VECTOR ANALYSIS

Yale Bicentennial Publications

With the approval of the President and Fellows of Yale University, a series of volumes has been prepared by a number of the Professors and Instructors, to be issued in connection with the Bicentennial Anniversary, as a partial indication of the character of the studies in which the University teachers are engaged.

This series of volumes is respectfully dedicated to

The Graduates of the University

VECTOR ANALYSIS

A TEXT-BOOK FOR THE USE OF STUDENTS
OF MATHEMATICS AND PHYSICS

FOUNDED UPON THE LECTURES OF

J. WILLARD GIBBS, PH.D., LL.D.

BY

EDWIN BIDWELL WILSON, PH.D.

DOVER PUBLICATIONS, INC.
NEW YORK NEW YORK

Published in the United Kingdom by Constable and Company Limited, 10 Orange Street, London W.C. 2.

This new Dover edition, first published in 1960, is an unabridged and corrected republication of the second edition published by Charles Scribner's Sons in 1909.

Manufactured in the United States of America

Dover Publications, Inc.
180 Varick Street
New York 14, N. Y.

PREFACE BY PROFESSOR GIBBS

SINCE the printing of a short pamphlet on the *Elements of Vector Analysis* in the years 1881–84, — never published, but somewhat widely circulated among those who were known to be interested in the subject, — the desire has been expressed in more than one quarter, that the substance of that treatise, perhaps in fuller form, should be made accessible to the public.

As, however, the years passed without my finding the leisure to meet this want, which seemed a real one, I was very glad to have one of the hearers of my course on Vector Analysis in the year 1899–1900 undertake the preparation of a text-book on the subject.

I have not desired that Dr. Wilson should aim simply at the reproduction of my lectures, but rather that he should use his own judgment in all respects for the production of a text-book in which the subject should be so illustrated by an adequate number of examples as to meet the wants of students of geometry and physics.

J. WILLARD GIBBS.

YALE UNIVERSITY, September, 1901.

GENERAL PREFACE

When I undertook to adapt the lectures of Professor Gibbs on Vector Analysis for publication in the Yale Bicentennial Series, Professor Gibbs himself was already so fully engaged upon his work to appear in the same series, *Elementary Principles in Statistical Mechanics*, that it was understood no material assistance in the composition of this book could be expected from him. For this reason he wished me to feel entirely free to use my own discretion alike in the selection of the topics to be treated and in the mode of treatment. It has been my endeavor to use the freedom thus granted only in so far as was necessary for presenting his method in text-book form.

By far the greater part of the material used in the following pages has been taken from the course of lectures on Vector Analysis delivered annually at the University by Professor Gibbs. Some use, however, has been made of the chapters on Vector Analysis in Mr. Oliver Heaviside's *Electromagnetic Theory* (Electrician Series, 1893) and in Professor Föppl's lectures on *Die Maxwell'sche Theorie der Electricität* (Teubner, 1894). My previous study of Quaternions has also been of great assistance.

The material thus obtained has been arranged in the way which seems best suited to easy mastery of the subject. Those Arts. which it seemed best to incorporate in the text but which for various reasons may well be omitted at the first reading have been marked with an asterisk (*). Numerous illustrative examples have been drawn from geometry, mechanics, and physics. Indeed, a large part of the text has to do with applications of the method. These applications have not been set apart in chapters by themselves, but have been distributed throughout the body of the book as fast as the analysis has been developed sufficiently for their adequate treatment. It is hoped that by this means the reader may be better enabled to make practical use of the book. Great care has been taken in avoiding the introduction of unnecessary ideas, and in so illustrating each idea that is introduced as to make its necessity evident and its meaning easy to grasp. Thus the book is not intended as a complete exposition of the theory of Vector Analysis, but as a text-book from which so much of the subject as may be required for practical applications may be learned. Hence a summary, including a list of the more important formulæ, and a number of exercises, have been placed at the end of each chapter, and many less essential points in the text have been indicated rather than fully worked out, in the hope that the reader will supply the details. The summary may be found useful in reviews and for reference.

The subject of Vector Analysis naturally divides itself into three distinct parts. First, that which concerns addition and the scalar and vector products of vectors. Second, that which concerns the differential and integral calculus in its relations to scalar and vector functions. Third, that which contains the theory of the linear vector function. The first part is a necessary introduction to both other parts. The second and third are mutually independent. Either may be taken up first. For practical purposes in mathematical physics the second must be regarded as more elementary than the third. But a student not primarily interested in physics would naturally pass from the first part to the third, which he would probably find more attractive and easy than the second.

Following this division of the subject, the main body of the book is divided into six chapters of which two deal with each of the three parts in the order named. Chapters I. and II. treat of addition, subtraction, scalar multiplication, and the scalar and vector products of vectors. The exposition has been made quite elementary. It can readily be understood by and is especially suited for such readers as have a knowledge of only the elements of Trigonometry and Analytic Geometry. Those who are well versed in Quaternions or allied subjects may perhaps need to read only the summaries. Chapters III. and IV. contain the treatment of those topics in Vector Analysis which, though of less value to the students of pure mathematics, are of the utmost importance to students of physics. Chapters V. and VI. deal with the linear vector function. To students of physics the linear vector function is of particular importance in the mathematical treatment of phenomena connected with non-isotropic media; and to the student of pure mathematics this part of the book will probably be the most interesting of all, owing to the fact that it leads to Multiple Algebra or the Theory of Matrices. A concluding chapter, VII., which contains the development of certain higher parts of the theory, a number of applications, and a short sketch of imaginary or complex vectors, has been added.

In the treatment of the integral calculus, Chapter IV., questions of mathematical rigor arise. Although modern theorists are devoting much time and thought to rigor, and although they will doubtless criticise this portion of the book adversely, it has been deemed best to give but little attention to the discussion of this subject. And the more so for the reason that whatever system of notation be employed questions of rigor are indissolubly associated with the *calculus* and occasion no new difficulty to the student of Vector Analysis, who must first learn what the facts are and may postpone until later the detailed consideration of the restrictions that are put upon those facts.

Notwithstanding the efforts which have been made during more than half a century to introduce Quaternions into physics the fact remains that they have not found wide favor. On the other hand there has been a growing tendency especially in the last decade toward the adoption of some form of Vector Analysis. The works of Heaviside and Föppl referred to before may be cited in evidence. As yet however no system of Vector Analysis which makes any claim to completeness has been published. In fact Heaviside says: "I am in hopes that the chapter which I now finish may serve as a stopgap till regular vectorial treatises come to be written suitable for physicists, based upon the vectorial treatment of vectors" (Electromagnetic Theory, Vol. I., p. 305). Elsewhere in the same chapter Heaviside has set forth the claims of vector analysis as against Quaternions, and others have expressed similar views.

The keynote, then, to any system of vector analysis must be its practical utility. This, I feel confident, was Professor Gibbs's point of view in building up his system. He uses it entirely in his courses on Electricity and Magnetism and on Electromagnetic Theory of Light. In writing this book I have tried to present the subject from this practical standpoint, and keep clearly before the reader's mind the questions: What combinations or functions of vectors occur in physics and geometry? And how may these be represented symbolically in the way best suited to facile analytic manipulation? The treatment of these questions in modern books on physics has been too much confined to the addition and subtraction of vectors. This is scarcely enough. It has been the aim here to give also an exposition of scalar and vector products, of the operator ∇, of divergence and curl which have gained such universal recognition since the appearance of Maxwell's *Treatise on Electricity and Magnetism*, of slope, potential, linear vector function, etc., such as shall be adequate for the needs of students of physics at the present day and adapted to them.

It has been asserted by some that Quaternions, Vector Analysis, and all such algebras are of little value for investigating questions in mathematical physics. Whether this assertion shall prove true or not, one may still maintain that vectors are to mathematical physics what invariants are to geometry. As every geometer must be thoroughly conversant with the ideas of invariants, so every student of physics should be able to *think* in terms of vectors. And there is no way in which he, especially at the beginning of his scientific studies, can come to so true an appreciation of the importance of vectors and of the ideas connected with them as by working in Vector Analysis and dealing directly with the vectors themselves. To those that hold these views the success of Professor Föppl's *Vorlesungen über Technische Mechanik* (four volumes, Teubner, 1897–1900, already in a second edition), in which the theory of mechanics is developed by means of a vector analysis, can be but an encouraging sign.

I take pleasure in thanking my colleagues, Dr. M. B. Porter and Prof. H. A. Bumstead, for assisting me with the manuscript. The good services of the latter have been particularly valuable in arranging Chapters III. and IV. in their present form and in suggesting many of the illustrations used in the work. I am also under obligations to my father, Mr. Edwin H. Wilson, for help in connection both with the proofs and the manuscript. Finally, I wish to express my deep indebtedness to Professor Gibbs. For although he has been so preoccupied as to be unable to read either manuscript or proof, he has always been ready to talk matters over with me, and it is he who has furnished me with inspiration sufficient to carry through the work.

EDWIN BIDWELL WILSON.

YALE UNIVERSITY, October, 1901.

PREFACE TO THE SECOND EDITION

THE only changes which have been made in this edition are a few corrections which my readers have been kind enough to point out to me.

E. B. W.

TABLE OF CONTENTS

CHAPTER III

THE DIFFERENTIAL CALCULUS OF VECTORS

CHAPTER IV

THE INTEGRAL CALCULUS OF VECTORS

CHAPTER V

LINEAR VECTOR FUNCTIONS

CHAPTER VI

ROTATIONS AND STRAINS

CHAPTER VII

MISCELLANEOUS APPLICATIONS

VECTOR ANALYSIS

VECTOR ANALYSIS

CHAPTER I

ADDITION AND SCALAR MULTIPLICATION

1.] In mathematics and especially in physics two very different kinds of quantity present themselves. Consider, for example, mass, time, density, temperature, force, displacement of a point, velocity, and acceleration. Of these quantities some can be represented adequately by a single number — temperature, by degrees on a thermometric scale; time, by years, days, or seconds; mass and density, by numerical values which are wholly determined when the unit of the scale is fixed. On the other hand the remaining quantities are not capable of such representation. Force to be sure is said to be of so many pounds or grams weight; velocity, of so many feet or centimeters per second. But in *addition* to this each of them must be considered as having *direction* as well as magnitude. A force points North, South, East, West, up, down, or in some intermediate direction. The same is true of displacement, velocity, and acceleration. No scale of numbers can represent them adequately. It can represent only their magnitude, not their direction.

2.] *Definition:* A *vector* is a quantity which is considered as possessing *direction* as well as *magnitude.*

Definition: A *scalar* is a quantity which is considered as possessing *magnitude* but no direction.

The positive and negative numbers of ordinary algebra are the typical scalars. For this reason the ordinary algebra is called *scalar* algebra when necessary to distinguish it from the *vector* algebra or analysis which is the subject of this book.

The typical vector is the displacement of translation in space. Consider first a point P (Fig. 1). Let P be displaced in a straight line and take a new position P'. This change of position is represented by the line $\overline{PP'}$. The magnitude of the displacement is the length of $\overline{PP'}$; the direction of it is the direction of the line $\overline{PP'}$ *from* P *to* P'. Next consider a displacement not of one, but of all the points in space. Let all the points move in straight lines in the same direction and for the same distance D. This is equivalent to shifting space as a rigid body in that direction through the distance D without rotation. Such a displacement is called a *translation*. It possesses direction and magnitude. When space undergoes a translation **T**, each point of space undergoes a displacement equal to **T** in magnitude and direction; and conversely if the displacement $\overline{PP'}$ which any one particular point P suffers in the translation **T** is known, then that of any other point Q is also known: for $\overline{QQ'}$ must be equal and parallel to $\overline{PP'}$.

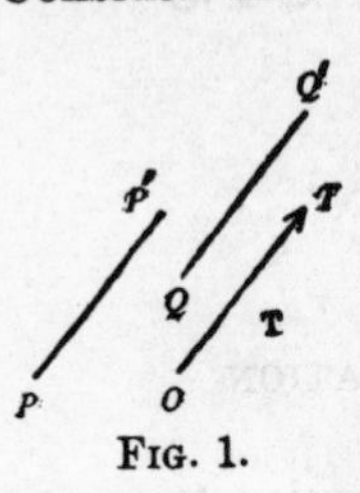

FIG. 1.

The translation **T** is represented geometrically or graphically by an arrow **T** (Fig. 1) of which the magnitude and direction are equal to those of the translation. The absolute position of this arrow in space is entirely immaterial. Technically the arrow is called a *stroke*. Its tail or initial point is its *origin*; and its head or final point, its *terminus*. In the figure the origin is designated by O and the terminus by T. This geometric quantity, a stroke, is used as the mathematical symbol for all vectors, just as the ordinary positive and negative numbers are used as the symbols for all scalars.

*3.] As examples of scalar quantities mass, time, density, and temperature have been mentioned. Others are distance, volume, moment of inertia, work, etc. Magnitude, however, is by no means the sole property of these quantities. Each implies something besides magnitude. Each has its own distinguishing characteristics, as an example of which its *dimensions* in the sense well known to physicists may be cited. A distance 3, a time 3, a work 3, etc., are very different. The magnitude 3 is, however, a property common to them all — perhaps the only one. Of all scalar quantities *pure number* is the simplest. It implies nothing but magnitude. It is the scalar *par excellence* and consequently it is used as the mathematical symbol for all scalars.

As examples of vector quantities force, displacement, velocity, and acceleration have been given. Each of these has other characteristics than those which belong to a vector pure and simple. The concept of *vector* involves two ideas and two alone — magnitude of the vector and direction of the vector. But force is more complicated. When it is applied to a rigid body the *line* in which it acts must be taken into consideration; magnitude and direction alone do not suffice. And in case it is applied to a non-rigid body the *point* of application of the force is as important as the magnitude or direction. Such is frequently true for vector quantities other than force. Moreover the question of dimensions is present as in the case of scalar quantities. The mathematical vector, the stroke, which is the primary object of consideration in this book, abstracts from all directed quantities their magnitude and direction and nothing but these; just as the mathematical scalar, pure number, abstracts the magnitude and that alone. Hence one must be on his guard lest from analogy he attribute some properties to the mathematical vector which do not belong to it; and he must be even more careful lest he obtain erroneous results by considering the vector quantities of physics as possessing no properties other than those of the mathematical vector. For example it would never do to consider force and its effects as unaltered by shifting it parallel to itself. This warning may not be necessary, yet it may possibly save some confusion.

4.] Inasmuch as, taken in its entirety, a vector or stroke is but a single concept, it may appropriately be designated by one letter. Owing however to the fundamental difference between scalars and vectors, it is necessary to distinguish carefully the one from the other. Sometimes, as in mathematical physics, the distinction is furnished by the physical interpretation. Thus if n be the index of refraction it must be scalar; m, the mass, and t, the time, are also scalars; but f, the force, and a, the acceleration, are vectors. When, however, the letters are regarded merely as symbols with no particular physical significance some typographical difference must be relied upon to distinguish vectors from scalars. Hence in this book **Clarendon type** is used for setting up vectors and ordinary type for scalars. This permits the use of the same letter differently printed to represent the vector and its scalar magnitude.[1] Thus if $\mathbf{C}$ be the electric current in magnitude and direction, C may be used to represent the magnitude of that current; if $\mathbf{g}$ be the vector acceleration due to gravity, g may be the scalar value of that acceleration; if $\mathbf{v}$ be the velocity of a moving mass, v may be the magnitude of that velocity. The use of Clarendons to denote vectors makes it possible to pass from directed quantities to their scalar magnitudes by a mere change in the appearance of a letter without any confusing change in the letter itself.

Definition: Two vectors are said to be *equal* when they have the same magnitude and the same direction.

[1] This convention, however, is by no means invariably followed. In some instances it would prove just as undesirable as it is convenient in others. It is chiefly valuable in the application of vectors to physics.

The equality of two vectors **A** and **B** is denoted by the usual sign =. Thus

$$\mathbf{A} = \mathbf{B}.$$

Evidently a vector or stroke is not altered by shifting it about *parallel* to itself in space. Hence any vector $\mathbf{A} = \overline{PP'}$ (Fig. 1) may be drawn from any assigned point O as origin; for the segment $\overline{PP'}$ may be moved parallel to itself until the point P falls upon the point O and P' upon some point T. Then

$$\mathbf{A} = \overline{PP'} = \overline{O\,T} = \mathbf{T}.$$

In this way all vectors in space may be replaced by directed segments radiating from one fixed point O. Equal vectors in space will of course coincide, when placed with their termini at the same point O. Thus (Fig. 1) $\mathbf{A} = \overline{PP'}$, and $\mathbf{B} = \overline{Q\,Q'}$, both fall upon $\mathbf{T} = \overline{O\,T}$.

For the numerical determination of a vector *three* scalars are necessary. These may be chosen in a variety of ways. If r, ϕ, θ be polar coördinates in space any vector **r** drawn with its origin at the origin of coördinates may be represented by the three scalars r, ϕ, θ which determine the terminus of the vector.

$$\mathbf{r} \sim (r, \phi, \theta).$$

Or if x, y, z be Cartesian coördinates in space a vector **r** may be considered as given by the differences of the coördinates x', y', z' of its terminus and those x, y, z of its origin.

$$\mathbf{r} \sim (x' - x, y' - y, z' - z).$$

If in particular the origin of the vector coincide with the origin of coördinates, the vector will be represented by the three coördinates of its terminus

$$\mathbf{r} \sim (x', y', z').$$

When two vectors are equal the three scalars which represent them must be equal respectively each to each. Hence one vector equality implies three scalar equalities.

Definition : A vector **A** is said to be equal to *zero* when its magnitude A is zero.

Such a vector **A** is called a *null* or *zero* vector and is written equal to naught in the usual manner. Thus

$$\mathbf{A} = 0 \text{ if } A = 0.$$

All null vectors are regarded as equal to each other without any considerations of direction.

In fact a null vector from a geometrical standpoint would be represented by a linear segment of length zero — that is to say, by a point. It consequently would have a wholly indeterminate direction or, what amounts to the same thing, none at all. If, however, it be regarded as the limit approached by a vector of finite length, it might be considered to have that direction which is the limit approached by the direction of the finite vector, when the length decreases indefinitely and approaches zero as a limit. The justification for disregarding this direction and looking upon all null vectors as equal is that when they are added (Art. 8) to other vectors no change occurs and when multiplied (Arts. 27, 31) by other vectors the product is zero.

5.] In extending to vectors the fundamental operations of algebra and arithmetic, namely, addition, subtraction, and multiplication, care must be exercised not only to avoid self-contradictory definitions but also to lay down useful ones. Both these ends may be accomplished most naturally and easily by looking to physics (for in that science vectors continually present themselves) and by observing how such quantities are treated there. If then **A** be a given displacement, force, or velocity, what is two, three, or in general x times **A**? What, the negative of **A**? And if **B** be another, what is the sum of **A** and **B**? That is to say, what is the equivalent of **A** and **B** taken together? The obvious answers to these questions suggest immediately the desired definitions.

Scalar Multiplication

6.] *Definition:* A vector is said to be multiplied by a positive scalar when its magnitude is multiplied by that scalar and its direction is left unaltered.

Thus if **v** be a velocity of nine knots East by North, $2\frac{1}{3}$ times **v** is a velocity of twenty-one knots with the direction still East by North. Or if **f** be the force exerted upon the scale-pan by a gram weight, 1000 times **f** is the force exerted by a kilogram. The direction in both cases is vertically downward.

If **A** be the vector and x the scalar the product of x and **A** is denoted as usual by

$$x\,\mathbf{A} \text{ or } \mathbf{A}\,x.$$

It is, however, more customary to place the scalar multiplier before the multiplicand **A**. This multiplication by a scalar is called *scalar multiplication*, and it follows the associative law

$$x\,(y\,\mathbf{A}) = (x\,y)\,\mathbf{A} = y\,(x\,\mathbf{A})$$

as in ordinary algebra and arithmetic. This statement is immediately obvious when the fact is taken into consideration that scalar multiplication does not alter direction but merely multiplies the length.

Definition: A *unit* vector is one whose magnitude is unity.

Any vector **A** may be looked upon as the product of a unit vector **a** in its direction by the positive scalar A, its magnitude.

$$\mathbf{A} = A\,\mathbf{a} = \mathbf{a}\,A.$$

The unit vector **a** may similarly be written as the product of **A** by $1/A$ or as the quotient of **A** and A.

$$\mathbf{a} = \frac{1}{A}\,\mathbf{A} = \frac{\mathbf{A}}{A}.$$

7.] *Definition:* The negative sign, $-$, prefixed to a vector *reverses* its direction but leaves its magnitude unchanged.

For example if **A** be a displacement for two feet to the right, $-$ **A** is a displacement for two feet to the left. Again if the stroke $\overline{AB}$ be **A**, the stroke $\overline{BA}$, which is of the same length as $\overline{AB}$ but which is in the direction from B to A instead of from A to B, will be $-$ **A**. Another illustration of the use of the negative sign may be taken from Newton's third law of motion. If **A** denote an "action," $-$ **A** will denote the "reaction." The positive sign, $+$, may be prefixed to a vector to call particular attention to the fact that the direction has not been reversed. The two signs $+$ and $-$ when used in connection with scalar multiplication of vectors follow the same laws of operation as in ordinary algebra. These are symbolically

$$++=+; \quad +-=-; \quad -+=-; \quad --=+;$$
$$-(m\,\mathbf{A}) = m\,(-\mathbf{A}).$$

The interpretation is obvious.

Addition and Subtraction

8.] The addition of two vectors or strokes may be treated most simply by regarding them as defining translations in space (Art. 2). Let **S** be one vector and **T** the other. Let P be a point of space (Fig. 2). The translation **S** carries P into P' such that the line $\overline{PP'}$ is equal to **S** in magnitude and direction. The transformation **T** will then carry P' into P'' — the line $\overline{P'P''}$ being parallel to **T** and equal to it in magnitude. Consequently the result of **S** followed by **T** is to carry the point P into the point P''. If now Q be any other point in space, **S** will carry Q into Q' such that $\overline{QQ'} = $ **S** and **T** will then carry Q' into Q''

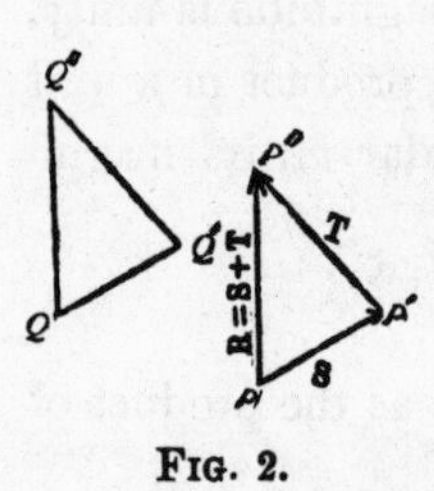

FIG. 2.

such that $\overline{Q' Q''} = \mathbf{T}$. Thus **S** followed by **T** carries Q into Q''. Moreover, the triangle $Q Q' Q''$ is equal to $P P' P''$. For the two sides $Q Q'$ and $Q' Q''$, being equal and parallel to **S** and **T** respectively, must be likewise parallel to $P P'$ and $P' P''$ respectively which are also parallel to **S** and **T**. Hence the third sides of the triangles must be equal and parallel That is

$$Q Q'' \text{ is equal and parallel to } P P''.$$

As Q is any point in space this is equivalent to saying that by means of **S** followed by **T** all points of space are displaced the same amount and in the same direction. This displacement is therefore a translation. Consequently the two translations **S** and **T** are equivalent to a single translation **R**. Moreover

$$\text{if} \qquad \mathbf{S} = \overline{P P'} \text{ and } \mathbf{T} = \overline{P' P''}, \text{ then } \mathbf{R} = \overline{P P''}.$$

The stroke **R** is called the *resultant* or *sum* of the two strokes **S** and **T** to which it is equivalent. This sum is denoted in the usual manner by

$$\mathbf{R} = \mathbf{S} + \mathbf{T}.$$

From analogy with the sum or resultant of two translations the following definition for the addition of any two vectors is laid down.

Definition: The sum or resultant of two vectors is found by placing the origin of the second upon the terminus of the first and drawing the vector from the origin of the first to the terminus of the second.

9.] *Theorem.* The order in which two vectors **S** and **T** are added does not affect the sum.

S followed by **T** gives precisely the same result as **T** followed by **S**. For let **S** carry P into P' (Fig. 3); and **T**, P' into P''. $\mathbf{S} + \mathbf{T}$ then carries P into P''. Suppose now that **T** carries P into P'''. The line PP''' is equal and parallel to $P' P''$. Consequently the points P, P', P'', and P''' lie at the vertices of a parallelogram. Hence $P''' P''$ is equal and parallel to $P P'$. Hence **S** carries P''' into P''. **T** followed by **S** therefore carries P into P'' through P', whereas **S** followed by **T** carries P into P'' through P'''. The final result is in either case the same. This may be designated symbolically by writing

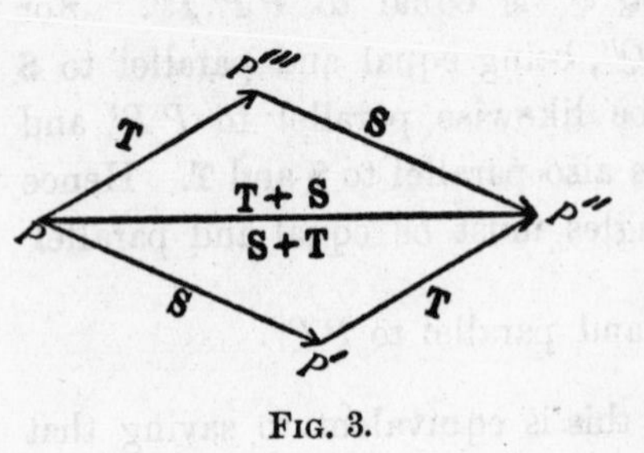

FIG. 3.

$$\mathbf{R} = \mathbf{S} + \mathbf{T} = \mathbf{T} + \mathbf{S}.$$

It is to be noticed that $\mathbf{S} = \overline{P P'}$ and $\mathbf{T} = \overline{P P'''}$ are the two sides of the parallelogram $P P' P'' P'''$ which have the point P as common origin; and that $\mathbf{R} = \overline{P P''}$ is the diagonal drawn through P. This leads to another very common way of stating the definition of the sum of two vectors.

If two vectors be drawn from the same origin and a parallelogram be constructed upon them as sides, their sum will be that diagonal which passes through their common origin.

This is the well-known "parallelogram law" according to which the physical vector quantities force, acceleration, velocity, and angular velocity are compounded. It is important to note that in case the vectors lie along the same line vector addition becomes equivalent to algebraic scalar addition. The lengths of the two vectors to be added are added if the vectors have the same direction; but subtracted if they have opposite directions. In either case the sum has the same direction as that of the greater vector.

10.] After the definition of the sum of two vectors has been laid down, the sum of several may be found by adding together the first two, to this sum the third, to this the fourth, and so on until all the vectors have been combined into a single one. The final result is the same as that obtained by placing the origin of each succeeding vector upon the terminus of the preceding one and then drawing at once the vector from the origin of the first to the terminus of the last. In case these two points coincide the vectors form a closed polygon and their sum is zero. Interpreted geometrically this states that if a number of displacements $\mathbf{R}, \mathbf{S}, \mathbf{T} \cdots$ are such that the strokes $\mathbf{R}, \mathbf{S}, \mathbf{T} \cdots$ form the sides of a closed polygon taken in order, then the effect of carrying out the displacements is *nil.* Each point of space is brought back to its starting point. Interpreted in mechanics it states that if any number of forces act at a point and if they form the sides of a closed polygon taken in order, then the resultant force is zero and the point is in equilibrium under the action of the forces.

The order of sequence of the vectors in a sum is of no consequence. This may be shown by proving that any two *adjacent* vectors may be interchanged without affecting the result.

To show

$$\mathbf{A} + \mathbf{B} + \mathbf{C} + \mathbf{D} + \mathbf{E} = \mathbf{A} + \mathbf{B} + \mathbf{D} + \mathbf{C} + \mathbf{E}.$$

Let $\quad \mathbf{A} = \overline{O\,A},\ \mathbf{B} = \overline{A\,B},\ \mathbf{C} = \overline{B\,C},\ \mathbf{D} = \overline{C\,D},\ \mathbf{E} = \overline{D\,E}.$
Then $\quad \overline{O\,E} = \mathbf{A} + \mathbf{B} + \mathbf{C} + \mathbf{D} + \mathbf{E}.$
Let now $\overline{B\,C'} = \mathbf{D}$. Then $C'\,B\,C\,D$ is a parallelogram and consequently $\overline{C'\,D} = \mathbf{C}$. Hence

$$\overline{O\,E} = \mathbf{A} + \mathbf{B} + \mathbf{D} + \mathbf{C} + \mathbf{E},$$

which proves the statement. Since any two adjacent vectors may be interchanged, and since the sum may be arranged in any order by successive interchanges of adjacent vectors, the order in which the vectors occur in the sum is immaterial.

11.] *Definition:* A vector is said to be subtracted when it is added after reversal of direction. Symbolically,

$$\mathbf{A} - \mathbf{B} = \mathbf{A} + (-\mathbf{B}).$$

By this means subtraction is reduced to addition and needs

no special consideration. There is however an interesting and important way of representing the difference of two vectors geometrically. Let $\mathbf{A} = \overline{OA}$, $\mathbf{B} = \overline{OB}$ (Fig. 4). Complete the parallelogram of which **A** and **B** are the sides. Then the diagonal $\overline{OC} = \mathbf{C}$ is the sum $\mathbf{A} + \mathbf{B}$ of the two vectors. Next complete the parallelogram of which **A** and $-\mathbf{B} = \overline{OB'}$ are the sides. Then the diagonal $\overline{OD} = \mathbf{D}$ will be the sum of **A** and the negative of **B**. But the segment $\overline{OD}$ is parallel and equal to $\overline{BA}$. Hence $\overline{BA}$ may be taken as the difference to the two vectors **A** and **B**. This leads to the following rule: The difference of two vectors which are drawn from the same origin is the vector drawn *from* the terminus of the vector to be subtracted *to* the terminus of the vector from which it is subtracted. Thus the two diagonals of the parallelogram, which is constructed upon **A** and **B** as sides, give the sum and difference of **A** and **B**.

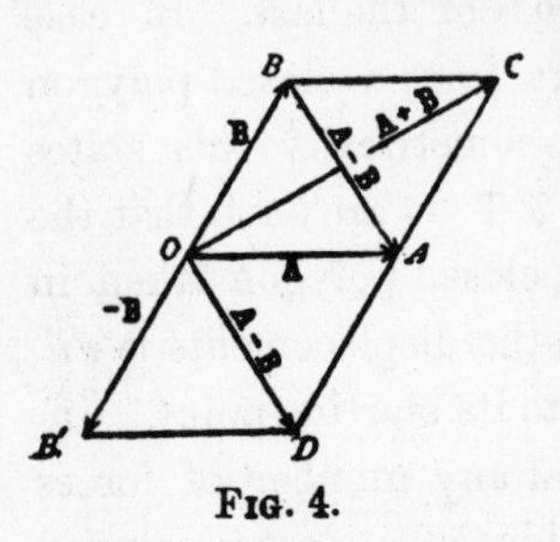

FIG. 4.

12.] In the foregoing paragraphs addition, subtraction, and scalar multiplication of vectors have been defined and interpreted. To make the development of vector algebra mathematically exact and systematic it would now become necessary to *demonstrate* that these three fundamental operations follow the same formal laws as in the ordinary scalar algebra, although from the standpoint of the physical and geometrical interpretation of vectors this may seem superfluous. These laws are

$$\mathrm{I}_a: \quad m\,(n\,\mathbf{A}) = n\,(m\,\mathbf{A}) = (m\,n)\,\mathbf{A},$$

$$\mathrm{I}_b: \quad (\mathbf{A} + \mathbf{B}) + \mathbf{C} = \mathbf{A} + (\mathbf{B} + \mathbf{C}),$$

$$\mathrm{II}: \quad \mathbf{A} + \mathbf{B} = \mathbf{B} + \mathbf{A},$$

$$\mathrm{III}_a: \quad (m + n)\,\mathbf{A} = m\,\mathbf{A} + n\,\mathbf{A},$$

$$\mathrm{III}_b: \quad m\,(\mathbf{A} + \mathbf{B}) = m\,\mathbf{A} + m\,\mathbf{B},$$

$$\mathrm{III}_c: \quad -(\mathbf{A} + \mathbf{B}) = -\mathbf{A} - \mathbf{B}.$$

I_a is the so-called law of association and commutation of the scalar factors in scalar multiplication.

I_b is the law of association for vectors in vector addition. It states that in adding vectors parentheses may be inserted at any points without altering the result.

II is the commutative law of vector addition.

III_a is the distributive law for scalars in scalar multiplication.

III_b is the distributive law for vectors in scalar multiplication.

III_c is the distributive law for the negative sign.

The proofs of these laws of operation depend upon those propositions in elementary geometry which have to deal with the first properties of the parallelogram and similar triangles. They will not be given here; but it is suggested that the reader work them out for the sake of fixing the fundamental ideas of addition, subtraction, and scalar multiplication more clearly in mind. The result of the laws may be summed up in the statement:

The laws which govern addition, subtraction, and scalar multiplication of vectors are identical with those governing these operations in ordinary scalar algebra.

It is precisely this identity of formal laws which justifies the extension of the use of the familiar signs $=$, $+$, and $-$ of arithmetic to the algebra of vectors and it is also this which ensures the correctness of results obtained by operating with those signs in the usual manner. One caution only need be mentioned. Scalars and vectors are entirely different sorts of quantity. For this reason they can never be equated to each other — except perhaps in the trivial case where each is zero. For the same reason they are not to be added together. So long as this is borne in mind no difficulty need be anticipated from dealing with vectors much as if they were scalars.

Thus from equations in which the vectors enter linearly with scalar coefficients unknown vectors may be eliminated or found by solution in the same way and with the same limitations as in ordinary algebra; for the eliminations and solutions depend solely on the scalar coefficients of the equations and not at all on what the variables represent. If for instance

$$a\,\mathbf{A} + b\,\mathbf{B} + c\,\mathbf{C} + d\,\mathbf{D} = 0,$$

then **A**, **B**, **C**, or **D** may be expressed in terms of the other three

as
$$\mathbf{D} = -\frac{1}{d}\,(a\,\mathbf{A} + b\,\mathbf{B} + c\,\mathbf{C}).$$

And two vector equations such as

$$3\,\mathbf{A} + 4\,\mathbf{B} = \mathbf{E}$$

and
$$2\,\mathbf{A} + 3\,\mathbf{B} = \mathbf{F}$$

yield by the usual processes the solutions

$$\mathbf{A} = 3\,\mathbf{E} - 4\,\mathbf{F}$$

and
$$\mathbf{B} = 3\,\mathbf{F} - 2\,\mathbf{E}.$$

Components of Vectors

13.] *Definition:* Vectors are said to be collinear when they are parallel to the same line; coplanar, when parallel to the same plane. Two or more vectors to which no line can be drawn parallel are said to be non-collinear. Three or more vectors to which no plane can be drawn parallel are said to be non-coplanar. Obviously any *two* vectors are coplanar.

Any vector **b** collinear with **a** may be expressed as the product of **a** and a positive or negative scalar which is the ratio of the magnitude of **b** to that of **a**. The sign is positive when **b** and **a** have the same direction; negative, when they have opposite directions. If then $\overline{OA} = \mathbf{a}$, the vector **r** drawn from the origin O to any point of the line OA produced in either direction is

$$\mathbf{r} = x\,\mathbf{a}. \tag{1}$$

If x be a variable scalar parameter this equation may therefore be regarded as the (vector) equation of all points in the line OA. Let now B be any point not upon the line OA or that line produced in either direction (Fig. 5).

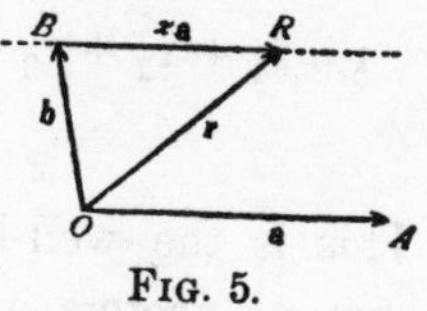

FIG. 5.

Let $\overline{OB} = \mathbf{b}$. The vector $\mathbf{b}$ is surely not of the form $x\,\mathbf{a}$. Draw through B a line parallel to OA and let R be any point upon it. The vector $\overline{BR}$ is collinear with $\mathbf{a}$ and is consequently expressible as $x\,\mathbf{a}$. Hence the vector drawn from O to R is

$$\overline{OR} = \overline{OB} + \overline{BR}$$

or

$$\mathbf{r} = \mathbf{b} + x\,\mathbf{a}. \tag{2}$$

This equation may be regarded as the (vector) equation of all the points in the line which is parallel to $\mathbf{a}$ and of which B is one point.

14.] Any vector $\mathbf{r}$ coplanar with two non-collinear vectors $\mathbf{a}$ and $\mathbf{b}$ may be resolved into two components parallel to $\mathbf{a}$ and $\mathbf{b}$ respectively. This resolution may be accomplished by constructing the parallelogram (Fig. 6) of which the sides are parallel to $\mathbf{a}$ and $\mathbf{b}$ and of which the diagonal is $\mathbf{r}$. Of these components one is $x\,\mathbf{a}$; the other, $y\,\mathbf{b}$. x and y are respectively the scalar ratios (taken with the proper sign) of the lengths of these components to the lengths of $\mathbf{a}$ and $\mathbf{b}$. Hence

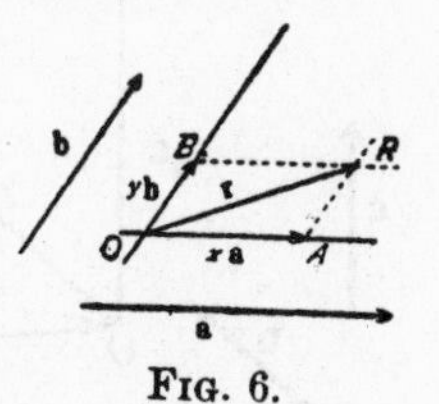

FIG. 6.

$$\mathbf{r} = x\,\mathbf{a} + y\,\mathbf{b} \tag{2'}$$

is a typical form for any vector coplanar with $\mathbf{a}$ and $\mathbf{b}$. If several vectors $\mathbf{r}_1, \mathbf{r}_2, \mathbf{r}_3 \cdots$ may be expressed in this form as

$$\mathbf{r}_1 = x_1\, \mathbf{a} + y_1\, \mathbf{b},$$
$$\mathbf{r}_2 = x_2\, \mathbf{a} + y_2\, \mathbf{b},$$
$$\mathbf{r}_3 = x_3\, \mathbf{a} + y_3\, \mathbf{b},$$
$$\cdots\cdots\cdots\cdots$$
$$\cdots\cdots\cdots\cdots$$

their sum $\mathbf{r}$ is then

$$\mathbf{r} = \mathbf{r}_1 + \mathbf{r}_2 + \mathbf{r}_3 + \cdots = (x_1 + x_2 + x_3 + \ldots)\, \mathbf{a} + (y_1 + y_2 + y_3 + \ldots)\, \mathbf{b}.$$

This is the well-known theorem that the components of a sum of vectors are the sums of the components of those vectors. If the vector $\mathbf{r}$ is zero each of its components must be zero. Consequently the one vector equation $\mathbf{r} = 0$ is equivalent to the two scalar equations

$$\left.\begin{array}{l} x_1 + x_2 + x_3 + \cdots = 0 \\ y_1 + y_2 + y_3 + \cdots = 0 \end{array}\right\rangle \mathbf{r} = 0. \qquad (3)$$

15.] Any vector $\mathbf{r}$ in space may be resolved into three components parallel to any three given non-coplanar vectors. Let the vectors be $\mathbf{a}$, $\mathbf{b}$, and $\mathbf{c}$. The resolution may then be accomplished by constructing the parallelopiped (Fig. 7) of which the edges are parallel to $\mathbf{a}$, $\mathbf{b}$, and $\mathbf{c}$ and of which the diagonal is $\mathbf{r}$. This parallelopiped may be drawn easily by passing three planes parallel respectively to $\mathbf{a}$ and $\mathbf{b}$, $\mathbf{b}$ and $\mathbf{c}$, $\mathbf{c}$ and $\mathbf{a}$ through the origin O of the vector $\mathbf{r}$; and a similar set of three planes through its terminus R. These six planes will then be parallel in pairs

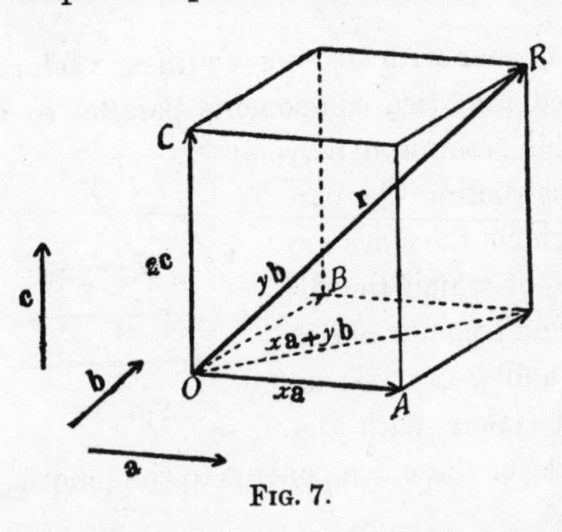

FIG. 7.

and hence form a parallelopiped. That the intersections of the planes are lines which are parallel to **a**, or **b**, or **c** is obvious. The three components of **r** are $x\,\mathbf{a}$, $y\,\mathbf{b}$, and $z\,\mathbf{c}$; where x, y, and z are respectively the scalar ratios (taken with the proper sign) of the lengths of these components to the length of **a**, **b**, and **c**. Hence

$$\mathbf{r} = x\,\mathbf{a} + y\,\mathbf{b} + z\,\mathbf{c} \tag{4}$$

is a typical form for any vector whatsoever in space. Several vectors $\mathbf{r}_1$, $\mathbf{r}_2$, $\mathbf{r}_3$ . . . may be expressed in this form as

$$\begin{aligned} \mathbf{r}_1 &= x_1\,\mathbf{a} + y_1\,\mathbf{b} + z_1\,\mathbf{c}, \\ \mathbf{r}_2 &= x_2\,\mathbf{a} + y_2\,\mathbf{b} + z_2\,\mathbf{c}, \\ \mathbf{r}_3 &= x_3\,\mathbf{a} + y_3\,\mathbf{b} + z_3\,\mathbf{c}, \\ &\cdots\cdots\cdots\cdots \\ &\cdots\cdots\cdots\cdots \end{aligned}$$

Their sum **r** is then

$$\begin{aligned} \mathbf{r} = \mathbf{r}_1 + \mathbf{r}_2 + \mathbf{r}_3 + \cdots = &(x_1 + x_2 + x_3 + \cdots)\,\mathbf{a} \\ &+ (y_1 + y_2 + y_3 + \cdots)\,\mathbf{b} \\ &+ (z_1 + z_2 + z_3 + \cdots)\,\mathbf{c}. \end{aligned}$$

If the vector **r** is zero each of its three components is zero. Consequently the one vector equation $\mathbf{r} = 0$ is equivalent to the three scalar equations

$$\left.\begin{aligned} x_1 + x_2 + x_3 + \cdots &= 0 \\ y_1 + y_2 + y_3 + \cdots &= 0 \\ z_1 + z_2 + z_3 + \cdots &= 0 \end{aligned}\right\rangle \mathbf{r} = 0. \tag{5}$$

Should the vectors all be coplanar with **a** and **b**, all the components parallel to **c** vanish. In this case therefore the above equations reduce to those given before.

16.] If two equal vectors are expressed in terms of the same three non-coplanar vectors, the corresponding scalar coefficients are equal.

Let $$\mathbf{r} = \mathbf{r}',$$

$$\mathbf{r} = x\,\mathbf{a} + y\,\mathbf{b} + z\,\mathbf{c},$$

$$\mathbf{r}' = x'\,\mathbf{a} + y'\,\mathbf{b} + z'\,\mathbf{c},$$

Then $$x = x', \quad y = y', \quad z = z'.$$

For $$\mathbf{r} - \mathbf{r}' = 0 = (x - x')\,\mathbf{a} + (y - y')\,\mathbf{b} + (z - z')\,\mathbf{c}.$$

Hence $$x - x' = 0, \quad y - y' = 0, \quad z - z' = 0.$$

But this would not be true if **a**, **b**, and **c** were coplanar. In that case one of the three vectors could be expressed in terms of the other two as

$$\mathbf{c} = m\,\mathbf{a} + n\,\mathbf{b}.$$

Then $$\mathbf{r} = x\,\mathbf{a} + y\,\mathbf{b} + z\,\mathbf{c} = (x + m\,z)\,\mathbf{a} + (y + n\,z)\,\mathbf{b},$$

$$\mathbf{r}' = x'\,\mathbf{a} + y'\,\mathbf{b} + z'\,\mathbf{c} = (x' + m\,z')\,\mathbf{a} + (y' + n\,z')\,\mathbf{b},$$

$$\mathbf{r} - \mathbf{r}' = [(x + m\,z) - (x' + m\,z')]\,\mathbf{a},$$

$$+ [(y + n\,z) - (y' + n\,z')]\,\mathbf{b} = 0.$$

Hence the individual components of $\mathbf{r} - \mathbf{r}'$ in the directions **a** and **b** (supposed different) are zero.

Hence $$x + m\,z = x' + m\,z'$$

$$y + n\,z = y' + n\,z'.$$

But this by no means necessitates x, y, z to be equal respectively to x', y', z'. In a similar manner if **a** and **b** were collinear it is impossible to infer that their coefficients vanish individually. The theorem may perhaps be stated as follows:

In case two equal vectors are expressed in terms of one vector, or two non-collinear *vectors, or three* non-coplanar *vectors, the corresponding scalar coefficients are equal. But this is* not *necessarily true if the two vectors be collinear ; or the three vectors, coplanar.* This principle will be used in the applications (Arts. 18 et seq.).

The Three Unit Vectors i, j, k.

17.] In the foregoing paragraphs the method of expressing vectors in terms of three given non-coplanar ones has been explained. The simplest set of three such vectors is the rectangular system familiar in Solid Cartesian Geometry. This rectangular system may however be either of two very distinct types. In one case (Fig. 8, first part) the Z-axis[1] lies upon that side of the XY- plane on which rotation through a right angle from the X-axis to the Y-axis appears *counterclockwise* or *positive* according to the convention adopted in Trigonometry. This relation may be stated in another form. If the X-axis be directed to the right and the Y-axis vertically, the Z-axis will be directed toward the observer. Or if the X-axis point toward the observer and the Y-axis to the right, the Z-axis will point upward. Still another method of state-

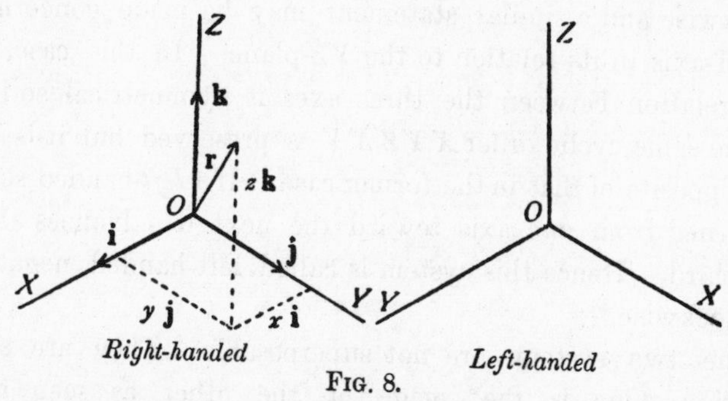

Fig. 8.

ment is common in mathematical physics and engineering. If a right-handed screw be turned from the X-axis to the Y-axis it will advance along the (*positive*) Z-*axis*. Such a system of axes is called right-handed, positive, or counterclockwise.[2] It is easy to see that the Y-axis lies upon that side of the ZX-plane on which rotation from the Z-axis to the X-axis is counterclockwise; and the X-axis, upon that side of

[1] By the X-, Y-, or Z-axis the *positive* half of that axis is meant. The XY-plane means the plane which contains the X- and Y-axis, *i.e.*, the plane $z = 0$.

[2] A convenient right-handed system and one which is always available consists of the thumb, first finger, and second finger of the right hand. If the thumb and first finger be stretched out from the palm perpendicular to each other, and if the second finger be bent over toward the palm at right angles to first finger, a right-handed system is formed by the fingers taken in the order thumb, first finger, second finger.

the YZ-plane on which rotation from the Y-axis to the Z-axis is counterclockwise. Thus it appears that the relation between the three axes is perfectly symmetrical so long as the same cyclic order $XYZXY$ is observed. If a right-handed screw is turned from one axis toward the next it advances along the third.

In the other case (Fig. 8, second part) the Z-axis lies upon that side of the XY-plane on which rotation through a right angle from the X-axis to the Y-axis appears *clockwise* or *negative.* The Y-axis then lies upon that side of the ZX-plane on which rotation from the Z-axis to the X-axis appears clockwise and a similar statement may be made concerning the X-axis in its relation to the YZ-plane. In this case, too, the relation between the three axes is symmetrical so long as the same cyclic order $XYZXY$ is preserved but it is just the opposite of that in the former case. If a *left*-handed screw is turned from one axis toward the next it advances along the third. Hence this system is called left-handed, negative, or clockwise.[1]

The two systems are not superposable. They are symmetric. One is the image of the other as seen in a mirror. If the X- and Y-axes of the two different systems be superimposed, the Z-axes will point in opposite directions. Thus one system may be obtained from the other by reversing the direction of *one* of the axes. A little thought will show that if *two* of the axes be reversed in direction the system will not be altered, but if all *three* be so reversed it will be.

Which of the two systems be used, matters little. But inasmuch as the formulæ of geometry and mechanics differ slightly in the matter of sign, it is advisable to settle once for all which shall be adopted. In this book the right-handed or counterclockwise system will be invariably employed.

[1] A left-handed system may be formed by the left hand just as a right-handed one was formed by the right.

Definition: The three letters **i**, **j**, **k** will be reserved to denote three vectors of unit length drawn respectively in the directions of the *X*-, *Y*-, and *Z*- axes of a right-handed rectangular system.

In terms of these vectors, any vector may be expressed as

$$\mathbf{r} = x\,\mathbf{i} + y\,\mathbf{j} + z\,\mathbf{k}. \qquad (6)$$

The coefficients x, y, z are the ordinary Cartesian coördinates of the terminus of **r** if its origin be situated at the origin of coördinates. The components of **r** parallel to the *X*-, *Y*-, and *Z*-axes are respectively

$$x\,\mathbf{i}, \quad y\,\mathbf{j}, \quad z\,\mathbf{k}.$$

The rotations about **i** from **j** to **k**, about **j** from **k** to **i**, and about **k** from **i** to **j** are all positive.

By means of these vectors **i**, **j**, **k** such a correspondence is established between vector analysis and the analysis in Cartesian coördinates that it becomes possible to pass at will from either one to the other. There is nothing contradictory between them. On the contrary it is often desirable or even necessary to translate the formulæ obtained by vector methods into Cartesian coördinates for the sake of comparing them with results already known and it is still more frequently convenient to pass from Cartesian analysis to vectors both on account of the brevity thereby obtained and because the vector expressions show forth the intrinsic meaning of the formulæ.

Applications

***18.**] Problems in plane geometry may frequently be solved easily by vector methods. Any two non-collinear vectors in the plane may be taken as the fundamental ones in terms of which all others in that plane may be expressed. The origin may also be selected at pleasure. Often it is possible to make such an advantageous choice of the origin and fundamental vectors that the analytic work of solution is materially simplified. The adaptability of the vector method is about the same as that of oblique Cartesian coördinates with different scales upon the two axes.

Example 1: The line which joins one vertex of a parallelogram to the middle point of an opposite side trisects the diagonal (Fig. 9).

Let $ABCD$ be the parallelogram, BE the line joining the vertex B to the middle point E of the side AD, R the point in which this line cuts the diagonal AC. To show AR is one third of AC. Choose A as origin, $\overline{AB}$ and $\overline{AD}$ as the two fundamental vectors $\mathbf{S}$ and $\mathbf{T}$. Then $\overline{AC}$ is the sum of $\mathbf{S}$ and $\mathbf{T}$. Let $\overline{AR} = \mathbf{R}$. To show

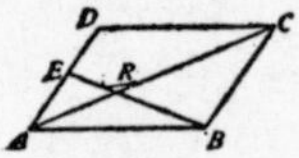

FIG. 9.

$$\mathbf{R} = \tfrac{1}{3}(\mathbf{S} + \mathbf{T}).$$

$$\mathbf{R} = \overline{AR} = \overline{AE} + \overline{ER} = \tfrac{1}{2}\,\mathbf{T} + x\,(\mathbf{S} - \tfrac{1}{2}\,\mathbf{T}),$$

where x is the ratio of ER to EB—an unknown scalar.

And $$\mathbf{R} = y\,(\mathbf{S} + \mathbf{T}),$$

where y is the scalar ratio of AR to AC to be shown equal to $\frac{1}{3}$.

Hence $$\tfrac{1}{2}\,\mathbf{T} + x\,(\mathbf{S} - \tfrac{1}{2}\,\mathbf{T}) = y\,(\mathbf{S} + \mathbf{T})$$

or $$x\,\mathbf{S} + \tfrac{1}{2}\,(1 - x)\,\mathbf{T} = y\,\mathbf{S} + y\,\mathbf{T}.$$

Hence, equating corresponding coefficients (Art. 16),

$$x = y,$$

$$\tfrac{1}{2}\,(1 - x) = y.$$

From which $$y = \frac{1}{3}.$$

Inasmuch as x is also $\frac{1}{3}$ the line EB must be trisected as well as the diagonal AC.

Example 2: If through any point within a triangle lines be drawn parallel to the sides the sum of the ratios of these lines to their corresponding sides is 2.

Let ABC be the triangle, R the point within it. Choose A as origin, AB and AC as the two fundamental vectors $\mathbf{S}$ and $\mathbf{T}$. Let

$$\overline{AR} = \mathbf{R} = m\,\mathbf{S} + n\,\mathbf{T}. \qquad (a)$$

$m\,\mathbf{S}$ is the fraction of $\overline{AB}$ which is cut off by the line through R parallel to AC. The remainder of AB must be the fraction $(1-m)\,\mathbf{S}$. Consequently by similar triangles the ratio of the line parallel to AC to the line AC itself is $(1-m)$. Similarly the ratio of the line parallel to AB to the line AB itself is $(1-n)$. Next express $\mathbf{R}$ in terms of $\mathbf{S}$ and $\mathbf{T}-\mathbf{S}$ the third side of the triangle. Evidently from (a)

$$\mathbf{R} = (m+n)\,\mathbf{S} + n\,(\mathbf{T}-\mathbf{S}).$$

Hence $(m+n)\,\mathbf{S}$ is the fraction of $\overline{AB}$ which is cut off by the line through R parallel to BC. Consequently by similar triangles the ratio of this line to BC itself is $(m+n)$. Adding the three ratios

$$(1-m) + (1-n) + (m+n) = 2,$$

and the theorem is proved.

Example 3: If from any point within a parallelogram lines be drawn parallel to the sides, the diagonals of the parallelograms thus formed intersect upon the diagonal of the given parallelogram.

Let $ABCD$ be a parallelogram, R a point within it, KM and LN two lines through R parallel respectively to AB and

AD, the points K, L, M, N lying upon the sides DA, AB, BC, CD respectively. To show that the diagonals KN and LM of the two parallelograms $KRND$ and $LBMR$ meet on AC. Choose A as origin, $\overline{AB}$ and $\overline{AD}$ as the two fundamental vectors $\mathbf{S}$ and $\mathbf{T}$. Let

$$\mathbf{R} = \overline{AR} = m\,\mathbf{S} + n\,\mathbf{T},$$

and let P be the point of intersection of KN with LM.

Then
$$\overline{KN} = \overline{KR} + \overline{RN} = m\,\mathbf{S} + (1-n)\,\mathbf{T},$$
$$\mathbf{P} = \overline{AP} = \overline{AK} + x\,\overline{KN},$$
$$\overline{LM} = (1-m)\,\mathbf{S} + n\,\mathbf{T},$$
$$\mathbf{P} = \overline{AP} = \overline{AL} + y\,\overline{LM}.$$

Hence
$$\mathbf{P} = n\,\mathbf{T} + x\,[m\,\mathbf{S} + (1-n)\,\mathbf{T}],$$
and
$$\mathbf{P} = m\,\mathbf{S} + y\,[(1-m)\,\mathbf{S} + n\,\mathbf{T}].$$

Equating coefficients,

$$x\,m = m + y\,(1-m)$$
$$y\,n = n + x\,(1-n)$$

By solution,
$$x = \frac{n}{m+n-1},$$
$$y = \frac{m}{m+n-1}.$$

Substituting either of these solutions in the expression for $\mathbf{P}$, the result is

$$\mathbf{P} = \frac{m\,n}{m+n-1}\,(\mathbf{S} + \mathbf{T}),$$

which shows that $\mathbf{P}$ is collinear with AC.

* **19.**] Problems in three dimensional geometry may be solved in essentially the same manner as those in two dimensions. In this case there are three fundamental vectors in terms of which all others can be expressed. The method of solution is analogous to that in the simpler case. Two

expressions for the same vector are usually found. The coefficients of the corresponding terms are equated. In this way the equations between three unknown scalars are obtained from which those scalars may be determined by solution and then substituted in either of the expressions for the required vector. The vector method has the same degree of adaptability as the Cartesian method in which oblique axes with different scales are employed. The following examples like those in the foregoing section are worked out not so much for their intrinsic value as for gaining a familiarity with vectors.

Example 1: Let $ABCD$ be a tetrahedron and P any point within it. Join the vertices to P and produce the lines until they intersect the opposite faces in A', B', C', D'. To show

$$\frac{PA'}{AA'} + \frac{PB'}{BB'} + \frac{PC'}{CC'} + \frac{PD'}{DD'} = 1.$$

Choose A as origin, and the edges $\overline{AB}$, $\overline{AC}$, $\overline{AD}$ as the three fundamental vectors $\mathbf{B}$, $\mathbf{C}$, $\mathbf{D}$. Let the vector $\overline{AP}$ be

$$\mathbf{P} = \overline{AP} = l\,\mathbf{B} + m\,\mathbf{C} + n\,\mathbf{D},$$

$$\mathbf{A}' = \overline{AA'} = k_1\,\mathbf{P} = k_1\,(l\,\mathbf{B} + m\,\mathbf{C} + n\,\mathbf{D}).$$

Also $$\mathbf{A}' = \overline{AA'} = \overline{AB} + \overline{BA'}.$$

The vector $\overline{BA'}$ is coplanar with $\overline{BC} = \mathbf{C} - \mathbf{B}$ and $\overline{BD} = \mathbf{D} - \mathbf{B}$. Hence it may be expressed in terms of them.

$$\mathbf{A}' = \mathbf{B} + x_1\,(\mathbf{C} - \mathbf{B}) + y_1\,(\mathbf{D} - \mathbf{B}).$$

Equating coefficients $$k_1\,m = x_1,$$

$$k_1\,n = y_1,$$

$$k_1\,l = 1 - x_1 - y_1.$$

Hence $$k_1 = \frac{1}{l + m + n}$$

and $$\frac{PA'}{AA'} = \frac{k_1 - 1}{k_1} = 1 - (l + m + n).$$

In like manner $\overline{A\,B'} = x_2\,\mathbf{C} + y_2\,\mathbf{D}$

and $\overline{AB'} = \overline{A\,B} + \overline{B\,B'} = \mathbf{B} + k_2\,(\mathbf{P} - \mathbf{B}).$

Hence $x_2\,\mathbf{C} + y_2\,\mathbf{D} = \mathbf{B} + k_2\,(l\,\mathbf{B} + m\,\mathbf{C} + n\,\mathbf{D} - \mathbf{B})$

and
$$0 = 1 + k_2\,(l - 1),$$
$$x_2 = k_2\,m,$$
$$y_2 = k_2\,n.$$

Hence
$$k_2 = \frac{1}{1 - l}$$

and
$$\frac{P\,B'}{B\,B'} = \frac{k_2 - 1}{k_2} = l.$$

In the same way it may be shown that

$$\frac{P\,C'}{C\,C'} = m \text{ and } \frac{P\,D'}{D\,D'} = n.$$

Adding the four ratios the result is

$$1 - (l + m + n) + l + m + n = 1.$$

Example 2: To find a line which passes through a given point and cuts two given lines in space.

Let the two lines be fixed respectively by two points A and B, C and D on each. Let O be the given point. Choose it as origin and let

$$\mathbf{A} = \overline{OA}, \quad \mathbf{B} = \overline{OB}, \quad \mathbf{C} = \overline{OC}, \quad \mathbf{D} = \overline{OD}.$$

Any point P of $A\,B$ may be expressed as

$$\mathbf{P} = \overline{O\,P} = \overline{O\,A} + x\,\overline{A\,B} = \mathbf{A} + x\,(\mathbf{B} - \mathbf{A}).$$

Any point Q of CD may likewise be written

$$\mathbf{Q} = \overline{O\,Q} = \overline{O\,C} + y\,\overline{C\,D} = \mathbf{C} + y\,(\mathbf{D} - \mathbf{C}).$$

If the points P and Q lie in the same line through O, $\mathbf{P}$ and $\mathbf{Q}$ are collinear. That is

$$\mathbf{P} = z\,\mathbf{Q}.$$

Before it is possible to equate coefficients one of the four vectors must be expressed in terms of the other three.

Let $$\mathbf{D} = l\,\mathbf{A} + m\,\mathbf{B} + n\,\mathbf{C}.$$

Then $$\mathbf{P} = \mathbf{A} + x\,(\mathbf{B} - \mathbf{A})$$
$$= z\,[\mathbf{C} + y\,(l\,\mathbf{A} + m\,\mathbf{B} + n\,\mathbf{C} - \mathbf{C})].$$

Hence $$1 - x = z\,y\,l,$$
$$x = z\,y\,m,$$
$$0 = z\,[1 + y\,(n-1)].$$

Hence $$x = \frac{m}{l+m},$$
$$y = \frac{1}{1-n},$$
$$z = \frac{1-n}{l+m}.$$

Substituting in **P** and **Q**

$$\mathbf{P} = \frac{l\,\mathbf{A} + m\,\mathbf{B}}{l+m},$$

$$\mathbf{Q} = \frac{n\,\mathbf{C} - \mathbf{D}}{n-1}.$$

Either of these may be taken as defining a line drawn from O and cutting AB and CD.

Vector Relations independent of the Origin

20.] *Example 1:* To divide a line AB in a given ratio $m : n$ (Fig. 10).

Choose any arbitrary point O as origin. Let $\overline{OA} = \mathbf{A}$ and $\overline{OB} = \mathbf{B}$. To find the vector $\mathbf{P} = \overline{OP}$ of which the terminus P divides AB in the ratio $m : n$.

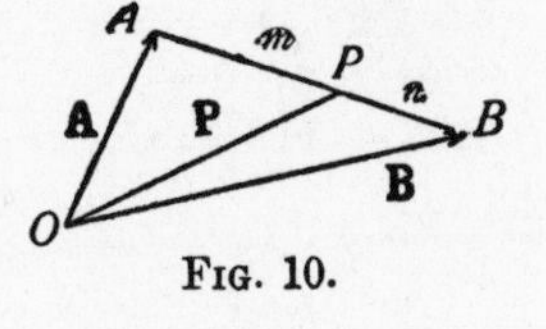

FIG. 10.

$$\mathbf{P} = \overline{OP} = \overline{OA} + \frac{m}{m+n}\,\overline{AB} = \mathbf{A} + \frac{m}{m+n}\,(\mathbf{B} - \mathbf{A}).$$

That is, $$\mathbf{P} = \frac{n\,\mathbf{A} + m\,\mathbf{B}}{m+n}. \tag{7}$$

The components of **P** parallel to **A** and **B** are in inverse ratio to the segments AP and PB into which the line AB is divided by the point P. If it should so happen that P divided the line AB externally, the ratio AP / PB would be negative, and the signs of m and n would be opposite, but the formula would hold without change if this difference of sign in m and n be taken into account.

Example 2: To find the point of intersection of the medians of a triangle.

Choose the origin O at random. Let ABC be the given triangle. Let $\overline{OA} = \mathbf{A}$, $\overline{OB} = \mathbf{B}$, and $\overline{OC} = \mathbf{C}$. Let A', B', C' be respectively the middle points of the sides opposite the vertices A, B, C. Let M be the point of intersection of the medians and $\mathbf{M} = OM$ the vector drawn to it. Then

$$\mathbf{M} = \overline{OM} = \overline{OA} + x\,\overline{AA'} = \mathbf{A} + x\left[\frac{(\mathbf{B} - \mathbf{A}) + (\mathbf{C} - \mathbf{A})}{2}\right]$$

and

$$\mathbf{M} = \overline{OM} = \overline{OB} + y\,\overline{BB'} = \mathbf{B} + y\left[\frac{(\mathbf{C} - \mathbf{B}) + (\mathbf{A} - \mathbf{B})}{2}\right].$$

Assuming that O has been chosen outside of the plane of the triangle so that **A, B, C** are non-coplanar, corresponding coefficients may be equated.

$$1 - x = \frac{1}{2}\,y,$$

$$\frac{1}{2}\,x = 1 - y,$$

$$\frac{1}{2}\,x = \frac{1}{2}\,y.$$

Hence
$$x = y = \frac{2}{3}.$$

Hence
$$\mathbf{M} = \frac{1}{3}\,(\mathbf{A} + \mathbf{B} + \mathbf{C}).$$

The vector drawn to the median point of a triangle is equal to one third of the sum of the vectors drawn to the vertices.

In the problems of which the solution has just been given the origin could be chosen arbitrarily and the result is independent of that choice. Hence it is even possible to disregard the origin entirely and replace the vectors **A**, **B**, **C**, etc., by their termini A, B, C, etc. Thus the points themselves become the subjects of analysis and the formulæ read

$$P = \frac{n\,A + m\,B}{m + n}$$

and

$$M = \frac{1}{3}\,(A + B + C).$$

This is typical of a whole class of problems soluble by vector methods. In fact any *purely geometric* relation between the different parts of a figure must necessarily be independent of the origin assumed for the analytic demonstration. In some cases, such as those in Arts. 18, 19, the position of the origin may be specialized with regard to some crucial point of the figure so as to facilitate the computation; but in many other cases the generality obtained by leaving the origin unspecialized and undetermined leads to a *symmetry* which renders the results just as easy to compute and more easy to remember.

Theorem: The necessary and sufficient condition that a vector equation represent a relation independent of the origin is that the sum of the scalar coefficients of the vectors on one side of the sign of equality is equal to the sum of the coefficients of the vectors upon the other side. Or if all the terms of a vector equation be transposed to one side leaving zero on the other, the sum of the scalar coefficients must be zero.

Let the equation written in the latter form be

$$a\,\mathbf{A} + b\,\mathbf{B} + c\,\mathbf{C} + d\,\mathbf{D} + \cdots = 0.$$

Change the origin from O to O' by adding a constant vector $\mathbf{R} = \overline{O\,O'}$ to each of the vectors **A**, **B**, **C**, **D**$\cdots$ The equation then becomes

$$a\,(\mathbf{A} + \mathbf{R}) + b\,(\mathbf{B} + \mathbf{R}) + c\,(\mathbf{C} + \mathbf{R}) + d\,(\mathbf{D} + \mathbf{R}) + \cdots = 0$$
$$= a\,\mathbf{A} + b\,\mathbf{B} + c\,\mathbf{C} + d\,\mathbf{D} + \cdots + \mathbf{R}\,(a + b + c + d + \ldots).$$

If this is to be independent of the origin the coefficient of **R** must vanish. Hence

$$a + b + c + d + \cdots = 0.$$

That this condition is fulfilled in the two examples cited is obvious.

If
$$\mathbf{P} = \frac{n\,\mathbf{A} + m\,\mathbf{B}}{m + n},$$
$$1 = \frac{n}{m + n} + \frac{m}{m + n}$$

If
$$\mathbf{M} = \frac{1}{3}\,(\mathbf{A} + \mathbf{B} + \mathbf{C}),$$
$$1 = \frac{1}{3} + \frac{1}{3} + \frac{1}{3}.$$

* **21.**] The necessary and sufficient condition that *two* vectors satisfy an equation, in which the sum of the scalar coefficients is zero, is that the vectors be equal in magnitude and in direction.

First let
$$a\,\mathbf{A} + b\,\mathbf{B} = 0$$
and
$$a + b = 0.$$

It is of course assumed that not both the coefficients a and b vanish. If they did the equation would mean nothing. Substitute the value of a obtained from the second equation into the first.

$$-b\,\mathbf{A} + b\,\mathbf{B} = 0.$$

Hence
$$\mathbf{A} = \mathbf{B}.$$

Secondly if $\mathbf{A}$ and $\mathbf{B}$ are equal in magnitude and direction the equation

$$\mathbf{A} - \mathbf{B} = 0$$

subsists between them. The sum of the coefficients is zero.

The necessary and sufficient condition that three vectors satisfy an equation, in which the sum of the scalar coefficients is zero, is that when drawn from a common origin they terminate in the same straight line.[1]

First let $$a\,\mathbf{A} + b\,\mathbf{B} + c\,\mathbf{C} = 0$$
and $$a + b + c = 0.$$

Not all the coefficients $a, b, c,$ vanish or the equations would be meaningless. Let c be a non-vanishing coefficient. Substitute the value of a obtained from the second equation into the first.

$$-(b + c)\,\mathbf{A} + b\,\mathbf{B} + c\,\mathbf{C} = 0,$$
or $$c\,(\mathbf{C} - \mathbf{A}) = b\,(\mathbf{A} - \mathbf{B}).$$

Hence the vector which joins the extremities of $\mathbf{C}$ and $\mathbf{A}$ is collinear with that which joins the extremities of $\mathbf{A}$ and $\mathbf{B}$. Hence those three points A, B, C lie on a line. Secondly suppose three vectors $\mathbf{A} = \overline{OA}, \mathbf{B} = \overline{OB}, \mathbf{C} = \overline{OC}$ drawn from the same origin O terminate in a straight line. Then the vectors

$$\overline{AB} = \mathbf{B} - \mathbf{A} \text{ and } \overline{AC} = \mathbf{C} - \mathbf{A}$$

are collinear. Hence the equation

$$(\mathbf{B} - \mathbf{A}) = x\,(\mathbf{C} - \mathbf{A})$$

subsists. The sum of the coefficients on the two sides is the same.

The necessary and sufficient condition that an equation, in which the sum of the scalar coefficients is zero, subsist

[1] Vectors which have a common origin and terminate in one line are called by Hamilton "*termino-collinear.*"

between four vectors, is that if drawn from a common origin they terminate in one plane.[1]

First let $$a\,\mathbf{A} + b\,\mathbf{B} + c\,\mathbf{C} + d\,\mathbf{D} = 0$$
and $$a + b + c + d = 0.$$

Let d be a non-vanishing coefficient. Substitute the value of a obtained from the last equation into the first.

$$-(b + c + d)\,\mathbf{A} + b\,\mathbf{B} + c\,\mathbf{C} + d\,\mathbf{D} = 0,$$
or $$d\,(\mathbf{D} - \mathbf{A}) = b\,(\mathbf{A} - \mathbf{B}) + c\,(\mathbf{A} - \mathbf{C}).$$

The line AD is coplanar with AB and AC. Hence all four termini A, B, C, D of $\mathbf{A}, \mathbf{B}, \mathbf{C}, \mathbf{D}$ lie in one plane. Secondly suppose that the termini of $\mathbf{A}, \mathbf{B}, \mathbf{C}, \mathbf{D}$ do lie in one plane. Then $\overline{AD} = \mathbf{D} - \mathbf{A}$, $\overline{AC} = \mathbf{C} - \mathbf{A}$, and $\overline{AB} = \mathbf{B} - \mathbf{A}$ are coplanar vectors. One of them may be expressed in terms of the other two. This leads to the equation

$$l\,(\mathbf{B} - \mathbf{A}) + m\,(\mathbf{C} - \mathbf{A}) + n\,(\mathbf{D} - \mathbf{A}) = 0,$$

where l, m, and n are certain scalars. The sum of the coefficients in this equation is zero.

Between any five vectors there exists one equation the sum of whose coefficients is zero.

Let $\mathbf{A}, \mathbf{B}, \mathbf{C}, \mathbf{D}, \mathbf{E}$ be the five given vectors. Form the differences

$$\mathbf{E} - \mathbf{A}, \quad \mathbf{E} - \mathbf{B}, \quad \mathbf{E} - \mathbf{C}, \quad \mathbf{E} - \mathbf{D}.$$

One of these may be expressed in terms of the other three — or what amounts to the same thing there must exist an equation between them.

$$k\,(\mathbf{E} - \mathbf{A}) + l\,(\mathbf{E} - \mathbf{B}) + m\,(\mathbf{E} - \mathbf{C}) + n\,(\mathbf{E} - \mathbf{D}) = 0.$$

The sum of the coefficients of this equation is zero.

[1] Vectors which have a common origin and terminate in one plane are called by Hamilton "*termino-complanar.*"

*22.] The results of the foregoing section afford simple solutions of many problems connected solely with the geometric properties of figures. Special theorems, the vector equations of lines and planes, and geometric nets in two and three dimensions are taken up in order.

Example 1: If a line be drawn parallel to the base of a triangle, the line which joins the opposite vertex to the intersection of the diagonals of the trapezoid thus formed bisects the base (Fig. 11).

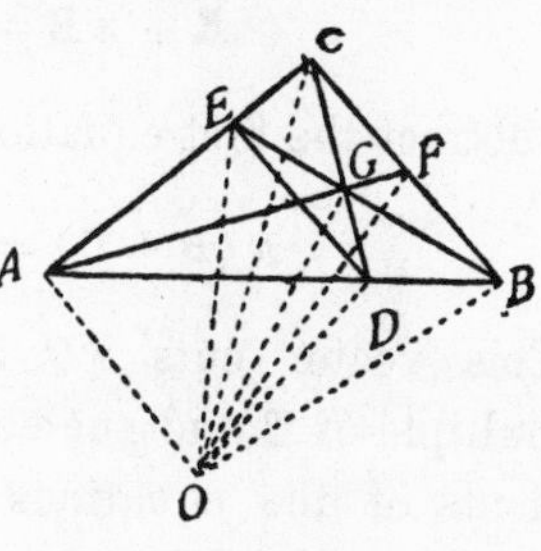

FIG. 11.

Let ABC be the triangle, ED the line parallel to the base CB, G the point of intersection of the diagonals EB and DC of the trapezoid $CBDE$, and F the intersection of AG with CB. To show that F bisects CB. Choose the origin at random. Let the vectors drawn from it to the various points of the figure be denoted by the corresponding Clarendons as usual. Then since ED is by hypothesis parallel to CB, the equation

$$\mathbf{E} - \mathbf{D} = n\,(\mathbf{C} - \mathbf{B})$$

holds true. The sum of the coefficients is evidently zero as it should be. Rearrange the terms so that the equation takes on the form

$$\mathbf{E} - n\,\mathbf{C} = \mathbf{D} - n\,\mathbf{B}.$$

The vector $\mathbf{E} - n\,\mathbf{C}$ is coplanar with $\mathbf{E}$ and $\mathbf{C}$. It must cut the line EC. The equal vector $\mathbf{D} - n\,\mathbf{B}$ is coplanar with $\mathbf{D}$ and $\mathbf{B}$. It must cut the line DB. Consequently the vector represented by either side of this equation must pass through the point A. Hence

$$\mathbf{E} - n\,\mathbf{C} = \mathbf{D} - n\,\mathbf{B} = x\,\mathbf{A}.$$

However the points E, C, and A lie upon the same straight line. Hence the equation which connects the vectors **E**, **C**, and **A** must be such that the sum of its coefficients is zero. This determines x as $1 - n$.

Hence $$\mathbf{E} - n\,\mathbf{C} = \mathbf{D} - n\,\mathbf{B} = (1 - n)\,\mathbf{A}.$$

By another rearrangement and similar reasoning

$$\mathbf{E} + n\,\mathbf{B} = \mathbf{D} + n\,\mathbf{C} = (1 + n)\,\mathbf{G}.$$

Subtract the first equation from the second:

$$n\,(\mathbf{B} + \mathbf{C}) = (1 + n)\,\mathbf{G} - (1 - n)\,\mathbf{A}.$$

This vector cuts $B\,C$ and $A\,G$. It must therefore be a multiple of **F** and such a multiple that the sum of the coefficients of the equations which connect **B**, **C**, and **F** or **G**, **A**, and **F** shall be zero.

Hence $$n\,(\mathbf{B} + \mathbf{C}) = (1 + n)\,\mathbf{G} - (1 - n)\,\mathbf{A} = 2\,n\,\mathbf{F}.$$

Hence $$\mathbf{F} = \frac{\mathbf{B} + \mathbf{C}}{2},$$

and the theorem has been proved. The proof has covered considerable space because each detail of the reasoning has been given. In reality, however, the actual analysis has consisted of just four equations obtained simply from the first.

Example 2: To determine the equations of the line and plane.

Let the line be fixed by two points A and B upon it. Let P be any point of the line. Choose an arbitrary origin. The vectors **A**, **B**, and **P** terminate in the same line. Hence

$$a\,\mathbf{A} + b\,\mathbf{B} + p\,\mathbf{P} = 0$$

and $$a + b + p = 0.$$

Therefore $$\mathbf{P} = \frac{a\,\mathbf{A} + b\,\mathbf{B}}{a + b}.$$

For different points P the scalars a and b have different values. They may be replaced by x and y, which are used more generally to represent variables. Then

$$\mathbf{P} = \frac{x\,\mathbf{A} + y\,\mathbf{B}}{x + y}.$$

Let a plane be determined by three points A, B, and C. Let P be any point of the plane. Choose an arbitrary origin. The vectors $\mathbf{A}$, $\mathbf{B}$, $\mathbf{C}$, and $\mathbf{P}$ terminate in one plane. Hence

$$a\,\mathbf{A} + b\,\mathbf{B} + c\,\mathbf{C} + p\,\mathbf{P} = 0$$

and
$$a + b + c + p = 0.$$

Therefore
$$\mathbf{P} = \frac{a\,\mathbf{A} + b\,\mathbf{B} + c\,\mathbf{C}}{a + b + c}.$$

As a, b, c, vary for different points of the plane, it is more customary to write in their stead x, y, z.

$$\mathbf{P} = \frac{x\,\mathbf{A} + y\,\mathbf{B} + z\,\mathbf{C}}{x + y + z}.$$

Example 3: The line which joins one vertex of a complete quadrilateral to the intersection of two diagonals divides the opposite sides harmonically (Fig. 12).

Let A, B, C, D be four vertices of a quadrilateral. Let AB meet CD in a fifth vertex E, and AD meet BC in the sixth vertex F. Let the two diagonals AC and BD intersect in G. To show that FG intersects AB in a point E' and CD in a point E'' such that the lines AB and CD are divided internally at E' and E'' in the same ratio as they are divided externally by E. That is to show that the cross ratios

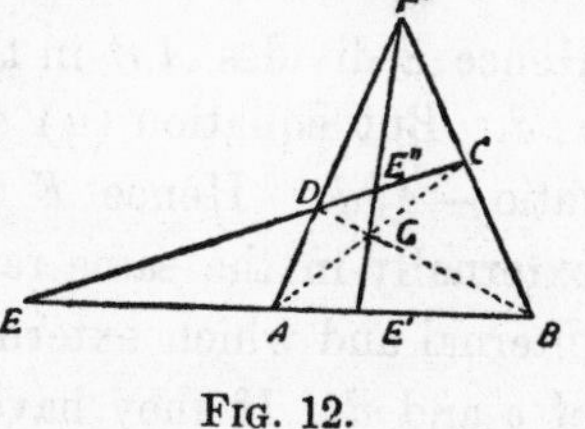

FIG. 12.

$$(AB \cdot EE') = (CD \cdot EE'') = -1.$$

Choose the origin at random. The four vectors **A, B, C, D** drawn from it to the points A, B, C, D terminate in one plane. Hence

$$a\,\mathbf{A} + b\,\mathbf{B} + c\,\mathbf{C} + d\,\mathbf{D} = 0$$

and

$$a + b + c + d = 0.$$

Separate the equations by transposing two terms:

$$a\,\mathbf{A} + c\,\mathbf{C} = -\,(b\,\mathbf{B} + d\,\mathbf{D}),$$
$$a + c = -\,(b + d).$$

Divide:

$$\mathbf{G} = \frac{a\,\mathbf{A} + c\,\mathbf{C}}{a + c} = \frac{b\,\mathbf{B} + d\,\mathbf{D}}{b + d}.$$

In like manner

$$\mathbf{F} = \frac{a\,\mathbf{A} + d\,\mathbf{D}}{a + d} = \frac{b\,\mathbf{B} + c\,\mathbf{C}}{b + c}.$$

Form:

$$\frac{(a + c)\,\mathbf{G} - (a + d)\,\mathbf{F}}{(a + c) - (a + d)} = \frac{c\,\mathbf{C} - d\,\mathbf{D}}{(a + c) - (a + d)}$$

or

$$\frac{(a + c)\,\mathbf{G} - (a + d)\,\mathbf{F}}{c - d} = \frac{c\,\mathbf{C} - d\,\mathbf{D}}{c - d} = \mathbf{E}''. \qquad (a)$$

Separate the equations again and divide:

$$\frac{a\,\mathbf{A} + b\,\mathbf{B}}{a + b} = \frac{c\,\mathbf{C} + d\,\mathbf{D}}{c + d} = \mathbf{E}. \qquad (b)$$

Hence E divides AB in the ratio $a : b$ and CD in the ratio $c : d$. But equation (a) shows that E'' divides CD in the ratio $-\,c : d$. Hence E and E'' divide CD internally and externally in the same ratio. Which of the two divisions is internal and which external depends upon the relative signs of c and d. If they have the same sign the internal point of division is E; if opposite signs, it is E''. In a similar way E' and E may be shown to divide AB harmonically.

Example 4: To discuss geometric nets.

By a geometric net in a plane is meant a figure composed of points and straight lines obtained in the following manner. Start with a certain number of points all of which lie in one plane. Draw all the lines joining these points in pairs. These lines will intersect each other in a number of points. Next draw all the lines which connect these points in pairs. This second set of lines will determine a still greater number of points which may in turn be joined in pairs and so on. The construction may be kept up indefinitely. At each step the number of points and lines in the figure increases. Probably the most interesting case of a plane geometric net is that in which *four* points are given to commence with. Joining these there are six lines which intersect in three points different from the given four. Three new lines may now be drawn in the figure. These cut out six new points. From these more lines may be obtained and so on.

To treat this net analytically write down the equations

$$a\,\mathbf{A} + b\,\mathbf{B} + c\,\mathbf{C} + d\,\mathbf{D} = 0 \qquad (c)$$

and

$$a + b + c + d = 0$$

which subsist between the four vectors drawn from an undetermined origin to the four given points. From these it is possible to obtain

$$\mathbf{E} = \frac{a\,\mathbf{A} + b\,\mathbf{B}}{a + b} = \frac{c\,\mathbf{C} + d\,\mathbf{D}}{c + d},$$

$$\mathbf{F} = \frac{a\,\mathbf{A} + c\,\mathbf{C}}{a + c} = \frac{b\,\mathbf{B} + d\,\mathbf{D}}{b + d},$$

$$\mathbf{G} = \frac{a\,\mathbf{A} + d\,\mathbf{D}}{a + d} = \frac{b\,\mathbf{B} + c\,\mathbf{C}}{b + c},$$

by splitting the equations into two parts and dividing. Next four vectors such as **A**, **D**, **E**, **F** may be chosen and the equation the sum of whose coefficients is zero may be determined. This would be

$$-\,a\,\mathbf{A} + d\,\mathbf{D} + (a + b)\,\mathbf{E} + (a + c)\,\mathbf{F} = 0.$$

By treating this equation as (c) was treated new points may be obtained.

$$\mathbf{H} = \frac{-a\,\mathbf{A} + d\,\mathbf{D}}{-a + d} = \frac{(a+b)\,\mathbf{E} + (a+c)\,\mathbf{F}}{2a + b + c},$$

$$\mathbf{I} = \frac{-a\,\mathbf{A} + (a+b)\,\mathbf{E}}{b} = \frac{d\,\mathbf{D} + (a+c)\,\mathbf{F}}{a + c + d},$$

$$\mathbf{K} = \frac{-a\mathbf{A} + (a+c)\mathbf{F}}{c} = \frac{d\,\mathbf{D} + (a+b)\,\mathbf{E}}{a + b + d}.$$

Equations between other sets of four vectors selected from **A, B, C, D, E, F, G** may be found; and from these more points obtained. The process of finding more points goes forward indefinitely. A fuller account of geometric nets may be found in Hamilton's "*Elements of Quaternions*," Book I.

As regards geometric nets in space just a word may be said. Five points are given. From these new points may be obtained by finding the intersections of planes passed through sets of three of the given points with lines connecting the remaining pairs. The construction may then be carried forward with the points thus obtained. The analytic treatment is similar to that in the case of plane nets. There are five vectors drawn from an undetermined origin to the given five points. Between these vectors there exists an equation the sum of whose coefficients is zero. This equation may be separated into parts as before and the new points may thus be obtained.

If
$$a\,\mathbf{A} + b\,\mathbf{B} + c\,\mathbf{C} + d\,\mathbf{D} + e\,\mathbf{E} = 0$$

and
$$a + b + c + d + e = 0,$$

then
$$\mathbf{F} = \frac{a\,\mathbf{A} + b\,\mathbf{B}}{a + b} = \frac{c\,\mathbf{C} + d\,\mathbf{D} + e\,\mathbf{E}}{c + d + e},$$

$$\mathbf{H} = \frac{a\,\mathbf{A} + c\,\mathbf{C}}{a + b} = \frac{b\,\mathbf{B} + d\,\mathbf{D} + e\,\mathbf{E}}{b + d + c},$$

are two of the points and others may be found in the same way. Nets in space are also discussed by Hamilton, *loc. cit.*

Centers of Gravity

*** 23.]** The *center of gravity* of a system of particles may be found very easily by vector methods. The two laws of physics which will be assumed are the following:

1°. The center of gravity of *two* masses (considered as situated at points) lies on the line connecting the two masses and divides it into two segments which are inversely proportional to the masses at the extremities.

2°. In finding the center of gravity of two systems of masses each system may be replaced by a single mass equal in magnitude to the sum of the masses in the system and situated at the center of gravity of the system.

Given two masses a and b situated at two points A and B. Their center of gravity G is given by

$$\mathbf{G} = \frac{a\,\mathbf{A} + b\,\mathbf{B}}{a + b}, \tag{8}$$

where the vectors are referred to any origin whatsoever. This follows immediately from law 1 and the formula (7) for division of a line in a given ratio.

The center of gravity of three masses a, b, c situated at the three points A, B, C may be found by means of law 2. The masses a and b may be considered as equivalent to a single mass $a + b$ situated at the point

$$\frac{a\,\mathbf{A} + b\,\mathbf{B}}{a + b}.$$

Then
$$\mathbf{G} = \frac{(a + b)\,\dfrac{a\,\mathbf{A} + b\,\mathbf{B}}{a + b} + c\,\mathbf{C}}{a + b + c}.$$

Hence
$$\mathbf{G} = \frac{a\,\mathbf{A} + b\,\mathbf{B} + c\,\mathbf{C}}{a + b + c}.$$

Evidently the center of gravity of any number of masses $a, b, c, d, \ldots$ situated at the points $A, B, C, D, \ldots$ may be found in a similar manner. The result is

$$\mathbf{G} = \frac{a\,\mathbf{A} + b\,\mathbf{B} + c\,\mathbf{C} + d\,\mathbf{D} + \cdots}{a + b + c + d + \cdots}. \tag{9}$$

Theorem 1: The lines which join the center of gravity of a triangle to the vertices divide it into three triangles which are proportional to the masses at the opposite vertices (Fig. 13). Let A, B, C be the vertices of a triangle weighted with masses a, b, c. Let G be the center of gravity. Join A, B, C to G and produce the lines until they intersect the opposite sides in A', B', C' respectively. To show that the areas

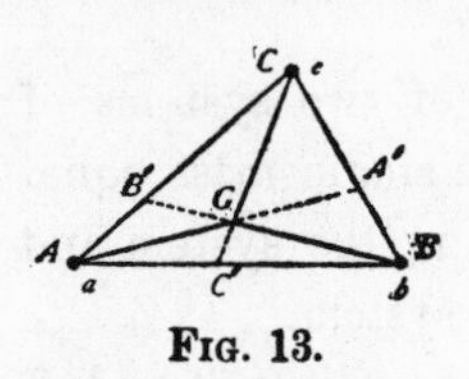

Fig. 13.

$$G\,B\,C : G\,C\,A : G\,A\,B : A\,B\,C = a : b : c : a + b + c.$$

The last proportion between $A\,B\,C$ and $a + b + c$ comes from compounding the first three. It is, however, useful in the demonstration.

$$\frac{A\,B\,C}{G\,B\,C} = \frac{A\,A'}{G\,A'} = \frac{A\,G}{G\,A'} + \frac{G\,A'}{G\,A'} = \frac{b + c}{a} + 1.$$

Hence $$\frac{A\,B\,C}{G\,B\,C} = \frac{a + b + c}{a}.$$

In a similar manner $$\frac{B\,C\,A}{G\,C\,A} = \frac{a + b + c}{b},$$

and $$\frac{C\,A\,B}{G\,A\,B} = \frac{a + b + c}{c}.$$

Hence the proportion is proved.

Theorem 2: The lines which join the center of gravity of a tetrahedron to the vertices divide the tetrahedron into four tetrahedra which are proportional to the masses at the opposite vertices.

Let A, B, C, D be the vertices of the tetrahedron weighted respectively with weights a, b, c, d. Let G be the center of gravity. Join A, B, C, D to G and produce the lines until they meet the opposite faces in A', B', C', D'. To show that the volumes

$$BCDG : CDAG : DABG : ABCG : ABCD$$
$$= a : b : c : d : a + b + c + d.$$

$$\frac{BCDA}{BCDG} = \frac{AA'}{GA'} = \frac{AG}{GA'} + \frac{GA'}{GA'} = \frac{b+c+d}{a} + 1$$
$$= \frac{a+b+c+d}{a}.$$

In like manner $\quad \dfrac{CDAG}{CDAB} = \dfrac{a+b+c+d}{b},$

and $\quad \dfrac{DABG}{DABC} = \dfrac{a+b+c+d}{c},$

and $\quad \dfrac{ABCG}{ABCD} = \dfrac{a+b+c+d}{d},$

which proves the proportion.

* **24.**] By a suitable choice of the three masses, a, b, c located at the vertices A, B, C, the center of gravity G may be made to coincide with any given point P of the triangle. If this be not obvious from physical considerations it certainly becomes so in the light of the foregoing theorems. For in order that the center of gravity fall at P, it is only necessary to choose the masses a, b, c proportional to the areas of the triangles PBC, PCA, and PAB respectively. Thus not merely one set of masses a, b, c may be found, but an infinite number of sets which differ from each other only by a common factor of proportionality. These quantities

a, b, c may therefore be looked upon as coördinates of the points P inside of the triangle ABC. To each set there corresponds a definite point P, and to each point P there corresponds an infinite number of sets of quantities, which however do not differ from one another except for a factor of proportionality.

To obtain the points P of the plane ABC which lie *outside* of the triangle ABC one may resort to the conception of negative weights or masses. The center of gravity of the masses 2 and -1 situated at the points A and B respectively would be a point G dividing the line AB *externally* in the ratio $1:2$. That is

$$GA : GB = 1 : 2.$$

Any point of the line AB produced may be represented by a suitable set of masses a, b which differ in sign. Similarly any point P of the plane ABC may be represented by a suitable set of masses a, b, c of which one will differ in sign from the other two if the point P lies outside of the triangle ABC. Inasmuch as only the ratios of a, b, and c are important two of the quantities may always be taken positive.

The idea of employing the masses situated at the vertices as coördinates of the center of gravity is due to Möbius and was published by him in his book entitled "*Der barycentrische Calcul*," in 1827. This may be fairly regarded as the starting point of modern analytic geometry.

The conception of negative masses which have no existence in nature may be avoided by replacing the masses at the vertices by the areas of the triangles GBC, GCA, and GAB to which they are proportional. The coördinates of a point P would then be three numbers proportional to the areas of the three triangles of which P is the common vertex; and the sides of a given triangle ABC, the bases. The sign of these areas is determined by the following definition.

Definition: The area ABC of a triangle is said to be positive when the vertices A, B, C follow each other in the positive or counterclockwise direction upon the circle described through them. The area is said to be negative when the points follow in the negative or clockwise direction.

Cyclic permutation of the letters therefore does not alter the sign of the area.

$$ABC = BCA = CAB.$$

Interchange of two letters which amounts to a reversal of the cyclic order changes the sign.

$$ACB = BAC = CBA = -ABC.$$

If P be any point within the triangle the equation

$$PAB + PBC + PCA = ABC$$

must hold. The same will also hold if P be outside of the triangle provided the *signs* of the areas be taken into consideration. The areas or three quantities proportional to them may be regarded as coördinates of the point P.

The extension of the idea of "*barycentric*" coördinates to space is immediate. The four points A, B, C, D situated at the vertices of a tetrahedron are weighted with mass a, b, c, d respectively. The center of gravity G is represented by these quantities or four others proportional to them. To obtain points outside of the tetrahedron negative masses may be employed. Or in the light of theorem 2, page 40, the masses may be replaced by the four tetrahedra which are proportional to them. Then the idea of negative volumes takes the place of that of negative weights. As this idea is of considerable importance later, a brief treatment of it here may not be out of place.

Definition: The volume $ABCD$ of a tetrahedron is said to be positive when the triangle ABC appears positive to the eye situated at the point D. The volume is negative if the area of the triangle appear negative.

To make the discussion of the signs of the various tetrahedra perfectly clear it is *almost necessary* to have a solid model. A *plane* drawing is scarcely sufficient. It is difficult to see from it which triangles appear positive and which negative. The following relations will be seen to hold if a model be examined.

The interchange of two letters in the tetrahedron $ABCD$ changes the sign.

$$ACBD = CBAD = BACD = DBCA$$
$$= ADCB = ABDC = -ABCD.$$

The sign of the tetrahedron for any given one of the possible twenty-four arrangements of the letters may be obtained by reducing that arrangement to the order $A\ B\ C\ D$ by means of a number of successive interchanges of two letters. If the number of interchanges is even the sign is the same as that of $ABCD$; if odd, opposite. Thus

$$CADB = -CABD = +ACBD = -ABCD.$$

If P is any point inside of the tetrahedron $ABCD$ the equation

$$ABCP - BCDP + CDAP - DABP = ABCD$$

holds good. It still is true if P be without the tetrahedron provided the signs of the volumes be taken into consideration. The equation may be put into a form more symmetrical and more easily remembered by transposing all the terms to one number. Then

$$ABCD + BCDP + CDPA + DPAB + PABC = 0.$$

The proportion in theorem 2, page 40, does not hold true if the signs of the tetrahedra be regarded. It should read

$$BCDG : CDGA : DGAB : GABC : ABCD$$
$$= a : b : c : d : a + b + c + d.$$

If the point G lies inside the tetrahedron a, b, c, d represent quantities proportional to the masses which must be located at the vertices A, B, C, D respectively if G is to be the center of gravity. If G lies outside of the tetrahedron they may still be regarded as masses some of which are negative — or perhaps better merely as four numbers whose ratios determine the position of the point G. In this manner a set of "*barycentric*" coördinates is established for space.

The vector **P** drawn from an indeterminate origin to any point of the plane ABC is (page 35)

$$\mathbf{P} = \frac{x\,\mathbf{A} + y\,\mathbf{B} + z\,\mathbf{C}}{x + y + z}.$$

Comparing this with the expression

$$\mathbf{G} = \frac{a\,\mathbf{A} + b\,\mathbf{B} + c\,\mathbf{C}}{a + b + c}$$

it will be seen that the quantities x, y, z are in reality nothing more nor less than the barycentric coördinates of the point P with respect to the triangle ABC. In like manner from equation

$$\mathbf{P} = \frac{x\,\mathbf{A} + y\,\mathbf{B} + z\,\mathbf{C} + w\,\mathbf{D}}{x + y + z + w}$$

which expresses any vector **P** drawn from an indeterminate origin in terms of four given vectors **A, B, C, D** drawn from the same origin, it may be seen by comparison with

$$\mathbf{G} = \frac{a\,\mathbf{A} + b\,\mathbf{B} + c\,\mathbf{C} + d\,\mathbf{D}}{a + b + c + d}$$

that the four quantities x, y, z, w are precisely the barycentric coördinates of P, the terminus of **P**, with respect to the tetrahedron $ABCD$. Thus the vector methods in which the origin is undetermined and the methods of the "*Barycentric Calculus*" are practically co-extensive.

It was mentioned before and it may be well to repeat here

that the origin may be left wholly out of consideration and the vectors replaced by their termini. The vector equations then become point equations

$$P = \frac{x\,A + y\,B + z\,C}{x + y + z}$$

and
$$P = \frac{x\,A + y\,B + z\,C + w\,D}{x + y + z + w}.$$

This step brings in the points themselves as the objects of analysis and leads still nearer to the "*Barycentrische Calcül*" of Möbius and the "*Ausdehnungslehre*" of Grassmann.

The Use of Vectors to denote Areas

25.] *Definition:* An area lying in one plane MN and bounded by a continuous curve PQR which nowhere cuts itself is said to appear positive from the point O when the letters PQR follow each other in the counterclockwise or positive order; negative, when they follow in the negative or clockwise order (Fig. 14).

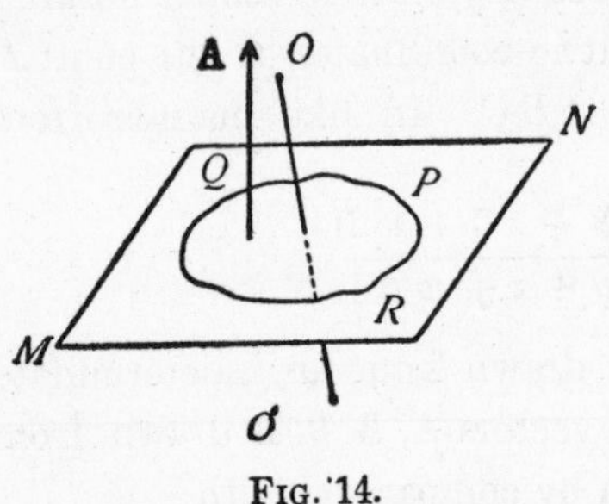

Fig. 14.

It is evident that an area can have no determined sign *per se*, but only in reference to that direction in which its boundary is supposed to be traced and to some point O outside of its plane. For the area $P\,R\,Q$ is negative relative to $P\,Q\,R$; and an area viewed from O is negative relative to the same area viewed from a point O' upon the side of the plane opposite to O. A circle lying in the XY-plane and described in the positive trigonometric order appears positive from every point on that side of the plane on which the positive Z-axis lies, but negative from all points on the side upon which the negative Z-axis lies. For this reason the point of view and the direction of description of the boundary must be kept clearly in mind.

Another method of stating the definition is as follows: If a person walking upon a plane traces out a closed curve, the area enclosed is said to be positive if it lies upon his left-hand side, negative if upon his right. It is clear that if two persons be considered to trace out together the same curve by walking upon opposite sides of the plane the area enclosed will lie upon the right hand of one and the left hand of the other. To one it will consequently appear positive; to the other, negative. That side of the plane upon which the area seems positive is called the positive side; the side upon which it appears negative, the negative side. This idea is familiar to students of electricity and magnetism. If an electric current flow around a closed plane curve the lines of magnetic force through the circuit pass from the negative to the positive side of the plane. A positive magnetic pole placed upon the positive side of the plane will be repelled by the circuit.

A plane area may be looked upon as possessing more than positive or negative magnitude. It may be considered to possess direction, namely, the direction of the normal to the positive side of the plane in which it lies. Hence a plane area is a vector quantity. The following theorems concerning areas when looked upon as vectors are important.

Theorem 1: If a plane area be denoted by a vector whose magnitude is the numerical value of that area and whose direction is the normal upon the positive side of the plane, then the orthogonal projection of that area upon a plane will be represented by the component of that vector in the direction normal to the plane of projection (Fig. 15).

Let the area A lie in the plane MN. Let it be projected orthogonally upon the plane $M'N'$. Let MN and $M'N'$ intersect in the line l and let the diedral angle between these two planes be x. Consider first a rectangle $PQRS$ in MN whose sides, PQ, RS and QR, SP are respectively parallel and perpendicular to the line l. This will project into a rectangle $P'Q'R'S'$ in $M'N'$. The sides $P'Q'$ and $R'S'$ will be equal to PQ and RS; but the sides $Q'R'$ and $S'P'$ will be equal to QR and SP multiplied by the cosine of x, the angle between the planes. Consequently the rectangle

$$P'Q'R'S' = PQRS \cos x.$$

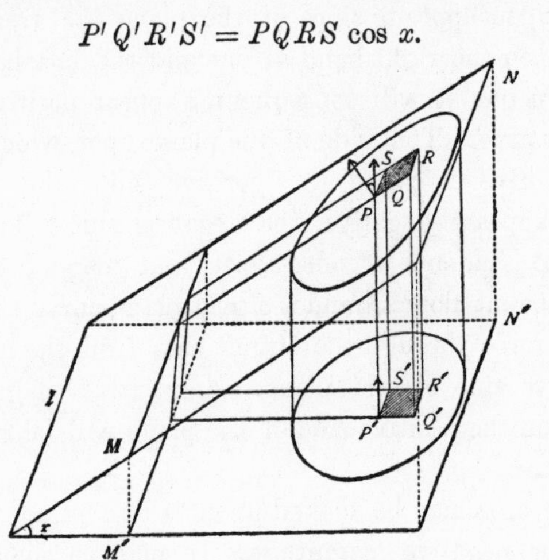

Fig. 15.

Hence rectangles, of which the sides are respectively parallel and perpendicular to l, the line of intersection of the two planes, project into rectangles whose sides are likewise respectively parallel and perpendicular to l and *whose area is equal to the area of the original rectangles multiplied by the cosine of the angle between the planes.*

From this it follows that *any* area A is projected into an area which is equal to the given area multiplied by the cosine of the angle between the planes. For any area A may be divided up into a large number of small rectangles by drawing a series of lines in MN parallel and perpendicular to the line l.

Each of these rectangles when projected is multiplied by the cosine of the angle between the planes and hence the total area is also multiplied by the cosine of that angle. On the other hand the component $\mathbf{A}'$ of the vector $\mathbf{A}$, which represents the given area, in the direction normal to the plane $M'N'$ of projection is equal to the total vector $\mathbf{A}$ multiplied by the cosine of the angle between its direction which is the normal to the plane MN and the normal to $M'N'$. This angle is x; for the angle between the normals to two planes is the same as the angle between the planes. The relation between the magnitudes of $\mathbf{A}$ and $\mathbf{A}'$ is therefore

$$A' = A \cos x,$$

which proves the theorem.

26.] *Definition:* Two plane areas regarded as vectors are said to be added when the vectors which represent them are added.

A vector area is consequently the sum of its three components obtainable by orthogonal projection upon three mutually perpendicular planes. Moreover in adding two areas each may be resolved into its three components, the corresponding components added as scalar quantities, and these sums compounded as vectors into the resultant area. A generalization of this statement to the case where the three planes are not mutually orthogonal and where the projection is oblique exists.

A surface made up of several plane areas may be represented by the vector which is the sum of all the vectors representing those areas. In case the surface be looked upon as forming the boundary or a portion of the boundary of a solid, those sides of the bounding planes which lie outside of the body are conventionally taken to be positive. The vectors which represent the faces of solids are always directed out from the solid, not into it.

Theorem 2: The vector which represents a *closed* polyhedral surface is zero.

This may be proved by means of certain considerations of hydrostatics. Suppose the polyhedron drawn in a body of fluid assumed to be free from all external forces, gravity included.[1] The fluid is in equilibrium under its own internal pressures. The portion of the fluid bounded by the closed surface moves neither one way nor the other. Upon each face of the surface the fluid exerts a definite force *proportional* to the *area* of the face and normal to it. The resultant of all these forces must be zero, as the fluid is in equilibrium. Hence the sum of all the vector areas in the closed surface is zero.

The proof may be given in a purely geometric manner. Consider the orthogonal projection of the closed surface upon any plane. This consists of a *double* area. The part of the surface farthest from the plane projects into positive area; the part nearest the plane, into negative area. Thus the surface projects into a certain portion of the plane which is covered twice, once with positive area and once with negative. These cancel each other. Hence the total projection of a closed surface upon a plane (if taken with regard to sign) is zero. But by theorem 1 the projection of an area upon a plane is equal to the component of the vector representing that area in the direction perpendicular to that plane. Hence the vector which represents a closed surface has no component along the line perpendicular to the plane of projection. This, however, was any plane whatsoever. Hence the vector is zero.

The theorem has been proved for the case in which the closed surface consists of planes. In case that surface be

[1] Such a state of affairs is realized to all practical purposes in the case of a polyhedron suspended in the atmosphere and consequently subjected to atmospheric pressure. The force of gravity acts but is counterbalanced by the tension in the suspending string.

curved it may be regarded as the limit of a polyhedral surface whose number of faces increases without limit. Hence the vector which represents any closed surface polyhedral or curved is zero. If the surface be not closed but be curved it may be represented by a vector just as if it were polyhedral. That vector is the limit [1] approached by the vector which represents that polyhedral surface of which the curved surface is the limit when the number of faces becomes indefinitely great.

Summary of Chapter I

A vector is a quantity considered as possessing magnitude and direction. Equal vectors possess the same magnitude and the same direction. A vector is not altered by shifting it parallel to itself. A null or zero vector is one whose magnitude is zero. To multiply a vector by a positive scalar multiply its length by that scalar and leave its direction unchanged. To multiply a vector by a negative scalar multiply its length by that scalar and reverse its direction.

Vectors add according to the parallelogram law. To subtract a vector reverse its direction and add. Addition, subtraction, and multiplication of vectors by a scalar follow the same laws as addition, subtraction, and multiplication in ordinary algebra. A vector may be resolved into three components parallel to any three non-coplanar vectors. This resolution can be accomplished in only one way.

$$\mathbf{r} = x\,\mathbf{a} + y\,\mathbf{b} + z\,\mathbf{c}. \tag{4}$$

The components of equal vectors, parallel to three given non-coplanar vectors, are equal, and conversely if the components are equal the vectors are equal. The three unit vectors $\mathbf{i}, \mathbf{j}, \mathbf{k}$ form a right-handed rectangular system. In

[1] This limit exists and is unique. It is independent of the method in which the polyhedral surface approaches the curved surface.

terms of them any vector may be expressed by means of the Cartesian coördinates x, y, z.

$$\mathbf{r} = x\,\mathbf{i} + y\,\mathbf{j} + z\,\mathbf{k}. \tag{6}$$

Applications. The point which divides a line in a given ratio $m : n$ is given by the formula

$$\mathbf{P} = \frac{n\,\mathbf{A} + m\,\mathbf{B}}{m + n}. \tag{7}$$

The necessary and sufficient condition that a vector equation represent a relation independent of the origin is that the sum of the scalar coefficients in the equation be zero. Between any four vectors there exists an equation with scalar coefficients. If the sum of the coefficients is zero the vectors are termino-coplanar. If an equation the sum of whose scalar coefficients is zero exists between three vectors they are termino-collinear. The center of gravity of a number of masses $a, b, c \cdots$ situated at the termini of the vectors $\mathbf{A}, \mathbf{B}, \mathbf{C} \cdots$ supposed to be drawn from a common origin is given by the formula

$$\mathbf{G} = \frac{a\,\mathbf{A} + b\,\mathbf{B} + c\,\mathbf{C} + \cdots}{a + b + c + \cdots}. \tag{9}$$

A vector may be used to denote an area. If the area is plane the magnitude of the vector is equal to the magnitude of the area, and the direction of the vector is the direction of the normal upon the positive side of the plane. The vector representing a *closed* surface is zero.

Exercises on Chapter I

1. Demonstrate the laws stated in Art. 12.

2. A triangle may be constructed whose sides are parallel and equal to the medians of any given triangle.

3. The six points in which the three diagonals of a complete quadrangle[1] meet the pairs of opposite sides lie three by three upon four straight lines.

4. If two triangles are so situated in space that the three points of intersection of corresponding sides lie on a line, then the lines joining the corresponding vertices pass through a common point and conversely.

5. Given a quadrilateral in space. Find the middle point of the line which joins the middle points of the diagonals. Find the middle point of the line which joins the middle points of two opposite sides. Show that these two points are the same and coincide with the center of gravity of a system of equal masses placed at the vertices of the quadrilateral.

6. If two opposite sides of a quadrilateral in space be divided proportionally and if two quadrilaterals be formed by joining the two points of division, then the centers of gravity of these two quadrilaterals lie on a line with the center of gravity of the original quadrilateral. By the center of gravity is meant the center of gravity of four equal masses placed at the vertices. Can this theorem be generalized to the case where the masses are not equal?

7. The bisectors of the angles of a triangle meet in a point.

8. If the edges of a hexahedron meet four by four in three points, the four diagonals of the hexahedron meet in a point. In the special case in which the hexahedron is a parallelopiped the three points are at an infinite distance.

9. Prove that the three straight lines through the middle points of the sides of any face of a tetrahedron, each parallel to the straight line connecting a fixed point P with the middle point of the opposite edge of the tetrahedron, meet in a

[1] A complete quadrangle consists of the six straight lines which may be passed through four points no three of which are collinear. The diagonals are the lines which join the points of intersection of pairs of sides.

point E and that this point is such that PE passes through and is bisected by the center of gravity of the tetrahedron.

10. Show that *without exception* there exists one vector equation with scalar coefficients between any four given vectors **A, B, C, D**.

11. Discuss the conditions imposed upon three, four, or five vectors if they satisfy *two* equations the sum of the coefficients in each of which is zero.

CHAPTER II

DIRECT AND SKEW PRODUCTS OF VECTORS

Products of Two Vectors

27.] THE operations of addition, subtraction, and scalar multiplication have been defined for vectors in the way suggested by physics and have been employed in a few applications. It now becomes necessary to introduce two new combinations of vectors. These will be called *products* because they obey the fundamental law of products; *i. e.*, the *distributive* law which states that the product of **A** into the sum of **B** and **C** is equal to the sum of the products of **A** into **B** and **A** into **C**.

Definition: The *direct product* of two vectors **A** and **B** is the *scalar* quantity obtained by multiplying the product of the magnitudes of the vectors by the *cosine* of the angle between them.

The direct product is denoted by writing the two vectors with a dot between them as

$$\mathbf{A} \cdot \mathbf{B}.$$

This is read **A** *dot* **B** and therefore may often be called the dot product instead of the direct product. It is also called the scalar product owing to the fact that its value is scalar. If A be the magnitude of **A** and B that of **B**, then by definition

$$\mathbf{A} \cdot \mathbf{B} = A\,B \cos(\mathbf{A}, \mathbf{B}). \tag{1}$$

Obviously the direct product follows the commutative law

$$\mathbf{A} \cdot \mathbf{B} = \mathbf{B} \cdot \mathbf{A}. \tag{2}$$

If either vector be multiplied by a scalar the product is multiplied by that scalar. That is

$$(x\,\mathbf{A})\cdot\mathbf{B} = \mathbf{A}\cdot(x\,\mathbf{B}) = x\,(\mathbf{A}\cdot\mathbf{B}).$$

In case the two vectors **A** and **B** are collinear the angle between them becomes zero or one hundred and eighty degrees and its cosine is therefore equal to unity with the positive or negative sign. Hence the scalar product of two parallel vectors is numerically equal to the product of their lengths. The sign of the product is positive when the directions of the vectors are the same, negative when they are opposite. The product of a vector by itself is therefore equal to the square of its length

$$\mathbf{A}\cdot\mathbf{A} = A^2. \tag{3}$$

Consequently if the product of a vector by itself vanish the vector is a null vector.

In case the two vectors **A** and **B** are perpendicular the angle between them becomes plus or minus ninety degrees and the cosine vanishes. Hence the product $\mathbf{A}\cdot\mathbf{B}$ vanishes. Conversely if the scalar product $\mathbf{A}\cdot\mathbf{B}$ vanishes, then

$$A\,B\cos(\mathbf{A},\mathbf{B}) = 0.$$

Hence either A or B or $\cos(\mathbf{A},\mathbf{B})$ is zero, and either the vectors are perpendicular or one of them is null. Thus *the condition for the perpendicularity of two vectors, neither of which vanishes, is* $\mathbf{A}\cdot\mathbf{B} = 0$.

28.] The scalar products of the three fundamental unit vectors **i**, **j**, **k** are evidently

$$\mathbf{i}\cdot\mathbf{i} = \mathbf{j}\cdot\mathbf{j} = \mathbf{k}\cdot\mathbf{k} = 1, \tag{4}$$
$$\mathbf{i}\cdot\mathbf{j} = \mathbf{j}\cdot\mathbf{k} = \mathbf{k}\cdot\mathbf{i} = 0.$$

If more generally **a** and **b** are any two unit vectors the product

$$\mathbf{a}\cdot\mathbf{b} = \cos(\mathbf{a},\mathbf{b}).$$

Thus the scalar product determines the cosine of the angle between two vectors and is in a certain sense equivalent to it. For this reason it might be better to give a purely geometric definition of the product rather than one which depends upon trigonometry. This is easily accomplished as follows: If $\mathbf{a}$ and $\mathbf{b}$ are two unit vectors, $\mathbf{a} \cdot \mathbf{b}$ is the length of the projection of either upon the other. If more generally $\mathbf{A}$ and $\mathbf{B}$ are any two vectors $\mathbf{A} \cdot \mathbf{B}$ is the product of the length of either by the length of projection of the other upon it. From these definitions the facts that the product of a vector by itself is the square of its length and the product of two perpendicular vectors is zero follow immediately. The trigonometric definition can also readily be deduced.

The scalar product of two vectors will appear whenever the cosine of the included angle is of importance. The following examples may be cited. The projection of a vector $\mathbf{B}$ upon a vector $\mathbf{A}$ is

$$\frac{\mathbf{A} \cdot \mathbf{B}}{\mathbf{A} \cdot \mathbf{A}} \mathbf{A} = \frac{A\,B}{A\,A} A\, \mathbf{a} \cos (\mathbf{A}, \mathbf{B}) = B \cos (\mathbf{A}, \mathbf{B})\, \mathbf{a}, \qquad (5)$$

where $\mathbf{a}$ is a unit vector in the direction of $\mathbf{A}$. If $\mathbf{A}$ is itself a unit vector the formula reduces to

$$(\mathbf{A} \cdot \mathbf{B})\, \mathbf{A} = B \cos (\mathbf{A}, \mathbf{B})\, \mathbf{A}.$$

If $\mathbf{A}$ be a constant force and $\mathbf{B}$ a displacement the work done by the force $\mathbf{A}$ during the displacement is $\mathbf{A} \cdot \mathbf{B}$. If $\mathbf{A}$ represent a plane area (Art. 25), and if $\mathbf{B}$ be a vector inclined to that plane, the scalar product $\mathbf{A} \cdot \mathbf{B}$ will be the volume of the cylinder of which the area $\mathbf{A}$ is the base and of which $\mathbf{B}$ is the directed slant height. For the volume (Fig. 16) is equal to the base A multiplied by the altitude h. This is the projection of $\mathbf{B}$ upon $\mathbf{A}$ or $B \cos (\mathbf{A}, \mathbf{B})$. Hence

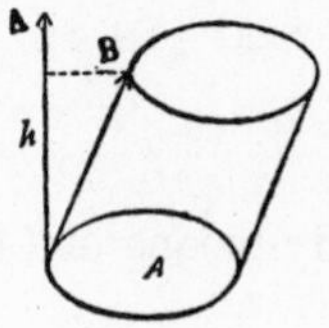

FIG. 16.

$$v = A\,h = A\,B \cos (\mathbf{A}, \mathbf{B}) = \mathbf{A} \cdot \mathbf{B}.$$

29.] The scalar or direct product follows the distributive law of multiplication. That is

$$(\mathbf{A} + \mathbf{B}) \cdot \mathbf{C} = \mathbf{A} \cdot \mathbf{C} + \mathbf{B} \cdot \mathbf{C}. \tag{6}$$

This may be proved by means of projections. Let **C** be equal to its magnitude C multiplied by a unit vector **c** in its direction. To show

$$(\mathbf{A} + \mathbf{B}) \cdot (C\,\mathbf{c}) = \mathbf{A} \cdot (C\,\mathbf{c}) + \mathbf{B} \cdot (C\,\mathbf{c})$$

or

$$(\mathbf{A} + \mathbf{B}) \cdot \mathbf{c} = \mathbf{A} \cdot \mathbf{c} + \mathbf{B} \cdot \mathbf{c}.$$

$\mathbf{A} \cdot \mathbf{c}$ is the projection of **A** upon **c**; $\mathbf{B} \cdot \mathbf{c}$, that of **B** upon **c**; $(\mathbf{A} + \mathbf{B}) \cdot \mathbf{c}$, that of $\mathbf{A} + \mathbf{B}$ upon **c**. But the projection of the sum $\mathbf{A} + \mathbf{B}$ is equal to the sum of the projections. Hence the relation (6) is proved. By an immediate generalization

$$\begin{aligned}(\mathbf{A} + \mathbf{B} + \cdots) \cdot (\mathbf{P} + \mathbf{Q} + \cdots) = {} & \mathbf{A} \cdot \mathbf{P} + \mathbf{A} \cdot \mathbf{Q} + \cdots \\ & + \mathbf{B} \cdot \mathbf{P} + \mathbf{B} \cdot \mathbf{Q} + \cdots \\ & + \cdot \;\cdot \;\cdot \;\cdot \;\cdot \;\cdot\end{aligned} \tag{6'}$$

The scalar product may be used just as the product in ordinary algebra. It has no peculiar difficulties.

If two vectors **A** and **B** are expressed in terms of the three unit vectors **i**, **j**, **k** as

$$\mathbf{A} = A_1\,\mathbf{i} + A_2\,\mathbf{j} + A_3\,\mathbf{k},$$

and

$$\mathbf{B} = B_1\,\mathbf{i} + B_2\,\mathbf{j} + B_3\,\mathbf{k},$$

then

$$\begin{aligned}\mathbf{A} \cdot \mathbf{B} &= (A_1\,\mathbf{i} + A_2\,\mathbf{j} + A_3\,\mathbf{k}) \cdot (B_1\,\mathbf{i} + B_2\,\mathbf{j} + B_3\,\mathbf{k}) \\ &= A_1 B_1\,\mathbf{i} \cdot \mathbf{i} + A_1 B_2\,\mathbf{i} \cdot \mathbf{j} + A_1 B_3\,\mathbf{i} \cdot \mathbf{k} \\ &\quad + A_2 B_1\,\mathbf{j} \cdot \mathbf{i} + A_2 B_2\,\mathbf{j} \cdot \mathbf{j} + A_2 B_3\,\mathbf{j} \cdot \mathbf{k} \\ &\quad + A_3 B_1\,\mathbf{k} \cdot \mathbf{j} + A_3 B_2\,\mathbf{k} \cdot \mathbf{j} + A_3 B_3\,\mathbf{k} \cdot \mathbf{k}.\end{aligned}$$

By means of (4) this reduces to

$$\mathbf{A} \cdot \mathbf{B} = A_1 B_1 + A_2 B_2 + A_3 B_3. \tag{7}$$

If in particular **A** and **B** are unit vectors, their components A_1, A_2, A_3 and B_1, B_2, B_3 are the direction cosines of the lines **A** and **B** referred to X, Y, Z.

$$A_1 = \cos(\mathbf{A}, X), \quad A_2 = \cos(\mathbf{A}, Y), \quad A_3 = \cos(\mathbf{A}, Z),$$
$$B_1 = \cos(\mathbf{B}, X), \quad B_2 = \cos(\mathbf{B}, Y), \quad B_3 = \cos(\mathbf{B}, Z).$$

Moreover $\mathbf{A} \cdot \mathbf{B}$ is the cosine of the included angle. Hence the equation becomes

$$\cos(\mathbf{A}, \mathbf{B}) = \cos(\mathbf{A}, X)\cos(\mathbf{B}, X) + \cos(\mathbf{A}, Y)\cos(\mathbf{B}, Y) + \cos(\mathbf{A}, Z)\cos(\mathbf{B}, Z).$$

In case $\mathbf{A}$ and $\mathbf{B}$ are perpendicular this reduces to the well-known relation

$$0 = \cos(\mathbf{A}, X)\cos(\mathbf{B}, X) + \cos(\mathbf{A}, Y)\cos(\mathbf{B}, Y) + \cos(\mathbf{A}, Z)\cos(\mathbf{B}, Z)$$

between the direction cosines of the line $\mathbf{A}$ and the line $\mathbf{B}$.

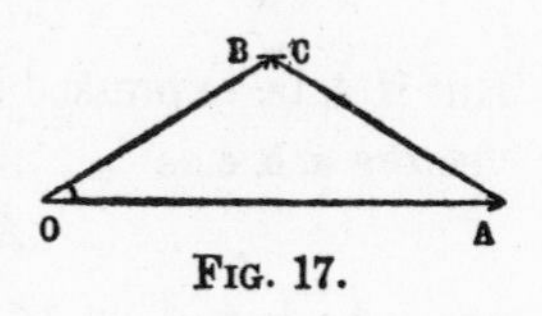

FIG. 17.

30.] If $\mathbf{A}$ and $\mathbf{B}$ are two sides OA and OB of a triangle OAB, the third side AB is $\mathbf{C} = \mathbf{B} - \mathbf{A}$ (Fig. 17).

$$\mathbf{C} \cdot \mathbf{C} = (\mathbf{B} - \mathbf{A}) \cdot (\mathbf{B} - \mathbf{A}) = \mathbf{B} \cdot \mathbf{B} + \mathbf{A} \cdot \mathbf{A} - 2\,\mathbf{A} \cdot \mathbf{B}$$

or

$$C^2 = A^2 + B^2 - 2\,A\,B\cos(\mathbf{A}\,\mathbf{B}).$$

That is, the square of one side of a triangle is equal to the sum of the squares of the other two sides diminished by twice their product times the cosine of the angle between them. Or, the square of one side of a triangle is equal to the sum of the squares of the other two sides diminished by twice the product of either of those sides by the projection of the other upon it — the generalized Pythagorean theorem.

If $\mathbf{A}$ and $\mathbf{B}$ are two sides of a parallelogram, $\mathbf{C} = \mathbf{A} + \mathbf{B}$ and $\mathbf{D} = \mathbf{A} - \mathbf{B}$ are the diagonals. Then

$$\mathbf{C} \cdot \mathbf{C} = (\mathbf{A} + \mathbf{B}) \cdot (\mathbf{A} + \mathbf{B}) = \mathbf{A} \cdot \mathbf{A} + 2\,\mathbf{A} \cdot \mathbf{B} + \mathbf{B} \cdot \mathbf{B},$$
$$\mathbf{D} \cdot \mathbf{D} = (\mathbf{A} - \mathbf{B}) \cdot (\mathbf{A} - \mathbf{B}) = \mathbf{A} \cdot \mathbf{A} - 2\,\mathbf{A} \cdot \mathbf{B} + \mathbf{B} \cdot \mathbf{B},$$
$$\mathbf{C} \cdot \mathbf{C} + \mathbf{D} \cdot \mathbf{D} = 2\,(\mathbf{A} \cdot \mathbf{A} + \mathbf{B} \cdot \mathbf{B}),$$

or

$$C^2 + D^2 = 2\,(A^2 + B^2).$$

That is, the sum of the squares of the diagonals of a parallelogram is equal to twice the sum of the squares of two sides.

In like manner also

$$\mathbf{C} \cdot \mathbf{C} - \mathbf{D} \cdot \mathbf{D} = 4\, \mathbf{A} \cdot \mathbf{B}$$

or

$$C^2 - D^2 = 4\, A\, B \cos (\mathbf{A}, \mathbf{B}).$$

That is, the difference of the squares of the diagonals of a parallelogram is equal to four times the product of one of the sides by the projection of the other upon it.

If $\mathbf{A}$ is any vector expressed in terms of $\mathbf{i}, \mathbf{j}, \mathbf{k}$ as

$$\mathbf{A} = A_1\, \mathbf{i} + A_2\, \mathbf{j} + A_3\, \mathbf{k},$$

then

$$\mathbf{A} \cdot \mathbf{A} = A^2 = A_1{}^2 + A_2{}^2 + A_3{}^2. \tag{8}$$

But if $\mathbf{A}$ be expressed in terms of any three non-coplanar unit vectors $\mathbf{a}, \mathbf{b}, \mathbf{c}$ as

$$\mathbf{A} = a\, \mathbf{a} + b\, \mathbf{b} + c\, \mathbf{c},$$

$$\mathbf{A} \cdot \mathbf{A} = A^2 = a^2\, \mathbf{a} \cdot \mathbf{a} + b^2\, \mathbf{b} \cdot \mathbf{b} + c^2\, \mathbf{c} \cdot \mathbf{c} + 2\, a\, b\, \mathbf{a} \cdot \mathbf{b} + 2\, b\, c\, \mathbf{b} \cdot \mathbf{c} + 2\, c\, a\, \mathbf{c} \cdot \mathbf{a}$$

$$A^2 = a^2 + b^2 + c^2 + 2\, a\, b \cos (\mathbf{a}, \mathbf{b}) + 2\, b\, c \cos (\mathbf{b}, \mathbf{c}) + 2\, c\, a \cos (\mathbf{c}, \mathbf{a}).$$

This formula is analogous to the one in Cartesian geometry which gives the distance between two points referred to oblique axes. If the points be x_1, y_1, z_1, and x_2, y_2, z_2 the distance squared is

$$\begin{aligned} \mathbf{D}^2 = {} & (x_2 - x_1)^2 + (y_2 - y_1)^2 + (z_2 - z_1)^2 \\ & + 2\,(x_2 - x_1)\,(y_2 - y_1) \cos (X, Y) \\ & + 2\,(y_2 - y_1)\,(z_2 - z_1) \cos (Y, Z) \\ & + 2\,(z_2 - z_1)\,(x_2 - x_1) \cos (Z, X). \end{aligned}$$

31.] *Definition:* The *skew product* of the vector $\mathbf{A}$ into the vector $\mathbf{B}$ is the *vector* quantity $\mathbf{C}$ whose direction is the normal upon that side of the plane of $\mathbf{A}$ and $\mathbf{B}$ on which rotation from **A** to **B** through an angle of less than one hundred and eighty degrees appears positive or counterclockwise; and whose magnitude is obtained by multiplying the product of the magnitudes of **A** and **B** by the *sine* of the angle from **A** to **B**.

The direction of $\mathbf{A} \times \mathbf{B}$ may also be defined as that in which an ordinary right-handed screw advances as it turns so as to carry **A** toward **B** (Fig. 18).

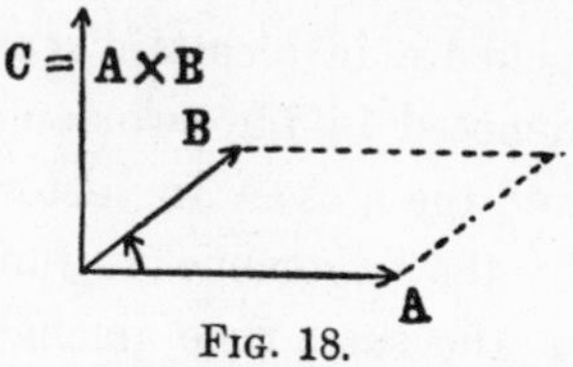

Fig. 18.

The skew product is denoted by a cross as the direct product was by a dot. It is written

$$\mathbf{C} = \mathbf{A} \times \mathbf{B}$$

and read **A** *cross* **B**. For this reason it is often called the *cross* product. More frequently, however, it is called the *vector* product, owing to the fact that it is a vector quantity and in contrast with the direct or scalar product whose value is scalar.

The vector product is by definition

$$\mathbf{C} = \mathbf{A} \times \mathbf{B} = A\,B \sin(\mathbf{A}, \mathbf{B})\,\mathbf{c}, \tag{9}$$

when A and B are the magnitudes of **A** and **B** respectively and where **c** is a unit vector in the direction of **C**. In case **A** and **B** are unit vectors the skew product $\mathbf{A} \times \mathbf{B}$ reduces to the unit vector **c** multiplied by the sine of the angle from **A** to **B**. Obviously also if either vector **A** or **B** is multiplied by a scalar x their product is multiplied by that scalar.

$$(x\,\mathbf{A}) \times \mathbf{B} = \mathbf{A} \times (x\,\mathbf{B}) = x\,\mathbf{C}.$$

If **A** and **B** are *parallel* the angle between them is either zero or one hundred and eighty degrees. In either case the sine vanishes and consequently the vector product $\mathbf{A} \times \mathbf{B}$ is a null vector. And conversely if $\mathbf{A} \times \mathbf{B}$ is zero

$$A\,B \sin(\mathbf{A}, \mathbf{B}) = 0.$$

Hence A or B or sin $(\mathbf{A}, \mathbf{B})$ is zero. Thus *the condition for parallelism of two vectors neither of which vanishes is* $\mathbf{A} \times \mathbf{B} = 0$. As a corollary the vector product of any vector into itself vanishes.

32.] The vector product of two vectors will appear wherever the sine of the included angle is of importance, just as the scalar product did in the case of the cosine. The two products are in a certain sense complementary. They have been denoted by the two common signs of multiplication, the dot and the cross. In vector analysis they occupy the place held by the trigonometric functions of scalar analysis. They are at the same time amenable to algebraic treatment, as will be seen later. At present a few uses of the vector product may be cited.

If $\mathbf{A}$ and $\mathbf{B}$ (Fig. 18) are the two adjacent sides of a parallelogram the vector product

$$\mathbf{C} = \mathbf{A} \times \mathbf{B} = A\, B \sin (\mathbf{A}, \mathbf{B})\, \mathbf{c}$$

represents the area of that parallelogram in magnitude and direction (Art. 25). This geometric representation of $\mathbf{A} \times \mathbf{B}$ is of such common occurrence and importance that it might well be taken as the definition of the product. From it the trigonometric definition follows at once. The vector product appears in mechanics in connection with couples. If $\mathbf{A}$ and $-\mathbf{A}$ are two forces forming a couple, the moment of the couple is $\mathbf{A} \times \mathbf{B}$ provided only that $\mathbf{B}$ is a vector drawn from any point of $\mathbf{A}$ to any point of $-\mathbf{A}$. The product makes its appearance again in considering the velocities of the individual particles of a body which is rotating with an angular velocity given in magnitude and direction by $\mathbf{A}$. If $\mathbf{R}$ be the radius vector drawn from any point of the axis of rotation $\mathbf{A}$ the product $\mathbf{A} \times \mathbf{R}$ will give the velocity of the extremity of $\mathbf{R}$ (Art. 51). This velocity is perpendicular alike to the axis of rotation and to the radius vector $\mathbf{R}$.

33.] The vector products $\mathbf{A} \times \mathbf{B}$ and $\mathbf{B} \times \mathbf{A}$ are not the same. They are in fact the negatives of each other. For if rotation from **A** to **B** appear positive on one side of the plane of **A** and **B**, rotation from **B** to **A** will appear positive on the other. Hence $\mathbf{A} \times \mathbf{B}$ is the normal to the plane of **A** and **B** upon that side opposite to the one upon which $\mathbf{B} \times \mathbf{A}$ is the normal. The magnitudes of $\mathbf{A} \times \mathbf{B}$ and $\mathbf{B} \times \mathbf{A}$ are the same. Hence

$$\mathbf{A} \times \mathbf{B} = - \mathbf{B} \times \mathbf{A}. \tag{10}$$

The factors in a vector product can be interchanged if and only if the sign of the product be reversed.

This is the first instance in which the laws of operation in vector analysis differ essentially from those of scalar analysis. It may be that at first this change of sign which must accompany the interchange of factors in a vector product will give rise to some difficulty and confusion. Changes similar to this are, however, very familiar. No one would think of interchanging the order of x and y in the expression $\sin (x - y)$ without prefixing the negative sign to the result. Thus

$$\sin (y - x) = - \sin (x - y),$$

although the sign is not required for the case of the cosine.

$$\cos (y - x) = \cos (x - y).$$

Again if the cyclic order of the letters $A\,B\,C$ in the area of a triangle be changed, the area will be changed in sign (Art. 25).

$$A\,B\,C = - A\,C\,B.$$

In the same manner this reversal of sign, which occurs when the order of the factors in a vector product is reversed, will appear after a little practice and acquaintance just as natural and convenient as it is necessary.

34.] The distributive law of multiplication holds in the case of vector products just as in ordinary algebra — *except* *that the order of the factors must be carefully maintained* when expanding.

$$(\mathbf{A} + \mathbf{B}) \times \mathbf{C} = \mathbf{A} \times \mathbf{C} + \mathbf{B} \times \mathbf{C}. \qquad (11)$$

A very simple proof may be given by making use of the ideas developed in Art. 26. Suppose that $\mathbf{C}$ is not coplanar with $\mathbf{A}$ and $\mathbf{B}$. Let $\mathbf{A}$ and $\mathbf{B}$ be two sides of a triangle taken in order. Then $-(\mathbf{A}+\mathbf{B})$ will be the third side (Fig. 19). Form the prism of which this triangle is the base and of which $\mathbf{C}$ is the slant height or edge. The areas of the lateral faces of this prism are

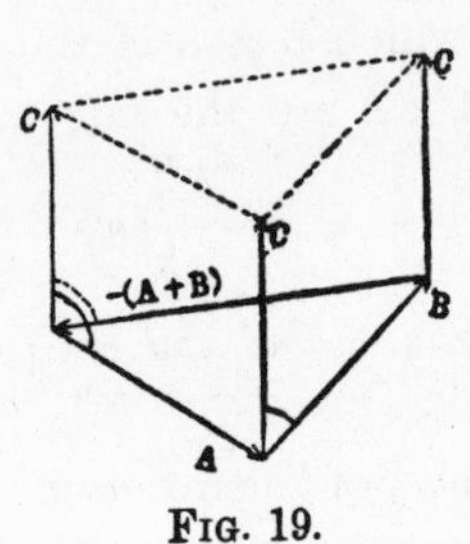

FIG. 19.

$$\mathbf{A} \times \mathbf{C}, \quad \mathbf{B} \times \mathbf{C}, \quad -(\mathbf{A} + \mathbf{B}) \times \mathbf{C}.$$

The areas of the bases are

$$\frac{1}{2}(\mathbf{A} \times \mathbf{B}) \text{ and } -\frac{1}{2}(\mathbf{A} \times \mathbf{B}).$$

But the sum of all the faces of the prism is zero; for the prism is a closed surface. Hence

$$\mathbf{A} \times \mathbf{C} + \mathbf{B} \times \mathbf{C} - (\mathbf{A} + \mathbf{B}) \times \mathbf{C} + \frac{1}{2}(\mathbf{A} \times \mathbf{B}) - \frac{1}{2}(\mathbf{A} \times \mathbf{B}) = 0,$$

$$\mathbf{A} \times \mathbf{C} + \mathbf{B} \times \mathbf{C} - (\mathbf{A} + \mathbf{B}) \times \mathbf{C} = 0,$$

or
$$\mathbf{A} \times \mathbf{C} + \mathbf{B} \times \mathbf{C} = (\mathbf{A} + \mathbf{B}) \times \mathbf{C}. \qquad (11)$$

The relation is therefore proved in case $\mathbf{C}$ is non-coplanar with $\mathbf{A}$ and $\mathbf{B}$. Should $\mathbf{C}$ be coplanar with $\mathbf{A}$ and $\mathbf{B}$, choose $\mathbf{D}$, any vector out of that plane. Then $\mathbf{C} + \mathbf{D}$ also will lie out of that plane. Hence by (11)

$$\mathbf{A} \times (\mathbf{C} + \mathbf{D}) + \mathbf{B} \times (\mathbf{C} + \mathbf{D}) = (\mathbf{A} + \mathbf{B}) \times (\mathbf{C} + \mathbf{D}).$$

Since the three vectors in each set $\mathbf{A}$, $\mathbf{C}$, $\mathbf{D}$, and $\mathbf{B}$, $\mathbf{C}$, $\mathbf{D}$, and $\mathbf{A} + \mathbf{B}$, $\mathbf{C}$, $\mathbf{D}$ will be non-coplanar if $\mathbf{D}$ is properly chosen, the products may be expanded.

$$\mathbf{A} \times \mathbf{C} + \mathbf{A} \times \mathbf{D} + \mathbf{B} \times \mathbf{C} + \mathbf{B} \times \mathbf{D}$$
$$= (\mathbf{A} + \mathbf{B}) \times \mathbf{C} + (\mathbf{A} + \mathbf{B}) \times \mathbf{D}.$$

But by (11) $\mathbf{A} \times \mathbf{D} + \mathbf{B} \times \mathbf{D} = (\mathbf{A} + \mathbf{B}) \times \mathbf{D}.$

Hence $\mathbf{A} \times \mathbf{C} + \mathbf{B} \times \mathbf{C} = (\mathbf{A} + \mathbf{B}) \times \mathbf{C}.$

This completes the demonstration. The distributive law holds for a vector product. The generalization is immediate.

$$(\mathbf{A} + \mathbf{B} + \cdots) \times (\mathbf{P} + \mathbf{Q} + \cdots) = \mathbf{A} \times \mathbf{P} + \mathbf{A} \times \mathbf{Q} + \cdots \qquad (11)'$$
$$+ \mathbf{B} \times \mathbf{P} + \mathbf{B} \times \mathbf{Q} + \cdots$$
$$+ \cdot \quad \cdot \quad \cdot \quad \cdot \quad \cdot \quad \cdot \quad \cdot \quad \cdot$$

35.] The vector products of the three unit vectors **i, j, k** are easily seen by means of Art. 17 to be

$$\begin{aligned} \mathbf{i} \times \mathbf{i} = \mathbf{j} \times \mathbf{j} = \mathbf{k} \times \mathbf{k} &= 0, \\ \mathbf{i} \times \mathbf{j} = - \mathbf{j} \times \mathbf{i} &= \mathbf{k}, \\ \mathbf{j} \times \mathbf{k} = - \mathbf{k} \times \mathbf{j} &= \mathbf{i}, \\ \mathbf{k} \times \mathbf{i} = - \mathbf{i} \times \mathbf{k} &= \mathbf{j}. \end{aligned} \qquad (12)$$

The skew product of two equal[1] vectors of the system **i, j, k** is zero. The product of two unequal vectors is the third taken with the positive sign if the vectors follow in the cyclic order **i j k** but with the negative sign if they do not.

If two vectors **A** and **B** are expressed in terms of **i, j, k**, their vector product may be found by expanding according to the distributive law and substituting.

$$\mathbf{A} = A_1 \mathbf{i} + A_2 \mathbf{j} + A_3 \mathbf{k},$$
$$\mathbf{B} = B_1 \mathbf{i} + B_2 \mathbf{j} + B_3 \mathbf{k},$$
$$\begin{aligned} \mathbf{A} \times \mathbf{B} &= (A_1 \mathbf{i} + A_2 \mathbf{j} + A_3 \mathbf{k}) \times (B_1 \mathbf{i} + B_2 \mathbf{j} + B_3 \mathbf{k}) \\ &= A_1 B_1 \mathbf{i} \times \mathbf{i} + A_1 B_2 \mathbf{i} \times \mathbf{j} + A_1 B_3 \mathbf{i} \times \mathbf{k} \\ &+ A_2 B_1 \mathbf{j} \times \mathbf{i} + A_2 B_2 \mathbf{j} \times \mathbf{j} + A_2 B_3 \mathbf{j} \times \mathbf{k}, \\ &+ A_3 B_1 \mathbf{k} \times \mathbf{i} + A_3 B_2 \mathbf{k} \times \mathbf{j} + A_3 B_3 \mathbf{k} \times \mathbf{k}. \end{aligned}$$

Hence $\mathbf{A} \times \mathbf{B} = (A_2 B_3 - A_3 B_2) \mathbf{i} + (A_3 B_1 - A_1 B_3) \mathbf{j}$
$$+ (A_1 B_2 - A_2 B_1) \mathbf{k}.$$

[1] This follows also from the fact that the sign is changed when the order of factors is reversed. Hence $\mathbf{j} \times \mathbf{j} = - \mathbf{j} \times \mathbf{j} = 0$.

This may be written in the form of a determinant as

$$\mathbf{A} \times \mathbf{B} = \begin{vmatrix} \mathbf{i} & \mathbf{j} & \mathbf{k} \\ A_1 & A_2 & A_3 \\ B_1 & B_2 & B_3 \end{vmatrix}.$$

The formulæ for the sine and cosine of the sum or difference of two angles follow immediately from the dot and cross products. Let **a** and **b** be two unit vectors lying in the **i j**-plane. If x be the angle that **a** makes with **i**, and y the angle **b** makes with **i**, then

$$\mathbf{a} = \cos x\, \mathbf{i} + \sin x\, \mathbf{j},$$

$$\mathbf{b} = \cos y\, \mathbf{i} + \sin y\, \mathbf{j},$$

$$\mathbf{a} \cdot \mathbf{b} = \cos(\mathbf{a}, \mathbf{b}) = \cos(y - x),$$

$$\mathbf{a} \cdot \mathbf{b} = \cos x \cos y + \sin x \sin y.$$

Hence $\quad \cos(y - x) = \cos y \cos x + \sin y \sin x.$

If $\quad \mathbf{b}' = \cos y\, \mathbf{i} - \sin y\, \mathbf{j},$

$$\mathbf{a} \cdot \mathbf{b}' = \cos(\mathbf{a}, \mathbf{b}') = \cos(y + x).$$

Hence $\quad \cos(y + x) = \cos y \cos x - \sin y \sin x.$

$$\mathbf{a} \times \mathbf{b} = \mathbf{k} \sin(\mathbf{a}, \mathbf{b}) = \mathbf{k} \sin(y - x),$$

$$\mathbf{a} \times \mathbf{b} = \mathbf{k} (\sin y \cos x - \sin x \cos y).$$

Hence $\quad \sin(y - x) = \sin y \cos x - \sin x \cos y.$

$$\mathbf{a} \times \mathbf{b}' = \mathbf{k} \sin(\mathbf{a}, \mathbf{b}') = \mathbf{k} \sin(y + x),$$

$$\mathbf{a} \times \mathbf{b}' = \mathbf{k} (\sin y \cos x + \sin x \cos y).$$

Hence $\quad \sin(y + x) = \sin y \cos x + \sin x \cos y.$

If l, m, n and l', m', n' are the direction cosines of two unit vectors **a** and **a**$'$ referred to X, Y, Z, then

$$\mathbf{a} = l\, \mathbf{i} + m\, \mathbf{j} + n\, \mathbf{k},$$

$$\mathbf{a}' = l'\, \mathbf{i} + m'\, \mathbf{j} + n'\, \mathbf{k},$$

$$\mathbf{a} \cdot \mathbf{a}' = \cos(\mathbf{a}, \mathbf{a}') = l\, l' + m\, m' + n\, n',$$

as has already been shown in Art. 29. The familiar formula for the square of the sine of the angle between **a** and **a**$'$ may be found.

$$\mathbf{a} \times \mathbf{a}' = \sin(\mathbf{a}, \mathbf{a}')\, \mathbf{e} = (m\, n' - m'\, n)\, \mathbf{i} + (n\, l' - n'\, l)\, \mathbf{j} + (l\, m' - l' m)\, \mathbf{k},$$

where **e** is a unit vector perpendicular to **a** and **a**′.

$$(\mathbf{a} \times \mathbf{a}') \cdot (\mathbf{a} \times \mathbf{a}') = \sin^2(\mathbf{a}, \mathbf{a}')\, \mathbf{e} \cdot \mathbf{e} = \sin^2(\mathbf{a}, \mathbf{a}').$$

$$\sin^2(\mathbf{a}, \mathbf{a}') = (m\, n' - m'\, n)^2 + (n\, l' - n'\, l)^2 + (l\, m' - l'\, m)^2.$$

This leads to an easy way of establishing the useful identity

$$(m\, n' - m'\, n)^2 + (n\, l' - n'\, l)^2 + (l\, m' - l'\, m)^2$$
$$= (l^2 + m^2 + n^2)(l'^2 + m'^2 + n'^2) - (l\, l' + m\, m' + n\, n')^2.$$

Products of More than Two Vectors

36.] Up to this point nothing has been said concerning products in which the number of vectors is greater than two. If three vectors are combined into a product the result is called a *triple* product. Next to the simple products **A·B** and **A×B** the triple products are the most important. All higher products may be reduced to them.

The simplest triple product is formed by multiplying the scalar product of two vectors **A** and **B** into a third **C** as

$$(\mathbf{A} \cdot \mathbf{B})\, \mathbf{C}.$$

This in reality does not differ essentially from scalar multiplication (Art. 6). The scalar in this case merely happens to be the scalar product of the two vectors **A** and **B**. Moreover inasmuch as two vectors cannot stand side by side in the form of a product as **B C** without either a dot or a cross to unite them, the parenthesis in (**A·B**) **C** is superfluous. The expression

$$\mathbf{A} \cdot \mathbf{B}\, \mathbf{C}$$

cannot be interpreted in any other way [1] than as the product of the vector **C** by the scalar **A·B**.

[1] Later (Chap. V.) the product **B C**, where no sign either dot or cross occurs, will be defined. But it will be seen there that (**A·B**) **C** and **A·**(**B C**) are identical and consequently no ambiguity can arise from the omission of the parenthesis.

37.] The second triple product is the scalar product of two vectors, of which one is itself a vector product, as

$$\mathbf{A}\cdot(\mathbf{B}\times\mathbf{C}) \text{ or } (\mathbf{A}\times\mathbf{B})\cdot\mathbf{C}.$$

This sort of product has a scalar value and consequently is often called the scalar triple product. Its properties are perhaps most easily deduced from its commonest geometrical interpretation. Let **A**, **B**, and **C** be any three vectors drawn from the same origin (Fig. 20). Then **B**×**C** is the area of the parallelogram of which **B** and **C** are two adjacent sides. The scalar

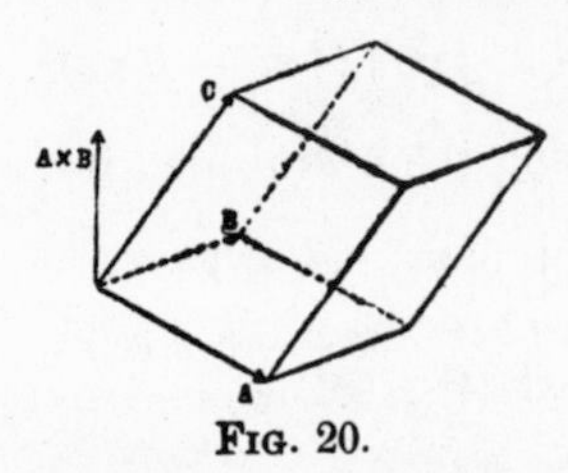

Fig. 20.

$$\mathbf{A}\cdot(\mathbf{B}\times\mathbf{C}) = v \tag{14}$$

will therefore be the volume of the parallelopiped of which **B**×**C** is the base and **A** the slant height or edge. See Art. 28. This volume v is positive if **A** and **B**×**C** lie upon the same side of the **BC**-plane; but negative if they lie on opposite sides. In other words if **A**, **B**, **C** form a right-handed or positive system of three vectors the scalar **A**·(**B**×**C**) is positive; but if they form a left-handed or negative system, it is negative.

In case **A**, **B**, and **C** are coplanar this volume will be neither positive nor negative but zero. And conversely if the volume is zero the three edges **A**, **B**, **C** of the parallelopiped must lie in one plane. Hence *the necessary and sufficient condition for the coplanarity of three vectors* **A**, **B**, **C** *none of which vanishes is* **A**·(**B**×**C**) = 0. As a corollary the scalar triple product of three vectors of which two are equal or collinear must vanish; for any two vectors are coplanar.

The two products **A**·(**B**×**C**) and (**A**×**B**)·**C** are equal to the same volume v of the parallelopiped whose concurrent edges are **A**, **B**, **C**. The sign of the volume is the same in both cases. Hence

$$(\mathbf{A}\times\mathbf{B})\cdot\mathbf{C} = \mathbf{A}\cdot(\mathbf{B}\times\mathbf{C}) = v. \tag{14}$$

This equality may be stated as a rule of operation. *The dot and the cross in a scalar triple product may be interchanged without altering the value of the product.*

It may also be seen that the vectors **A**, **B**, **C** may be permuted cyclicly without altering the product.

$$\mathbf{A}\cdot(\mathbf{B}\times\mathbf{C}) = \mathbf{B}\cdot(\mathbf{C}\times\mathbf{A}) = \mathbf{C}\cdot(\mathbf{A}\times\mathbf{B}). \qquad (15)$$

For each of the expressions gives the volume of the same parallelopiped and that volume will have in each case the same sign, because if **A** is upon the positive side of the **B C**-plane, **B** will be on the positive side of the **C A**-plane and **C** upon the positive side of the **A B**-plane. The triple product may therefore have any one of six equivalent forms

$$\mathbf{A}\cdot(\mathbf{B}\times\mathbf{C}) = \mathbf{B}\cdot(\mathbf{C}\times\mathbf{A}) = \mathbf{C}\cdot(\mathbf{A}\times\mathbf{B}) \qquad (15)'$$
$$= (\mathbf{A}\times\mathbf{B})\cdot\mathbf{C} = (\mathbf{B}\times\mathbf{C})\cdot\mathbf{A} = (\mathbf{C}\times\mathbf{A})\cdot\mathbf{B}.$$

If however the cyclic order of the letters is changed the product will change sign.

$$\mathbf{A}\cdot(\mathbf{B}\times\mathbf{C}) = -\,\mathbf{A}\cdot(\mathbf{C}\times\mathbf{B}). \qquad (16)$$

This may be seen from the figure or from the fact that

$$\mathbf{B}\times\mathbf{C} = -\,\mathbf{C}\times\mathbf{B}.$$

Hence: *A scalar triple product is not altered by interchanging the dot or the cross or by permuting cyclicly the order of the vectors, but it is reversed in sign if the cyclic order be changed.*

38.] A word is necessary upon the subject of parentheses in this triple product. Can they be omitted without ambiguity? They can. The expression

$$\mathbf{A}\cdot\mathbf{B}\times\mathbf{C}$$

can have only the one interpretation

$$\mathbf{A}\cdot(\mathbf{B}\times\mathbf{C}).$$

For the expression $(\mathbf{A}\cdot\mathbf{B})\times\mathbf{C}$ is meaningless. It is impossible to form the skew product of a scalar $\mathbf{A}\cdot\mathbf{B}$ and a vector

C. Hence as there is only one way in which $\mathbf{A \cdot B \times C}$ may be interpreted, no confusion can arise from omitting the parentheses. Furthermore owing to the fact that there are six scalar triple products of **A**, **B**, and **C** which have the same value and are consequently generally not worth distinguishing the one from another, it is often convenient to use the symbol

$$[\mathbf{A\ B\ C}]$$

to denote any one of the six equal products.

$$\begin{aligned} [\mathbf{A\ B\ C}] &= \mathbf{A \cdot B \times C} = \mathbf{B \cdot C \times A} = \mathbf{C \cdot A \times B} \\ &= \mathbf{A \times B \cdot C} = \mathbf{B \times C \cdot A} = \mathbf{C \times A \cdot B} \end{aligned} \tag{15)'$$

then

$$[\mathbf{A\ B\ C}] = -[\mathbf{A\ C\ B}]. \tag{16)'$$

The scalar triple products of the three unit vectors **i**, **j**, **k** all vanish except the two which contain the three different vectors.

$$[\mathbf{i\ j\ k}] = -[\mathbf{i\ k\ j}] = 1. \tag{17}$$

Hence if three vectors **A**, **B**, **C** be expressed in terms of **i**, **j**, **k** as

$$\begin{aligned} \mathbf{A} &= A_1\,\mathbf{i} + A_2\,\mathbf{j} + A_3\,\mathbf{k}, \\ \mathbf{B} &= B_1\,\mathbf{i} + B_2\,\mathbf{j} + B_3\,\mathbf{k}, \\ \mathbf{C} &= C_1\,\mathbf{i} + C_2\,\mathbf{j} + C_3\,\mathbf{k}, \end{aligned}$$

then

$$\begin{aligned} [\mathbf{A\ B\ C}] &= A_1\,B_2\,C_3 + B_1\,C_2\,A_3 + C_1\,A_2\,B_3 \\ &\quad - A_1\,B_3\,C_2 - B_1\,C_3\,A_2 - C_1\,A_3\,B_2. \end{aligned} \tag{18}$$

This may be obtained by actually performing the multiplications which are indicated in the triple product. The result may be written in the form of a determinant.[1]

$$[\mathbf{A\ B\ C}] = \begin{vmatrix} A_1 & A_2 & A_3 \\ B_1 & B_2 & B_3 \\ C_1 & C_2 & C_3 \end{vmatrix} \tag{18)'$$

[1] This is the formula given in solid analytic geometry for the volume of a tetrahedron one of whose vertices is at the origin. For a more general formula see exercises.

If more generally **A**, **B**, **C** are expressed in terms of any three non-coplanar vectors **a**, **b**, **c** which are not necessarily unit vectors,

$$\mathbf{A} = a_1\, \mathbf{a} + a_2\, \mathbf{b} + a_3\, \mathbf{c}$$
$$\mathbf{B} = b_1\, \mathbf{a} + b_2\, \mathbf{b} + b_3\, \mathbf{c}$$
$$\mathbf{C} = c_1\, \mathbf{a} + c_2\, \mathbf{b} + c_3\, \mathbf{c}$$

where a_1, a_2, a_3; b_1, b_2, b_3; and c_1, c_2, c_3 are certain constants, then

$$[\mathbf{A\ B\ C}] = (a_1\, b_2\, c_3 + b_1\, c_2\, a_3 + c_1\, a_2\, b_3 - a_1\, b_3\, c_2 - b_1\, c_3\, a_2 - c_1\, a_3\, b_2)\, [\mathbf{a\ b\ c}]. \tag{19}$$

or

$$[\mathbf{A\ B\ C}] = \begin{vmatrix} a_1 & a_2 & a_3 \\ b_1 & b_2 & b_3 \\ c_1 & c_2 & c_3 \end{vmatrix} [\mathbf{a\ b\ c}] \tag{19$'$}$$

39.] The third type of triple product is the vector product of two vectors of which one is itself a vector product. Such are

$$\mathbf{A}\times(\mathbf{B}\times\mathbf{C}) \text{ and } (\mathbf{A}\times\mathbf{B})\times\mathbf{C}.$$

The vector $\mathbf{A}\times(\mathbf{B}\times\mathbf{C})$ is perpendicular to **A** and to $(\mathbf{B}\times\mathbf{C})$. But $(\mathbf{B}\times\mathbf{C})$ is perpendicular to the plane of **B** and **C**. Hence $\mathbf{A}\times(\mathbf{B}\times\mathbf{C})$, being perpendicular to $(\mathbf{B}\times\mathbf{C})$ must lie in the plane of **B** and **C** and thus take the form

$$\mathbf{A}\times(\mathbf{B}\times\mathbf{C}) = x\, \mathbf{B} + y\, \mathbf{C},$$

where x and y are two scalars. In like manner also the vector $(\mathbf{A}\times\mathbf{B})\times\mathbf{C}$, being perpendicular to $(\mathbf{A}\times\mathbf{B})$ must lie in the plane of **A** and **B**. Hence it will be of the form

$$(\mathbf{A}\times\mathbf{B})\times\mathbf{C} = m\, \mathbf{A} + n\, \mathbf{B}$$

where m and n are two scalars. From this it is evident that in general

$$(\mathbf{A}\times\mathbf{B})\times\mathbf{C} \textit{ is not equal to } \mathbf{A}\times(\mathbf{B}\times\mathbf{C}).$$

The parentheses therefore cannot be removed or interchanged. It is essential to know which cross product is formed first and which second. This product is termed the vector triple product in contrast to the scalar triple product.

The vector triple product may be used to express that *component* of a vector **B** which is perpendicular to a given vector **A**. This geometric use of the product is valuable not only in itself but for the light it sheds upon the properties of the product. Let **A** (Fig. 21) be a given vector and **B** another vector whose components parallel and perpendicular to **A** are to be found. Let the components of **B** parallel and perpendicular to **A** be $\mathbf{B}'$ and $\mathbf{B}''$ respectively. Draw **A** and **B** from a common origin. The product $\mathbf{A}\times\mathbf{B}$ is perpendicular to the plane of **A** and **B**. The product $\mathbf{A}\times(\mathbf{A}\times\mathbf{B})$ lies in the plane of **A** and **B**. It is furthermore perpendicular to **A**. Hence it is collinear with $\mathbf{B}''$. An examination of the figure will show that the direction of $\mathbf{A}\times(\mathbf{A}\times\mathbf{B})$ is opposite to that of $\mathbf{B}''$. Hence

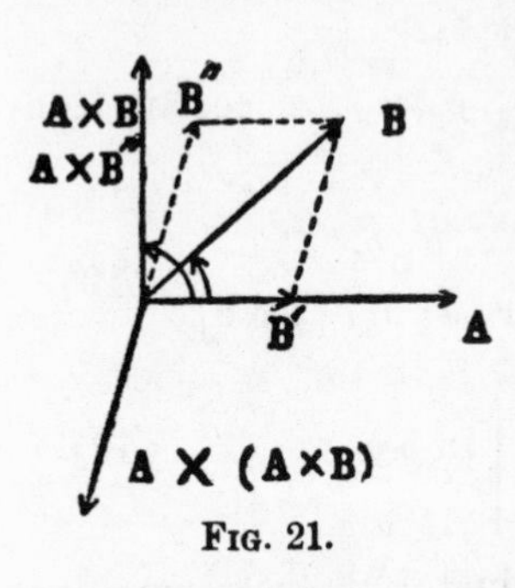

Fig. 21.

$$\mathbf{A}\times(\mathbf{A}\times\mathbf{B}) = -\,c\,\mathbf{B}'',$$

where c is some scalar constant.

Now $$\mathbf{A}\times(\mathbf{A}\times\mathbf{B}) = -\,A^2\,B \sin(\mathbf{A},\mathbf{B})\,\mathbf{b}''$$

but $$-\,c\,\mathbf{B}'' = -\,c\,B \sin(\mathbf{A},\mathbf{B})\,\mathbf{b}'',$$

if $\mathbf{b}''$ be a unit vector in the direction of $\mathbf{B}''$.

Hence $$c = A^2 = \mathbf{A}\cdot\mathbf{A}.$$

Hence $$\mathbf{B}'' = -\,\frac{\mathbf{A}\times(\mathbf{A}\times\mathbf{B})}{\mathbf{A}\cdot\mathbf{A}}. \qquad (20)$$

The component of **B** perpendicular to **A** has been expressed in terms of the vector triple product of **A**, **A**, and **B**. The component $\mathbf{B}'$ parallel to **A** was found in Art. 28 to be

$$\mathbf{B}' = \frac{\mathbf{A}\cdot\mathbf{B}}{\mathbf{A}\cdot\mathbf{A}}\,\mathbf{A} \tag{21}$$

$$\mathbf{B} = \mathbf{B}' + \mathbf{B}'' = \frac{\mathbf{A}\cdot\mathbf{B}}{\mathbf{A}\cdot\mathbf{A}}\,\mathbf{A} - \frac{\mathbf{A}\times(\mathbf{A}\times\mathbf{B})}{\mathbf{A}\cdot\mathbf{A}}. \tag{22}$$

40.] The vector triple product $\mathbf{A}\times(\mathbf{B}\times\mathbf{C})$ may be expressed as the sum of two terms as

$$\mathbf{A}\times(\mathbf{B}\times\mathbf{C}) = \mathbf{A}\cdot\mathbf{C}\ \mathbf{B} - \mathbf{A}\cdot\mathbf{B}\ \mathbf{C}$$

In the first place consider the product when two of the vectors are the same. By equation (22)

$$\mathbf{A}\cdot\mathbf{A}\ \mathbf{B} = \mathbf{A}\cdot\mathbf{B}\ \mathbf{A} - \mathbf{A}\times(\mathbf{A}\times\mathbf{B}) \tag{22}$$

or

$$\mathbf{A}\times(\mathbf{A}\times\mathbf{B}) = \mathbf{A}\cdot\mathbf{B}\ \mathbf{A} - \mathbf{A}\cdot\mathbf{A}\ \mathbf{B} \tag{23}$$

This proves the formula in case two vectors are the same.

To prove it in general express $\mathbf{A}$ in terms of the three non-coplanar vectors $\mathbf{B}$, $\mathbf{C}$, and $\mathbf{B}\times\mathbf{C}$.

$$\mathbf{A} = b\,\mathbf{B} + c\,\mathbf{C} + a\,(\mathbf{B}\times\mathbf{C}), \tag{I}$$

where a, b, c are scalar constants. Then

$$\begin{aligned}\mathbf{A}\times(\mathbf{B}\times\mathbf{C}) &= b\,\mathbf{B}\times(\mathbf{B}\times\mathbf{C}) + c\,\mathbf{C}\times(\mathbf{B}\times\mathbf{C}) \\ &\quad + a\,(\mathbf{B}\times\mathbf{C})\times(\mathbf{B}\times\mathbf{C}).\end{aligned} \tag{II}$$

The vector product of any vector by itself is zero. Hence

$$(\mathbf{B}\times\mathbf{C})\times(\mathbf{B}\times\mathbf{C}) = 0$$

$$\mathbf{A}\times(\mathbf{B}\times\mathbf{C}) = b\,\mathbf{B}\times(\mathbf{B}\times\mathbf{C}) + c\,\mathbf{C}\times(\mathbf{B}\times\mathbf{C}). \tag{II$'$}$$

By (23)

$$\mathbf{B}\times(\mathbf{B}\times\mathbf{C}) = \mathbf{B}\cdot\mathbf{C}\ \mathbf{B} - \mathbf{B}\cdot\mathbf{B}\ \mathbf{C}$$

$$\mathbf{C}\times(\mathbf{B}\times\mathbf{C}) = -\,\mathbf{C}\times(\mathbf{C}\times\mathbf{B}) = -\,\mathbf{C}\cdot\mathbf{B}\ \mathbf{C} + \mathbf{C}\cdot\mathbf{C}\ \mathbf{B}.$$

Hence $\mathbf{A}\times(\mathbf{B}\times\mathbf{C}) = [(b\,\mathbf{B}\cdot\mathbf{C} + c\,\mathbf{C}\cdot\mathbf{C})\mathbf{B} - (b\,\mathbf{B}\cdot\mathbf{B} + c\,\mathbf{C}\cdot\mathbf{B})\mathbf{C}]$. (II)$''$

But from (I) $\mathbf{A}\cdot\mathbf{B} = b\,\mathbf{B}\cdot\mathbf{B} + c\,\mathbf{C}\cdot\mathbf{B} + a\,(\mathbf{B}\times\mathbf{C})\cdot\mathbf{B}$

and $\mathbf{A}\cdot\mathbf{C} = b\,\mathbf{B}\cdot\mathbf{C} + c\,\mathbf{C}\cdot\mathbf{C} + a\,(\mathbf{B}\times\mathbf{C})\cdot\mathbf{C}$.

By Art. 37 $(\mathbf{B}\times\mathbf{C})\cdot\mathbf{B} = 0$ and $(\mathbf{B}\times\mathbf{C})\cdot\mathbf{C} = 0$.

Hence

$$\mathbf{A}\cdot\mathbf{B} = b\,\mathbf{B}\cdot\mathbf{B} + c\,\mathbf{C}\cdot\mathbf{B},$$

$$\mathbf{A}\cdot\mathbf{C} = b\,\mathbf{B}\cdot\mathbf{C} + c\,\mathbf{C}\cdot\mathbf{C}.$$

Substituting these values in (II)″,

$$\mathbf{A}\times(\mathbf{B}\times\mathbf{C}) = \mathbf{A}\cdot\mathbf{C}\ \mathbf{B} - \mathbf{A}\cdot\mathbf{B}\ \mathbf{C}. \tag{24}$$

The relation is therefore proved for any three vectors **A**, **B**, **C**.

Another method of giving the demonstration is as follows. It was shown that the vector triple product $\mathbf{A}\times(\mathbf{B}\times\mathbf{C})$ was of the form

$$\mathbf{A}\times(\mathbf{B}\times\mathbf{C}) = x\,\mathbf{B} + y\,\mathbf{C}.$$

Since $\mathbf{A}\times(\mathbf{A}\times\mathbf{C})$ is perpendicular to **A**, the direct product of it by **A** is zero. Hence

$$\mathbf{A}\cdot[\mathbf{A}\times(\mathbf{B}\times\mathbf{C})] = x\,\mathbf{A}\cdot\mathbf{B} + y\,\mathbf{A}\cdot\mathbf{C} = 0$$

and $$x : y = \mathbf{A}\cdot\mathbf{C} : -\mathbf{A}\cdot\mathbf{B}.$$

Hence $$\mathbf{A}\times(\mathbf{B}\times\mathbf{C}) = n\,(\mathbf{A}\cdot\mathbf{C}\ \mathbf{B} - \mathbf{A}\cdot\mathbf{B}\ \mathbf{C}),$$

where n is a scalar constant. It remains to show $n = 1$. Multiply by **B**.

$$\mathbf{A}\times(\mathbf{B}\times\mathbf{C})\cdot\mathbf{B} = n\,(\mathbf{A}\cdot\mathbf{C}\ \mathbf{B}\cdot\mathbf{B} - \mathbf{A}\cdot\mathbf{B}\ \mathbf{C}\cdot\mathbf{B}).$$

The scalar triple product allows an interchange of dot and cross. Hence

$$\mathbf{A}\times(\mathbf{B}\times\mathbf{C})\cdot\mathbf{B} = \mathbf{A}\cdot(\mathbf{B}\times\mathbf{C})\times\mathbf{B} = -\mathbf{A}\cdot[\mathbf{B}\times(\mathbf{B}\times\mathbf{C})],$$

if the order of the factors $(\mathbf{B}\times\mathbf{C})$ and **B** be inverted.

$$-\mathbf{A}\cdot[\mathbf{B}\times(\mathbf{B}\times\mathbf{C})] = -\mathbf{A}\cdot[\mathbf{B}\cdot\mathbf{C}\ \mathbf{B} - \mathbf{B}\cdot\mathbf{B}\ \mathbf{C}]$$
$$= -\mathbf{B}\cdot\mathbf{C}\ \mathbf{A}\cdot\mathbf{B} + \mathbf{B}\cdot\mathbf{B}\ \mathbf{A}\cdot\mathbf{C}.$$

Hence $n = 1$ and $$\mathbf{A}\times(\mathbf{B}\times\mathbf{C}) = \mathbf{A}\cdot\mathbf{C}\ \mathbf{B} - \mathbf{A}\cdot\mathbf{B}\ \mathbf{C}. \tag{24}$$

From the three letters **A**, **B**, **C** by different arrangements, four allied products in each of which **B** and **C** are included in parentheses may be formed. These are

$$\mathbf{A}\times(\mathbf{B}\times\mathbf{C}),\quad \mathbf{A}\times(\mathbf{C}\times\mathbf{B}),\quad (\mathbf{C}\times\mathbf{B})\times\mathbf{A},\quad (\mathbf{B}\times\mathbf{C})\times\mathbf{A}.$$

As a vector product changes its sign whenever the order of two factors is interchanged, the above products evidently satisfy the equations

$$\mathbf{A}\times(\mathbf{B}\times\mathbf{C}) = -\mathbf{A}\times(\mathbf{C}\times\mathbf{B}) = (\mathbf{C}\times\mathbf{B})\times\mathbf{A} = -(\mathbf{B}\times\mathbf{C})\times\mathbf{A}.$$

The expansion for a vector triple product in which the parenthesis comes first may therefore be obtained directly from that already found when the parenthesis comes last.

$$(\mathbf{A}\times\mathbf{B})\times\mathbf{C} = -\mathbf{C}\times(\mathbf{A}\times\mathbf{B}) = -\mathbf{C}\cdot\mathbf{B}\;\mathbf{A} + \mathbf{C}\cdot\mathbf{A}\;\mathbf{B}.$$

The formulæ then become

$$\mathbf{A}\times(\mathbf{B}\times\mathbf{C}) = \mathbf{A}\cdot\mathbf{C}\;\mathbf{B} - \mathbf{A}\cdot\mathbf{B}\;\mathbf{C} \qquad (24)$$

and

$$(\mathbf{A}\times\mathbf{B})\times\mathbf{C} = \mathbf{A}\cdot\mathbf{C}\;\mathbf{B} - \mathbf{C}\cdot\mathbf{B}\;\mathbf{A}. \qquad (24)'$$

These reduction formulæ are of such constant occurrence and great importance that they should be committed to memory. Their content may be stated in the following rule. *To expand a vector triple product first multiply the exterior factor into the remoter term in the parenthesis to form a scalar coefficient for the nearer one, then multiply the exterior factor into the nearer term in the parenthesis to form a scalar coefficient for the remoter one, and subtract this result from the first.*

41.] As far as the practical applications of vector analysis are concerned, one can generally get along without any formulæ more complicated than that for the vector triple product. But it is frequently more convenient to have at hand other reduction formulæ of which all may be derived simply by making use of the expansion for the triple product $\mathbf{A}\times(\mathbf{B}\times\mathbf{C})$ and of the rules of operation with the triple product $\mathbf{A}\cdot\mathbf{B}\times\mathbf{C}$.

To reduce a scalar product of two vectors each of which is itself a vector product of two vectors, as

$$(\mathbf{A}\times\mathbf{B})\cdot(\mathbf{C}\times\mathbf{D}).$$

Let this be regarded as a scalar triple product of the three vectors **A**, **B**, and $\mathbf{C}\times\mathbf{D}$ — thus

$$\mathbf{A}\times\mathbf{B}\cdot(\mathbf{C}\times\mathbf{D}).$$

Interchange the dot and the cross.

$$\mathbf{A}\times\mathbf{B}\cdot(\mathbf{C}\times\mathbf{D}) = \mathbf{A}\cdot\mathbf{B}\times(\mathbf{C}\times\mathbf{D})$$

$$\mathbf{B}\times(\mathbf{C}\times\mathbf{D}) = \mathbf{B}\cdot\mathbf{D}\ \mathbf{C} - \mathbf{B}\cdot\mathbf{C}\ \mathbf{D}.$$

Hence
$$(\mathbf{A}\times\mathbf{B})\cdot(\mathbf{C}\times\mathbf{D}) = \mathbf{A}\cdot\mathbf{C}\ \mathbf{B}\cdot\mathbf{D} - \mathbf{A}\cdot\mathbf{D}\ \mathbf{B}\cdot\mathbf{C}. \qquad (25)$$

This may be written in determinantal form.

$$(\mathbf{A}\times\mathbf{B})\cdot(\mathbf{C}\times\mathbf{D}) = \begin{vmatrix} \mathbf{A}\cdot\mathbf{C} & \mathbf{A}\cdot\mathbf{D} \\ \mathbf{B}\cdot\mathbf{C} & \mathbf{B}\cdot\mathbf{D} \end{vmatrix} \qquad (25)'$$

If **A** and **D** be called the extremes; **B** and **C** the means; **A** and **C** the antecedents; **B** and **D** the consequents in this product according to the familiar usage in proportions, then the expansion may be stated in words. The scalar product of two vector products is equal to the (scalar) product of the antecedents times the (scalar) product of the consequents diminished by the (scalar) product of the means times the (scalar) product of the extremes.

To reduce a vector product of two vectors each of which is itself a vector product of two vectors, as

$$(\mathbf{A}\times\mathbf{B})\times(\mathbf{C}\times\mathbf{D}).$$

Let $\mathbf{C}\times\mathbf{D} = \mathbf{E}$. The product becomes

$$(\mathbf{A}\times\mathbf{B})\times\mathbf{E} = \mathbf{A}\cdot\mathbf{E}\ \mathbf{B} - \mathbf{B}\cdot\mathbf{E}\ \mathbf{A}.$$

Substituting the value of **E** back into the equation:

$$(\mathbf{A}\times\mathbf{B})\times(\mathbf{C}\times\mathbf{D}) = (\mathbf{A}\cdot\mathbf{C}\times\mathbf{D})\mathbf{B} - (\mathbf{B}\cdot\mathbf{C}\times\mathbf{D})\ \mathbf{A}. \qquad (26)$$

Let $\mathbf{F} = \mathbf{A}\times\mathbf{B}$. The product then becomes

$$\mathbf{F}\times(\mathbf{C}\times\mathbf{D}) = \mathbf{F}\cdot\mathbf{D}\ \mathbf{C} - \mathbf{F}\cdot\mathbf{C}\ \mathbf{D}$$

$$(\mathbf{A}\times\mathbf{B})\times(\mathbf{C}\times\mathbf{D}) = (\mathbf{A}\times\mathbf{B}\cdot\mathbf{D})\mathbf{C} - (\mathbf{A}\times\mathbf{B}\cdot\mathbf{C})\ \mathbf{D}. \qquad (26)'$$

By equating these two equivalent results and transposing all the terms to one side of the equation,

$$[\mathbf{B}\ \mathbf{C}\ \mathbf{D}]\ \mathbf{A} - [\mathbf{C}\ \mathbf{D}\ \mathbf{A}]\ \mathbf{B} + [\mathbf{D}\ \mathbf{A}\ \mathbf{B}]\ \mathbf{C} - [\mathbf{A}\ \mathbf{B}\ \mathbf{C}]\ \mathbf{D} = 0. \qquad (27)$$

This is an equation with scalar coefficients between the four vectors **A, B, C, D.** There is in general only one such equation, because any one of the vectors can be expressed in only one way in terms of the other three: thus the scalar coefficients of that equation which exists between four vectors are found to be nothing but the four scalar triple products of those vectors taken three at a time. The equation may also be written in the form

$$[\mathbf{A}\,\mathbf{B}\,\mathbf{C}\,\mathbf{D}] = [\mathbf{B}\,\mathbf{C}\,\mathbf{D}]\,\mathbf{A} + [\mathbf{C}\,\mathbf{A}\,\mathbf{D}]\,\mathbf{B} + [\mathbf{A}\,\mathbf{B}\,\mathbf{D}]\,\mathbf{C}. \qquad (27)'$$

More examples of reduction formulæ, of which some are important, are given among the exercises at the end of the chapter. In view of these it becomes fairly obvious that the combination of any number of vectors connected in any legitimate way by dots and crosses or the product of any number of such combinations can be ultimately reduced to a sum of terms each of which contains only one cross at most. The proof of this theorem depends solely upon analyzing the possible combinations of vectors and showing that they all fall under the reduction formulæ in such a way that the crosses may be removed two at a time until not more than one remains.

* **42.**] The formulæ developed in the foregoing article have interesting geometric interpretations. They also afford a simple means of deducing the formulæ of Spherical Trigonometry. These do not occur in the vector analysis proper. Their place is taken by the two quadruple products,

$$(\mathbf{A}\times\mathbf{B})\cdot(\mathbf{C}\times\mathbf{D}) = \mathbf{A}\cdot\mathbf{C}\;\mathbf{B}\cdot\mathbf{D} - \mathbf{B}\cdot\mathbf{C}\;\mathbf{A}\cdot\mathbf{D} \qquad (25)$$

and

$$(\mathbf{A}\times\mathbf{B})\times(\mathbf{C}\times\mathbf{D}) = [\mathbf{A}\mathbf{C}\mathbf{D}]\;\mathbf{B} - [\mathbf{B}\mathbf{C}\mathbf{D}]\;\mathbf{A}$$

$$= [\mathbf{A}\mathbf{B}\mathbf{D}]\;\mathbf{C} - [\mathbf{A}\mathbf{B}\mathbf{C}]\;\mathbf{D}, \qquad (26)$$

which are now to be interpreted.

Let a unit sphere (Fig. 22) be given. Let the vectors **A**, **B**, **C**, **D** be unit vectors drawn from a common origin, the centre of the sphere, and terminating in the surface of the sphere at the points *A*, *B*, *C*, *D*. The great circular arcs

AB, AC, etc., give the angles between the vectors **A** and **B**, **A** and **C**, etc. The points A, B, C, D determine a quadrilateral upon the sphere. AC and BD are one pair of opposite sides; AD and BC, the other. AB and CD are the diagonals.

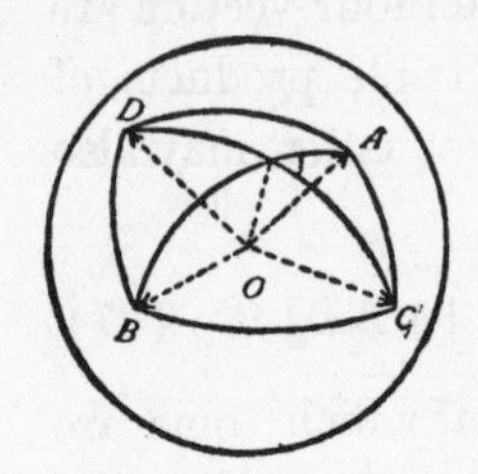

FIG. 22.

$$(\mathbf{A}\times\mathbf{B})\cdot(\mathbf{C}\times\mathbf{D}) = \mathbf{A}\cdot\mathbf{C}\ \mathbf{B}\cdot\mathbf{D} - \mathbf{A}\cdot\mathbf{D}\ \mathbf{B}\cdot\mathbf{C}$$

$$\mathbf{A}\times\mathbf{B} = \sin(\mathbf{A}, \mathbf{B}), \quad \mathbf{C}\times\mathbf{D} = \sin(\mathbf{C}, \mathbf{D}).$$

The angle between $\mathbf{A}\times\mathbf{B}$ and $\mathbf{C}\times\mathbf{D}$ is the angle between the normals to the **AB**- and **CD**-planes. This is the same as the angle between the planes themselves. Let it be denoted by x. Then

$$(\mathbf{A}\times\mathbf{B})\cdot(\mathbf{C}\times\mathbf{D}) = \sin(\mathbf{A}, \mathbf{B}) \sin(\mathbf{C}, \mathbf{D}) \cos x.$$

The angles $(\mathbf{A}, \mathbf{B})$, $(\mathbf{C}, \mathbf{D})$ may be replaced by the great circular arcs AB, CD which measure them. Then

$$(\mathbf{A}\times\mathbf{B})\cdot(\mathbf{C}\times\mathbf{D}) = \sin AB \sin CD \cos x,$$

$$\mathbf{A}\cdot\mathbf{C}\ \mathbf{B}\cdot\mathbf{D} - \mathbf{A}\cdot\mathbf{D}\ \mathbf{B}\cdot\mathbf{C} = \cos AC \cos BD - \cos AD \cos BC.$$

Hence

$$\sin AB \sin CD \cos x = \cos AC \cos BD - \cos AD \cos BC.$$

In words: The product of the cosines of two opposite sides of a spherical quadrilateral less the product of the cosines of the other two opposite sides is equal to the product of the sines of the diagonals multiplied by the cosine of the angle between them. This theorem is credited to Gauss.

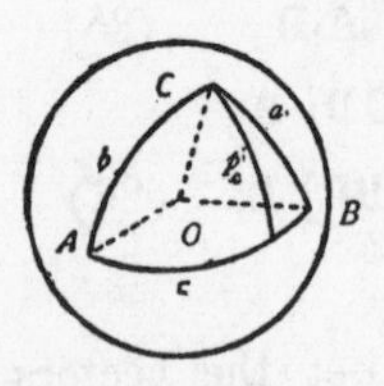

FIG. 23.

Let A, B, C (Fig. 23) be a spherical triangle, the sides of which are arcs of great circles. Let the sides be denoted by a, b, c respectively. Let **A**, **B**, **C** be the unit vectors drawn from the center of the sphere to the points A, B, C. Furthermore let p_a, p_b, p_c be the great circular arcs dropped perpendicularly from the vertices A, B, C to the sides a, b, c. Interpret the formula

$$(\mathbf{A}\times\mathbf{B})\cdot(\mathbf{C}\times\mathbf{A}) = \mathbf{A}\cdot\mathbf{C}\ \mathbf{B}\cdot\mathbf{A} - \mathbf{B}\cdot\mathbf{C}\ \mathbf{A}\cdot\mathbf{A}.$$

$$(\mathbf{A}\times\mathbf{B}) = \sin(\mathbf{A}, \mathbf{B}) = \sin c, \quad (\mathbf{C}\times\mathbf{A}) = \sin(\mathbf{C}, \mathbf{A}) = \sin b.$$

Then $$(\mathbf{A}\times\mathbf{B})\cdot(\mathbf{C}\times\mathbf{A}) = \sin c \sin b \cos x,$$

where x is the angle between $\mathbf{A}\times\mathbf{B}$ and $\mathbf{C}\times\mathbf{A}$. This angle is equal to the angle between the plane of $\mathbf{A}$, $\mathbf{B}$ and the plane of $\mathbf{C}$, $\mathbf{A}$. It is, however, not the interior angle A which is one of the angles of the triangle: but it is the exterior angle $180° - A$, as an examination of the figure will show. Hence

$$(\mathbf{A}\times\mathbf{B})\cdot(\mathbf{C}\times\mathbf{A}) = \sin c \sin b \cos(180° - A)$$
$$= -\sin c \sin b \cos A$$
$$\mathbf{A}\cdot\mathbf{C}\ \mathbf{B}\cdot\mathbf{A} - \mathbf{B}\cdot\mathbf{C}\ \mathbf{A}\cdot\mathbf{A} = \cos b \cos c - \cos a\ 1.$$

By equating the results and transposing,

$$\cos a = \cos b \cos c - \sin b \sin c \cos A$$
$$\cos b = \cos c \cos a - \sin c \sin a \cos B$$
$$\cos c = \cos a \cos b - \sin a \sin b \cos C.$$

The last two may be obtained by cyclic permutation of the letters or from the identities

$$(\mathbf{B}\times\mathbf{C})\cdot(\mathbf{A}\times\mathbf{B}) = \mathbf{B}\cdot\mathbf{A}\ \mathbf{C}\cdot\mathbf{B} - \mathbf{C}\cdot\mathbf{A},$$
$$(\mathbf{C}\times\mathbf{A})\cdot(\mathbf{B}\times\mathbf{C}) = \mathbf{C}\cdot\mathbf{B}\ \mathbf{A}\cdot\mathbf{C} - \mathbf{B}\cdot\mathbf{C}.$$

Next interpret the identity $(\mathbf{A}\times\mathbf{B})\times(\mathbf{C}\times\mathbf{D})$ in the special cases in which one of the vectors is repeated.

$$(\mathbf{A}\times\mathbf{B})\times(\mathbf{A}\times\mathbf{C}) = [\mathbf{A}\,\mathbf{B}\,\mathbf{C}]\ \mathbf{A}.$$

Let the three vectors $\mathbf{a}$, $\mathbf{b}$, $\mathbf{c}$ be unit vectors in the direction of $\mathbf{B}\times\mathbf{C}$, $\mathbf{C}\times\mathbf{A}$, $\mathbf{A}\times\mathbf{B}$ respectively. Then

$$\mathbf{A}\times\mathbf{B} = \mathbf{c}\sin c, \qquad \mathbf{A}\times\mathbf{C} = -\mathbf{b}\sin b$$
$$(\mathbf{A}\times\mathbf{B})\times(\mathbf{A}\times\mathbf{C}) = -\mathbf{c}\times\mathbf{b}\sin c \sin b = \mathbf{A}\sin c \sin b \sin A$$
$$[\mathbf{A}\,\mathbf{B}\,\mathbf{C}] = (\mathbf{A}\times\mathbf{B})\cdot\mathbf{C} = \mathbf{c}\cdot\mathbf{C}\sin c = \cos(90° - p_c)\sin c$$
$$[\mathbf{A}\,\mathbf{B}\,\mathbf{C}]\ \mathbf{A} = \sin c \sin p_c\ \mathbf{A}.$$

By equating the results and cancelling the common factor,

$$\sin p_c = \sin b \sin A$$
$$\sin p_a = \sin c \sin B$$
$$\sin p_b = \sin a \sin C.$$

The last two may be obtained by cyclic permutation of the letters. The formulæ give the sines of the altitudes of the triangle in terms of the sines of the angle and sides. Again write

$$(\mathbf{A}\times\mathbf{B})\times(\mathbf{A}\times\mathbf{C}) = [\mathbf{A}\,\mathbf{B}\,\mathbf{C}]\,\mathbf{A}$$
$$(\mathbf{B}\times\mathbf{C})\times(\mathbf{B}\times\mathbf{A}) = [\mathbf{B}\,\mathbf{C}\,\mathbf{A}]\,\mathbf{B}$$
$$(\mathbf{C}\times\mathbf{A})\times(\mathbf{C}\times\mathbf{B}) = [\mathbf{C}\,\mathbf{A}\,\mathbf{B}]\,\mathbf{C}.$$

Hence
$$\sin c \sin b \sin A = [\mathbf{A}\,\mathbf{B}\,\mathbf{C}]$$
$$\sin a \sin c \sin B = [\mathbf{B}\,\mathbf{C}\,\mathbf{A}]$$
$$\sin b \sin a \sin C = [\mathbf{C}\,\mathbf{A}\,\mathbf{B}].$$

The expressions $[\mathbf{A}\,\mathbf{B}\,\mathbf{C}]$, $[\mathbf{B}\,\mathbf{C}\,\mathbf{A}]$, $[\mathbf{C}\,\mathbf{A}\,\mathbf{B}]$ are equal. Equate the results in pairs and the formulæ

$$\sin b \sin A = \sin a \sin B$$
$$\sin c \sin B = \sin b \sin C$$
$$\sin a \sin C = \sin c \sin A$$

are obtained. These may be written in a single line.

$$\frac{\sin A}{\sin a} = \frac{\sin B}{\sin b} = \frac{\sin C}{\sin c}.$$

The formulæ of Plane Trigonometry are even more easy to obtain. If $A\,B\,C$ be a triangle, the sum of the sides taken as vectors is zero — for the triangle is a closed polygon. From this equation

$$\mathbf{a} + \mathbf{b} + \mathbf{c} = 0$$

almost all the elementary formulæ follow immediately. It is to be noticed that the angles from **a** to **b**, from **b** to **c**, from

$\mathbf{c}$ to $\mathbf{a}$ are not the interior angles A, B, C, but the exterior angles $180° - A$, $180° - B$, $180° - C$.

$$-\mathbf{a} = \mathbf{b} + \mathbf{c}$$

$$\mathbf{a}\cdot\mathbf{a} = (\mathbf{b} + \mathbf{c})\cdot(\mathbf{b} + \mathbf{c}) = \mathbf{b}\cdot\mathbf{b} + \mathbf{c}\cdot\mathbf{c} + 2\,\mathbf{b}\cdot\mathbf{c}.$$

If a, b, c be the length of the sides $\mathbf{a}$, $\mathbf{b}$, $\mathbf{c}$, this becomes

$$a^2 = b^2 + c^2 - 2\,b\,c\cos A$$
$$b^2 = c^2 + a^2 - 2\,c\,a\cos B$$
$$c^2 = a^2 + b^2 - 2\,a\,b\cos C.$$

The last two are obtained in a manner similar to the first one or by cyclic permutation of the letters.

The area of the triangle is

$$\frac{1}{2}\mathbf{a} \times \mathbf{b} = \frac{1}{2}\mathbf{b} \times \mathbf{c} = \frac{1}{2}\mathbf{c} \times \mathbf{a} =$$
$$\frac{1}{2}\,a\,b\sin C = \frac{1}{2}\,b\,c\sin A = \frac{1}{2}\,c\,a\sin B.$$

If each of the last three equalities be divided by the product $\frac{1}{2}\,a\,b\,c$, the fundamental relation

$$\frac{\sin A}{a} = \frac{\sin B}{b} = \frac{\sin C}{c}$$

is obtained. Another formula for the area may be found from the product

$$(\mathbf{b}\times\mathbf{c})\cdot(\mathbf{b}\times\mathbf{c}) = (\mathbf{c}\times\mathbf{a})\cdot(\mathbf{a}\times\mathbf{b})$$

$$2\text{ Area }(b\,c\sin A) = (c\,a\sin B)\,(a\,b\sin C)$$

$$2\text{ Area} = \frac{a^2\sin B\sin C}{\sin A}.$$

Reciprocal Systems of Three Vectors. Solution of Equations

43.] The problem of expressing any vector $\mathbf{r}$ in terms of three non-coplanar vectors $\mathbf{a}$, $\mathbf{b}$, $\mathbf{c}$ may be solved as follows. Let

$$\mathbf{r} = a\,\mathbf{a} + b\,\mathbf{b} + c\,\mathbf{c}$$

where a, b, c are three scalar constants to be determined. Multiply by $\cdot\, \mathbf{b} \times \mathbf{c}$.

$$\mathbf{r} \cdot \mathbf{b} \times \mathbf{c} = a\, \mathbf{a} \cdot \mathbf{b} \times \mathbf{c} + b\, \mathbf{b} \cdot \mathbf{b} \times \mathbf{c} + c\, \mathbf{c} \cdot \mathbf{b} \times \mathbf{c}$$

or

$$[\mathbf{r}\, \mathbf{b}\, \mathbf{c}] = a\, [\mathbf{a}\, \mathbf{b}\, \mathbf{c}].$$

In like manner by multiplying the equation by $\cdot\, \mathbf{c} \times \mathbf{a}$ and $\cdot\, \mathbf{a} \times \mathbf{b}$ the coefficients b and c may be found.

$$[\mathbf{r}\, \mathbf{c}\, \mathbf{a}] = b\, [\mathbf{b}\, \mathbf{c}\, \mathbf{a}]$$

$$[\mathbf{r}\, \mathbf{a}\, \mathbf{b}] = c\, [\mathbf{c}\, \mathbf{a}\, \mathbf{b}]$$

Hence

$$\mathbf{r} = \frac{[\mathbf{r}\, \mathbf{b}\, \mathbf{c}]}{[\mathbf{a}\, \mathbf{b}\, \mathbf{c}]}\, \mathbf{a} + \frac{[\mathbf{r}\, \mathbf{c}\, \mathbf{a}]}{[\mathbf{b}\, \mathbf{c}\, \mathbf{a}]}\, \mathbf{b} + \frac{[\mathbf{r}\, \mathbf{a}\, \mathbf{b}]}{[\mathbf{c}\, \mathbf{a}\, \mathbf{b}]}\, \mathbf{c}. \qquad (28)$$

The denominators are all equal. Hence this gives the equation

$$[\mathbf{a}\, \mathbf{b}\, \mathbf{c}]\, \mathbf{r} - [\mathbf{b}\, \mathbf{c}\, \mathbf{r}]\, \mathbf{a} + [\mathbf{c}\, \mathbf{r}\, \mathbf{a}]\, \mathbf{b} - [\mathbf{r}\, \mathbf{a}\, \mathbf{b}]\, \mathbf{c} = 0$$

which must exist between the four vectors $\mathbf{r}$, $\mathbf{a}$, $\mathbf{b}$, $\mathbf{c}$.

The equation may also be written

$$\mathbf{r} = \frac{\mathbf{r} \cdot \mathbf{b} \times \mathbf{c}}{[\mathbf{a}\, \mathbf{b}\, \mathbf{c}]}\, \mathbf{a} + \frac{\mathbf{r} \cdot \mathbf{c} \times \mathbf{a}}{[\mathbf{a}\, \mathbf{b}\, \mathbf{c}]}\, \mathbf{b} + \frac{\mathbf{r} \cdot \mathbf{a} \times \mathbf{b}}{[\mathbf{a}\, \mathbf{b}\, \mathbf{c}]}\, \mathbf{c}$$

or

$$\mathbf{r} = \mathbf{r} \cdot \frac{\mathbf{b} \times \mathbf{c}}{[\mathbf{a}\, \mathbf{b}\, \mathbf{c}]}\, \mathbf{a} + \mathbf{r} \cdot \frac{\mathbf{c} \times \mathbf{a}}{[\mathbf{a}\, \mathbf{b}\, \mathbf{c}]}\, \mathbf{b} + \mathbf{r} \cdot \frac{\mathbf{a} \times \mathbf{b}}{[\mathbf{a}\, \mathbf{b}\, \mathbf{c}]}\, \mathbf{c}.$$

The three vectors which appear here multiplied by $\mathbf{r}\cdot$, namely

$$\frac{\mathbf{b} \times \mathbf{c}}{[\mathbf{a}\, \mathbf{b}\, \mathbf{c}]}, \quad \frac{\mathbf{c} \times \mathbf{a}}{[\mathbf{a}\, \mathbf{b}\, \mathbf{c}]}, \quad \frac{\mathbf{a} \times \mathbf{b}}{[\mathbf{a}\, \mathbf{b}\, \mathbf{c}]}$$

are very important. They are perpendicular respectively to the planes of $\mathbf{b}$ and $\mathbf{c}$, $\mathbf{c}$ and $\mathbf{a}$, $\mathbf{a}$ and $\mathbf{b}$. They occur over and over again in a large number of important relations. For this reason they merit a distinctive name and notation.

Definition : The system of three vectors

$$\frac{\mathbf{b} \times \mathbf{c}}{[\mathbf{a}\, \mathbf{b}\, \mathbf{c}]}, \quad \frac{\mathbf{c} \times \mathbf{a}}{[\mathbf{a}\, \mathbf{b}\, \mathbf{c}]}, \quad \frac{\mathbf{a} \times \mathbf{b}}{[\mathbf{a}\, \mathbf{b}\, \mathbf{c}]}$$

which are found by dividing the three vector products $\mathbf{b} \times \mathbf{c}$, $\mathbf{c} \times \mathbf{a}$, $\mathbf{a} \times \mathbf{b}$ of three *non-coplanar* vectors **a**, **b**, **c** by the scalar product $[\mathbf{a}\ \mathbf{b}\ \mathbf{c}]$ is called the *reciprocal system* to **a**, **b**, **c**.

The word *non-coplanar* is important. If **a**, **b**, **c** were coplanar the scalar triple product $[\mathbf{a}\ \mathbf{b}\ \mathbf{c}]$ would vanish and consequently the fractions

$$\frac{\mathbf{b} \times \mathbf{c}}{[\mathbf{a}\,\mathbf{b}\,\mathbf{c}]}, \quad \frac{\mathbf{c} \times \mathbf{a}}{[\mathbf{a}\,\mathbf{b}\,\mathbf{c}]}, \quad \frac{\mathbf{a} \times \mathbf{b}}{[\mathbf{a}\,\mathbf{b}\,\mathbf{c}]}$$

would all become meaningless. Three coplanar vectors have no reciprocal system. This must be carefully remembered. Hereafter when the term *reciprocal system* is used, it will be understood that the three vectors **a**, **b**, **c** are not coplanar.

The system of three vectors reciprocal to system **a**, **b**, **c** will be denoted by primes as $\mathbf{a}'$, $\mathbf{b}'$, $\mathbf{c}'$.

$$\mathbf{a}' = \frac{\mathbf{b} \times \mathbf{c}}{[\mathbf{a}\,\mathbf{b}\,\mathbf{c}]}, \quad \mathbf{b}' = \frac{\mathbf{c} \times \mathbf{a}}{[\mathbf{a}\,\mathbf{b}\,\mathbf{c}]}, \quad \mathbf{c}' = \frac{\mathbf{a} \times \mathbf{b}}{[\mathbf{a}\,\mathbf{b}\,\mathbf{c}]}. \tag{29}$$

The expression for **r** reduces then to the very simple form

$$\mathbf{r} = \mathbf{r}\cdot\mathbf{a}'\,\mathbf{a} + \mathbf{r}\cdot\mathbf{b}'\,\mathbf{b} + \mathbf{r}\cdot\mathbf{c}'\,\mathbf{c}. \tag{30}$$

The vector **r** may be expressed in terms of the reciprocal system $\mathbf{a}'$, $\mathbf{b}'$, $\mathbf{c}'$ instead of in terms of **a**, **b**, **c**. In the first place it is necessary to note that if **a**, **b**, **c** are non-coplanar, $\mathbf{a}'$, $\mathbf{b}'$, $\mathbf{c}'$ which are the normals to the planes of **b** and **c**, **c** and **a**, **a** and **b** must also be non-coplanar. Hence **r** may be expressed in terms of them by means of proper scalar coefficients x, y, z.

$$\mathbf{r} = x\,\mathbf{a}' + y\,\mathbf{b}' + z\,\mathbf{c}'$$

or

$$[\mathbf{a}\,\mathbf{b}\,\mathbf{c}]\,\mathbf{r} = x\,\mathbf{b} \times \mathbf{c} + y\,\mathbf{c} \times \mathbf{a} + z\,\mathbf{a} \times \mathbf{b}.$$

Multiply successively by $\cdot\mathbf{a}$, $\cdot\mathbf{b}$, $\cdot\mathbf{c}$. This gives

$$[\mathbf{a}\,\mathbf{b}\,\mathbf{c}]\,\mathbf{r}\cdot\mathbf{a} = x\,[\mathbf{b}\,\mathbf{c}\,\mathbf{a}], \qquad x = \mathbf{r}\cdot\mathbf{a}$$
$$[\mathbf{a}\,\mathbf{b}\,\mathbf{c}]\,\mathbf{r}\cdot\mathbf{b} = y\,[\mathbf{c}\,\mathbf{a}\,\mathbf{b}], \qquad y = \mathbf{r}\cdot\mathbf{b}$$
$$[\mathbf{a}\,\mathbf{b}\,\mathbf{c}]\,\mathbf{r}\cdot\mathbf{c} = z\,[\mathbf{a}\,\mathbf{b}\,\mathbf{c}], \qquad z = \mathbf{r}\cdot\mathbf{c}$$

Hence

$$\mathbf{r} = \mathbf{r}\cdot\mathbf{a}\,\mathbf{a}' + \mathbf{r}\cdot\mathbf{b}\,\mathbf{b}' + \mathbf{r}\cdot\mathbf{c}\,\mathbf{c}'. \tag{31}$$

44.] If $\mathbf{a}'$, $\mathbf{b}'$, $\mathbf{c}'$ be the system reciprocal to $\mathbf{a}$, $\mathbf{b}$, $\mathbf{c}$ the scalar product of any vector of the reciprocal system into the corresponding vector of the given system is unity; but the product of two non-corresponding vectors is zero. That is

$$\mathbf{a}' \cdot \mathbf{a} = \mathbf{b}' \cdot \mathbf{b} = \mathbf{c}' \cdot \mathbf{c} = 1 \tag{32}$$

$$\mathbf{a}' \cdot \mathbf{b} = \mathbf{a}' \cdot \mathbf{c} = \mathbf{b}' \cdot \mathbf{a} = \mathbf{b}' \cdot \mathbf{c} = \mathbf{c}' \cdot \mathbf{a} = \mathbf{c}' \cdot \mathbf{b} = 0.$$

This may be seen most easily by expressing $\mathbf{a}'$, $\mathbf{b}'$, $\mathbf{c}'$ in terms of themselves according to the formula (31)

$$\mathbf{r} = \mathbf{r} \cdot \mathbf{a}\, \mathbf{a}' + \mathbf{r} \cdot \mathbf{b}\, \mathbf{b}' + \mathbf{r} \cdot \mathbf{c}\, \mathbf{c}'.$$

Hence

$$\mathbf{a}' = \mathbf{a}' \cdot \mathbf{a}\, \mathbf{a}' + \mathbf{a}' \cdot \mathbf{b}\, \mathbf{b}' + \mathbf{a}' \cdot \mathbf{c}\, \mathbf{c}'$$

$$\mathbf{b}' = \mathbf{b}' \cdot \mathbf{a}\, \mathbf{a}' + \mathbf{b}' \cdot \mathbf{b}\, \mathbf{b}' + \mathbf{b}' \cdot \mathbf{c}\, \mathbf{c}'$$

$$\mathbf{c}' = \mathbf{c}' \cdot \mathbf{a}\, \mathbf{a}' + \mathbf{c}' \cdot \mathbf{b}\, \mathbf{b}' + \mathbf{c}' \cdot \mathbf{c}\, \mathbf{c}'.$$

Since $\mathbf{a}'$, $\mathbf{b}'$, $\mathbf{c}'$ are non-coplanar the corresponding coefficients on the two sides of each of these three equations must be equal. Hence from the first

$$1 = \mathbf{a}' \cdot \mathbf{a} \quad 0 = \mathbf{a}' \cdot \mathbf{b} \quad 0 = \mathbf{a}'\, \mathbf{c}.$$

From the second $\quad 0 = \mathbf{b}' \cdot \mathbf{a} \quad 1 = \mathbf{b}' \cdot \mathbf{b} \quad 0 = \mathbf{b}' \cdot \mathbf{c}.$

From the third $\quad 0 = \mathbf{c}' \cdot \mathbf{a} \quad 0 = \mathbf{c}' \cdot \mathbf{b} \quad 1 = \mathbf{c}' \cdot \mathbf{c}.$

This proves the relations. They may also be proved directly from the definitions of $\mathbf{a}'$, $\mathbf{b}'$, $\mathbf{c}'$.

$$\mathbf{a}' \cdot \mathbf{a} = \frac{\mathbf{b} \times \mathbf{c}}{[\mathbf{a}\,\mathbf{b}\,\mathbf{c}]} \cdot \mathbf{a} = \frac{\mathbf{b} \times \mathbf{c} \cdot \mathbf{a}}{[\mathbf{a}\,\mathbf{b}\,\mathbf{c}\,]} = \frac{[\mathbf{b}\,\mathbf{c}\,\mathbf{a}]}{[\mathbf{a}\,\mathbf{b}\,\mathbf{c}]} = 1$$

$$\mathbf{a}' \cdot \mathbf{b} = \frac{\mathbf{b} \times \mathbf{c}}{[\mathbf{a}\,\mathbf{b}\,\mathbf{c}]} \cdot \mathbf{b} = \frac{\mathbf{b} \times \mathbf{c} \cdot \mathbf{b}}{[\mathbf{a}\,\mathbf{b}\,\mathbf{c}]} = \frac{0}{[\mathbf{a}\,\mathbf{b}\,\mathbf{c}]} = 0$$

and so forth.

Conversely if two sets of three vectors each, say $\mathbf{A}$, $\mathbf{B}$, $\mathbf{C}$, and $\mathbf{a}$, $\mathbf{b}$, $\mathbf{c}$, satisfy the relations

$$\mathbf{A} \cdot \mathbf{a} = \mathbf{B} \cdot \mathbf{b} = \mathbf{C} \cdot \mathbf{c} = 1$$

$$\mathbf{A} \cdot \mathbf{b} = \mathbf{A} \cdot \mathbf{c} = \mathbf{B} \cdot \mathbf{a} = \mathbf{B} \cdot \mathbf{c} = \mathbf{C} \cdot \mathbf{a} = \mathbf{C} \cdot \mathbf{b} = 0$$

then the set $\mathbf{A}$, $\mathbf{B}$, $\mathbf{C}$ is the system reciprocal to $\mathbf{a}$, $\mathbf{b}$, $\mathbf{c}$. By reasoning similar to that before

$$\mathbf{A} = \mathbf{A}\cdot\mathbf{a}\,\mathbf{a}' + \mathbf{A}\cdot\mathbf{b}\,\mathbf{b}' + \mathbf{A}\cdot\mathbf{c}\,\mathbf{c}'$$
$$\mathbf{B} = \mathbf{B}\cdot\mathbf{a}\,\mathbf{a}' + \mathbf{B}\cdot\mathbf{b}\,\mathbf{b}' + \mathbf{B}\cdot\mathbf{c}\,\mathbf{c}'$$
$$\mathbf{C} = \mathbf{C}\cdot\mathbf{a}\,\mathbf{a}' + \mathbf{C}\cdot\mathbf{b}\,\mathbf{b}' + \mathbf{C}\cdot\mathbf{c}\,\mathbf{c}'.$$

Substituting in these equations the given relations the result is

$$\mathbf{A} = \mathbf{a}', \quad \mathbf{B} = \mathbf{b}', \quad \mathbf{C} = \mathbf{c}'.$$

Hence

Theorem: The necessary and sufficient conditions that the set of vectors $\mathbf{a}'$, $\mathbf{b}'$, $\mathbf{c}'$ be the reciprocals of $\mathbf{a}$, $\mathbf{b}$, $\mathbf{c}$ is that they satisfy the equations

$$\mathbf{a}'\cdot\mathbf{a} = \mathbf{b}'\cdot\mathbf{b} = \mathbf{c}'\cdot\mathbf{c} = 1 \tag{32}$$
$$\mathbf{a}'\cdot\mathbf{b} = \mathbf{a}'\cdot\mathbf{c} = \mathbf{b}'\cdot\mathbf{a} = \mathbf{b}'\cdot\mathbf{c} = \mathbf{c}'\cdot\mathbf{a} = \mathbf{c}'\cdot\mathbf{b} = 0.$$

As these equations are perfectly symmetrical with respect to $\mathbf{a}'$, $\mathbf{b}'$, $\mathbf{c}'$ and $\mathbf{a}$, $\mathbf{b}$, $\mathbf{c}$ it is evident that the system $\mathbf{a}$, $\mathbf{b}$, $\mathbf{c}$ may be looked upon as the reciprocal of the system $\mathbf{a}'$, $\mathbf{b}'$, $\mathbf{c}'$ just as the system $\mathbf{a}'$, $\mathbf{b}'$, $\mathbf{c}'$ may be regarded as the reciprocal of $\mathbf{a}$, $\mathbf{b}$, $\mathbf{c}$. That is to say,

Theorem: If $\mathbf{a}'$, $\mathbf{b}'$, $\mathbf{c}'$ be the reciprocal system of $\mathbf{a}$, $\mathbf{b}$, $\mathbf{c}$, then $\mathbf{a}$, $\mathbf{b}$, $\mathbf{c}$ will be the reciprocal system of $\mathbf{a}'$, $\mathbf{b}'$, $\mathbf{c}'$.

$$\mathbf{a} = \frac{\mathbf{b}' \times \mathbf{c}'}{[\mathbf{a}'\,\mathbf{b}'\,\mathbf{c}']}, \quad \mathbf{b} = \frac{\mathbf{c}' \times \mathbf{a}'}{[\mathbf{a}'\,\mathbf{b}'\,\mathbf{c}']}, \quad \mathbf{c} = \frac{\mathbf{a}' \times \mathbf{b}'}{[\mathbf{a}'\,\mathbf{b}'\,\mathbf{c}']}. \tag{29$'$}$$

These relations may be demonstrated directly from the definitions of $\mathbf{a}'$, $\mathbf{b}'$, $\mathbf{c}'$. The demonstration is straightforward, but rather long and tedious as it depends on complicated reduction formulæ. The proof given above is as short as could be desired. The relations between $\mathbf{a}'$, $\mathbf{b}'$, $\mathbf{c}'$ and $\mathbf{a}$, $\mathbf{b}$, $\mathbf{c}$ are symmetrical and hence if $\mathbf{a}'$, $\mathbf{b}'$, $\mathbf{c}'$ is the reciprocal system of $\mathbf{a}$, $\mathbf{b}$, $\mathbf{c}$, then $\mathbf{a}$, $\mathbf{b}$, $\mathbf{c}$ must be the reciprocal system of $\mathbf{a}'$, $\mathbf{b}'$, $\mathbf{c}'$.

45.] *Theorem:* If $\mathbf{a}'$, $\mathbf{b}'$, $\mathbf{c}'$ and $\mathbf{a}$, $\mathbf{b}$, $\mathbf{c}$ be reciprocal systems the scalar triple products $[\mathbf{a}'\,\mathbf{b}'\,\mathbf{c}']$ and $[\mathbf{a}\,\mathbf{b}\,\mathbf{c}]$ are numerical reciprocals. That is

$$[\mathbf{a}'\,\mathbf{b}'\,\mathbf{c}']\;[\mathbf{a}\,\mathbf{b}\,\mathbf{c}] = 1 \tag{33}$$

$$[\mathbf{a}'\,\mathbf{b}'\,\mathbf{c}'] = \left[\frac{\mathbf{b}\times\mathbf{c}}{[\mathbf{a}\,\mathbf{b}\,\mathbf{c}]}\quad\frac{\mathbf{c}\times\mathbf{a}}{[\mathbf{a}\,\mathbf{b}\,\mathbf{c}]}\quad\frac{\mathbf{a}\times\mathbf{b}}{[\mathbf{a}\,\mathbf{b}\,\mathbf{c}]}\right]$$

$$= \frac{1}{[\mathbf{a}\,\mathbf{b}\,\mathbf{c}]^3}\,[\mathbf{b}\times\mathbf{c}\;\;\mathbf{c}\times\mathbf{a}\;\;\mathbf{a}\times\mathbf{b}].$$

$$[\mathbf{b}\times\mathbf{c}\;\;\mathbf{c}\times\mathbf{a}\;\;\mathbf{a}\times\mathbf{b}] = (\mathbf{b}\times\mathbf{c})\times(\mathbf{c}\times\mathbf{a})\cdot(\mathbf{a}\times\mathbf{b}).$$

But
$$(\mathbf{b}\times\mathbf{c})\times(\mathbf{c}\times\mathbf{a}) = [\mathbf{a}\,\mathbf{b}\,\mathbf{c}]\mathbf{c}.$$

Hence
$$[\mathbf{b}\times\mathbf{c}\;\;\mathbf{c}\times\mathbf{a}\;\;\mathbf{a}\times\mathbf{b}] = [\mathbf{a}\,\mathbf{b}\,\mathbf{c}]\;\mathbf{c}\cdot\mathbf{a}\times\mathbf{b} = [\mathbf{a}\,\mathbf{b}\,\mathbf{c}]^2.$$

Hence
$$[\mathbf{a}'\,\mathbf{b}'\,\mathbf{c}'] = \frac{1}{[\mathbf{a}\,\mathbf{b}\,\mathbf{c}]^3}\,[\mathbf{a}\,\mathbf{b}\,\mathbf{c}]^2 = \frac{1}{[\mathbf{a}\,\mathbf{b}\,\mathbf{c}]}\cdot \tag{33}'$$

By means of this relation between $[\mathbf{a}'\,\mathbf{b}'\,\mathbf{c}']$ and $[\mathbf{a}\,\mathbf{b}\,\mathbf{c}]$ it is possible to prove an important reduction formula,

$$(\mathbf{P}\cdot\mathbf{Q}\times\mathbf{R})(\mathbf{A}\cdot\mathbf{B}\times\mathbf{C}) = \begin{vmatrix} \mathbf{P}\cdot\mathbf{A} & \mathbf{P}\cdot\mathbf{B} & \mathbf{P}\cdot\mathbf{C} \\ \mathbf{Q}\cdot\mathbf{A} & \mathbf{Q}\cdot\mathbf{B} & \mathbf{Q}\cdot\mathbf{C} \\ \mathbf{R}\cdot\mathbf{A} & \mathbf{R}\cdot\mathbf{B} & \mathbf{R}\cdot\mathbf{C} \end{vmatrix}, \tag{34}$$

which replaces the two scalar triple products by a sum of nine terms each of which is the product of three direct products. Thus the two crosses which occur in the two scalar products are removed. To give the proof let $\mathbf{P}$, $\mathbf{Q}$, $\mathbf{R}$ be expressed as

$$\mathbf{P} = \mathbf{P}\cdot\mathbf{A}\;\mathbf{A}' + \mathbf{P}\cdot\mathbf{B}\;\mathbf{B}' + \mathbf{P}\cdot\mathbf{C}\;\mathbf{C}'$$
$$\mathbf{Q} = \mathbf{Q}\cdot\mathbf{A}\;\mathbf{A}' + \mathbf{Q}\cdot\mathbf{B}\;\mathbf{B}' + \mathbf{Q}\cdot\mathbf{C}\;\mathbf{C}'$$
$$\mathbf{R} = \mathbf{R}\cdot\mathbf{A}\;\mathbf{A}' + \mathbf{R}\cdot\mathbf{B}\;\mathbf{B}' + \mathbf{R}\cdot\mathbf{C}\;\mathbf{C}'.$$

Then
$$[\mathbf{P}\,\mathbf{Q}\,\mathbf{R}] = \begin{vmatrix} \mathbf{P}\cdot\mathbf{A} & \mathbf{P}\cdot\mathbf{B} & \mathbf{P}\cdot\mathbf{C} \\ \mathbf{Q}\cdot\mathbf{A} & \mathbf{Q}\cdot\mathbf{B} & \mathbf{Q}\cdot\mathbf{C} \\ \mathbf{R}\cdot\mathbf{A} & \mathbf{R}\cdot\mathbf{B} & \mathbf{R}\cdot\mathbf{C} \end{vmatrix}\,[\mathbf{A}'\,\mathbf{B}'\,\mathbf{C}'].$$

But
$$[\mathbf{A}'\,\mathbf{B}'\,\mathbf{C}'] = \frac{1}{[\mathbf{A}\,\mathbf{B}\,\mathbf{C}]}\cdot$$

Hence
$$[\mathbf{P}\,\mathbf{Q}\,\mathbf{R}]\,[\mathbf{A}\,\mathbf{B}\,\mathbf{C}] = \begin{vmatrix} \mathbf{P}\cdot\mathbf{A} & \mathbf{P}\cdot\mathbf{B} & \mathbf{P}\cdot\mathbf{C} \\ \mathbf{Q}\cdot\mathbf{A} & \mathbf{Q}\cdot\mathbf{B} & \mathbf{Q}\cdot\mathbf{C} \\ \mathbf{R}\cdot\mathbf{A} & \mathbf{R}\cdot\mathbf{B} & \mathbf{R}\cdot\mathbf{C} \end{vmatrix}.$$

The system of three unit vectors **i, j, k** *is its own reciprocal system.*

$$\mathbf{i}' = \frac{\mathbf{j}\times\mathbf{k}}{[\mathbf{i}\,\mathbf{j}\,\mathbf{k}]} = \frac{\mathbf{i}}{1} = \mathbf{i},\quad \mathbf{j}' = \frac{\mathbf{k}\times\mathbf{i}}{[\mathbf{i}\,\mathbf{j}\,\mathbf{k}]} = \mathbf{j},\quad \mathbf{k}' = \frac{\mathbf{i}\times\mathbf{j}}{[\mathbf{i}\,\mathbf{j}\,\mathbf{k}]} = \mathbf{k}. \quad (35)$$

For this reason the primes $\mathbf{i}'$, $\mathbf{j}'$, $\mathbf{k}'$ are not needed to denote a system of vectors reciprocal to **i, j, k.** The primes will therefore be used in the future to denote another set of rectangular axes **i, j, k**, just as X', Y', Z' are used to denote a set of axes different from X, Y, Z.

The only systems of three vectors which are their own reciprocals are the right-handed and left-handed systems of three unit vectors. That is the system **i, j, k** and the system **i, j, – k.**

Let **A, B, C** be a set of vectors which is its own reciprocal. Then by (32)

$$\mathbf{A}\cdot\mathbf{A} = \mathbf{B}\cdot\mathbf{B} = \mathbf{C}\cdot\mathbf{C} = 1.$$

Hence the vectors are all unit vectors.

$$\mathbf{A}\cdot\mathbf{B} = \mathbf{A}\cdot\mathbf{C} = 0.$$

Hence **A** is perpendicular to **B** and **C.**

$$\mathbf{B}\cdot\mathbf{A} = \mathbf{B}\cdot\mathbf{C} = 0.$$

Hence **B** is perpendicular to **A** and **C.**

$$\mathbf{C}\cdot\mathbf{A} = \mathbf{C}\cdot\mathbf{B} = 0.$$

Hence **C** is perpendicular to **A** and **B.**

Hence **A, B, C** must be a system like **i, j, k** or like **i, j, – k.**

* **46.**] A scalar equation of the first degree in a vector **r** is an equation in each term of which **r** occurs not more than once. The value of each term must be scalar. As an example of such an equation the following may be given.

$$a\ \mathbf{a}\cdot\mathbf{b}\times\mathbf{r} + b(\mathbf{c}\times\mathbf{d})\cdot(\mathbf{e}\times\mathbf{r}) + c\ \mathbf{f}\cdot\mathbf{r} + d = 0,$$

where **a**, **b**, **c**, **d**, **e**, **f** are known vectors; and a, b, c, d, known scalars. Obviously any scalar equation of the first degree in an unknown vector **r** may be reduced to the form

$$\mathbf{r}\cdot\mathbf{A} = a$$

where **A** is a known vector; and a, a known scalar. To accomplish this result in the case of the given equation proceed as follows.

$$a\,\mathbf{a}\times\mathbf{b}\cdot\mathbf{r} + b\,(\mathbf{c}\times\mathbf{d})\times\mathbf{e}\cdot\mathbf{r} + c\,\mathbf{f}\cdot\mathbf{r} + d = 0$$

$$\{a\,\mathbf{a}\times\mathbf{b} + b\,(\mathbf{c}\times\mathbf{d})\times\mathbf{e} + c\,\mathbf{f}\}\cdot\mathbf{r} = -\,d.$$

In more complicated forms it may be necessary to make use of various reduction formulæ before the equation can be made to take the desired form,

$$\mathbf{r}\cdot\mathbf{A} = a.$$

As a vector has three degrees of freedom it is clear that one scalar equation is insufficient to determine a vector. Three scalar equations are necessary.

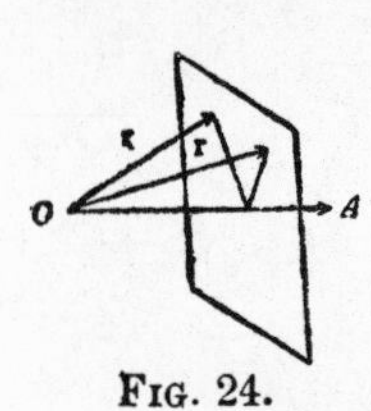

FIG. 24.

The geometric interpretation of the equation

$$\mathbf{r}\cdot\mathbf{A} = a \qquad (36)$$

is interesting. Let **r** be a variable vector (Fig. 24) drawn from a fixed origin. Let **A** be a fixed vector drawn from the same origin. The equation then becomes

$$r\,A\cos(\mathbf{r},\mathbf{A}) = a,$$

or

$$r\cos(\mathbf{r},\mathbf{A}) = \frac{a}{A},$$

if r be the magnitude of **r**; and A that of **A**. The expression

$$r\cos(\mathbf{r},\mathbf{A})$$

is the projection of **r** upon **A**. The equation therefore states that the projection of **r** upon a certain fixed vector **A** must always be constant and equal to a/A. Consequently the terminus of **r** must trace out a plane perpendicular to the vector **A** at a distance equal to a/A from the origin. The projection upon **A** of any radius vector drawn from the origin to a point of this plane is constant and equal to a/A. This gives the following theorem.

Theorem: A scalar equation in an unknown vector may be regarded as the equation of a plane, which is the locus of the terminus of the unknown vector if its origin be fixed.

It is easy to see why three scalar equations in an unknown vector determine the vector completely. Each equation determines a plane in which the terminus of **r** must lie. The three planes intersect in one common point. Hence one vector **r** is determined. The analytic solution of three scalar equations is extremely easy. If the equations are

$$\begin{aligned} \mathbf{r}\cdot\mathbf{A} &= a \\ \mathbf{r}\cdot\mathbf{B} &= b \\ \mathbf{r}\cdot\mathbf{C} &= c, \end{aligned} \tag{37}$$

it is only necessary to call to mind the formula

$$\mathbf{r} = \mathbf{r}\cdot\mathbf{A}\,\mathbf{A}' + \mathbf{r}\cdot\mathbf{B}\,\mathbf{B}' + \mathbf{r}\cdot\mathbf{C}\,\mathbf{C}'.$$

Hence
$$\mathbf{r} = a\,\mathbf{A}' + b\,\mathbf{B}' + c\,\mathbf{C}'. \tag{38}$$

The solution is therefore accomplished. It is expressed in terms **A**′, **B**′, **C**′ which is the reciprocal system to **A**, **B**, **C**. One caution must however be observed. The vectors **A**, **B**, **C** will have no reciprocal system if they are coplanar. Hence the solution will fail. In this case, however, the three planes determined by the three equations will be parallel to a line. They will therefore either not intersect (as in the case of the lateral faces of a triangular prism) or they will intersect in a common line. Hence there will be either no solution for **r** or there will be an infinite number.

From four scalar equations

$$\begin{aligned} \mathbf{r}\cdot\mathbf{A} &= a \\ \mathbf{r}\cdot\mathbf{B} &= b \\ \mathbf{r}\cdot\mathbf{C} &= c \\ \mathbf{r}\cdot\mathbf{D} &= d \end{aligned} \tag{39}$$

the vector **r** may be entirely eliminated. To accomplish this solve three of the equations and substitute the value in the fourth.

$$\mathbf{r} = a\,\mathbf{A}' + b\,\mathbf{B}' + c\,\mathbf{C}'$$

$$a\,\mathbf{A}'\cdot\mathbf{D} + b\,\mathbf{B}'\cdot\mathbf{D} + c\,\mathbf{C}'\cdot\mathbf{D} = d$$

or $$a\,[\mathbf{B}\,\mathbf{C}\,\mathbf{D}] + b\,[\mathbf{C}\,\mathbf{A}\,\mathbf{D}] + c\,[\mathbf{A}\,\mathbf{B}\,\mathbf{D}] = d\,[\mathbf{A}\,\mathbf{B}\,\mathbf{C}]. \tag{40}$$

* **47.**] A vector equation of the first degree in an unknown vector is an equation each term of which is a vector quantity containing the unknown vector not more than once. Such an equation is

$$(\mathbf{A}\times\mathbf{B})\times(\mathbf{C}\times\mathbf{r}) + \mathbf{D}\,\mathbf{E}\cdot\mathbf{r} + n\,\mathbf{r} + \mathbf{F} = 0,$$

where **A**, **B**, **C**, **D**, **E**, **F** are known vectors, n a known scalar, and **r** the unknown vector. One such equation may in general be solved for **r**. That is to say, one vector equation is in general sufficient to determine the unknown vector which is contained in it to the first degree.

The method of solving a vector equation is to multiply it with a dot successively by three arbitrary known non-coplanar vectors. Thus three scalar equations are obtained. These may be solved by the methods of the foregoing article. In the first place let the equation be

$$\mathbf{A}\,\mathbf{a}\cdot\mathbf{r} + \mathbf{B}\,\mathbf{b}\cdot\mathbf{r} + \mathbf{C}\,\mathbf{c}\cdot\mathbf{r} = \mathbf{D},$$

where **A**, **B**, **C**, **D**, **a**, **b**, **c** are known vectors. No scalar coefficients are written in the terms, for they may be incorporated in the vectors. Multiply the equation successively by **A**′, **B**′, **C**′. It is understood of course that **A**, **B**, **C** are non-coplanar.

$$\mathbf{a \cdot r} = \mathbf{D \cdot A'}$$
$$\mathbf{b \cdot r} = \mathbf{D \cdot B'}$$
$$\mathbf{c \cdot r} = \mathbf{D \cdot C'}.$$

But $$\mathbf{r} = \mathbf{a'}\, \mathbf{a \cdot r} + \mathbf{b'}\, \mathbf{b \cdot r} + \mathbf{c'}\, \mathbf{c \cdot r}.$$

Hence $$\mathbf{r} = \mathbf{D \cdot A'}\, \mathbf{a'} + \mathbf{D \cdot B'}\, \mathbf{b'} + \mathbf{D \cdot C'}\, \mathbf{c'}.$$

The solution is therefore accomplished in case $\mathbf{A}$, $\mathbf{B}$, $\mathbf{C}$ are non-coplanar and $\mathbf{a}$, $\mathbf{b}$, $\mathbf{c}$ also non-coplanar. The special cases in which either of these sets of three vectors is coplanar will not be discussed here.

The most general vector equation of the first degree in an unknown vector $\mathbf{r}$ contains terms of the types

$$\mathbf{A}\, \mathbf{a \cdot r}, \quad n\, \mathbf{r}, \quad \mathbf{E \times r}, \quad \mathbf{D}.$$

That is it will contain terms which consist of a known vector multiplied by the scalar product of another known vector and the unknown vector; terms which are scalar multiples of the unknown vector; terms which are the vector product of a known and the unknown vector; and constant terms. The terms of the type $\mathbf{A}\, \mathbf{a \cdot r}$ may always be reduced to three in number. For the vectors $\mathbf{a}$, $\mathbf{b}$, $\mathbf{c}$, $\cdots$ which are multiplied into $\mathbf{r}$ may all be expressed in terms of three non-coplanar vectors. Hence all the products $\mathbf{a \cdot r}$, $\mathbf{b \cdot r}$, $\mathbf{c \cdot r}$, $\cdots$ may be expressed in terms of three. The sum of all terms of the type $\mathbf{A}\, \mathbf{a \cdot r}$ therefore reduces to an expression of three terms, as

$$\mathbf{A}\, \mathbf{a \cdot r} + \mathbf{B}\, \mathbf{b \cdot r} + \mathbf{C}\, \mathbf{c \cdot r}.$$

The terms of the types $n\, \mathbf{r}$ and $\mathbf{E \times r}$ may also be expressed in this form.

$$n\, \mathbf{r} = n\, \mathbf{a'}\, \mathbf{a \cdot r} + n\, \mathbf{b'}\, \mathbf{b \cdot r} + n\, \mathbf{c'}\, \mathbf{c \cdot r}$$
$$\mathbf{E \times r} = \mathbf{E \times a'}\, \mathbf{a \cdot r} + \mathbf{E \times b'}\, \mathbf{b \cdot r} + \mathbf{E \times c'}\, \mathbf{c \cdot r}.$$

Adding all these terms together the whole equation reduces to the form

$$\mathbf{L}\, \mathbf{a \cdot r} + \mathbf{M}\, \mathbf{b \cdot r} + \mathbf{N}\, \mathbf{c \cdot r} = \mathbf{K}.$$

This has already been solved as

$$\mathbf{r} = \mathbf{K}\cdot\mathbf{L}'\,\mathbf{a}' + \mathbf{K}\cdot\mathbf{M}'\,\mathbf{b}' + \mathbf{K}\cdot\mathbf{N}'\,\mathbf{c}'.$$

The solution is in terms of three non-coplanar vectors $\mathbf{a}'$, $\mathbf{b}'$, $\mathbf{c}'$. These form the system reciprocal to $\mathbf{a}$, $\mathbf{b}$, $\mathbf{c}$ in terms of which the products containing the unknown vector $\mathbf{r}$ were expressed.

* Sundry Applications of Products

Applications to Mechanics

48.] In the mechanics of a rigid body a force is *not* a vector in the sense understood in this book. See Art. 3. A force has magnitude and direction; but it has also a line of application. Two forces which are alike in magnitude and direction, but which lie upon different lines in the body do not produce the same effect. Nevertheless vectors are sufficiently like forces to be useful in treating them.

If a number of forces $\mathbf{f}_1$, $\mathbf{f}_2$, $\mathbf{f}_3$, $\cdots$ act on a body at the same point O, the sum of the forces added as vectors is called the *resultant* $\mathbf{R}$.

$$\mathbf{R} = \mathbf{f}_1 + \mathbf{f}_2 + \mathbf{f}_3 + \cdots$$

In the same way if $\mathbf{f}_1$, $\mathbf{f}_2$, $\mathbf{f}_3 \cdots$ do not act at the same point the term resultant is still applied to the sum of these forces added just as if they were vectors.

$$\mathbf{R} = \mathbf{f}_1 + \mathbf{f}_2 + \mathbf{f}_3 + \cdots \qquad (41)$$

The idea of the resultant therefore does not introduce the line of action of a force. As far as the resultant is concerned a force does not differ from a vector.

Definition: The *moment* of a force $\mathbf{f}$ about the point O is equal to the product of the force by the perpendicular distance from O to the line of action of the force. The moment however is best looked upon as a vector quantity. Its magnitude is as defined above. Its direction is usually taken to be the normal on that side of the plane passed through the point O and the line **f** upon which the force appears to produce a tendency to rotation about the point O in the positive trigonometric direction. Another method of defining the moment of a force $\mathbf{f} = \overline{PQ}$ about the point O is as follows: The moment of the force $\mathbf{f} = \overline{PQ}$ about the point O is equal to twice the area of the triangle OPQ. This includes at once both the magnitude and direction of the moment (Art. 25). The point P is supposed to be the origin; and the point Q, the terminus of the arrow which represents the force **f**. The letter **M** will be used to denote the moment. A subscript will be attached to designate the point about which the moment is taken.

$$\mathbf{M}_O \{\mathbf{f}\} = 2\, O P Q.$$

The moment of a number of forces $\mathbf{f}_1, \mathbf{f}_2, \cdots$ is the (vector) sum of the moments of the individual forces.

If
$$\mathbf{f}_1 = \overline{P_1 Q_1}, \quad \mathbf{f}_2 = \overline{P_2 Q_2} \cdots$$
$$\mathbf{M}_O \{\mathbf{f}_1, \mathbf{f}_2, \cdots\} = 2\,(O P_1 Q_1 + O P_2 Q_2 + \cdots).$$

This is known as the total or resultant moment of the forces $\mathbf{f}_1, \mathbf{f}_2, \cdots$.

49.] If **f** be a force acting on a body and if **d** be the vector drawn from the point O to any point in the line of action of the force, the moment of the force about the point O is the vector product of **d** into **f**.

$$\mathbf{M}_O \{\mathbf{f}\} = \mathbf{d} \times \mathbf{f} \tag{42}$$

For
$$\mathbf{d} \times \mathbf{f} = d\, f \sin(\mathbf{d}, \mathbf{f})\, \mathbf{e},$$

if **e** be a unit vector in the direction of $\mathbf{d} \times \mathbf{f}$.

$$\mathbf{d} \times \mathbf{f} = d \sin(\mathbf{d}, \mathbf{f})\, f\, \mathbf{e}.$$

Now $d \sin(\mathbf{d}, \mathbf{f})$ is the perpendicular distance from O to **f**. The magnitude of $\mathbf{d} \times \mathbf{f}$ is accordingly equal to this perpendicular distance multiplied by f, the magnitude of the force.

This is the magnitude of the moment $\mathbf{M}_O\,\{\mathbf{f}\}$. The direction of $\mathbf{d}\times\mathbf{f}$ is the same as the direction of the moment. Hence the relation is proved.

$$\mathbf{M}_O\,\{\mathbf{f}\} = \mathbf{d}\times\mathbf{f}.$$

The sum of the moments about O of a number of forces $\mathbf{f}_1, \mathbf{f}_2, \cdots$ acting at the *same point* P is equal to the moment of the resultant $\mathbf{R}$ of the forces acting at that point. For let $\mathbf{d}$ be the vector from O to P. Then

$$\mathbf{M}_O\,\{\mathbf{f}_1\} = \mathbf{d}\times\mathbf{f}_1$$

$$\mathbf{M}_O\,\{\mathbf{f}_2\} = \mathbf{d}\times\mathbf{f}_2$$

$$\vdots \qquad \vdots$$

$$\mathbf{M}_O\,\{\mathbf{f}_1\} + \mathbf{M}_O\,\{\mathbf{f}_2\} + \cdots = \mathbf{d}\times\mathbf{f}_1 + \mathbf{d}\times\mathbf{f}_2 + \cdots \tag{43}$$

$$= \mathbf{d}\times(\mathbf{f}_1 + \mathbf{f}_2 + \cdots) = \mathbf{d}\times\mathbf{R}$$

The total moment about O' of any number of forces $\mathbf{f}_1, \mathbf{f}_2, \cdots$ acting on a *rigid body* is equal to the total moment of those forces about O increased by the moment about O' of the resultant $\mathbf{R}_O$ considered as acting at O.

$$\mathbf{M}_{O'}\,\{\mathbf{f}_1, \mathbf{f}_2, \cdots\} = \mathbf{M}_O\,\{\mathbf{f}_1, \mathbf{f}_2, \cdots\} + \mathbf{M}_{O'}\,\{\mathbf{R}_O\}. \tag{44}$$

Let $\mathbf{d}_1, \mathbf{d}_2, \cdots$ be vectors drawn from O to any point in $\mathbf{f}_1, \mathbf{f}_2, \cdots$ respectively. Let $\mathbf{d}_1', \mathbf{d}_2', \cdots$ be the vectors drawn from O' to the same points in $\mathbf{f}_1, \mathbf{f}_2, \cdots$ respectively. Let $\mathbf{c}$ be the vector from O to O'. Then

$$\mathbf{d}_1 = \mathbf{d}_1' + \mathbf{c}, \qquad \mathbf{d}_2 = \mathbf{d}_2' + \mathbf{c}, \cdots$$

$$\mathbf{M}_O\,\{\mathbf{f}_1, \mathbf{f}_2, \cdots\} = \mathbf{d}_1\times\mathbf{f}_1 + \mathbf{d}_2\times\mathbf{f}_2 + \cdots$$

$$\mathbf{M}_{O'}\,\{\mathbf{f}_1, \mathbf{f}_2, \cdots\} = \mathbf{d}_1'\times\mathbf{f}_1 + \mathbf{d}_2'\times\mathbf{f}_2 + \cdots$$

$$= (\mathbf{d}_1 - \mathbf{c})\times\mathbf{f}_1 + (\mathbf{d}_2 - \mathbf{c})\times\mathbf{f}_2 + \cdots$$

$$= \mathbf{d}_1\times\mathbf{f}_1 + \mathbf{d}_2\times\mathbf{f}_2 + \cdots - \mathbf{c}\times(\mathbf{f}_1 + \mathbf{f}_2 + \cdots)$$

But $-\mathbf{c}$ is the vector drawn from O' to O. Hence $-\mathbf{c}\times\mathbf{f}_1$ is the moment about O' of a force equal in magnitude and parallel in direction to $\mathbf{f}_1$ but situated at O. Hence

$$-\mathbf{c}\times(\mathbf{f}_1+\mathbf{f}_2+\cdots) = -\,\mathbf{c}\times\mathbf{R}_O = \mathbf{M}_{O'}\,\{\mathbf{R}_O\}.$$

Hence $\mathbf{M}_{O'}\,\{\mathbf{f}_1, \mathbf{f}_2, \cdots\} = \mathbf{M}_O\,\{\mathbf{f}_1, \mathbf{f}_2, \cdots\} + \mathbf{M}_{O'}\,\{\mathbf{R}_O\}.$ (44)

The theorem is therefore proved.

The resultant $\mathbf{R}$ is of course the same at all points. The subscript O is attached merely to show at what point it is *supposed* to act when the moment about O' is taken. For the point of application of $\mathbf{R}$ affects the value of that moment.

The scalar product of the total moment and the resultant is the same no matter about what point the moment be taken. In other words the product of the total moment, the resultant, and the cosine of the angle between them is invariant for all points of space.

$$\mathbf{R}\cdot\mathbf{M}_{O'}\,\{\mathbf{f}_1, \mathbf{f}_2, \cdots\} = \mathbf{R}\cdot\mathbf{M}_O\,\{\mathbf{f}_1, \mathbf{f}_2, \cdots\}$$

where O' and O are any two points in space. This important relation follows immediately from the equation

$$\mathbf{M}_{O'}\,\{\mathbf{f}_1, \mathbf{f}_2, \cdots\} = \mathbf{M}_O\,\{\mathbf{f}_1, \mathbf{f}_2, \cdots\} + \mathbf{M}_{O'}\,\{\mathbf{R}_O\}.$$

For $\mathbf{R}\cdot\mathbf{M}_{O'}\,\{\mathbf{f}_1, \mathbf{f}_2, \cdots\} = \mathbf{R}\cdot\mathbf{M}_O\,\{\mathbf{f}_1, \mathbf{f}_2, \cdots\} + \mathbf{R}\cdot\mathbf{M}_{O'}\,\{\mathbf{R}_O\}.$

But the moment of $\mathbf{R}$ is perpendicular to $\mathbf{R}$ no matter what the point O of application be. Hence

$$\mathbf{R}\cdot\mathbf{M}_{O'}\,\{\mathbf{R}_O\} = 0$$

and the relation is proved. The variation in the total moment due to a variation of the point about which the moment is taken is always perpendicular to the resultant.

50.] A point O' may be found such that the total moment about it is parallel to the resultant. The condition for parallelism is

$$\mathbf{R}\times\mathbf{M}_{O'}\,\{\mathbf{f}_1, \mathbf{f}_2, \cdots\} = 0$$

$$\mathbf{R}\times\mathbf{M}_{O'}\,\{\mathbf{f}_1, \mathbf{f}_2, \cdots\} = \mathbf{R}\times\mathbf{M}_O\,\{\mathbf{f}_1, \mathbf{f}_2, \cdots\} + \mathbf{R}\times\mathbf{M}_{O'}\,\{\mathbf{R}_O\} = 0$$

where O is any point chosen at random. Replace $\mathbf{M}_{O'}\{\mathbf{R}_O\}$ by its value and for brevity omit to write the $\mathbf{f}_1, \mathbf{f}_2, \cdots$ in the braces $\{\ \}$. Then

$$\mathbf{R}\times\mathbf{M}_{O'} = \mathbf{R}\times\mathbf{M}_O - \mathbf{R}\times(\mathbf{c}\times\mathbf{R}) = 0.$$

The problem is to solve this equation for $\mathbf{c}$.

$$\mathbf{R}\times\mathbf{M}_O - \mathbf{R}\cdot\mathbf{R}\ \mathbf{c} + \mathbf{R}\cdot\mathbf{c}\ \mathbf{R} = 0.$$

Now $\mathbf{R}$ is a known quantity. $\mathbf{M}_O$ is also supposed to be known. Let $\mathbf{c}$ be chosen in the plane through O perpendicular to $\mathbf{R}$. Then $\mathbf{R}\cdot\mathbf{c} = 0$ and the equation reduces to

$$\mathbf{R}\times\mathbf{M}_O = \mathbf{R}\cdot\mathbf{R}\ \mathbf{c}$$

$$\mathbf{c} = \frac{\mathbf{R}\times\mathbf{M}_O}{\mathbf{R}\cdot\mathbf{R}}.$$

If $\mathbf{c}$ be chosen equal to this vector the total moment about the point O', which is at a vector distance from O equal to $\mathbf{c}$, will be parallel to $\mathbf{R}$. Moreover, since the scalar product of the total moment and the resultant is constant and since the resultant itself is constant it is clear that in the case where they are parallel the numerical value of the total moment will be a minimum.

The total moment is unchanged by displacing the point about which it is taken in the direction of the resultant.

For $$\mathbf{M}_{O'}\{\mathbf{f}_1, \mathbf{f}_2, \cdots\} = \mathbf{M}_O\{\mathbf{f}_1, \mathbf{f}_2, \cdots\} - \mathbf{c}\times\mathbf{R}.$$

If $\mathbf{c} = \overline{O\,O'}$ is parallel to $\mathbf{R}$, $\mathbf{c}\times\mathbf{R}$ vanishes and the moment about O' is equal to that about O. Hence it is possible to find not merely *one* point O' about which the total moment is parallel to the resultant; but the total moment about any point in the line drawn through O' parallel to $\mathbf{R}$ is parallel to $\mathbf{R}$. Furthermore the solution found in equation for $\mathbf{c}$ is the only one which exists in the plane perpendicular to $\mathbf{R}$— unless the resultant $\mathbf{R}$ vanishes. The results that have been obtained may be summed up as follows:

If any system of forces $\mathbf{f}_1, \mathbf{f}_2, \cdots$ whose resultant is not zero act upon a rigid body, then there exists in space one and only one line such that the total moment about any point of it is parallel to the resultant. This line is itself parallel to the resultant. The total moment about all points of it is the same and is numerically less than that about any other point in space.

This theorem is equivalent to the one which states that any system of forces acting upon a rigid body is equivalent to a single force (the resultant) acting in a definite line and a couple of which the plane is perpendicular to the resultant and of which the moment is a minimum. A system of forces may be reduced to a single force (the resultant) acting at any desired point O of space and a couple the moment of which (regarded as a vector quantity) is equal to the total moment about O of the forces acting on the body. But in general the plane of this couple will not be perpendicular to the resultant, nor will its moment be a minimum.

Those who would pursue the study of systems of forces acting on a rigid body further and more thoroughly may consult the *Traité de Mécanique Rationnelle* [1] by P. APPELL. The first chapter of the first volume is entirely devoted to the discussion of systems of forces. Appell defines a vector as a quantity possessing magnitude, direction, and point of application. His vectors are consequently not the same as those used in this book. The treatment of his vectors is carried through in the Cartesian coördinates. Each step however may be easily converted into the notation of vector analysis. A number of exercises is given at the close of the chapter.

51.] Suppose a body be rotating about an axis with a constant angular velocity a. The points in the body describe circles concentric with the axis in planes perpendicular to

[1] Paris, Gauthier-Villars et Fils, 1893.

the axis. The velocity of any point in its circle is equal to the product of the angular velocity and the radius of the circle. It is therefore equal to the product of the angular velocity and the perpendicular distance from the point to the axis. The direction of the velocity is perpendicular to the axis and to the radius of the circle described by the point.

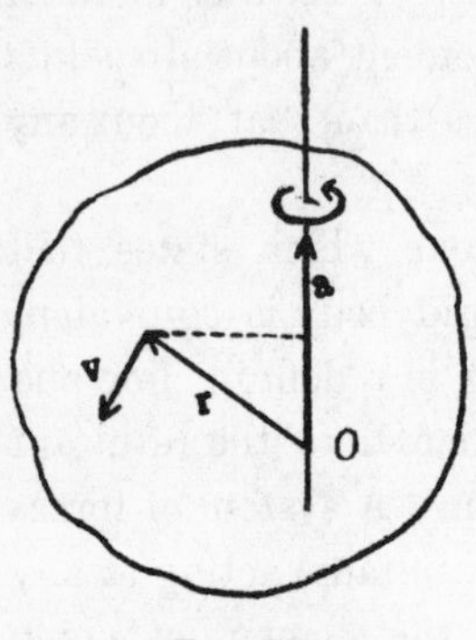

FIG. 25.

Let **a** (Fig. 25) be a vector drawn along the axis of rotation in that direction in which a right-handed screw would advance if turned in the direction in which the body is rotating. Let the magnitude of **a** be a, the angular velocity. The vector **a** may be taken to represent the rotation of the body. Let **r** be a radius vector drawn from any point of the axis of rotation to a point in the body. The vector product

$$\mathbf{a} \times \mathbf{r} = a\, r \sin(\mathbf{a}, \mathbf{r})$$

is equal in magnitude and direction to the velocity **v** of the terminus of **r**. For its direction is perpendicular to **a** and **r** and its magnitude is the product of a and the perpendicular distance $r \sin(\mathbf{a}, \mathbf{r})$ from the point to the line **a**. That is

$$\mathbf{v} = \mathbf{a} \times \mathbf{r}. \tag{45}$$

If the body be rotating simultaneously about several axes $\mathbf{a}_1, \mathbf{a}_2, \mathbf{a}_3 \cdots$ which pass through the same point as in the case of the gyroscope, the velocities due to the various rotations are

$$\begin{aligned} \mathbf{v}_1 &= \mathbf{a}_1 \times \mathbf{r}_1 \\ \mathbf{v}_2 &= \mathbf{a}_2 \times \mathbf{r}_2 \\ \mathbf{v}_3 &= \mathbf{a}_3 \times \mathbf{r}_3 \\ \vdots \quad & \quad \vdots \quad \vdots \end{aligned}$$

where $\mathbf{r}_1, \mathbf{r}_2, \mathbf{r}_3, \cdots$ are the radii vectores drawn from points on the axis $\mathbf{a}_1, \mathbf{a}_2, \mathbf{a}_3, \cdots$ to the same point of the body. Let the vectors $\mathbf{r}_1, \mathbf{r}_2, \mathbf{r}_3, \cdots$ be drawn from the common point of intersection of the axes. Then

$$\mathbf{r}_1 = \mathbf{r}_2 = \mathbf{r}_3 = \cdots = \mathbf{r}$$

and

$$\mathbf{v} = \mathbf{v}_1 + \mathbf{v}_2 + \mathbf{v}_3 + \cdots = \mathbf{a}_1 \times \mathbf{r} + \mathbf{a}_2 \times \mathbf{r} + \mathbf{a}_3 \times \mathbf{r} + \cdots$$
$$= (\mathbf{a}_1 + \mathbf{a}_2 + \mathbf{a}_3 + \cdots) \times \mathbf{r}.$$

This shows that the body moves as if rotating with the angular velocity which is the vector sum of the angular velocities $\mathbf{a}_1, \mathbf{a}_2, \mathbf{a}_3, \cdots$ This theorem is sometimes known as the parallelogram law of angular velocities.

It will be shown later (Art.) 60 that the motion of any rigid body one point of which is fixed is at each instant of time a rotation about some axis drawn through that point. This axis is called the instantaneous axis of rotation. The axis is not the same for all time, but constantly changes its position. The motion of a rigid body one point of which is fixed is therefore represented by

$$\mathbf{v} = \mathbf{a} \times \mathbf{r} \qquad (45)$$

where $\mathbf{a}$ is the instantaneous angular velocity; and $\mathbf{r}$, the radius vector drawn from the fixed point to any point of the body.

The most general motion of a rigid body no point of which is fixed may be treated as follows. Choose an arbitrary point O. At any instant this point will have a velocity $\mathbf{v}_o$. *Relative to the point* O the body will have a motion of rotation about some axis drawn through O. Hence the velocity $\mathbf{v}$ of any point of the body may be represented by the sum of $\mathbf{v}_o$ the velocity of O and $\mathbf{a} \times \mathbf{r}$ the velocity of that point relative to O.

$$\mathbf{v} = \mathbf{v}_o + \mathbf{a} \times \mathbf{r}. \qquad (46)$$

In case $\mathbf{v}_0$ is parallel to $\mathbf{a}$, the body moves around $\mathbf{a}$ and along $\mathbf{a}$ simultaneously. This is precisely the motion of a screw advancing along $\mathbf{a}$. In case $\mathbf{v}_0$ is perpendicular to $\mathbf{a}$, it is possible to find a point, given by the vector $\mathbf{r}$, such that its velocity is zero. That is

$$\mathbf{a}\times\mathbf{r} = -\mathbf{v}_0.$$

This may be done as follows. Multiply by $\times\mathbf{a}$.

$$(\mathbf{a}\times\mathbf{r})\times\mathbf{a} = -\mathbf{v}_0\times\mathbf{a}$$

or

$$\mathbf{a}\cdot\mathbf{a}\,\mathbf{r} - \mathbf{a}\cdot\mathbf{r}\,\mathbf{a} = -\mathbf{v}_0\times\mathbf{a}.$$

Let $\mathbf{r}$ be chosen perpendicular to $\mathbf{a}$. Then $\mathbf{a}\cdot\mathbf{r}$ is zero and

$$\mathbf{a}\cdot\mathbf{a}\,\mathbf{r} = -\mathbf{v}_0\times\mathbf{a}$$

$$\mathbf{r} = \frac{-\mathbf{v}_0\times\mathbf{a}}{\mathbf{a}\cdot\mathbf{a}}.$$

The point $\mathbf{r}$, thus determined, has the property that its velocity is zero. If a line be drawn through this point parallel to $\mathbf{a}$, the motion of the body is one of instantaneous rotation about this new axis.

In case $\mathbf{v}_0$ is neither parallel nor perpendicular to $\mathbf{a}$ it may be resolved into two components

$$\mathbf{v}_0 = \mathbf{v}_0' + \mathbf{v}_0''$$

which are respectively parallel and perpendicular to $\mathbf{a}$.

$$\mathbf{v} = \mathbf{v}_0' + \mathbf{v}_0'' + \mathbf{a}\times\mathbf{r}$$

A point may now be found such that

$$\mathbf{v}_0'' = -\mathbf{a}\times\mathbf{r}.$$

Let the different points of the body referred to this point be denoted by $\mathbf{r}'$. Then the equation becomes

$$\mathbf{v} = \mathbf{v}_0' + \mathbf{a}\times\mathbf{r}'. \qquad (46)'$$

The motion here expressed consists of rotation about an axis $\mathbf{a}$ and translation along that axis. It is therefore seen that the most general motion of a rigid body is at any instant the motion of a screw advancing at a certain rate along a definite axis **a** in space. The axis of the screw and its rate of advancing per unit of rotation (*i. e.* its pitch) change from instant to instant.

52.] The conditions for equilibrium as obtained by the principle of virtual velocities may be treated by vector methods. Suppose any system of forces $\mathbf{f}_1, \mathbf{f}_2, \cdots$ act on a rigid body. If the body be displaced through a vector distance **D** whether this distance be finite or infinitesimal the work done by the forces is

$$\mathbf{D}\cdot\mathbf{f}_1, \mathbf{D}\cdot\mathbf{f}_2, \cdots$$

The total work done is therefore

$$W = \mathbf{D}\cdot\mathbf{f}_1 + \mathbf{D}\cdot\mathbf{f}_2 + \cdots$$

If the body be in equilibrium under the action of the forces the work done must be zero.

$$W = \mathbf{D}\cdot\mathbf{f}_1 + \mathbf{D}\cdot\mathbf{f}_2 + \cdots = \mathbf{D}\cdot(\mathbf{f}_1 + \mathbf{f}_2 + \cdots) = \mathbf{D}\cdot\mathbf{R} = 0.$$

The work done by the forces is equal to the work done by their resultant. This must be zero for *every* displacement **D**. The equation

$$\mathbf{D}\cdot\mathbf{R} = 0$$

holds for all vectors **D**. Hence

$$\mathbf{R} = 0.$$

The total resultant must be zero if the body be in equilibrium.

The work done by a force **f** when the rigid body is displaced by a rotation of angular velocity **a** for an infinitesimal time t is approximately

$$\mathbf{a}\cdot\mathbf{d}\times\mathbf{f}\ t,$$

where **d** is a vector drawn from any point of the axis of rotation **a** to any point of **f**. To prove this break up **f** into two components $\mathbf{f}', \mathbf{f}''$ parallel and perpendicular respectively to **a**.

$$\mathbf{a}\cdot\mathbf{d}\times\mathbf{f} = \mathbf{a}\cdot\mathbf{d}\times\mathbf{f}' + \mathbf{a}\cdot\mathbf{d}\times\mathbf{f}''.$$

As $\mathbf{f}'$ is parallel to $\mathbf{a}$ the scalar product $[\mathbf{a}\;\mathbf{d}\;\mathbf{f}']$ vanishes.

$$\mathbf{a}\cdot\mathbf{d}\times\mathbf{f} = \mathbf{a}\cdot\mathbf{d}\times\mathbf{f}''.$$

On the other hand the work done by $\mathbf{f}''$ is equal to the work done by $\mathbf{f}$ during the displacement. For $\mathbf{f}'$ being parallel to $\mathbf{a}$ is perpendicular to its line of action. If $\mathbf{h}$ be the common vector perpendicular from the line $\mathbf{a}$ to the force $\mathbf{f}''$, the work done by $\mathbf{f}''$ during a rotation of angular velocity $\mathbf{a}$ for time t is approximately

$$W = h\,f''\,a\,t = \mathbf{a}\cdot\mathbf{h}\times\mathbf{f}''\,t.$$

The vector $\mathbf{d}$ drawn from any point of $\mathbf{a}$ to any point of $\mathbf{f}$ may be broken up into three components of which one is $\mathbf{h}$, another is parallel to $\mathbf{a}$, and the third is parallel to $\mathbf{f}''$. In the scalar triple product $[\mathbf{a}\;\mathbf{d}\;\mathbf{f}'']$ only that component of $\mathbf{d}$ which is perpendicular alike to $\mathbf{a}$ and $\mathbf{f}''$ has any effect. Hence

$$W = \mathbf{a}\cdot\mathbf{h}\times\mathbf{f}''\,t = \mathbf{a}\cdot\mathbf{d}\times\mathbf{f}'\,t' = \mathbf{a}\cdot\mathbf{d}\times\mathbf{f}\,t.$$

If a rigid body upon which the forces $\mathbf{f}_1, \mathbf{f}_2, \cdots$ act be displaced by an angular velocity $\mathbf{a}$ for an infinitesimal time t and if $\mathbf{d}_1, \mathbf{d}_2, \cdots$ be the vectors drawn from any point O of $\mathbf{a}$ to any points of $\mathbf{f}_1, \mathbf{f}_2, \cdots$ respectively, then the work done by the forces $\mathbf{f}_1, \mathbf{f}_2, \cdots$ will be approximately

$$\begin{aligned} W &= (\mathbf{a}\cdot\mathbf{d}_1\times\mathbf{f}_1 + \mathbf{a}\cdot\mathbf{d}_2\times\mathbf{f}_2 + \cdots)\,t \\ &= \mathbf{a}\cdot(\mathbf{d}_1\times\mathbf{f}_1 + \mathbf{d}_2\times\mathbf{f}_2 + \cdots)\,t \\ &= \mathbf{a}\cdot\mathbf{M}_O\,\{\mathbf{f}_1, \mathbf{f}_2, \cdots\}\,t. \end{aligned}$$

If the body be in equilibrium this work must be zero.

Hence $$\mathbf{a}\cdot\mathbf{M}_O\,\{\mathbf{f}_1, \mathbf{f}_2, \cdots\}\,t = 0.$$

The scalar product of the angular velocity $\mathbf{a}$ and the total moment of the forces $\mathbf{f}_1, \mathbf{f}_2, \cdots$ about any point O must be zero. As $\mathbf{a}$ may be any vector whatsoever the moment itself must vanish.

$$\mathbf{M}_O\,\{\mathbf{f}_1, \mathbf{f}_2, \cdots\} = 0.$$

The necessary conditions that a rigid body be in equilibrium under the action of a system of forces is that the resultant of those forces and the total moment about any point in space shall vanish.

Conversely if the resultant of a system of forces and the moment of those forces about any one particular point in space vanish simultaneously, the body will be in equilibrium.

If $\mathbf{R} = 0$, then for any displacement of translation $\mathbf{D}$

$$\mathbf{D} \cdot \mathbf{R} = 0.$$

$$W = \mathbf{D} \cdot \mathbf{f}_1 + \mathbf{D} \cdot \mathbf{f}_2 + \cdots = 0$$

and the total work done is zero, when the body suffers any displacement of translation.

Let $\mathbf{M}_O \{\mathbf{f}_1, \mathbf{f}_2, \cdots\}$ be zero for a given point O. Then for any other point O'

$$\mathbf{M}_{O'} \{\mathbf{f}_1, \mathbf{f}_2, \cdots\} = \mathbf{M}_O \{\mathbf{f}_1, \mathbf{f}_2, \cdots\} + \mathbf{M}_{O'} \{\mathbf{R}_O\}.$$

But by hypothesis $\mathbf{R}$ is also zero. Hence

$$\mathbf{M}_{O'} \{\mathbf{f}_1, \mathbf{f}_2, \cdots\} = 0.$$

Hence $$\mathbf{a} \cdot \mathbf{M}_{O'} \{\mathbf{f}_1, \mathbf{f}_2, \cdots\}\, t = 0$$

where $\mathbf{a}$ is any vector whatsoever. But this expression is equal to the work done by the forces when the body is rotated for a time t with an angular velocity a about the line $\mathbf{a}$ passing through the point O'. This work is zero.

Any displacement of a rigid body may be regarded as a translation through a distance $\mathbf{D}$ combined with a rotation for a time t with angular velocity a about a suitable line $\mathbf{a}$ in space. It has been proved that the total work done by the forces during this displacement is zero. Hence the forces must be in equilibrium. The theorem is proved.

Applications to Geometry

53.] Relations between two right-handed systems of three mutually perpendicular unit vectors. — Let $\mathbf{i}, \mathbf{j}, \mathbf{k}$ and $\mathbf{i}', \mathbf{j}', \mathbf{k}'$ be two such systems. They form their own reciprocal systems. Hence

$$\mathbf{r} = \mathbf{r}\cdot\mathbf{i}\,\mathbf{i} + \mathbf{r}\cdot\mathbf{j}\,\mathbf{j} + \mathbf{r}\cdot\mathbf{k}\,\mathbf{k}$$

and

$$\mathbf{r} = \mathbf{r}\cdot\mathbf{i}'\,\mathbf{i}' + \mathbf{r}\cdot\mathbf{j}'\,\mathbf{j}' + \mathbf{r}\cdot\mathbf{k}'\,\mathbf{k}'. \tag{47}$$

From this

$$\begin{cases} \mathbf{i}' = \mathbf{i}'\cdot\mathbf{i}\,\mathbf{i} + \mathbf{i}'\cdot\mathbf{j}\,\mathbf{j} + \mathbf{i}'\cdot\mathbf{k}\,\mathbf{k} = a_1\,\mathbf{i} + a_2\,\mathbf{j} + a_3\,\mathbf{k} \\ \mathbf{j}' = \mathbf{j}'\cdot\mathbf{i}\,\mathbf{i} + \mathbf{j}'\cdot\mathbf{j}\,\mathbf{j} + \mathbf{j}'\cdot\mathbf{k}\,\mathbf{k} = b_1\,\mathbf{i} + b_2\,\mathbf{j} + b_3\,\mathbf{k} \\ \mathbf{k}' = \mathbf{k}'\cdot\mathbf{i}\,\mathbf{i} + \mathbf{k}'\cdot\mathbf{j}\,\mathbf{j} + \mathbf{k}'\cdot\mathbf{k}\,\mathbf{k} = c_1\,\mathbf{i} + c_2\,\mathbf{j} + c_3\,\mathbf{k}. \end{cases} \tag{47'}$$

The scalars a_1, a_2, a_3; b_1, b_2, b_3; c_1, c_2, c_3 are respectively the direction cosines of $\mathbf{i}'$; $\mathbf{j}'$; $\mathbf{k}'$ with respect to $\mathbf{i}, \mathbf{j}, \mathbf{k}$. That is

$$\begin{matrix} a_1 = \cos\,(\mathbf{i}', \mathbf{i}) & a_2 = \cos\,(\mathbf{i}', \mathbf{j}) & a_3 = \cos\,(\mathbf{i}', \mathbf{k}) \\ b_1 = \cos\,(\mathbf{j}', \mathbf{i}) & b_2 = \cos\,(\mathbf{j}', \mathbf{j}) & b_3 = \cos\,(\mathbf{j}', \mathbf{k}) \\ c_1 = \cos\,(\mathbf{k}', \mathbf{i}) & c_2 = \cos\,(\mathbf{k}', \mathbf{j}) & c_3 = \cos\,(\mathbf{k}', \mathbf{k}). \end{matrix} \tag{48}$$

In the same manner

$$\begin{cases} \mathbf{i} = \mathbf{i}\cdot\mathbf{i}'\,\mathbf{i}' + \mathbf{i}\cdot\mathbf{j}'\,\mathbf{j}' + \mathbf{i}\cdot\mathbf{k}'\,\mathbf{k}' = a_1\,\mathbf{i}' + b_1\,\mathbf{j}' + c_1\,\mathbf{k}' \\ \mathbf{j} = \mathbf{j}\cdot\mathbf{i}'\,\mathbf{i}' + \mathbf{j}\cdot\mathbf{j}'\,\mathbf{j}' + \mathbf{j}\cdot\mathbf{k}'\,\mathbf{k}' = a_2\,\mathbf{i}' + b_2\,\mathbf{j}' + c_2\,\mathbf{k}' \\ \mathbf{k} = \mathbf{k}\cdot\mathbf{i}'\,\mathbf{i}' + \mathbf{k}\cdot\mathbf{j}'\,\mathbf{j}' + \mathbf{k}\cdot\mathbf{k}'\,\mathbf{k}' = a_3\,\mathbf{i}' + b_3\,\mathbf{j}' + c_3\,\mathbf{k}' \end{cases} \tag{47''}$$

$$\begin{cases} \mathbf{i}'\cdot\mathbf{i}' = 1 = a_1^2 + a_2^2 + a_3^2 \\ \mathbf{j}'\cdot\mathbf{j}' = 1 = b_1^2 + b_2^2 + b_3^2 \\ \mathbf{k}'\cdot\mathbf{k}' = 1 = c_1^2 + c_2^2 + c_3^2 \end{cases} \tag{49}$$

and

$$\begin{cases} \mathbf{i}\cdot\mathbf{i} = 1 = a_1^2 + b_1^2 + c_1^2 \\ \mathbf{j}\cdot\mathbf{j} = 1 = a_2^2 + b_2^2 + c_2^2 \\ \mathbf{k}\cdot\mathbf{k} = 1 = a_3^2 + b_3^2 + c_3^2 \end{cases} \tag{49'}$$

and

$$\begin{cases} \mathbf{i}'\cdot\mathbf{j}' = 0 = a_1\,b_1 + a_2\,b_2 + a_3\,b_3 \\ \mathbf{j}'\cdot\mathbf{k}' = 0 = b_1\,c_1 + b_2\,c_2 + b_3\,c_3 \\ \mathbf{k}'\cdot\mathbf{i}' = 0 = c_1\,a_1 + c_2\,a_2 + c_3\,a_3 \end{cases} \tag{50}$$

and
$$\left\{\begin{array}{l}\mathbf{i}\cdot\mathbf{j} = 0 = a_1\, a_2 + b_1\, b_2 + c_1\, c_2 \\ \mathbf{j}\cdot\mathbf{k} = 0 = a_2\, a_3 + b_2\, b_3 + c_2\, c_3 \\ \mathbf{k}\cdot\mathbf{i} = 0 = a_3\, a_1 + b_3\, b_1 + c_3\, c_1\end{array}\right. \tag{50}'$$

and
$$[\mathbf{i}\,\mathbf{j}\,\mathbf{k}] = [\mathbf{i}'\,\mathbf{j}'\,\mathbf{k}'] = 1 = \begin{vmatrix} a_1 & a_2 & a_3 \\ b_1 & b_2 & b_3 \\ c_1 & c_2 & c_3 \end{vmatrix} \tag{51}$$

$$\mathbf{k}' = \mathbf{i}'\times\mathbf{j}' = (a_2\, b_3 - a_3\, b_2)\,\mathbf{i} + (a_3\, b_1 - a_1\, b_3)\,\mathbf{j} + (a_1\, b_2 - a_2\, b_1)\,\mathbf{k}.$$

But
$$\mathbf{k}' = c_1\,\mathbf{i} + c_2\,\mathbf{j} + c_3\,\mathbf{k}.$$

Hence
$$\left\{\begin{array}{l} c_1 = (a_2\, b_3 - a_3\, b_2), \\ c_2 = (a_3\, b_1 - a_1\, b_3), \\ c_3 = (a_1\, b_2 - a_2\, b_1). \end{array}\right. \tag{52}$$

Or
$$c_1 = \begin{vmatrix} a_2 & a_3 \\ b_2 & b_3 \end{vmatrix}, \quad c_2 = \begin{vmatrix} a_3 & a_1 \\ b_3 & b_1 \end{vmatrix}, \quad c_3 = \begin{vmatrix} a_1 & a_2 \\ b_1 & b_2 \end{vmatrix},$$

and similar relations may be found for the other six quantities a_1, a_2, a_3; b_1, b_2, b_3. All these scalar relations between the coefficients of a transformation which expresses one set of orthogonal axes X', Y', Z' in terms of another set X, Y, Z are important and well known to students of Cartesian methods. The ease with which they are obtained here may be noteworthy.

A number of *vector* relations, which are perhaps not so well known, but nevertheless important, may be found by multiplying the equations

$$\mathbf{i}' = a_1\,\mathbf{i} + a_2\,\mathbf{j} + a_3\,\mathbf{k}$$
$$a_1\,\mathbf{i}' + b_1\,\mathbf{j}' + c_1\,\mathbf{k}' = \mathbf{i}$$

in vector multiplication.

$$b_1\,\mathbf{k}' - c_1\,\mathbf{j}' = a_3\,\mathbf{j} - a_2\,\mathbf{k}. \tag{53}$$

The quantity on either side of this equality is a vector. From its form upon the right it is seen to possess no component in the **i** direction but to lie wholly in the **jk**-plane; and from its form upon the left it is seen to lie in the **j**′**k**′-plane. Hence it must be the line of intersection of those two planes. Its magnitude is $\sqrt{a_2^2 + a_3^2}$ or $\sqrt{b_1^2 + c_1^2}$. This gives the scalar relations

$$a_2^2 + a_3^2 = b_1^2 + c_1^2 = 1 - a_1^2.$$

The magnitude $1 - a_1^2$ is the square of the sine of the angle between the vectors **i** and **i**′. Hence the vector

$$b_1\,\mathbf{k}' - c_1\,\mathbf{j}' = a_3\,\mathbf{j} - a_2\,\mathbf{k} \qquad (53)$$

is the line of intersection of the **j**′**k**′- and **j k**-planes, and its magnitude is the sine of the angle between the planes. Eight other similar vectors may be found, each of which gives one of the nine lines of intersection of the two sets of mutually orthogonal planes. The magnitude of the vector is in each case the sine of the angle between the planes.

54.] Various examples in Plane and Solid Geometry may be solved by means of products.

Example 1: The perpendiculars from the vertices of a triangle to the opposite sides meet in a point. Let ABC be the triangle. Let the perpendiculars from A to BC and from B to CA meet in the point O. To show OC is perpendicular to AB. Choose O as origin and let $\overline{OA} = \mathbf{A}$, $\overline{OB} = \mathbf{B}$, and $\overline{OC} = \mathbf{C}$. Then

$$\overline{BC} = \mathbf{C} - \mathbf{B}, \quad \overline{CA} = \mathbf{A} - \mathbf{C}, \quad \overline{AB} = \mathbf{B} - \mathbf{A}.$$

By hypothesis

$$\mathbf{A}\cdot(\mathbf{C} - \mathbf{B}) = 0$$

and

$$\mathbf{B}\cdot(\mathbf{A} - \mathbf{C}) = 0.$$

Add;

$$\mathbf{C}\cdot(\mathbf{B} - \mathbf{A}) = 0,$$

which proves the theorem.

Example 2: To find the vector equation of a line drawn through the point B parallel to a given vector **A**.

Let O be the origin and $\mathbf{B}$ the vector $\overline{OB}$. Let $\mathbf{R}$ be the radius vector from O to any point of the required line. Then $\mathbf{R} - \mathbf{B}$ is parallel to $\mathbf{A}$. Hence the vector product vanishes.

$$\mathbf{A}\times(\mathbf{R} - \mathbf{B}) = 0.$$

This is the desired equation. It is a vector equation in the unknown vector $\mathbf{R}$. The equation of a plane was seen (page 88) to be a scalar equation such as

$$\mathbf{R}\cdot\mathbf{C} = c$$

in the unknown vector $\mathbf{R}$.

The point of intersection of a line and a plane may be found at once. The equations are

$$\begin{cases} \mathbf{A}\times(\mathbf{R} - \mathbf{B}) = 0 \\ \mathbf{R}\cdot\mathbf{C} = c \end{cases}$$

$$\mathbf{A}\times\mathbf{R} = \mathbf{A}\times\mathbf{B}$$

$$(\mathbf{A}\times\mathbf{R})\times\mathbf{C} = (\mathbf{A}\times\mathbf{B})\times\mathbf{C}$$

$$\mathbf{A}\cdot\mathbf{C}\,\mathbf{R} - \mathbf{C}\cdot\mathbf{R}\,\mathbf{A} = (\mathbf{A}\times\mathbf{B})\times\mathbf{C}$$

$$\mathbf{A}\cdot\mathbf{C}\,\mathbf{R} - c\,\mathbf{A} = (\mathbf{A}\times\mathbf{B})\times\mathbf{C}$$

Hence
$$\mathbf{R} = \frac{(\mathbf{A}\times\mathbf{B})\times\mathbf{C} + c\,\mathbf{A}}{\mathbf{A}\cdot\mathbf{C}}.$$

The solution evidently fails when $\mathbf{A}\cdot\mathbf{C} = 0$. In this case however the line is parallel to the plane and there is no solution; or, if it lies in the plane, there are an infinite number of solutions.

Example 3: The introduction of vectors to represent planes.

Heretofore vectors have been used to denote plane areas of definite extent. The direction of the vector was normal to the plane and the magnitude was equal to the area to be represented. But it is possible to use vectors to denote not a plane area but the entire plane itself, just as a vector represents a point. The result is analogous to the *plane* coördinates of analytic geometry. Let O be an assumed origin. Let MN be a plane in space. The plane MN is to be denoted by a vector whose direction is the direction of the perpendicular dropped upon the plane from the origin O and whose magnitude is the *reciprocal* of the length of that perpendicular. Thus the nearer a plane is to the origin the longer will be the vector which represents it.

If $\mathbf{r}$ be any radius vector drawn from the origin to a point in the plane and if $\mathbf{p}$ be the vector which denotes the plane, then

$$\mathbf{r} \cdot \mathbf{p} = 1$$

is the equation of the plane. For

$$\mathbf{r} \cdot \mathbf{p} = r \cos(\mathbf{r}, \mathbf{p})\, p.$$

Now p, the length of $\mathbf{p}$ is the reciprocal of the perpendicular distance from O to the plane. On the other hand $r \cos(\mathbf{r}, \mathbf{p})$ is that perpendicular distance. Hence $\mathbf{r} \cdot \mathbf{p}$ must be unity.

If $\mathbf{r}$ and $\mathbf{p}$ be expressed in terms of $\mathbf{i}$, $\mathbf{j}$, $\mathbf{k}$

$$\mathbf{r} = x\,\mathbf{i} + y\,\mathbf{j} + z\,\mathbf{k}$$

$$\mathbf{p} = u\,\mathbf{i} + v\,\mathbf{j} + w\,\mathbf{k}$$

Hence $$\mathbf{r} \cdot \mathbf{p} = x\,u + y\,v + z\,w = 1.$$

The quantities u, v, w are the reciprocals of the intercepts of the plane $\mathbf{p}$ upon the axes.

The relation between $\mathbf{r}$ and $\mathbf{p}$ is symmetrical. It is a relation of duality. If in the equation

$$\mathbf{r} \cdot \mathbf{p} = 1$$

$\mathbf{r}$ be regarded as variable, the equation represents a plane $\mathbf{p}$ which is the locus of all points given by $\mathbf{r}$. If however $\mathbf{p}$ be regarded as variable and $\mathbf{r}$ as constant, the equation represents a point $\mathbf{r}$ through which all the planes $\mathbf{p}$ pass. The development of the idea of duality will not be carried out. It is familiar to all students of geometry. The use of vectors to denote planes will scarcely be alluded to again until Chapter VII.

SUMMARY OF CHAPTER II

The scalar product of two vectors is equal to the product of their lengths multiplied by the cosine of the angle between them.

$$\mathbf{A} \cdot \mathbf{B} = A\,B \cos\,(\mathbf{A}, \mathbf{B}) \tag{1}$$

$$\mathbf{A} \cdot \mathbf{B} = \mathbf{B} \cdot \mathbf{A} \tag{2}$$

$$\mathbf{A} \cdot \mathbf{A} = A^2. \tag{3}$$

The necessary and sufficient condition for the perpendicularity of two vectors neither of which vanishes is that their scalar product vanishes. The scalar products of the vectors **i, j, k** are

$$\begin{aligned} \mathbf{i} \cdot \mathbf{i} = \mathbf{j} \cdot \mathbf{j} = \mathbf{k} \cdot \mathbf{k} = 1 \\ \mathbf{i} \cdot \mathbf{j} = \mathbf{j} \cdot \mathbf{k} = \mathbf{k} \cdot \mathbf{i} = 0 \end{aligned} \tag{4}$$

$$\mathbf{A} \cdot \mathbf{B} = A_1\,B_1 + A_2\,B_2 + A_3\,B_3 \tag{7}$$

$$\mathbf{A} \cdot \mathbf{A} = A^2 = A_1^{\,2} + A_2^{\,2} + A_3^{\,2}. \tag{8}$$

If the projection of a vector **B** upon a vector **A** is **B′**,

$$\mathbf{B}' = \frac{\mathbf{A} \cdot \mathbf{B}}{\mathbf{A} \cdot \mathbf{A}}\,\mathbf{A}. \tag{5}$$

The vector product of two vectors is equal in magnitude to the product of their lengths multiplied by the sine of the angle between them. The direction of the vector product is the normal to the plane of the two vectors on that side on which a rotation of less than 180° from the first vector to the second appears positive.

$$\mathbf{A} \times \mathbf{B} = A\,B \sin\,(\mathbf{A}, \mathbf{B})\;\mathbf{c}. \tag{9}$$

The vector product is equal in magnitude and direction to the vector which represents the parallelogram of which **A** and **B** are the two adjacent sides. The necessary and sufficient condition for the parallelism of two vectors neither of which vanishes is that their vector product vanishes. The commutative laws do not hold.

$$\mathbf{A}\times\mathbf{B} = -\mathbf{B}\times\mathbf{A} \tag{10}$$

$$\begin{aligned} \mathbf{i}\times\mathbf{i} &= \mathbf{j}\times\mathbf{j} = \mathbf{k}\times\mathbf{k} = 0 \\ \mathbf{i}\times\mathbf{j} &= -\mathbf{j}\times\mathbf{i} = \mathbf{k} \\ \mathbf{j}\times\mathbf{k} &= -\mathbf{k}\times\mathbf{j} = \mathbf{i} \\ \mathbf{k}\times\mathbf{i} &= -\mathbf{i}\times\mathbf{k} = \mathbf{j} \end{aligned} \tag{12}$$

$$\mathbf{A}\times\mathbf{B} = (A_2\, B_3 - A_3\, B_2)\,\mathbf{i} + (A_3\, B_1 - A_1\, B_3)\mathbf{j} + (A_1\, B_2 - A_2\, B_1)\,\mathbf{k} \tag{13}$$

$$\mathbf{A}\times\mathbf{B} = \begin{vmatrix} \mathbf{i} & \mathbf{j} & \mathbf{k} \\ A_1 & A_2 & A_3 \\ B_1 & B_2 & B_3 \end{vmatrix} \tag{13$'$}$$

The scalar triple product of three vectors $[\mathbf{A}\ \mathbf{B}\ \mathbf{C}]$ is equal to the volume of the parallelopiped of which **A**, **B**, **C** are three edges which meet in a point.

$$\begin{aligned} [\mathbf{A}\,\mathbf{B}\,\mathbf{C}] &= \mathbf{A}\cdot\mathbf{B}\times\mathbf{C} = \mathbf{B}\cdot\mathbf{C}\times\mathbf{A} = \mathbf{C}\cdot\mathbf{A}\times\mathbf{B} \\ &= \mathbf{A}\times\mathbf{B}\cdot\mathbf{C} = \mathbf{B}\times\mathbf{C}\cdot\mathbf{A} = \mathbf{C}\times\mathbf{A}\cdot\mathbf{B} \end{aligned} \tag{15$'$}$$

$$[\mathbf{A}\,\mathbf{B}\,\mathbf{C}] = -[\mathbf{A}\,\mathbf{C}\,\mathbf{B}]. \tag{16$'$}$$

The dot and the cross in a scalar triple product may be interchanged and the order of the letters may be permuted *cyclicly* without altering the value of the product; but a change of cyclic order changes the sign.

$$[\mathbf{A}\,\mathbf{B}\,\mathbf{C}] = \begin{vmatrix} A_1 & A_2 & A_3 \\ B_1 & B_2 & B_3 \\ C_1 & C_2 & C_3 \end{vmatrix} \tag{18$'$}$$

$$[\mathbf{A}\,\mathbf{B}\,\mathbf{C}] = \begin{vmatrix} a_1 & a_2 & a_3 \\ b_1 & b_2 & b_3 \\ c_1 & c_2 & c_3 \end{vmatrix} [\mathbf{a}\,\mathbf{b}\,\mathbf{c}] \tag{19$'$}$$

If the component of **B** perpendicular to **A** be **B**″,

$$\mathbf{B}'' = -\frac{\mathbf{A}\times(\mathbf{A}\times\mathbf{B})}{\mathbf{A}\cdot\mathbf{A}} \qquad (20)$$

$$\mathbf{A}\times(\mathbf{B}\times\mathbf{C}) = \mathbf{A}\cdot\mathbf{C}\,\mathbf{B} - \mathbf{A}\cdot\mathbf{B}\,\mathbf{C} \qquad (24)$$

$$(\mathbf{A}\times\mathbf{B})\times\mathbf{C} = \mathbf{A}\cdot\mathbf{C}\,\mathbf{B} - \mathbf{C}\cdot\mathbf{B}\,\mathbf{A} \qquad (24)'$$

$$(\mathbf{A}\times\mathbf{B})\cdot(\mathbf{C}\times\mathbf{D}) = \mathbf{A}\cdot\mathbf{C}\,\mathbf{B}\cdot\mathbf{D} - \mathbf{A}\cdot\mathbf{D}\,\mathbf{B}\cdot\mathbf{C} \qquad (25)$$

$$\begin{aligned}(\mathbf{A}\times\mathbf{B})\times(\mathbf{C}\times\mathbf{D}) &= [\mathbf{A}\,\mathbf{C}\,\mathbf{D}]\,\mathbf{B} - [\mathbf{B}\,\mathbf{C}\,\mathbf{D}]\,\mathbf{A} \\ &= [\mathbf{A}\,\mathbf{B}\,\mathbf{D}]\,\mathbf{C} - [\mathbf{A}\,\mathbf{B}\,\mathbf{C}]\,\mathbf{D}.\end{aligned} \qquad (26)$$

The equation which subsists between four vectors **A, B, C, D** is

$$[\mathbf{B}\,\mathbf{C}\,\mathbf{D}]\,\mathbf{A} - [\mathbf{C}\,\mathbf{D}\,\mathbf{A}]\,\mathbf{B} + [\mathbf{D}\,\mathbf{A}\,\mathbf{B}]\,\mathbf{C} - [\mathbf{A}\,\mathbf{B}\,\mathbf{C}]\,\mathbf{D} = 0. \qquad (27)$$

Application of formulæ of vector analysis to obtain the formulæ of Plane and Spherical Trigonometry.

The system of vectors **a**′, **b**′, **c**′ is said to be reciprocal to the system of three non-coplanar vectors **a, b, c**

$$\text{when} \qquad \mathbf{a}' = \frac{\mathbf{b}\times\mathbf{c}}{[\mathbf{a}\,\mathbf{b}\,\mathbf{c}]}, \quad \mathbf{b}' = \frac{\mathbf{c}\times\mathbf{a}}{[\mathbf{a}\,\mathbf{b}\,\mathbf{c}]}, \quad \mathbf{c}' = \frac{\mathbf{a}\times\mathbf{b}}{[\mathbf{a}\,\mathbf{b}\,\mathbf{c}]}. \qquad (29)$$

A vector **r** may be expressed in terms of a set of vectors and its reciprocal in two similar ways

$$\mathbf{r} = \mathbf{r}\cdot\mathbf{a}'\,\mathbf{a} + \mathbf{r}\cdot\mathbf{b}'\,\mathbf{b} + \mathbf{r}\cdot\mathbf{c}'\,\mathbf{c} \qquad (30)$$

or

$$\mathbf{r} = \mathbf{r}\cdot\mathbf{a}\,\mathbf{a}' + \mathbf{r}\cdot\mathbf{b}\,\mathbf{b}' + \mathbf{r}\cdot\mathbf{c}\,\mathbf{c}'. \qquad (31)$$

The necessary and sufficient conditions that the two systems of non-coplanar vectors **a, b, c** and **a**′, **b**′, **c**′ be reciprocals is that

$$\begin{aligned}\mathbf{a}'\cdot\mathbf{a} = \mathbf{b}'\cdot\mathbf{b} = \mathbf{c}'\cdot\mathbf{c} = 1 \\ \mathbf{a}'\cdot\mathbf{b} = \mathbf{a}'\cdot\mathbf{c} = \mathbf{b}'\cdot\mathbf{c} = \mathbf{b}'\cdot\mathbf{a} = \mathbf{c}'\cdot\mathbf{a} = \mathbf{c}'\cdot\mathbf{b} = 0.\end{aligned} \qquad (32)$$

If **a**′, **b**′, **c**′ form a system reciprocal to **a, b, c**; then **a, b, c** will form a system reciprocal to **a**′, **b**′, **c**′.

$$[\mathbf{a}'\,\mathbf{b}'\,\mathbf{c}'] = \frac{1}{[\mathbf{a}\,\mathbf{b}\,\mathbf{c}]} \qquad (33)'$$

$$[\mathbf{P}\ \mathbf{Q}\ \mathbf{R}]\,[\mathbf{A}\ \mathbf{B}\ \mathbf{C}] = \begin{vmatrix} \mathbf{P}\cdot\mathbf{A} & \mathbf{P}\cdot\mathbf{B} & \mathbf{P}\cdot\mathbf{C} \\ \mathbf{Q}\cdot\mathbf{A} & \mathbf{Q}\cdot\mathbf{B} & \mathbf{Q}\cdot\mathbf{C} \\ \mathbf{R}\cdot\mathbf{A} & \mathbf{R}\cdot\mathbf{B} & \mathbf{R}\cdot\mathbf{C} \end{vmatrix} \tag{34}$$

The system $\mathbf{i}, \mathbf{j}, \mathbf{k}$ is its own reciprocal and if conversely a system be its own reciprocal it must be a right or left handed system of three mutually perpendicular unit vectors. Application of the theory of reciprocal systems to the solution of scalar and vector equations of the first degree in an unknown vector. The vector equation of a plane is

$$\mathbf{r}\cdot\mathbf{A} = a. \tag{36}$$

Applications of the methods developed in Chapter II., to the treatment of a system of forces acting on a rigid body and in particular to the reduction of any system of forces to a single force and a couple of which the plane is perpendicular to that force. Application of the methods to the treatment of instantaneous motion of a rigid body obtaining

$$\mathbf{v} = \mathbf{v}_0 + \mathbf{a} \times \mathbf{r} \tag{46}$$

where $\mathbf{v}$ is the velocity of any point, $\mathbf{v}_0$ a translational velocity in the direction $\mathbf{a}$, and $\mathbf{a}$ the vector angular velocity of rotation. Further application of the methods to obtain the conditions for equilibrium by making use of the principle of virtual velocities. Applications of the method to obtain the relations which exist between the nine direction cosines of the angles between two systems of mutually orthogonal axes. Application to special problems in geometry including the form under which plane coördinates make their appearance in vector analysis and the method by which planes (as distinguished from finite plane areas) may be represented by vectors.

EXERCISES ON CHAPTER II

Prove the following reduction formulæ

1. $\mathbf{A}\times\{\mathbf{B}\times(\mathbf{C}\times\mathbf{D})\} = [\mathbf{A\,C\,D}]\,\mathbf{B} - \mathbf{A}\cdot\mathbf{B}\;\mathbf{C}\times\mathbf{D}$
$= \mathbf{B}\cdot\mathbf{D}\;\mathbf{A}\times\mathbf{C} - \mathbf{B}\cdot\mathbf{C}\;\mathbf{A}\times\mathbf{D}.$

2. $[\mathbf{A}\times\mathbf{B}\;\;\mathbf{C}\times\mathbf{D}\;\;\mathbf{E}\times\mathbf{F}] = [\mathbf{A\,B\,D}]\,[\mathbf{C\,E\,F}] - [\mathbf{A\,B\,C}]\,[\mathbf{D\,E\,F}]$
$= [\mathbf{A\,B\,E}]\,[\mathbf{F\,C\,D}] - [\mathbf{A\,B\,F}]\,[\mathbf{E\,C\,D}]$
$= [\mathbf{C\,D\,A}]\,[\mathbf{B\,E\,F}] - [\mathbf{C\,D\,B}]\,[\mathbf{A\,E\,F}].$

3. $[\mathbf{A}\times\mathbf{B}\;\;\mathbf{B}\times\mathbf{C}\;\;\mathbf{C}\times\mathbf{A}] = [\mathbf{A\,B\,C}]^2.$

4. $$[\mathbf{P\,Q\,R}]\,(\mathbf{A}\times\mathbf{B}) = \begin{vmatrix} \mathbf{P}\cdot\mathbf{A} & \mathbf{P}\cdot\mathbf{B} & \mathbf{P} \\ \mathbf{Q}\cdot\mathbf{A} & \mathbf{Q}\cdot\mathbf{B} & \mathbf{Q} \\ \mathbf{R}\cdot\mathbf{A} & \mathbf{R}\cdot\mathbf{B} & \mathbf{R} \end{vmatrix}.$$

5. $\mathbf{A}\times(\mathbf{B}\times\mathbf{C}) + \mathbf{B}\times(\mathbf{C}\times\mathbf{A}) + \mathbf{C}\times(\mathbf{A}\times\mathbf{B}) = 0.$

6. $[\mathbf{A}\times\mathbf{P}\;\;\mathbf{B}\times\mathbf{Q}\;\;\mathbf{C}\times\mathbf{R}] + [\mathbf{A}\times\mathbf{Q}\;\;\mathbf{B}\times\mathbf{R}\;\;\mathbf{C}\times\mathbf{P}]$
$+ [\mathbf{A}\times\mathbf{R}\;\;\mathbf{B}\times\mathbf{P}\;\;\mathbf{C}\times\mathbf{Q}] = 0.$

7. Obtain formula (34) in the text by expanding

$$[(\mathbf{A}\times\mathbf{B})\times\mathbf{P}]\cdot[\mathbf{C}\times(\mathbf{Q}\times\mathbf{R})]$$

in two different ways and equating the results.

8. Demonstrate directly by the above formulæ that if $\mathbf{a}'$, $\mathbf{b}'$, $\mathbf{c}'$ form a reciprocal system to $\mathbf{a}$, $\mathbf{b}$, $\mathbf{c}$; then $\mathbf{a}$, $\mathbf{b}$, $\mathbf{c}$ form a system reciprocal to $\mathbf{a}'$, $\mathbf{b}'$, $\mathbf{c}'$.

9. Show the connection between reciprocal systems of vectors and polar triangles upon a sphere. Obtain some of the geometrical formulæ connected with polar triangles by interpreting vector formulæ such as (3) in the above list.

10. The perpendicular bisectors of the sides of a triangle meet in a point.

11. Find an expression for the common perpendicular to two lines not lying in the same plane.

12. Show by vector methods that the formulæ for the volume of a tetrahedron whose four vertices are

$$(x_1, y_1, z_1) \quad (x_2, y_2, z_2) \quad (x_3, y_3, z_3) \quad (x_4, y_4, z_4)$$

is

$$\frac{1}{6}\begin{vmatrix} x_1 & y_1 & z_1 & 1 \\ x_2 & y_2 & z_2 & 1 \\ x_3 & y_3 & z_3 & 1 \\ x_4 & y_4 & z_4 & 1 \end{vmatrix}$$

13. Making use of formula (34) of the text show that

$$[\mathbf{a}\,\mathbf{b}\,\mathbf{c}] = a\,b\,c\sqrt{\begin{matrix} 1 & n & m \\ n & 1 & l \\ m & l & 1 \end{matrix}}$$

where a, b, c are the lengths of **a**, **b**, **c** respectively and where

$$l = \cos(\mathbf{b}, \mathbf{c}), \quad m = \cos(\mathbf{c}, \mathbf{a}), \quad n = \cos(\mathbf{a}, \mathbf{b}).$$

14. Determine the perpendicular (as a vector quantity) which is dropped from the origin upon a plane determined by the termini of the vectors **a**, **b**, **c**. Use the method of solution given in Art. 46.

15. Show that the volume of a tetrahedron is equal to one sixth of the product of two opposite edges by the perpendicular distance between them and the sine of the included angle.

16. If a line is drawn in each face plane of any triedral angle through the vertex and perpendicular to the third edge, the three lines thus obtained lie in a plane.

CHAPTER III

THE DIFFERENTIAL CALCULUS OF VECTORS

Differentiation of Functions of One Scalar Variable

55.] IF a vector varies and changes from $\mathbf{r}$ to $\mathbf{r}'$ the increment of $\mathbf{r}$ will be the difference between $\mathbf{r}'$ and $\mathbf{r}$ and will be denoted as usual by $\Delta\,\mathbf{r}$.

$$\Delta\,\mathbf{r} = \mathbf{r}' - \mathbf{r}, \tag{1}$$

where $\Delta\,\mathbf{r}$ must be a vector quantity. If the variable $\mathbf{r}$ be unrestricted the increment $\Delta\,\mathbf{r}$ is of course also unrestricted: it may have any magnitude and any direction. If, however, the vector $\mathbf{r}$ be regarded as a function (a vector function) of a single scalar variable t the value of $\Delta\,\mathbf{r}$ will be completely determined when the two values t and t' of t, which give the two values $\mathbf{r}$ and $\mathbf{r}'$, are known.

To obtain a clearer conception of the quantities involved it will be advantageous to think of the vector $\mathbf{r}$ as drawn from a fixed origin O (Fig. 26). When the independent variable t changes its value the vector $\mathbf{r}$ will change, and as t possesses one degree of freedom $\mathbf{r}$ will vary in such a way that its terminus describes a curve in space. $\mathbf{r}$ will be the radius vector of one point P of the curve; $\mathbf{r}'$, of a neighboring point P'. $\Delta\,\mathbf{r}$ will be the chord $\overline{PP'}$ of the curve. The ratio

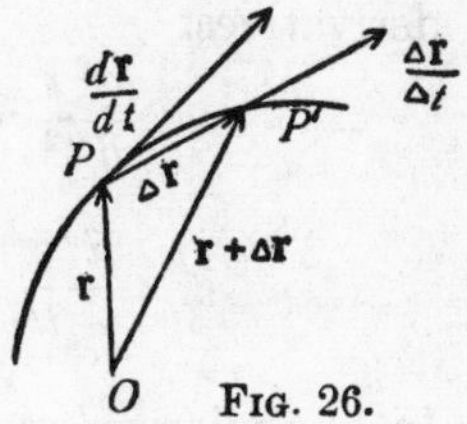

FIG. 26.

$$\frac{\Delta\,\mathbf{r}}{\Delta\,t}$$

will be a vector collinear with the chord $\overline{PP'}$ but magnified in the ratio $1 : \Delta t$. When Δt approaches zero P' will approach P, the chord $\overline{PP'}$ will approach the tangent at P, and the vector

$$\frac{\Delta \mathbf{r}}{\Delta t} \text{ will approach } \frac{d\mathbf{r}}{dt}$$

which is a vector tangent to the curve at P directed in that sense in which the variable t increases along the curve.

If $\mathbf{r}$ be expressed in terms of $\mathbf{i}$, $\mathbf{j}$, $\mathbf{k}$ as

$$\mathbf{r} = r_1 \mathbf{i} + r_2 \mathbf{j} + r_3 \mathbf{k}$$

the components r_1, r_2, r_3 will be functions of the scalar t.

$$\mathbf{r}' = (r_1 + \Delta r_1) \mathbf{i} + (r_2 + \Delta r_2) \mathbf{j} + (r_3 + \Delta r_3) \mathbf{k}$$

$$\Delta \mathbf{r} = \mathbf{r}' - \mathbf{r} = \Delta r_1 \mathbf{i} + \Delta r_2 \mathbf{j} + \Delta r_3 \mathbf{k}$$

$$\frac{\Delta \mathbf{r}}{\Delta t} = \frac{\Delta r_1}{\Delta t} \mathbf{i} + \frac{\Delta r_2}{\Delta t} \mathbf{j} + \frac{\Delta r_3}{\Delta t} \mathbf{k}$$

and
$$\frac{d\mathbf{r}}{dt} = \frac{d r_1}{dt} \mathbf{i} + \frac{d r_2}{dt} \mathbf{j} + \frac{d r_3}{dt} \mathbf{k}. \tag{2}$$

Hence the components of the first derivative of $\mathbf{r}$ with respect to t are the first derivatives with respect to t of the components of $\mathbf{r}$. The same is true for the second and higher derivatives.

$$\frac{d^2 \mathbf{r}}{dt^2} = \frac{d^2 r_1}{dt^2} \mathbf{i} + \frac{d^2 r_2}{dt^2} \mathbf{j} + \frac{d^2 r_3}{dt^2} \mathbf{k},$$

$$\frac{d^n \mathbf{r}}{dt^n} = \frac{d^n r_1}{dt^n} \mathbf{i} + \frac{d^n r_2}{dt^n} \mathbf{j} + \frac{d^n r_3}{dt^n} \mathbf{k}. \tag{2)'$$

In a similar manner if $\mathbf{r}$ be expressed in terms of any three non-coplanar vectors $\mathbf{a}$, $\mathbf{b}$, $\mathbf{c}$ as

$$\mathbf{r} = a \mathbf{a} + b \mathbf{b} + c \mathbf{c}$$

$$\frac{d^n \mathbf{r}}{dt^n} = \frac{d^n a}{dt^n} \mathbf{a} + \frac{d^n b}{dt^n} \mathbf{b} + \frac{d^n c}{dt^n} \mathbf{c}.$$

Example: Let $\mathbf{r} = \mathbf{a} \cos t + \mathbf{b} \sin t.$

The vector $\mathbf{r}$ will then describe an ellipse of which $\mathbf{a}$ and $\mathbf{b}$ are two conjugate diameters. This may be seen by assuming a set of oblique Cartesian axes X, Y coincident with $\mathbf{a}$ and $\mathbf{b}$. Then

$$X = a \cos t, \qquad Y = b \sin t,$$

$$\frac{X^2}{a^2} + \frac{Y^2}{b^2} = 1,$$

which is the equation of an ellipse referred to a pair of conjugate diameters of lengths a and b respectively.

$$\frac{d\,\mathbf{r}}{d\,t} = - \mathbf{a} \sin t + \mathbf{b} \cos t.$$

Hence $$\frac{d\,\mathbf{r}}{d\,t} = \mathbf{a} \cos (t + 90°) + \mathbf{b} \sin (t + 90°).$$

The tangent to the curve is parallel to the radius vector for $(t + 90°)$.

$$\frac{d^2\,\mathbf{r}}{d\,t^2} = - (\mathbf{a} \cos t + \mathbf{b} \sin t).$$

The second derivative is the negative of $\mathbf{r}$. Hence

$$\frac{d^2\,\mathbf{r}}{d\,t^2} = - \mathbf{r}$$

is evidently a differential equation satisfied by the ellipse.

Example: Let $\mathbf{r} = \mathbf{a} \cosh t + \mathbf{b} \sinh t.$

The vector $\mathbf{r}$ will then describe an hyperbola of which $\mathbf{a}$ and $\mathbf{b}$ are two conjugate diameters.

$$\frac{d\,\mathbf{r}}{d\,t} = \mathbf{a} \sinh t + \mathbf{b} \cosh t,$$

and $$\frac{d^2\,\mathbf{r}}{d\,t^2} = \mathbf{a} \cosh t + \mathbf{b} \sinh t.$$

Hence $$\frac{d^2\,\mathbf{r}}{d\,t^2} = \mathbf{r}$$

is a differential equation satisfied by the hyperbola.

56.] A combination of vectors all of which depend on the same scalar variable t may be differentiated very much as in ordinary calculus.

$$\frac{d}{dt}(\mathbf{a}\cdot\mathbf{b}) = \mathbf{a}\cdot\left(\frac{d\,\mathbf{b}}{d\,t}\right) + \left(\frac{d\,\mathbf{a}}{d\,t}\right)\cdot\mathbf{b}.$$

For

$$(\mathbf{a} + \Delta\,\mathbf{b})\cdot(\mathbf{b} + \Delta\,\mathbf{b}) = \mathbf{a}\cdot\mathbf{b} + \mathbf{a}\cdot\Delta\,\mathbf{b} + \Delta\,\mathbf{a}\cdot\mathbf{b} + \Delta\,\mathbf{a}\cdot\Delta\,\mathbf{b}$$

$$\Delta\,(\mathbf{a}\cdot\mathbf{b}) = (\mathbf{a} + \Delta\,\mathbf{a})\cdot(\mathbf{b} + \Delta\,\mathbf{b}) - \mathbf{a}\cdot\mathbf{b}$$

$$= \mathbf{a}\cdot\Delta\,\mathbf{b} + \Delta\,\mathbf{a}\cdot\mathbf{b} + \Delta\,\mathbf{a}\cdot\Delta\,\mathbf{b}$$

$$\frac{\Delta\,(\mathbf{a}\cdot\mathbf{b})}{\Delta\,t} = \mathbf{a}\cdot\frac{\Delta\,\mathbf{b}}{\Delta\,t} + \frac{\Delta\,\mathbf{a}}{\Delta\,t}\cdot\mathbf{b} + \frac{\Delta\,\mathbf{a}\cdot\Delta\,\mathbf{b}}{\Delta\,t}.$$

Hence in the limit when $\Delta\,t = 0$,

$$\frac{d}{d\,t}(\mathbf{a}\cdot\mathbf{b}) = \mathbf{a}\cdot\frac{d\,\mathbf{b}}{d\,t} + \frac{d\,\mathbf{a}}{d\,t}\cdot\mathbf{b} \qquad (3)$$

$$\frac{d}{d\,t}(\mathbf{a}\times\mathbf{b}) = \mathbf{a}\times\left(\frac{d\,\mathbf{b}}{d\,t}\right) + \left(\frac{d\,\mathbf{a}}{d\,t}\right)\times\mathbf{b} \qquad (4)$$

$$\frac{d}{d\,t}(\mathbf{a}\cdot\mathbf{b}\times\mathbf{c}) = \mathbf{a}\cdot\mathbf{b}\times\left(\frac{d\,\mathbf{c}}{d\,t}\right) + \mathbf{a}\cdot\left(\frac{d\,\mathbf{b}}{d\,t}\right)\times\mathbf{c} + \left(\frac{d\,\mathbf{a}}{d\,t}\right)\cdot\mathbf{b}\times\mathbf{c}. \qquad (5)$$

$$\frac{d}{d\,t}(\mathbf{a}\times[\mathbf{b}\times\mathbf{c}]) = \mathbf{a}\times\left[\mathbf{b}\times\left(\frac{d\,\mathbf{c}}{d\,t}\right)\right] + \mathbf{a}\times\left[\left(\frac{d\,\mathbf{b}}{d\,t}\right)\times\mathbf{c}\right] + \left(\frac{d\,\mathbf{a}}{d\,t}\right)\times[\mathbf{b}\times\mathbf{c}]. \qquad (6)$$

The last three of these formulæ may be demonstrated exactly as the first was.

The formal process of differentiation in vector analysis differs in no way from that in scalar analysis except in this one point in which vector analysis always differs from scalar analysis, namely: The order of the factors in a vector product cannot be changed without changing the sign of the product. Hence of the two formulæ

$$\frac{d}{dt}(\mathbf{a} \times \mathbf{b}) = \left(\frac{d\,\mathbf{a}}{d\,t}\right) \times \mathbf{b} + \left(\frac{d\,\mathbf{b}}{d\,t}\right) \times \mathbf{a}$$

and $$\frac{d}{dt}(\mathbf{a} \times \mathbf{b}) = \left(\frac{d\,\mathbf{a}}{d\,t}\right) \times \mathbf{b} + \mathbf{a} \times \left(\frac{d\,\mathbf{b}}{d\,t}\right)$$

the first is evidently incorrect, but the second correct. In other words, scalar differentiation must take place without altering the order of the factors of a vector product. The factors must be differentiated *in situ.* This of course was to be expected.

In case the vectors depend upon more than one variable the results are practically the same. In place of total derivatives with respect to the scalar variables, partial derivatives occur. Suppose **a** and **b** are two vectors which depend on three scalar variables x, y, z. The scalar product $\mathbf{a}\cdot\mathbf{b}$ will depend upon these three variables, and it will have three partial derivatives of the first order.

$$\begin{aligned}
\frac{\partial}{\partial x}(\mathbf{a} \cdot \mathbf{b}) &= \left(\frac{\partial \mathbf{a}}{\partial x}\right) \cdot \mathbf{b} + \mathbf{a} \cdot \left(\frac{\partial \mathbf{b}}{\partial x}\right) \\
\frac{\partial}{\partial y}(\mathbf{a} \cdot \mathbf{b}) &= \left(\frac{\partial \mathbf{a}}{\partial y}\right) \cdot \mathbf{b} + \mathbf{a} \cdot \left(\frac{\partial \mathbf{b}}{\partial y}\right) \qquad (7) \\
\frac{\partial}{\partial z}(\mathbf{a} \cdot \mathbf{b}) &= \left(\frac{\partial \mathbf{a}}{\partial z}\right) \cdot \mathbf{b} + \mathbf{a} \cdot \left(\frac{\partial \mathbf{b}}{\partial z}\right) \cdot
\end{aligned}$$

The second partial derivatives are formed in the same way.

$$\begin{aligned}
\frac{\partial^2}{\partial x\,\partial y}(\mathbf{a} \cdot \mathbf{b}) = \left(\frac{\partial^2 \mathbf{a}}{\partial x\,\partial y}\right) \cdot \mathbf{b} + \left(\frac{\partial \mathbf{a}}{\partial x}\right) \cdot \left(\frac{\partial \mathbf{b}}{\partial y}\right) \\
+ \left(\frac{\partial \mathbf{a}}{\partial y}\right) \cdot \left(\frac{\partial \mathbf{b}}{\partial x}\right) + \mathbf{a} \cdot \left(\frac{\partial^2 \mathbf{b}}{\partial x\,\partial y}\right) \cdot
\end{aligned}$$

Often it is more convenient to use not the derivatives but the differentials. This is particularly true when dealing with *first* differentials. The formulæ (3), (4) become

$$d\,(\mathbf{a}\cdot\mathbf{b}) = d\,\mathbf{a}\cdot\mathbf{b} + \mathbf{a}\cdot d\,\mathbf{b}, \tag{3)'$$

$$d\,(\mathbf{a}\times\mathbf{b}) = d\,\mathbf{a}\times\mathbf{b} + \mathbf{a}\times d\,\mathbf{b}, \tag{4)'$$

and so forth. As an illustration consider the following example. If $\mathbf{r}$ be a unit vector

$$\mathbf{r}\cdot\mathbf{r} = 1.$$

The locus of the terminus of $\mathbf{r}$ is a spherical surface of unit radius described about the origin. $\mathbf{r}$ depends upon two variables. Differentiate the equation.

$$(d\,\mathbf{r})\cdot\mathbf{r} + \mathbf{r}\cdot(d\,\mathbf{r}) = 2\,\mathbf{r}\cdot(d\,\mathbf{r}) = 0.$$

Hence
$$\mathbf{r}\cdot d\,\mathbf{r} = 0.$$

Hence the increment $d\mathbf{r}$ of a *unit* vector is perpendicular to the vector. This can be seen geometrically. If $\mathbf{r}$ traces a sphere the variation $d\,\mathbf{r}$ must be at each point in the tangent plane and hence perpendicular to $\mathbf{r}$.

***57.]** Vector methods may be employed advantageously in the discussion of curvature and torsion of curves. Let $\mathbf{r}$ denote the radius vector of a curve

$$\mathbf{r} = \mathbf{f}\,(t),$$

where $\mathbf{f}$ is some vector function of the scalar t. In most applications in physics and mechanics t represents the time. Let s be the length of arc measured from some definite point of the curve as origin. The increment $\Delta\,\mathbf{r}$ is the chord of the curve. Hence $\Delta\,\mathbf{r}\,/\,\Delta\,s$ is approximately equal in magnitude to unity and approaches unity as its limit when $\Delta\,s$ becomes infinitesimal. Hence $d\,\mathbf{r}\,/\,d\,s$ will be a *unit* vector tangent to the curve and will be directed toward that portion of the curve along which s is increasing (Fig. 27). Let **t** be the unit tangent

$$\frac{d\mathbf{r}}{ds} = \mathbf{t}. \qquad (8)$$

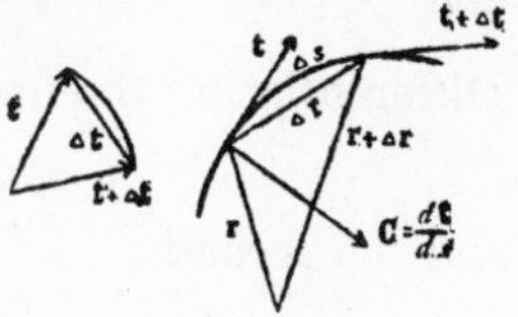

FIG. 27.

The curvature of the curve is the limit of the ratio of the angle through which the tangent turns to the length of the arc. The tangent changes by the increment $\Delta\mathbf{t}$. As **t** is of unit length, the length of $\Delta\mathbf{t}$ is approximately the angle through which the tangent has turned measured in circular measure. Hence the directed curvature **C** is

$$\mathbf{C} = \operatorname*{Lim}_{\Delta s \doteq 0} \frac{\Delta\mathbf{t}}{\Delta s} = \frac{d\mathbf{t}}{ds} = \frac{d^2\mathbf{r}}{ds^2}. \qquad (9)$$

The vector **C** is collinear with $\Delta\mathbf{t}$ and hence perpendicular to **t**; for inasmuch as **t** is a unit vector $\Delta\mathbf{t}$ is perpendicular to **t**.

The tortuosity of a curve is the limit of the ratio of the angle through which the osculating plane turns to the length of the arc. The osculating plane is the plane of the tangent vector **t** and the curvature vector **C**. The normal to this plane is

$$\mathbf{N} = \mathbf{t} \times \mathbf{C}.$$

If **c** be a unit vector collinear with **C**

$$\mathbf{n} = \mathbf{t} \times \mathbf{c}$$

will be a unit normal (Fig. 28) to the osculating plane and the three vectors **t**, **c**, **n** form an **i**, **j**, **k** system, that is, a right-handed rectangular system. Then the angle through which the osculating plane turns will be given approximately by $\Delta\mathbf{n}$ and hence the tortuosity is by definition $d\mathbf{n}/ds$.

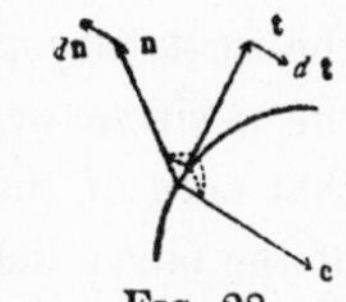

FIG. 28.

From the fact that **t**, **c**, **n** form an **i**, **j**, **k** system of unit vectors

$$\mathbf{t}\cdot\mathbf{t}=\mathbf{c}\cdot\mathbf{c}=\mathbf{n}\cdot\mathbf{n}=1$$

and $$\mathbf{t}\cdot\mathbf{c}=\mathbf{c}\cdot\mathbf{n}=\mathbf{n}\cdot\mathbf{t}=0.$$

Differentiating the first set

$$\mathbf{t}\cdot d\,\mathbf{t}=\mathbf{c}\cdot d\,\mathbf{c}=\mathbf{n}\cdot d\,\mathbf{n}=0,$$

and the second

$$\mathbf{t}\cdot d\mathbf{c}+d\mathbf{t}\cdot\mathbf{c}=\mathbf{c}\cdot d\,\mathbf{n}+d\,\mathbf{c}\cdot\mathbf{n}=\mathbf{n}\cdot d\,\mathbf{t}+d\,\mathbf{n}\cdot\mathbf{t}=0.$$

But $d\,\mathbf{t}$ is parallel to $\mathbf{c}$ and consequently perpendicular to $\mathbf{n}$.

$$\mathbf{n}\cdot d\mathbf{t}=0.$$

Hence $$d\,\mathbf{n}\cdot\mathbf{t}=0.$$

The increment of $\mathbf{n}$ is perpendicular to $\mathbf{t}$. But the increment of $\mathbf{n}$ is also perpendicular to $\mathbf{n}$. It is therefore parallel to $\mathbf{c}$. As the tortuosity is $\mathbf{T}=d\,\mathbf{n}/d\,s$, it is parallel to $d\,\mathbf{n}$ and hence to $\mathbf{c}$.

The tortuosity $\mathbf{T}$ is

$$\mathbf{T}=\frac{d}{d\,s}\,(\mathbf{t}\times\mathbf{c})=\frac{d}{d\,s}\left(\frac{d\,\mathbf{r}}{d\,s}\times\frac{d^2\,\mathbf{r}}{d\,s^2}\,\frac{1}{\sqrt{\mathbf{C}\cdot\mathbf{C}}}\right) \qquad (11)$$

$$\mathbf{T}=\frac{d^2\,\mathbf{r}}{d\,s^2}\times\frac{d^2\,\mathbf{r}}{d\,s^2}\,\frac{1}{\sqrt{\mathbf{C}\cdot\mathbf{C}}}+\frac{d\,\mathbf{r}}{d\,s}\times\frac{d^3\,\mathbf{r}}{d\,s^3}\,\frac{1}{\sqrt{\mathbf{C}\cdot\mathbf{C}}}$$

$$+\frac{d\,\mathbf{r}}{d\,s}\times\frac{d^2\,\mathbf{r}}{d\,s^2}\,\frac{d}{d\,s}\,\frac{1}{\sqrt{\mathbf{C}\cdot\mathbf{C}}}\,.$$

The first term of this expression vanishes. $\mathbf{T}$ moreover has been seen to be parallel to $\mathbf{C}=d^2\,\mathbf{r}/d\,s^2$. Consequently the magnitude of $\mathbf{T}$ is the scalar product of $\mathbf{T}$ by the unit vector $\mathbf{c}$ in the direction of $\mathbf{C}$. It is desirable however to have the tortuosity positive when the normal $\mathbf{n}$ appears to turn in the positive or counterclockwise direction if viewed from that side of the $\mathbf{n}\,\mathbf{c}$-plane upon which $\mathbf{t}$ or the positive part of the curve lies. With this convention $d\,\mathbf{n}$ appears to move in the direction $-\mathbf{c}$ when the tortuosity is positive, that is, $\mathbf{n}$ turns away from $\mathbf{c}$. The scalar value of the tortuosity will therefore be given by $-\mathbf{c}\cdot\mathbf{T}$.

$$-\mathbf{c}\cdot\mathbf{T} = -\mathbf{c}\cdot\frac{d\,\mathbf{r}}{d\,s}\times\frac{d^3\,\mathbf{r}}{d\,s^3}\frac{1}{\sqrt{\mathbf{C}\cdot\mathbf{C}}} - \mathbf{c}\cdot\frac{d\,\mathbf{r}}{d\,s}\times\frac{d^2\,\mathbf{r}}{d\,s^2}\frac{d}{d\,s}\frac{1}{\sqrt{\mathbf{C}\cdot\mathbf{C}}}.$$

But $\mathbf{c}$ is parallel to the vector $d^2\,\mathbf{r}/d\,s^2$. Hence

$$\mathbf{c}\cdot\frac{d\,\mathbf{r}}{d\,s}\times\frac{d^2\,\mathbf{r}}{d\,s^2} = 0.$$

And $\mathbf{c}$ is a unit vector in the direction $\mathbf{C}$. Hence

$$\mathbf{c} = \frac{\mathbf{C}}{\sqrt{\mathbf{C}\cdot\mathbf{C}}} = \frac{d^2\,\mathbf{r}}{d\,s^2}\frac{1}{\sqrt{\mathbf{C}\cdot\mathbf{C}}}.$$

Hence $$T = -\mathbf{c}\cdot\mathbf{T} = -\frac{d^2\,\mathbf{r}}{d\,s^2}\cdot\frac{d\,\mathbf{r}}{d\,s}\times\frac{d^3\,\mathbf{r}}{d\,s^3}\frac{1}{\mathbf{C}\cdot\mathbf{C}}. \tag{12}$$

Or $$T = \frac{\left[\dfrac{d\,\mathbf{r}}{d\,s}\;\dfrac{d^2\,\mathbf{r}}{d\,s^2}\;\dfrac{d^3\,\mathbf{r}}{d\,s^3}\right]}{\dfrac{d^2\,\mathbf{r}}{d\,s^2}\cdot\dfrac{d^2\,\mathbf{r}}{d\,s^2}}. \tag{13}$$

The tortuosity may be obtained by another method which is somewhat shorter if not quite so straightforward.

$$\mathbf{t}\cdot\mathbf{c} = \mathbf{c}\cdot\mathbf{n} = \mathbf{n}\cdot\mathbf{t} = 0.$$

Hence
$$d\,\mathbf{t}\cdot\mathbf{c} = -\,d\,\mathbf{c}\cdot\mathbf{t}$$
$$d\,\mathbf{c}\cdot\mathbf{n} = -\,d\,\mathbf{n}\cdot\mathbf{c}$$
$$d\,\mathbf{n}\cdot\mathbf{t} = -\,d\,\mathbf{t}\cdot\mathbf{n}.$$

Now $d\,\mathbf{t}$ is parallel to $\mathbf{c}$; hence perpendicular to $\mathbf{n}$. Hence $d\,\mathbf{t}\cdot\mathbf{n} = 0$. Hence $d\,\mathbf{n}\cdot\mathbf{t} = 0$. But $d\,\mathbf{n}$ is perpendicular to $\mathbf{n}$. Hence $d\,\mathbf{n}$ must be parallel to $\mathbf{c}$. The tortuosity is the magnitude of $d\,\mathbf{n}/d\,s$ taken however with the negative sign because $d\,\mathbf{n}$ appears clockwise from the positive direction of the curve. Hence the scalar tortuosity T may be given by

$$T = -\frac{d\,\mathbf{n}}{d\,s}\cdot\mathbf{c} = \mathbf{n}\cdot\frac{d\,\mathbf{c}}{d\,s}, \tag{14}$$

$$T = \mathbf{t}\times\mathbf{c}\cdot\frac{d\,\mathbf{c}}{d\,s}, \tag{14'}$$

$$\mathbf{c} = \frac{\mathbf{C}}{\sqrt{\mathbf{C}\cdot\mathbf{C}}},$$

$$\frac{d\,\mathbf{c}}{d\,s} = \frac{\sqrt{\mathbf{C}\cdot\mathbf{C}}\,\frac{d\,\mathbf{C}}{d\,s} - \mathbf{C}\,\frac{d}{d\,s}\sqrt{\mathbf{C}\cdot\mathbf{C}}}{\mathbf{C}\cdot\mathbf{C}},$$

$$\mathbf{t}\times\mathbf{c}\cdot\frac{d\,\mathbf{c}}{d\,s} = \mathbf{t}\times\mathbf{c}\cdot\frac{d\,\mathbf{C}}{d\,s}\sqrt{\mathbf{C}\cdot\mathbf{C}} - \mathbf{t}\times\mathbf{c}\cdot\mathbf{C}\,\frac{d}{d\,s}\sqrt{\mathbf{C}\cdot\mathbf{C}}.$$

But $$\mathbf{t}\times\mathbf{c}\cdot\mathbf{C} = 0.$$

$$T = \frac{\mathbf{t}\times\mathbf{c}\cdot\frac{d\,\mathbf{C}}{d\,s}\sqrt{\mathbf{C}\cdot\mathbf{C}}}{\mathbf{C}\cdot\mathbf{C}},$$

$$T = \frac{\mathbf{t}\times\mathbf{C}\cdot\frac{d\,\mathbf{C}}{d\,s}}{\mathbf{C}\cdot\mathbf{C}},$$

$$T = \frac{\left[\frac{d\,\mathbf{r}}{d\,s}\quad\frac{d\,\mathbf{r}^2}{d\,s^2}\quad\frac{d^3\,\mathbf{r}}{d\,s^3}\right]}{\frac{d^2\,\mathbf{r}}{d\,s^2}\cdot\frac{d^2\,\mathbf{r}}{d\,s^2}}. \tag{13}$$

In Cartesian coördinates this becomes

$$T = \frac{\begin{vmatrix} \frac{d\,x}{d\,s} & \frac{d\,y}{d\,s} & \frac{d\,z}{d\,s} \\ \frac{d^2\,x}{d\,s^2} & \frac{d^2\,y}{d\,s^2} & \frac{d^2\,z}{d\,s^2} \\ \frac{d^3\,x}{d\,s^3} & \frac{d^3\,y}{d\,s^3} & \frac{d^3\,z}{d\,s^3} \end{vmatrix}}{\left(\frac{d^2\,x}{d\,s^2}\right)^2 + \left(\frac{d^2\,y}{d\,s^2}\right)^2 + \left(\frac{d^2\,z}{d\,s^2}\right)^2}. \tag{13)'$$

Those who would pursue the study of twisted curves and surfaces in space further from the standpoint of vectors will find the book "*Application de la Méthode Vectorielle de Grassmann à la Géométrie Infinitésimale*"[1] by FEHR extremely

[1] Paris, Carré et Naud, 1899.

helpful. He works with vectors constantly. The treatment is elegant. The notation used is however slightly different from that used by the present writer. The fundamental points of difference are exhibited in this table

$$\mathbf{a}_1 \times \mathbf{a}_2 \qquad \sim [a_1\, a_2]$$

$$\mathbf{a}_1 \cdot \mathbf{a}_2 \qquad \sim [a_1 \mid a_2]$$

$$\mathbf{a}_1 \cdot \mathbf{a}_2 \times \mathbf{a}_3 = [\mathbf{a}_1\, \mathbf{a}_2\, \mathbf{a}_3] \sim [a_1\, a_2\, a_3].$$

One used to either method need have no difficulty with the other. All the important elementary properties of curves and surfaces are there treated. They will not be taken up here.

* *Kinematics*

58.] Let $\mathbf{r}$ be a radius vector drawn from a fixed origin to a moving point or particle. Let t be the time. The equation of the path is then

$$\mathbf{r} = \mathbf{f}\,(t).$$

The *velocity* of the particle is its rate of change of position. This is the limit of the increment $\Delta\, \mathbf{r}$ to the increment $\Delta\, t$.

$$\mathbf{v} = \underset{\Delta t \doteq 0}{\text{Lim}} \left[\frac{\Delta\, \mathbf{r}}{\Delta\, t}\right] = \frac{d\,\mathbf{r}}{d\,t}. \tag{15}$$

This velocity is a vector quantity. Its direction is the direction of the tangent of the curve described by the particle. The term *speed* is used frequently to denote merely the scalar value of the velocity. This convention will be followed here. Then

$$v = \frac{d\,s}{d\,t}; \tag{16}$$

if s be the length of the arc measured from some fixed point of the curve. It is found convenient in mechanics to denote differentiations with respect to the time by dots placed over the quantity differentiated. This is the old *fluxional* notation introduced by Newton. It will also be convenient to denote the unit tangent to the curve by $\mathbf{t}$. The equations become

$$\mathbf{v} = \dot{\mathbf{r}} = \frac{d\,\mathbf{r}}{d\,t} \tag{15}$$

$$v = \dot{s} = \frac{d\,s}{d\,t} \tag{16}$$

$$\mathbf{v} = v\,\mathbf{t}. \tag{17}$$

The *acceleration* is the rate of change of velocity. It is a vector quantity. Let it be denoted by $\mathbf{A}$. Then by definition

$$\mathbf{A} = \underset{\Delta t \doteq 0}{\text{Lim}} \frac{\Delta\,\mathbf{v}}{\Delta\,t} = \frac{d\,\mathbf{v}}{d\,t} = \dot{\mathbf{v}}$$

and
$$\mathbf{A} = \dot{\mathbf{v}} = \frac{d\,\mathbf{v}}{d\,t} = \frac{d}{d\,t}\left(\frac{d\,\mathbf{r}}{d\,t}\right) = \frac{d^2\,\mathbf{r}}{d\,t^2} = \ddot{\mathbf{r}}. \tag{18}$$

Differentiate the expression $\mathbf{v} = v\,\mathbf{t}$.

$$\mathbf{A} = \frac{d\,\mathbf{v}}{d\,t} = \frac{d\,(v\,\mathbf{t})}{d\,t} = \frac{d\,v}{d\,t}\,\mathbf{t} + v\,\frac{d\,\mathbf{t}}{d\,t},$$

$$\frac{d\,v}{d\,t} = \frac{d^2\,s}{d\,t^2} = \ddot{s},$$

$$\frac{d\,\mathbf{t}}{d\,t} = \frac{d\,\mathbf{t}}{d\,s}\,\frac{d\,s}{d\,t} = \mathbf{C}\,v,$$

where $\mathbf{C}$ is the (vector) curvature of the curve and v is the speed in the curve. Substituting these values in the equation the result is

$$\mathbf{A} = \ddot{s}\,\mathbf{t} + v^2\,\mathbf{C}.$$

The acceleration of a particle moving in a curve has therefore been broken up into two components of which one is *parallel* to the tangent $\mathbf{t}$ and of which the other is parallel to the curvature $\mathbf{C}$, that is, *perpendicular* to the tangent. That this resolution has been accomplished would be unimportant were it not for the remarkable fact which it brings to light. The component of the acceleration parallel to the tangent is equal in magnitude to the rate of change of *speed*. It is entirely independent of what sort of curve the particle is describing. It would be the same if the particle described a right line with the same speed as it describes the curve. On the other hand the component of the acceleration normal to the tangent is equal in magnitude to the product of the square of the speed of the particle and the curvature of the curve. The sharper the curve, the greater this component. The greater the speed of the particle, the greater the component. But the rate of change of *speed* in path has no effect at all on this normal component of the acceleration.

If **r** be expressed in terms of **i**, **j**, **k** as

$$\mathbf{r} = x\,\mathbf{i} + y\,\mathbf{j} + z\,\mathbf{k},$$

$$\mathbf{v} = \dot{\mathbf{r}} = \dot{x}\,\mathbf{i} + \dot{y}\,\mathbf{j} + \dot{z}\,\mathbf{k}, \qquad (15)'$$

$$v = \sqrt{\dot{x}^2 + \dot{y}^2 + \dot{z}^2}, \qquad (16)'$$

$$\mathbf{A} = \dot{\mathbf{v}} = \ddot{\mathbf{r}} = \ddot{x}\,\mathbf{i} + \ddot{y}\,\mathbf{j} + \ddot{z}\,\mathbf{k}, \qquad (18)'$$

$$A = \dot{v} = \ddot{s} = \frac{\dot{x}\,\ddot{x} + \dot{y}\,\ddot{y} + \dot{z}\,\ddot{z}}{\sqrt{\dot{x}^2 + \dot{y}^2 + \dot{z}^2}}.$$

From these formulæ the difference between $\ddot{s}$, the rate of change of speed, and $\mathbf{A} = \ddot{\mathbf{r}}$, the rate of change of velocity, is apparent. Just when this difference first became clearly recognized would be hard to say. But certain it is that Newton must have had it in mind when he stated his second law of motion. The rate of change of velocity is proportional to the impressed force; but rate of change of speed is not.

59.] The *hodograph* was introduced by Hamilton as an aid to the study of the curvilinear motion of a particle. With any assumed origin the vector velocity $\dot{\mathbf{r}}$ is laid off. The locus of its terminus is the hodograph. In other words, the radius vector in the hodograph gives the velocity of the particle in magnitude and direction at any instant. It is possible to proceed one step further and construct the hodograph of the hodograph. This is done by laying off the vector acceleration $\mathbf{A} = \ddot{\mathbf{r}}$ from an assumed origin. The radius vector in the hodograph of the hodograph therefore gives the acceleration at each instant.

Example 1: Let a particle revolve in a circle (Fig. 29) of radius r with a uniform angular velocity $\mathbf{a}$. The speed of the particle will then be equal to

$$v = a\,r.$$

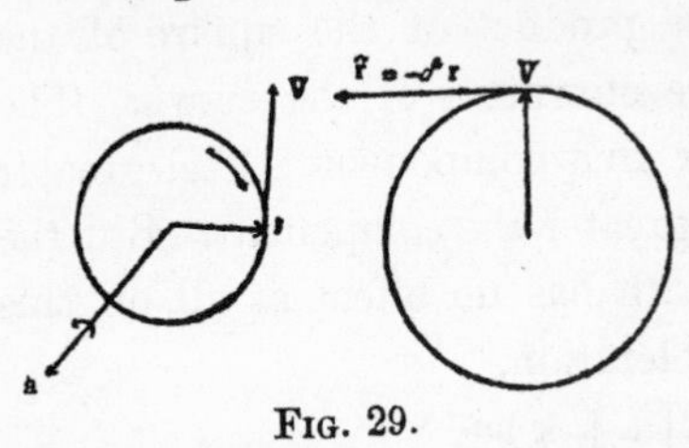

Fig. 29.

Let $\mathbf{r}$ be the radius vector drawn to the particle. The velocity $\mathbf{v}$ is perpendicular to $\mathbf{r}$ and to $\mathbf{a}$. It is

$$\dot{\mathbf{r}} = \mathbf{v} = \mathbf{a} \times \mathbf{r}.$$

The vector $\mathbf{v}$ is always perpendicular and of constant magnitude. The hodograph is therefore a circle of radius $v = a\,r$. The radius vector $\dot{\mathbf{r}}$ in this circle is just ninety degrees in advance of the radius vector $\mathbf{r}$ in its circle, and it consequently describes the circle with the same angular velocity $\mathbf{a}$. The acceleration $\mathbf{A}$ which is the rate of change of $\mathbf{v}$ is always perpendicular to $\mathbf{v}$ and equal in magnitude to

$$A = a\,v = a^2\,r.$$

The acceleration $\mathbf{A}$ may be given by the formula

$$\ddot{\mathbf{r}} = \mathbf{A} = \mathbf{a}\times\mathbf{v} = \mathbf{a}\times(\mathbf{a}\times\mathbf{r}) = \mathbf{a}\cdot\mathbf{r}\;\mathbf{a} - \mathbf{a}\cdot\mathbf{a}\;\mathbf{r}.$$

But as $\mathbf{a}$ is perpendicular to the plane in which $\mathbf{r}$ lies, $\mathbf{a} \cdot \mathbf{r} = 0$. Hence

$$\ddot{\mathbf{r}} = \mathbf{A} = -\,\mathbf{a}\cdot\mathbf{a}\;\mathbf{r} = -\,a^2\,\mathbf{r}.$$

The acceleration due to the uniform motion of a particle in a circle is directed toward the centre and is equal in magnitude to the square of the angular velocity multiplied by the radius of the circle.

Example 2: Consider the motion of a projectile. The acceleration in this case is the acceleration $\mathbf{g}$ due to gravity.

$$\ddot{\mathbf{r}} = \mathbf{A} = \mathbf{g}.$$

The hodograph of the hodograph reduces to a constant vector. The curve is merely a point. It is easy to find the hodograph. Let $\mathbf{v}_0$ be the velocity of the projectile in path at any given instant. At a later instant the velocity will be

$$\mathbf{v} = \mathbf{v}_0 + t\,\mathbf{g}.$$

Thus the hodograph is a straight line parallel to $\mathbf{g}$ and passing through the extremity of $\mathbf{v}_0$. The hodograph of a particle moving under the influence of gravity is hence a straight line. The path is well known to be a parabola.

Example 3: In case a particle move under any central acceleration

$$\ddot{\mathbf{r}} = \mathbf{A} = \mathbf{f}(r).$$

The tangents to the hodograph of $\mathbf{r}$ are the accelerations $\ddot{\mathbf{r}}$. But these tangents are approximately collinear with the chords between two successive values $\mathbf{r}$ and $\dot{\mathbf{r}}_0$ of the radius vector in the hodograph. That is approximately

$$\ddot{\mathbf{r}} = \frac{\dot{\mathbf{r}} - \dot{\mathbf{r}}_0}{\Delta t}.$$

Multiply by $\mathbf{r}\times$.
$$\mathbf{r} \times \ddot{\mathbf{r}} = \mathbf{r} \times \frac{(\dot{\mathbf{r}} - \dot{\mathbf{r}}_0)}{\Delta t}.$$

Since $\mathbf{r}$ and $\ddot{\mathbf{r}}$ are parallel

$$\mathbf{r} \times (\dot{\mathbf{r}} - \dot{\mathbf{r}}_0) = 0.$$

Hence
$$\mathbf{r} \times \dot{\mathbf{r}} = \mathbf{r} \times \dot{\mathbf{r}}_0.$$

But $\frac{1}{2}\,\mathbf{r} \times \dot{\mathbf{r}}$ is the rate of description of area. Hence the equation states that when a particle moves under an acceleration directed towards the centre, equal areas are swept over in equal times by the radius vector.

Perhaps it would be well to go a little more carefully into this question. If $\mathbf{r}$ be the radius vector of the particle in its path at one instant, the radius vector at the next instant is $\mathbf{r} + \Delta\,\mathbf{r}$. The area of the vector of which $\mathbf{r}$ and $\mathbf{r} + \Delta\,\mathbf{r}$ are the bounding radii is approximately equal to the area of the triangle enclosed by $\mathbf{r}$, $\mathbf{r} + \Delta\,\mathbf{r}$, and the chord $\Delta\,\mathbf{r}$. This area is

$$\frac{1}{2}\mathbf{r} \times (\mathbf{r} + \Delta\,\mathbf{r}) = \frac{1}{2}\mathbf{r} \times \mathbf{r} + \frac{1}{2}\mathbf{r} \times \Delta\,\mathbf{r} = \frac{1}{2}\mathbf{r} \times \Delta\,\mathbf{r}.$$

The rate of description of area by the radius vector is consequently

$$\underset{\Delta t \doteq 0}{\text{Lim}}\ \frac{1}{2}\frac{\mathbf{r} \times (\mathbf{r} + \Delta\,\mathbf{r})}{\Delta\,t} = \underset{\Delta t \doteq 0}{\text{Lim}}\ \frac{1}{2}\,\mathbf{r} \times \frac{\Delta\,\mathbf{r}}{\Delta\,t} = \frac{1}{2}\mathbf{r} \times \dot{\mathbf{r}}.$$

Let $\dot{\mathbf{r}}$ and $\dot{\mathbf{r}}_0$ be two values of the velocity at two points P and P_0 which are near together. The acceleration $\ddot{\mathbf{r}}_0$ at P_0 is the limit of

$$\frac{\dot{\mathbf{r}} - \dot{\mathbf{r}}_0}{\Delta\,t} = \frac{\Delta\,\dot{\mathbf{r}}}{\Delta\,t}.$$

Break up the vector $\dfrac{\Delta\,\dot{\mathbf{r}}}{\Delta\,t} = \dfrac{\dot{\mathbf{r}} - \dot{\mathbf{r}}_0}{\Delta\,t}$ into two components one parallel and the other perpendicular to the acceleration $\ddot{\mathbf{r}}_0$.

$$\frac{\Delta\,\dot{\mathbf{r}}}{\Delta\,t} = x\,\ddot{\mathbf{r}}_0 + y\,\mathbf{n},$$

if $\mathbf{n}$ be a normal to the vector $\ddot{\mathbf{r}}_0$. The quantity x approaches unity when $\Delta\,t$ approaches zero. The quantity y approaches zero when $\Delta\,t$ approaches zero.

$$\Delta\,\dot{\mathbf{r}} = \dot{\mathbf{r}} - \dot{\mathbf{r}}_0 = x\,\Delta\,t\ \ddot{\mathbf{r}}_0 + y\,\Delta\,t\ \mathbf{n}.$$

Hence $\quad \mathbf{r} \times (\dot{\mathbf{r}} - \dot{\mathbf{r}}_0) = x\,\Delta\,t\ \mathbf{r} \times \ddot{\mathbf{r}}_0 + y\,\Delta\,t\ \mathbf{r} \times \mathbf{n}.$

$$\mathbf{r} \times (\dot{\mathbf{r}} - \dot{\mathbf{r}}_0) = \mathbf{r} \times \dot{\mathbf{r}} - \left(\mathbf{r}_0 + \frac{\Delta\,\mathbf{r}}{\Delta\,t}\,\Delta\,t\right) \times \dot{\mathbf{r}}_0.$$

Hence

$$\mathbf{r} \times \dot{\mathbf{r}} - \mathbf{r}_0 \times \dot{\mathbf{r}}_0 = \frac{\Delta \mathbf{r}}{\Delta t} \times \dot{\mathbf{r}}_0 \, \Delta t + x \, \Delta t \; \mathbf{r} \times \ddot{\mathbf{r}}_0 + y \, \Delta t \; \mathbf{r} \times \mathbf{n}.$$

But each of the three terms upon the right-hand side is an infinitesimal of the *second* order. Hence the rates of description of area at P and P_0 differ by an infinitesimal of the second order with respect to the time. This is true for any point of the curve. Hence the rates must be exactly equal at all points. This proves the theorem.

60.] The motion of a rigid body one point of which is fixed is at any instant a rotation about an instantaneous axis passing through the fixed point.

Let **i**, **j**, **k** be three axes fixed in the body but moving in space. Let the radius vector **r** be drawn from the fixed point to any point of the body. Then

$$\mathbf{r} = x \, \mathbf{i} + y \, \mathbf{j} + z \, \mathbf{k},$$

$$d \, \mathbf{r} = x \, d \, \mathbf{i} + y \, d \, \mathbf{j} + z \, d \, \mathbf{k}.$$

But $$d \, \mathbf{r} = (d \, \mathbf{r} \cdot \mathbf{i}) \, \mathbf{i} + (d \, \mathbf{r} \cdot \mathbf{j}) \, \mathbf{j} + (d \, \mathbf{r} \cdot \mathbf{k}) \, \mathbf{k}.$$

Substituting the values of $d \, \mathbf{r} \cdot \mathbf{i}$, $d \, \mathbf{r} \cdot \mathbf{j}$, $d \, \mathbf{r} \cdot \mathbf{k}$ obtained from the second equation

$$\begin{aligned} d \, \mathbf{r} = & (x \, \mathbf{i} \cdot d \, \mathbf{i} + y \, \mathbf{i} \cdot d \, \mathbf{j} + z \, \mathbf{i} \cdot d \, \mathbf{k}) \, \mathbf{i} \\ & + (x \, \mathbf{j} \cdot d \, \mathbf{i} + y \, \mathbf{j} \cdot d \, \mathbf{j} + z \, \mathbf{j} \cdot d \, \mathbf{k}) \, \mathbf{j} \\ & + (x \, \mathbf{k} \cdot d \, \mathbf{i} + y \, \mathbf{k} \cdot d \, \mathbf{j} + z \, \mathbf{k} \cdot d \, \mathbf{k}) \, \mathbf{k}. \end{aligned}$$

But $$\mathbf{i} \cdot \mathbf{j} = \mathbf{j} \cdot \mathbf{k} = \mathbf{k} \cdot \mathbf{i} = 0.$$

Hence $$\begin{aligned} \mathbf{i} \cdot d \mathbf{j} + \mathbf{j} \cdot d \mathbf{i} = 0 \quad & \text{or} \quad \mathbf{j} \cdot d \mathbf{i} = - \mathbf{i} \cdot d \mathbf{j} \\ \mathbf{j} \cdot d \mathbf{k} + \mathbf{k} \cdot d \mathbf{j} = 0 \quad & \text{or} \quad \mathbf{k} \cdot d \mathbf{j} = - \mathbf{j} \cdot d \mathbf{k} \\ \mathbf{k} \cdot d \mathbf{i} + \mathbf{i} \cdot d \mathbf{k} = 0 \quad & \text{or} \quad \mathbf{i} \cdot d \mathbf{k} = - \mathbf{k} \cdot d \mathbf{i}. \end{aligned}$$

Moreover $$\mathbf{i} \cdot \mathbf{i} = \mathbf{j} \cdot \mathbf{j} = \mathbf{k} \cdot \mathbf{k} = 1.$$

Hence $$\mathbf{i} \cdot d \mathbf{i} = \mathbf{j} \cdot d \mathbf{j} = \mathbf{k} \cdot d \mathbf{k} = 0.$$

Substituting these values in the expression for $d\,\mathbf{r}$.

$$d\,\mathbf{r} = (z\,\mathbf{i}\cdot d\,\mathbf{k} - y\,\mathbf{j}\cdot d\,\mathbf{i})\,\mathbf{i} + (x\,\mathbf{j}\cdot d\,\mathbf{i} - z\,\mathbf{k}\cdot d\,\mathbf{j})\,\mathbf{j} + (y\,\mathbf{k}\cdot d\,\mathbf{j} - x\,\mathbf{i}\cdot d\,\mathbf{k})\,\mathbf{k}.$$

This is a vector product.

$$d\,\mathbf{r} = (\mathbf{k}\cdot d\mathbf{j}\,\mathbf{i} + \mathbf{i}\cdot d\mathbf{k}\,\mathbf{j} + \mathbf{j}\cdot d\mathbf{i}\,\mathbf{k})\times(x\,\mathbf{i} + y\,\mathbf{j} + z\,\mathbf{k}).$$

Let
$$\mathbf{a} = \mathbf{k}\cdot\frac{d\,\mathbf{j}}{d\,t}\,\mathbf{i} + \mathbf{i}\cdot\frac{d\,\mathbf{k}}{d\,t}\,\mathbf{j} + \mathbf{j}\cdot\frac{d\,\mathbf{i}}{d\,t}\,\mathbf{k}.$$

Then
$$\dot{\mathbf{r}} = \frac{d\,\mathbf{r}}{d\,t} = \mathbf{a}\times\mathbf{r}.$$

This shows that the instantaneous motion of the body is one of rotation with the angular velocity a about the line $\mathbf{a}$. This angular velocity changes from instant to instant. The proof of this theorem fills the lacuna in the work in Art. 51.

Two infinitesimal rotations may be added like vectors. Let $\mathbf{a}_1$ and $\mathbf{a}_2$ be two angular velocities. The displacements due to them are

$$d_1\,\mathbf{r} = \mathbf{a}_1\times\mathbf{r}\,d\,t,$$

$$d_2\,\mathbf{r} = \mathbf{a}_2\times\mathbf{r}\,d\,t.$$

If $\mathbf{r}$ be displaced by $\mathbf{a}$, it becomes

$$\mathbf{r} + d_1\,\mathbf{r} = \mathbf{r} + \mathbf{a}_1\times\mathbf{r}\,d\,t.$$

If it then be displaced by $\mathbf{a}_2$, it becomes

$$\mathbf{r} + d\,\mathbf{r} = \mathbf{r} + d_1\,\mathbf{r} + \mathbf{a}_2\times[\mathbf{r} + (\mathbf{a}_1\times\mathbf{r})\,d\,t]\,d\,t.$$

Hence $\quad d\,\mathbf{r} = \mathbf{a}_1\times\mathbf{r}\,d\,t + \mathbf{a}_2\times\mathbf{r}\,d\,t + \mathbf{a}_2\times(\mathbf{a}_1\times\mathbf{r})\,(d\,t)^2.$

If the infinitesimals $(d\,t)^2$ of order higher than the first be neglected,

$$d\,\mathbf{r} = \mathbf{a}_1\times\mathbf{r}\,d\,t + \mathbf{a}_2\times\mathbf{r}\,d\,t = (\mathbf{a}_1 + \mathbf{a}_2)\times\mathbf{r}\,d\,t,$$

which proves the theorem. If both sides be divided by $d\,t$

$$\dot{\mathbf{r}} = \frac{d\,\mathbf{r}}{d\,t} = (\mathbf{a}_1 + \mathbf{a}_2)\times\mathbf{r}.$$

This is the parallelogram law for angular velocities. It was obtained before (Art. 51) in a different way.

In case the direction of **a**, the instantaneous axis, is constant, the motion reduces to one of steady rotation about **a**.

$$d\,\mathbf{r} = \mathbf{a} \times \mathbf{r}\, d\,t,$$

$$\dot{\mathbf{r}} = \mathbf{a} \times \mathbf{r}.$$

The acceleration $\ddot{\mathbf{r}} = \dot{\mathbf{a}} \times \mathbf{r} + \mathbf{a} \times \dot{\mathbf{r}} = \dot{\mathbf{a}} \times \mathbf{r} + \mathbf{a} \times (\mathbf{a} \times \mathbf{r})$.

As **a** does not change its direction $\dot{\mathbf{a}}$ must be collinear with **a** and hence $\dot{\mathbf{a}} \times \mathbf{r}$ is parallel to $\mathbf{a} \times \mathbf{r}$. That is, it is perpendicular to **r**. On the other hand $\mathbf{a} \times (\mathbf{a} \times \mathbf{r})$ is parallel to **r**. Inasmuch as all points of the rotating body move in concentric circles about **a** in planes perpendicular to **a**, it is unnecessary to consider more than one such plane.

The part of the acceleration of a particle toward the centre of the circle in which it moves is

$$\mathbf{a} \times (\mathbf{a} \times \mathbf{r}).$$

This is equal in magnitude to the square of the angular velocity multiplied by the radius of the circle. It does not depend upon the angular acceleration $\dot{\mathbf{a}}$ at all. It corresponds to what is known as centrifugal force. On the other hand the acceleration normal to the radius of the circle is

$$\dot{\mathbf{a}} \times \mathbf{r}.$$

This is equal in magnitude to the rate of change of angular velocity multiplied by the radius of the circle. It does not depend in any way upon the angular velocity itself but only upon its rate of change.

61.] The subject of *integration* of vector equations in which the differentials depend upon scalar variables needs but a word. It is precisely like integration in ordinary calculus. If then

$$d\,\mathbf{r} = d\,\mathbf{s},$$

$$\mathbf{r} = \mathbf{s} + \mathbf{C},$$

where $\mathbf{C}$ is some constant vector. To accomplish the integration in any particular case may be a matter of some difficulty just as it is in the case of ordinary integration of scalars.

Example 1: Integrate the equation of motion of a projectile.

The equation of motion is simply

$$\ddot{\mathbf{r}} = \mathbf{g},$$

which expresses the fact that the acceleration is always vertically downward and due to gravity.

$$\dot{\mathbf{r}} = \mathbf{g}\, t + \mathbf{b},$$

where $\mathbf{b}$ is a constant of integration. It is evidently the velocity at the time $t = 0$.

$$\mathbf{r} = \frac{1}{2}\,\mathbf{g}\, t^2 + \mathbf{b}\, t + \mathbf{c}.$$

$\mathbf{c}$ is another constant of integration. It is the position vector of the point at time $t = 0$. The path which is given by this last equation is a parabola. That this is so may be seen by expressing it in terms of x and y and eliminating t.

Example 2: The rate of description of areas when a particle moves under a central acceleration is constant.

$$\ddot{\mathbf{r}} = \mathbf{f}\,(r).$$

Since the acceleration is parallel to the radius,

$$\mathbf{r} \times \ddot{\mathbf{r}} = 0.$$

But
$$\mathbf{r} \times \ddot{\mathbf{r}} = \frac{d}{d\,t}\,(\mathbf{r} \times \dot{\mathbf{r}}).$$

For
$$\frac{d}{d\,t}\,(\mathbf{r} \times \dot{\mathbf{r}}) = \dot{\mathbf{r}} \times \dot{\mathbf{r}} + \mathbf{r} \times \ddot{\mathbf{r}}.$$

Hence
$$\frac{d}{d\,t}\,(\mathbf{r} \times \dot{\mathbf{r}}) = 0$$

and
$$\mathbf{r} \times \dot{\mathbf{r}} = \mathbf{C},$$

which proves the statement.

Example 3: Integrate the equation of motion for a particle moving with an acceleration toward the centre and equal to a constant multiple of the inverse square of the distance from the centre.

Given $$\ddot{\mathbf{r}} = -\frac{c^2}{r^3}\mathbf{r}.$$

Then $$\mathbf{r} \times \ddot{\mathbf{r}} = 0.$$

Hence $$\mathbf{r} \times \dot{\mathbf{r}} = \mathbf{C}.$$

Multiply the equations together with $\times$.

$$\frac{\ddot{\mathbf{r}} \times \mathbf{C}}{c^2} = \frac{-1}{r^3}\,\mathbf{r} \times (\mathbf{r} \times \dot{\mathbf{r}}) = \frac{-1}{r^3}\,\{\mathbf{r}\cdot\dot{\mathbf{r}}\,\mathbf{r} - \mathbf{r}\cdot\mathbf{r}\,\dot{\mathbf{r}}\}.$$

$$\mathbf{r} \cdot \mathbf{r} = r^2.$$

Differentiate. Then $$\mathbf{r} \cdot \dot{\mathbf{r}} = r\,\dot{r}.$$

Hence $$\frac{\ddot{\mathbf{r}} \times \mathbf{C}}{c^2} = \frac{\dot{\mathbf{r}}}{r} - \frac{\dot{r}}{r^2}\mathbf{r}.$$

Each side of this equality is a perfect differential.

$$d\left(\frac{\dot{\mathbf{r}} \times \mathbf{C}}{c^2}\right) = d\left(\frac{\mathbf{r}}{r}\right).$$

Integrate. Then $$\frac{\dot{\mathbf{r}} \times \mathbf{C}}{c^2} = \frac{\mathbf{r}}{r} + e\,\mathbf{I},$$

where $e\,\mathbf{I}$ is the vector constant of integration. e is its magnitude and $\mathbf{I}$ a unit vector in its direction. Multiply the equation by $\mathbf{r}\,\cdot$

$$\frac{\mathbf{r} \cdot \dot{\mathbf{r}} \times \mathbf{C}}{c^2} = \frac{\mathbf{r} \cdot \mathbf{r}}{r} + e\,\mathbf{r} \cdot \mathbf{I}.$$

But $$\frac{\mathbf{r} \cdot \dot{\mathbf{r}} \times \mathbf{C}}{c^2} = \frac{\mathbf{r} \times \dot{\mathbf{r}} \cdot \mathbf{C}}{c^2} = \frac{\mathbf{C} \cdot \mathbf{C}}{c^2}.$$

Let
$$p = \frac{\mathbf{C} \cdot \mathbf{C}}{c^2} \quad \text{and} \quad \cos u = \cos (\mathbf{r}, \mathbf{I}).$$

Then
$$p = r + e\, r \cos u.$$

Or
$$r = \frac{p}{1 + e \cos u}.$$

This is the equation of the ellipse of which e is the eccentricity. The vector **I** is drawn in the direction of the major axis. The length of this axis is

$$a = \frac{p}{1 - e^2}.$$

It is possible to carry the integration further and obtain the *time*. So far merely the path has been found.

Scalar Functions of Position in Space. The Operator ∇

62.] A function $V(x, y, z)$ which takes on a definite scalar value for each set of coördinates x, y, z in space is called a scalar function of position in space. Such a function, for example, is

$$V(x, y, z) = x^2 + y^2 + z^2 = r^2.$$

This function gives the square of the distance of the point (x, y, z) from the origin. The function V will be supposed to be in general continuous and single-valued. In physics scalar functions of position are of constant occurrence. In the theory of heat the temperature T at any point of a body is a scalar function of the position of that point. In mechanics and theories of attraction the potential is the all-important function. This, too, is a scalar function of position.

If a scalar function V be set equal to a constant, the equation

$$V(x, y, z) = c. \tag{20}$$

defines a surface in space such that at every point of it the function V has the same value c. In case V be the temperature, this is a surface of constant temperature. It is called an *isothermal* surface. In case V be the potential, this surface of constant potential is known as an *equipotential* surface. As the potential is a typical scalar function of position in space, and as it is perhaps the most important of all such functions owing to its manifold applications, the surface

$$V(x, y, z) = c$$

obtained by setting V equal to a constant is frequently spoken of as an equipotential surface even in the case where V has no connection with the potential, but is *any* scalar function of positions in space.

The rate at which the function V increases in the X direction — that is, when x changes to $x + \Delta x$ and y and z remain constant — is

$$\underset{\Delta x \doteq 0}{\text{Lim}} \left[\frac{V(x + \Delta x, y, z) - V(x, y, z)}{\Delta x} \right].$$

This is the partial derivative of V with respect to x. Hence the rates at which V increases in the directions of the three axes X, Y, Z are respectively

$$\frac{\partial V}{\partial x}, \quad \frac{\partial V}{\partial y}, \quad \frac{\partial V}{\partial z}.$$

Inasmuch as these are rates in a certain direction, they may be written appropriately as vectors. Let **i**, **j**, **k** be a system of unit vectors coincident with the rectangular system of axes X, Y, Z. The rates of increase of V are

$$\mathbf{i} \frac{\partial V}{\partial x}, \quad \mathbf{j} \frac{\partial V}{\partial y}, \quad \mathbf{k} \frac{\partial V}{\partial z}.$$

The sum of these three vectors would therefore appear to be a vector which represents both in magnitude and direction the *resultant* or *most rapid* rate of increase of V. That this is actually the case will be shown later (Art. 64).

63.] The vector sum which is the resultant rate of increase of V is denoted by ∇V.

$$\nabla V = \mathbf{i} \frac{\partial V}{\partial x} + \mathbf{j} \frac{\partial V}{\partial y} + \mathbf{k} \frac{\partial V}{\partial z}. \qquad (21)$$

∇V represents a directed rate of change of V — a directed or vector derivative of V, so to speak. For this reason ∇V will be called the *derivative* of V; and V, the *primitive* of ∇V. The terms *gradient* and *slope* of V are also used for ∇V. It is customary to regard ∇ as an operator which obtains a vector ∇V from a scalar function V of position in space.

$$\nabla V = \left(\mathbf{i} \frac{\partial}{\partial x} + \mathbf{j} \frac{\partial}{\partial y} + \mathbf{k} \frac{\partial}{\partial z} \right) V \qquad (21)'$$

$$\nabla = \mathbf{i} \frac{\partial}{\partial x} + \mathbf{j} \frac{\partial}{\partial y} + \mathbf{k} \frac{\partial}{\partial z}. \qquad (22)$$

This symbolic operator ∇ was introduced by Sir W. R. Hamilton and is now in universal employment. There seems, however, to be no universally recognized name[1] for it, although owing to the frequent occurrence of the symbol some name is a practical necessity. It has been found by experience that the monosyllable *del* is so short and easy to pronounce that even in complicated formulæ in which ∇ occurs a number of times no inconvenience to the speaker or hearer arises from the repetition. ∇V is read simply as "del V."

Although this operator ∇ has been defined as

$$\nabla = \mathbf{i} \frac{\partial}{\partial x} + \mathbf{j} \frac{\partial}{\partial y} + \mathbf{k} \frac{\partial}{\partial z},$$

[1] Some use the term *Nabla* owing to its fancied resemblance to an Assyrian harp. Others have noted its likeness to an inverted Δ and have consequently coined the none too euphonious name *Atled* by inverting the order of the letters in the word *Delta*. Föppl in his *Einführung in die Maxwell'sche Theorie der Electricität* avoids any special designation and refers to the symbol as "*die Operation* ∇." How this is to be read is not divulged. Indeed, for printing no particular name is necessary, but for lecturing and purposes of instruction something is required — something too that does not confuse the speaker or hearer even when often repeated.

so that it appears to depend upon the choice of the axes, it is in reality independent of them. This would be surmised from the interpretation of ∇ as the magnitude and direction of the most rapid increase of V. To demonstrate the independence take another set of axes, $\mathbf{i}'$, $\mathbf{j}'$, $\mathbf{k}'$ and a new set of variables x', y', z' referred to them. Then ∇ referred to this system is

$$\nabla' = \mathbf{i}' \frac{\partial}{\partial x'} + \mathbf{j}' \frac{\partial}{\partial y'} + \mathbf{k}' \frac{\partial}{\partial z'}. \qquad (22)'$$

By making use of the formulæ (47)′ and (47)″, Art. 53, page 104, for transformation of axes from $\mathbf{i}$, $\mathbf{j}$, $\mathbf{k}$ to $\mathbf{i}'$, $\mathbf{j}'$, $\mathbf{k}'$ and by actually carrying out the differentiations and finally by taking into account the identities (49) and (50), ∇' may actually be transformed into ∇.

$$\nabla' = \nabla.$$

The details of the proof are omitted here, because another shorter method of demonstration is to be given.

64.] Consider two surfaces (Fig. 30)

$$V(x, y, z) = c$$

and
$$V(x, y, z) = c + d\,c,$$

upon which V is constant and which are moreover infinitely near together. Let x, y, z be a given point upon the surface $V = c$. Let $\mathbf{r}$ denote the radius vector drawn to this point from any fixed origin. Then any point near by in the neighboring surface $V = c + d\,c$ may be represented by the radius vector $\mathbf{r} + d\,\mathbf{r}$. The *actual* increase of V from the first surface to the second is a fixed quantity $d\,c$. The *rate* of increase is a variable

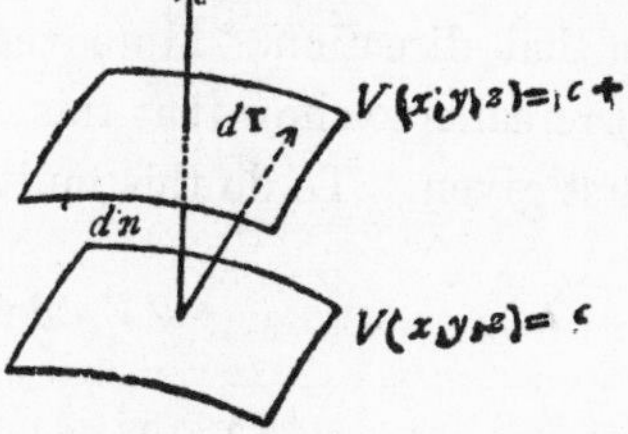

Fig. 30.

quantity and depends upon the direction $d\mathbf{r}$ which is followed when passing from one surface to the other. The rate of increase will be the quotient of the actual increase dc and the distance $\sqrt{d\mathbf{r}\cdot d\mathbf{r}}$ between the surfaces at the point x, y, z in the direction $d\mathbf{r}$. Let $\mathbf{n}$ be a unit normal to the surfaces and dn the segment of that normal intercepted between the surfaces, $\mathbf{n}\,dn$ will then be the least value for $d\mathbf{r}$. The quotient

$$\frac{dc}{\sqrt{d\mathbf{r}\cdot d\mathbf{r}}}$$

will therefore be a maximum when $d\mathbf{r}$ is parallel to $\mathbf{n}$ and equal in magnitude of dn. The expression

$$\frac{dc}{dn}\mathbf{n} \tag{23}$$

is therefore a vector of which the direction is the direction of most rapid increase of V and of which the magnitude is the rate of that increase. This vector is entirely independent of the axes X, Y, Z. Let dc be replaced by its equal dV which is the increment of V in passing from the first surface to the second. Then let ∇V be *defined* again as

$$\nabla V = \frac{dV}{dn}\mathbf{n}. \tag{24}$$

From this definition, ∇V is certainly the vector which gives the direction of most rapid increase of V and the rate in that direction. Moreover ∇V is independent of the axes. It remains to show that this definition is equivalent to the one first given. To do this multiply by $\cdot\, d\mathbf{r}$.

$$\nabla V\cdot d\mathbf{r} = \frac{dV}{dn}\mathbf{n}\cdot d\mathbf{r}. \tag{25}$$

$\mathbf{n}$ is a unit normal. Hence $\mathbf{n}\cdot d\mathbf{r}$ is the projection of $d\mathbf{r}$ on $\mathbf{n}$ and must be equal to the perpendicular distance dn between the surfaces.

$$\nabla V \cdot d\mathbf{r} = \frac{dV}{dn}\, dn = dV. \tag{25)'$$

But
$$dV = \frac{\partial V}{\partial x}\, dx + \frac{\partial V}{\partial y}\, dy + \frac{\partial V}{\partial z}\, dz,$$

where
$$(dx)^2 + (dy)^2 + (dz)^2 = d\mathbf{r} \cdot d\mathbf{r}.$$

If $d\mathbf{r}$ takes on successively the values $\mathbf{i}\, dx$, $\mathbf{j}\, dy$, $\mathbf{k}\, dz$ the equation (25)′ takes on the values

$$\nabla V \cdot \mathbf{i}\, dx = \frac{\partial V}{\partial x}\, dx$$

$$\nabla V \cdot \mathbf{j}\, dy = \frac{\partial V}{\partial y}\, dy \tag{26}$$

$$\nabla V \cdot \mathbf{k}\, dz = \frac{\partial V}{\partial z}\, dz.$$

If the factors dx, dy, dz be cancelled these equations state that the components $\nabla V \cdot \mathbf{i}$, $\nabla V \cdot \mathbf{j}$, $\nabla V \cdot \mathbf{k}$ of ∇V in the $\mathbf{i}$, $\mathbf{j}$, $\mathbf{k}$ directions respectively are equal to

$$\frac{\partial V}{\partial x}, \qquad \frac{\partial V}{\partial y}, \qquad \frac{\partial V}{\partial z}.$$

$$\nabla V = (\nabla V \cdot \mathbf{i})\, \mathbf{i} + (\nabla V \cdot \mathbf{j})\, \mathbf{j} + (\nabla V \cdot \mathbf{k})\, \mathbf{k}.$$

Hence by (26)
$$\nabla V = \mathbf{i}\, \frac{\partial V}{\partial x} + \mathbf{j}\, \frac{\partial V}{\partial y} + \mathbf{k}\, \frac{\partial V}{\partial z}. \tag{21}$$

The second definition (24) has been reduced to the first and consequently is equivalent to it.

***65.**] The equation (25)′ found above is often taken as a definition of ∇V. According to ordinary calculus the derivative $\frac{dy}{dx}$ satisfies the equation

$$dx\, \frac{dy}{dx} = dy.$$

Moreover this equation defines dy/dx. In a similar manner it is possible to lay down the following definition.

Definition: The derivative ∇V of a scalar function of position in space shall satisfy the equation

$$d\mathbf{r} \cdot \nabla V = dV$$

for all values of $d\mathbf{r}$.

This definition is certainly the most natural and important from theoretical considerations. But for practical purposes either of the definitions before given seems to be better. They are more tangible. The real significance of this last definition cannot be appreciated until the subject of linear vector functions has been treated. See Chapter VII.

The computation of the derivative ∇ of a function is most frequently carried on by means of the ordinary partial differentiation.

Example 1: Let $V(x, y, z) = r = \sqrt{x^2 + y^2 + z^2}$.

$$\nabla r = \mathbf{i}\frac{\partial r}{\partial x} + \mathbf{j}\frac{\partial r}{\partial y} + \mathbf{k}\frac{\partial r}{\partial z}.$$

$$\nabla r = \mathbf{i}\frac{x}{\sqrt{x^2 + y^2 + z^2}} + \mathbf{j}\frac{y}{\sqrt{x^2 + y^2 + z^2}} + \mathbf{k}\frac{z}{\sqrt{x^2 + y^2 + z^2}}.$$

Hence
$$\nabla r = \frac{1}{\sqrt{x^2 + y^2 + z^2}}(\mathbf{i}x + \mathbf{j}y + \mathbf{k}z)$$

and
$$\nabla r = \frac{\mathbf{r}}{\sqrt{\mathbf{r} \cdot \mathbf{r}}} = \frac{\mathbf{r}}{r}.$$

The derivative of r is a unit vector in the direction of $\mathbf{r}$. This is evidently the direction of most rapid increase of r and the rate of that increase.

Example 2: Let
$$V(x, y, z) = \frac{1}{r} = \frac{1}{\sqrt{x^2 + y^2 + z^2}}.$$

$$\nabla \frac{1}{r} = -\mathbf{i} \frac{x}{(x^2 + y^2 + z^2)^{\frac{3}{2}}} - \mathbf{j} \frac{y}{(x^2 + y^2 + z^2)^{\frac{3}{2}}}$$
$$- \mathbf{k} \frac{z}{(x^2 + y^2 + z^2)^{\frac{3}{2}}}.$$

Hence
$$\nabla \frac{1}{r} = \frac{1}{(x^2 + y^2 + z^2)^{\frac{3}{2}}} (-\mathbf{i}\, x - \mathbf{j}\, y - \mathbf{k}\, z)$$

and
$$\nabla \frac{1}{r} = \frac{-\mathbf{r}}{(\mathbf{r} \cdot \mathbf{r})^{\frac{3}{2}}} = \frac{-\mathbf{r}}{r^3}.$$

The derivative of $1/r$ is a vector whose direction is that of $-\mathbf{r}$, and whose magnitude is equal to the reciprocal of the square of the length r.

Example 3:
$$\nabla r^n = n\, r^{n-2}\, \mathbf{r} = n\, r^n \frac{\mathbf{r}}{\mathbf{r} \cdot \mathbf{r}}.$$
The proof is left to the reader.

Example 4: Let
$$V(x, y, z) = \log \sqrt{x^2 + y^2}.$$

$$\nabla \log \sqrt{x^2 + y^2} = \mathbf{i} \frac{x}{x^2 + y^2} + \mathbf{j} \frac{y}{x^2 + y^2} + 0\, \mathbf{k}$$
$$= \frac{1}{x^2 + y^2} (\mathbf{i}\, x + \mathbf{j}\, y).$$

If $\mathbf{r}$ denote the vector drawn from the origin to the point (x, y, z) of space, the function V may be written as
$$V(x, y, z) = \log \sqrt{\mathbf{r} \cdot \mathbf{r} - (\mathbf{k} \cdot \mathbf{r})^2}$$

and
$$\mathbf{i}\, x + \mathbf{j}\, y = \mathbf{r} - \mathbf{k}\, \mathbf{k} \cdot \mathbf{r}.$$

Hence
$$\nabla \log \sqrt{x^2 + y^2} = \frac{\mathbf{r} - \mathbf{k}\, \mathbf{k} \cdot \mathbf{r}}{\mathbf{r} \cdot \mathbf{r} - (\mathbf{k} \cdot \mathbf{r})^2}$$
$$= \frac{\mathbf{r} - \mathbf{k}\, \mathbf{k} \cdot \mathbf{r}}{(\mathbf{r} - \mathbf{k}\, \mathbf{k} \cdot \mathbf{r}) \cdot (\mathbf{r} - \mathbf{k}\, \mathbf{k} \cdot \mathbf{r})}.$$

There is another method of computing ∇ which is based upon the identity

$$d\mathbf{r} \cdot \nabla V = dV.$$

Example 1: Let $$V = \sqrt{\mathbf{r} \cdot \mathbf{r}} = r.$$

$$dV = \frac{d\mathbf{r} \cdot \mathbf{r}}{\sqrt{\mathbf{r} \cdot \mathbf{r}}} = d\mathbf{r} \cdot \frac{\mathbf{r}}{\sqrt{\mathbf{r} \cdot \mathbf{r}}} = d\mathbf{r} \cdot \nabla V.$$

Hence $$\nabla V = \frac{\mathbf{r}}{\sqrt{\mathbf{r} \cdot \mathbf{r}}} = \frac{\mathbf{r}}{r}.$$

Example 2: Let $V = \mathbf{r} \cdot \mathbf{a}$, where $\mathbf{a}$ is a constant vector.

$$dV = d\mathbf{r} \cdot \mathbf{a} = d\mathbf{r} \cdot \nabla V.$$

Hence $$\nabla V = \mathbf{a}.$$

Example 3: Let $V = (\mathbf{r} \times \mathbf{a}) \cdot (\mathbf{r} \times \mathbf{b})$, where $\mathbf{a}$ and $\mathbf{b}$ are constant vectors.

$$V = \mathbf{r} \cdot \mathbf{r}\ \mathbf{a} \cdot \mathbf{b} - \mathbf{r} \cdot \mathbf{a}\ \mathbf{r} \cdot \mathbf{b}.$$

$$dV = 2\, d\mathbf{r} \cdot \mathbf{r}\ \mathbf{a} \cdot \mathbf{b} - d\mathbf{r} \cdot \mathbf{a}\ \mathbf{r} \cdot \mathbf{b} - d\mathbf{r} \cdot \mathbf{b}\ \mathbf{r} \cdot \mathbf{a} = d\mathbf{r} \cdot \nabla V.$$

Hence $$\nabla V = 2\, \mathbf{r}\ \mathbf{a} \cdot \mathbf{b} - \mathbf{a}\ \mathbf{r} \cdot \mathbf{b} - \mathbf{b}\ \mathbf{r} \cdot \mathbf{a}$$

$$\nabla V = (\mathbf{r}\ \mathbf{a} \cdot \mathbf{b} - \mathbf{a}\ \mathbf{r} \cdot \mathbf{b}) + (\mathbf{r}\ \mathbf{a} \cdot \mathbf{b} - \mathbf{b}\ \mathbf{r} \cdot \mathbf{a})$$

$$= \mathbf{b} \times (\mathbf{r} \times \mathbf{a}) + \mathbf{a} \times (\mathbf{r} \times \mathbf{b}).$$

Which of these two methods for computing ∇ shall be applied in a particular case depends entirely upon their relative ease of execution in that case. The latter method is independent of the coördinate axes and may therefore be preferred. It is also shorter in case the function V can be expressed easily in terms of $\mathbf{r}$. But when V cannot be so expressed the former method has to be resorted to.

***66.**] The great importance of the operator ∇ in mathematical physics may be seen from a few illustrations. Suppose $T\ (x, y, z)$ be the temperature at the point x, y, z of a heated body. That direction in which the temperature *decreases* most rapidly gives the direction of the flow of heat. ∇T, as has been seen, gives the direction of most rapid *increase* of temperature. Hence the flow of heat $\mathbf{f}$ is

$$\mathbf{f} = -k \nabla T,$$

where k is a constant depending upon the material of the body. Suppose again that V be the gravitational potential due to a fixed body. The force acting upon a unit mass at the point (x, y, z) is in the direction of most rapid increase of potential and is in magnitude equal to the rate of increase per unit length in that direction. Let $\mathbf{F}$ be the force per unit mass. Then

$$\mathbf{F} = \nabla V.$$

As different writers use different conventions as regards the *sign* of the gravitational potential, it might be well to state that the potential V referred to here has the *opposite* sign to the potential energy. If W denoted the potential energy of a mass m situated at x, y, z, the force acting upon that mass would be

$$\mathbf{F} = -\nabla W.$$

In case V represent the electric or magnetic potential due to a definite electric charge or to a definite magnetic pole respectively the force $\mathbf{F}$ acting upon a unit charge or unit pole as the case might be is

$$\mathbf{F} = -\nabla V.$$

The force is in the direction of most rapid decrease of potential. In dealing with electricity and magnetism potential and potential energy have the *same* sign; whereas in attraction problems they are generally considered to have opposite signs. The direction of the force in either case is in the direction of most rapid *decrease* of potential energy. The difference between potential and potential energy is this.

Potential in electricity or magnetism is the potential energy per unit charge or pole; and potential in attraction problems is potential energy per unit mass taken, however, with the negative sign.

***67.**] It is often convenient to treat an *operator* as a quantity provided it obeys the same formal laws as that quantity. Consider for example the partial differentiators

$$\frac{\partial}{\partial x}, \frac{\partial}{\partial y}, \frac{\partial}{\partial z}.$$

As far as combinations of these are concerned, the formal laws are precisely what they would be if instead of differentiators three true scalars

$$a, \quad b, \quad c$$

were given. For instance

the commutative law

$$\frac{\partial}{\partial x}\frac{\partial}{\partial y} = \frac{\partial}{\partial y}\frac{\partial}{\partial x} \sim a\,b = b\,a,$$

the associative law

$$\frac{\partial}{\partial x}\left(\frac{\partial}{\partial y}\frac{\partial}{\partial z}\right) = \left(\frac{\partial}{\partial x}\frac{\partial}{\partial y}\right)\frac{\partial}{\partial z} \sim a\,(b\,c) = (a\,b)\,c,$$

and the distributive law

$$\frac{\partial}{\partial x}\left(\frac{\partial}{\partial y} + \frac{\partial}{\partial z}\right) = \frac{\partial}{\partial x}\frac{\partial}{\partial y} + \frac{\partial}{\partial x}\frac{\partial}{\partial z} \sim a\,(b + c) = a\,b + a\,c$$

hold for the differentiators just as for scalars. Of course such formulæ as

$$u\,\frac{\partial}{\partial x} = \frac{\partial}{\partial x}\,u,$$

where u is a function of x cannot hold on account of the properties of differentiators. A scalar function u cannot be placed under the influence of the sign of differentiators. Such a patent error may be avoided by remembering that an operand must be understood upon which $\partial / \partial x$ is to operate.

In the same way a great advantage may be obtained by looking upon

$$\nabla = \mathbf{i}\frac{\partial}{\partial x} + \mathbf{j}\frac{\partial}{\partial y} + \mathbf{k}\frac{\partial}{\partial z}$$

as a vector. It is not a true vector, for the coefficients

$$\frac{\partial}{\partial x},\ \frac{\partial}{\partial y},\ \frac{\partial}{\partial z}$$

are not true scalars. It is a vector differentiator and of course an operand is always implied with it. As far as formal operations are concerned it behaves like a vector. For instance

$$\nabla(u+v) = \nabla u + \nabla v,$$

$$\nabla(u\,v) = (\nabla u)\,v + u\,(\nabla v),$$

$$c\,\nabla u = \nabla(c\,u),$$

if u and v are any two scalar functions of the scalar variables x, y, z and if c be a scalar independent of the variables with regard to which the differentiations are performed.

68.] If **A** represent any vector the formal combination $\mathbf{A}\cdot\nabla$ is

$$\mathbf{A}\cdot\nabla = A_1\frac{\partial}{\partial x} + A_2\frac{\partial}{\partial y} + A_3\frac{\partial}{\partial z}, \tag{27}$$

provided $\mathbf{A} = A_1\,\mathbf{i} + A_2\,\mathbf{j} + A_3\,\mathbf{k}.$

This operator $\mathbf{A}\cdot\nabla$ is a scalar differentiator. When applied to a scalar function $V(x, y, z)$ it gives a scalar.

$$(\mathbf{A}\cdot\nabla)\,V = A_1\frac{\partial V}{\partial x} + A_2\frac{\partial V}{\partial y} + A_3\frac{\partial V}{\partial z}. \tag{28}$$

Suppose for convenience that **A** is a unit vector **a**.

$$(\mathbf{a}\cdot\nabla)\,V = a_1\frac{\partial V}{\partial x} + a_2\frac{\partial V}{\partial y} + a_3\frac{\partial V}{\partial z} \tag{29}$$

where a_1, a_2, a_3 are the direction cosines of the line **a** referred to the axes X, Y, Z. Consequently $(\mathbf{a} \cdot \nabla) V$ appears as the well-known *directional* derivative of V in the direction **a**.

This is often written

$$\frac{\partial V}{\partial s} = a_1 \frac{\partial V}{\partial x} + a_2 \frac{\partial V}{\partial y} + a_3 \frac{\partial V}{\partial z}. \qquad (29)'$$

It expresses the magnitude of the rate of increase of V in the direction **a**. In the particular case where this direction is the normal **n** to a surface of constant value of V, this relation becomes the normal derivative.

$$(\mathbf{n} \cdot \nabla) V = \frac{\partial V}{\partial n} = n_1 \frac{\partial V}{\partial x} + n_2 \frac{\partial V}{\partial y} + n_3 \frac{\partial V}{\partial z}, \qquad (29)''$$

if n_1, n_2, n_3 be the direction cosines of the normal.

The operator $\mathbf{a} \cdot \nabla$ applied to a scalar function of position V yields the same result as the direct product of **a** and the vector ∇V.

$$(\mathbf{a} \cdot \nabla) V = \mathbf{a} \cdot (\nabla V). \qquad (30)$$

For this reason either operation may be denoted simply by

$$\mathbf{a} \cdot \nabla V$$

without parentheses and no ambiguity can result from the omission. The two different forms $(\mathbf{a} \cdot \nabla) V$ and $\mathbf{a} \cdot (\nabla V)$ may however be interpreted in an important theorem. $(\mathbf{a} \cdot \nabla) V$ is the directional derivative of V in the direction **a**. On the other hand $\mathbf{a} \cdot (\nabla V)$ is the component of ∇V in the direction **a**. Hence: The directional derivative of V in any direction is equal to the component of the derivative ∇V in that direction. If V denote gravitational potential the theorem becomes: The directional derivative of the potential in any direction gives the component of the force per unit mass in that direction. In case V be electric or magnetic potential a difference of sign must be observed.

Vector Functions of Position in Space

69.] A *vector* function of position in space is a function

$$\mathbf{V}\,(x, y, z)$$

which associates with each point x, y, z in space a definite vector. The function may be broken up into its three components

$$\mathbf{V}\,(x, y, z) = V_1\,(x, y, z)\,\mathbf{i} + V_2\,(x, y, z)\,\mathbf{j} + V_3\,(x, y, z)\,\mathbf{k}.$$

Examples of vector functions are very numerous in physics. Already the function ∇V has occurred. At each point of space ∇V has in general a definite vector value. In mechanics of rigid bodies the velocity of each point of the body is a vector function of the position of the point. Fluxes of heat, electricity, magnetic force, fluids, etc., are all vector functions of position in space.

The scalar operator $\mathbf{a} \cdot \nabla$ may be applied to a vector function $\mathbf{V}$ to yield another vector function.

Let $\quad \mathbf{V} = V_1\,(x, y, z)\,\mathbf{i} + V_2\,(x, y, z)\,\mathbf{j} + V_3\,(x, y, z)\,\mathbf{k}$

and $$\mathbf{a} = a_1\,\mathbf{i} + a_2\,\mathbf{j} + a_3\,\mathbf{k}.$$

Then $$\mathbf{a} \cdot \nabla = a_1 \frac{\partial}{\partial x} + a_2 \frac{\partial}{\partial y} + a_3 \frac{\partial}{\partial z}$$

$$(\mathbf{a} \cdot \nabla)\,\mathbf{V} = (\mathbf{a} \cdot \nabla)\,V_1\,\mathbf{i} + (\mathbf{a} \cdot \nabla)\,V_2\,\mathbf{j} + (\mathbf{a} \cdot \nabla)\,V_3\,\mathbf{k}$$

and $$(\mathbf{a} \cdot \nabla)\,\mathbf{V} = \left(a_1 \frac{\partial V_1}{\partial x} + a_2 \frac{\partial V_1}{\partial y} + a_3 \frac{\partial V_1}{\partial z}\right)\mathbf{i}$$

$$+ \left(a_1 \frac{\partial V_2}{\partial x} + a_2 \frac{\partial V_2}{\partial y} + a_3 \frac{\partial V_2}{\partial z}\right)\mathbf{j} \qquad (31)$$

$$+ \left(a_1 \frac{\partial V_3}{\partial x} + a_2 \frac{\partial V_3}{\partial y} + a_3 \frac{\partial V_3}{\partial z}\right)\mathbf{k}.$$

This may be written in the form

$$(\mathbf{a} \cdot \nabla) \mathbf{V} = \frac{\partial V_1}{\partial s} \mathbf{i} + \frac{\partial V_2}{\partial s} \mathbf{j} + \frac{\partial V_3}{\partial s} \mathbf{k}. \qquad (31)'$$

Hence $(\mathbf{a} \cdot \nabla) \mathbf{V}$ is the directional derivative of the vector function $\mathbf{V}$ in the direction $\mathbf{a}$. It is possible to write

$$(\mathbf{a} \cdot \nabla) \mathbf{V} = \mathbf{a} \cdot \nabla \mathbf{V}$$

without parentheses. For the meaning of the vector symbol ∇ when applied to a vector function $\mathbf{V}$ has not yet been defined. Hence from the present standpoint the expression $\mathbf{a} \cdot \nabla \mathbf{V}$ can have but the one interpretation given to it by $(\mathbf{a} \cdot \nabla) \mathbf{V}$.

70.] Although the operation $\nabla \mathbf{V}$ has not been defined and cannot be at present,[1] two formal combinations of the vector operator ∇ and a vector function $\mathbf{V}$ may be treated. These are the (formal) scalar product and the (formal) vector product of ∇ into $\mathbf{V}$. They are

$$\nabla \cdot \mathbf{V} = \left(\mathbf{i} \frac{\partial}{\partial x} + \mathbf{j} \frac{\partial}{\partial y} + \mathbf{k} \frac{\partial}{\partial z} \right) \cdot \mathbf{V} \qquad (32)$$

and

$$\nabla \times \mathbf{V} = \left(\mathbf{i} \frac{\partial}{\partial x} + \mathbf{j} \frac{\partial}{\partial y} + \mathbf{k} \frac{\partial}{\partial z} \right) \times \mathbf{V}. \qquad (33)$$

$\nabla \cdot \mathbf{V}$ is read *del dot* $\mathbf{V}$; and $\nabla \times \mathbf{V}$, *del cross* $\mathbf{V}$.

The differentiators $\frac{\partial}{\partial x}, \frac{\partial}{\partial y}, \frac{\partial}{\partial z}$, being scalar operators, pass by the dot and the cross. That is

$$\nabla \cdot \mathbf{V} = \mathbf{i} \cdot \frac{\partial \mathbf{V}}{\partial x} + \mathbf{j} \cdot \frac{\partial \mathbf{V}}{\partial y} + \mathbf{k} \cdot \frac{\partial \mathbf{V}}{\partial z} \qquad (32)'$$

$$\nabla \times \mathbf{V} = \mathbf{i} \times \frac{\partial \mathbf{V}}{\partial x} + \mathbf{j} \times \frac{\partial \mathbf{V}}{\partial y} + \mathbf{k} \times \frac{\partial \mathbf{V}}{\partial z}. \qquad (33)'$$

These may be expressed in terms of the components V_1, V_2, V_3 of $\mathbf{V}$.

[1] A definition of $\nabla \mathbf{V}$ will be given in Chapter VII.

Now
$$\frac{\partial \mathbf{V}}{\partial x} = \frac{\partial V_1}{\partial x}\mathbf{i} + \frac{\partial V_2}{\partial x}\mathbf{j} + \frac{\partial V_3}{\partial x}\mathbf{k},$$
$$\frac{\partial \mathbf{V}}{\partial y} = \frac{\partial V_1}{\partial y}\mathbf{i} + \frac{\partial V_2}{\partial y}\mathbf{j} + \frac{\partial V_3}{\partial y}\mathbf{k}, \qquad (34)$$
$$\frac{\partial \mathbf{V}}{\partial z} = \frac{\partial V_1}{\partial z}\mathbf{i} + \frac{\partial V_2}{\partial z}\mathbf{j} + \frac{\partial V_3}{\partial z}\mathbf{k}.$$

Then
$$\mathbf{i} \cdot \frac{\partial \mathbf{V}}{\partial x} = \frac{\partial V_1}{\partial x},$$
$$\mathbf{j} \cdot \frac{\partial \mathbf{V}}{\partial y} = \frac{\partial V_2}{\partial y},$$
$$\mathbf{k} \cdot \frac{\partial \mathbf{V}}{\partial z} = \frac{\partial V_3}{\partial z}.$$

Hence
$$\nabla \cdot \mathbf{V} = \frac{\partial V_1}{\partial x} + \frac{\partial V_2}{\partial y} + \frac{\partial V_3}{\partial z}. \qquad (32)''$$

Moreover
$$\mathbf{i} \times \frac{\partial \mathbf{V}}{\partial x} = \mathbf{k}\frac{\partial V_2}{\partial x} - \mathbf{j}\frac{\partial V_3}{\partial x},$$
$$\mathbf{j} \times \frac{\partial \mathbf{V}}{\partial y} = \mathbf{i}\frac{\partial V_3}{\partial y} - \mathbf{k}\frac{\partial V_1}{\partial y},$$
$$\mathbf{k} \times \frac{\partial \mathbf{V}}{\partial z} = \mathbf{j}\frac{\partial V_1}{\partial z} - \mathbf{i}\frac{\partial V_2}{\partial z}.$$

Hence
$$\nabla \times \mathbf{V} = \mathbf{i}\left(\frac{\partial V_3}{\partial y} - \frac{\partial V_2}{\partial z}\right) + \mathbf{j}\left(\frac{\partial V_1}{\partial z} - \frac{\partial V_3}{\partial x}\right) + \mathbf{k}\left(\frac{\partial V_2}{\partial x} - \frac{\partial V_1}{\partial y}\right). \qquad (33)''$$

This may be written in the form of a determinant
$$\nabla \times \mathbf{V} = \begin{vmatrix} \mathbf{i} & \mathbf{j} & \mathbf{k} \\ \frac{\partial}{\partial x} & \frac{\partial}{\partial y} & \frac{\partial}{\partial z} \\ V_1 & V_2 & V_3 \end{vmatrix}. \qquad (33)'''$$

It is to be understood that the operators are to be applied to the functions V_1, V_2, V_3 when expanding the determinant.

From some standpoints objections may be brought forward against treating ∇ as a symbolic vector and introducing $\nabla \cdot \mathbf{V}$ and $\nabla \times \mathbf{V}$ respectively as the symbolic scalar and vector products of ∇ into $\mathbf{V}$. These objections may be avoided by simply laying down the definition that the symbols $\nabla \cdot$ and $\nabla \times$, which may be looked upon as entirely new operators quite distinct from ∇, shall be

$$\nabla \cdot \mathbf{V} = \mathbf{i} \cdot \frac{\partial \mathbf{V}}{\partial x} + \mathbf{j} \cdot \frac{\partial \mathbf{V}}{\partial y} + \mathbf{k} \cdot \frac{\partial \mathbf{V}}{\partial z} \tag{32)'}$$

and $$\nabla \times \mathbf{V} = \mathbf{i} \times \frac{\partial \mathbf{V}}{\partial x} + \mathbf{j} \times \frac{\partial \mathbf{V}}{\partial y} + \mathbf{k} \times \frac{\partial \mathbf{V}}{\partial z}. \tag{33)'}$$

But for practical purposes and for remembering formulæ it seems by all means advisable to regard

$$\nabla = \mathbf{i} \frac{\partial}{\partial x} + \mathbf{j} \frac{\partial}{\partial y} + \mathbf{k} \frac{\partial}{\partial z}$$

as a symbolic vector differentiator. This symbol obeys the same laws as a vector just in so far as the differentiators $\frac{\partial}{\partial x}, \frac{\partial}{\partial y}, \frac{\partial}{\partial z}$ obey the same laws as ordinary scalar quantities.

71.] That the two functions $\nabla \cdot \mathbf{V}$ and $\nabla \times \mathbf{V}$ have very important physical meanings in connection with the vector function $\mathbf{V}$ may be easily recognized. By the straightforward proof indicated in Art. 63 it was seen that the operator ∇ is independent of the choice of axes. From this fact the inference is immediate that $\nabla \cdot \mathbf{V}$ and $\nabla \times \mathbf{V}$ represent intrinsic properties of $\mathbf{V}$ invariant of choice of axes. In order to perceive these properties it is convenient to attribute to the function $\mathbf{V}$ some definite physical meaning such as flux or flow of a fluid substance. Let therefore the vector $\mathbf{V}$ denote at each point of space the direction and the magnitude of the flow of some fluid. This may be a material fluid as water or gas, or a fictitious one as heat or electricity. To obtain as great clearness as possible let the fluid be material but not necessarily restricted to incompressibility like water.

Then $$\nabla \cdot \mathbf{V} = \mathbf{i} \cdot \frac{\partial \mathbf{V}}{\partial x} + \mathbf{j} \cdot \frac{\partial \mathbf{V}}{\partial y} + \mathbf{k} \cdot \frac{\partial \mathbf{V}}{\partial z}$$

is called the divergence of $\mathbf{V}$ and is often written

$$\nabla \cdot \mathbf{V} = \text{div } \mathbf{V}.$$

The reason for this term is that $\nabla \cdot \mathbf{V}$ gives at each point the rate per unit volume per unit time at which fluid is leaving that point — the rate of diminution of density. To prove this consider a small cube of matter (Fig. 31). Let the edges of the cube be dx, dy, and dz respectively. Let

$$\mathbf{V}(x, y, z) = V_1(x, y, z)\, \mathbf{i} + V_2(x, y, z)\, \mathbf{j} + V_3(x, y, z)\, \mathbf{k}.$$

Consider the amount of fluid which passes through those faces of the cube which are parallel to the YZ-plane, *i. e.* perpendicular to the X axis. The normal to the face whose x coördinate is the lesser, that is, the normal to the left-hand face of the cube is $-\mathbf{i}$. The flux of substance through this face is

$$-\mathbf{i} \cdot \mathbf{V}(x, y, z)\, dy\, dz.$$

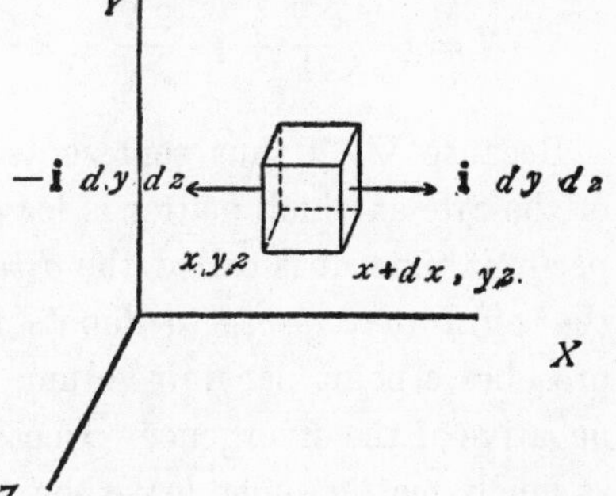

FIG. 31.

The normal to the opposite face, the face whose x coördinate is greater by the amount dx, is $+\mathbf{i}$ and the flux through it is therefore

$$\mathbf{i} \cdot \mathbf{V}\,(x + dx, y, z)\,dy\,dz = \mathbf{i} \cdot \left[\mathbf{V}(x, y, z) + \frac{\partial \mathbf{V}}{\partial x} dx \right] dy\,dz$$

$$= \mathbf{i} \cdot \mathbf{V}\,(x, y, z)\,dy\,dz + \mathbf{i} \cdot \frac{\partial \mathbf{V}}{\partial x}\,dx\,dy\,dz.$$

The total flux outward from the cube through these two faces is therefore the algebraic sum of these quantities. This is simply

$$\mathbf{i} \cdot \frac{\partial \mathbf{V}}{\partial x}\,dx\,dy\,dz = \frac{\partial V_1}{\partial x}\,dx\,dy\,dz.$$

In like manner the fluxes through the other pairs of faces of the cube are

$$\mathbf{j} \cdot \frac{\partial \mathbf{V}}{\partial y}\,dx\,dy\,dz \text{ and } \mathbf{k} \cdot \frac{\partial \mathbf{V}}{\partial z}\,dx\,dy\,dz.$$

The total flux out from the cube is therefore

$$\left(\mathbf{i} \cdot \frac{\partial \mathbf{V}}{\partial x} + \mathbf{j} \cdot \frac{\partial \mathbf{V}}{\partial y} + \mathbf{k} \cdot \frac{\partial \mathbf{V}}{\partial z} \right) dx\,dy\,dz.$$

This is the net quantity of fluid which leaves the cube per unit time. The quotient of this by the volume $dx\,dy\,dz$ of the cube gives the rate of diminution of density. This is

$$\nabla \cdot \mathbf{V} = \mathbf{i} \cdot \frac{\partial \mathbf{V}}{\partial x} + \mathbf{j} \cdot \frac{\partial \mathbf{V}}{\partial y} + \mathbf{k} \cdot \frac{\partial \mathbf{V}}{\partial z} = \frac{\partial V_1}{\partial x} + \frac{\partial V_2}{\partial y} + \frac{\partial V_3}{\partial z}.$$

Because $\nabla \cdot \mathbf{V}$ thus represents the diminution of density or the rate at which matter is leaving a point per unit volume per unit time, it is called the *divergence.* Maxwell employed the term *convergence* to denote the rate at which fluid approaches a point per unit volume per unit time. This is the negative of the divergence. In case the fluid is *incompressible,* as much matter must leave the cube as enters it. The total change of contents must therefore be zero. For this reason the characteristic differential equation which any incompressible fluid must satisfy is

$$\nabla \cdot \mathbf{V} = 0$$

where $\mathbf{V}$ is the flux of the fluid. This equation is often known as the *hydrodynamic equation.* It is satisfied by any flow of water, since water is practically incompressible. The great importance of the equation for work in electricity is due to the fact that according to Maxwell's hypothesis electric displacement obeys the same laws as an incompressible fluid. If then $\mathbf{D}$ be the electric displacement,

$$\text{div}\,\mathbf{D} = \nabla \cdot \mathbf{D} = 0.$$

72.] To the operator $\nabla \times$ Maxwell gave the name *curl.* This nomenclature has become widely accepted.

$$\nabla \times \mathbf{V} = \text{curl}\,\mathbf{V}.$$

The curl of a vector function $\mathbf{V}$ is itself a vector function of position in space. As the name indicates, it is closely connected with the angular velocity or spin of the flux at each point. But the interpretation of the curl is neither so easily obtained nor so simple as that of the divergence.

Consider as before that $\mathbf{V}$ represents the flux of a fluid. Take at a definite instant an infinitesimal sphere about any point (x, y, z). At the next instant what has become of the sphere? In the first place it may have moved off as a whole in a certain direction by an amount $d\mathbf{r}$. In other words it may have a translational velocity of $d\mathbf{r}/dt$. In addition to this it may have undergone such a deformation that it is no longer a sphere. It may have been subjected to a *strain* by virtue of which it becomes slightly ellipsoidal in shape. Finally it may have been rotated as a whole about some axis through an angle $d\omega$. That is to say, it may have an angular velocity the magnitude of which is $d\omega/dt$. An infinitesimal sphere therefore may have any one of three distinct types of motion or all of them combined. First, a translation with definite velocity. Second, a strain with three definite rates of elongation along the axes of an ellipsoid.

Third, an angular velocity about a definite axis. It is this third type of motion which is given by the curl. In fact, the *curl* of the flux $\mathbf{V}$ is a vector which has at each point of space the direction of the instantaneous axis of rotation at that point and a magnitude equal to twice the instantaneous angular velocity about that axis.

The analytic discussion of the motion of a fluid presents more difficulties than it is necessary to introduce in treating the curl. The motion of a rigid body is sufficiently complex to give an adequate idea of the operation. It was seen (Art. 51) that the velocity of the particles of a rigid body at any instant is given by the formula

$$\mathbf{v} = \mathbf{v}_0 + \mathbf{a} \times \mathbf{r}.$$

$$\text{curl } \mathbf{v} = \nabla \times \mathbf{v} = \nabla \times \mathbf{v}_0 + \nabla \times (\mathbf{a} \times \mathbf{r}).$$

Let
$$\mathbf{a} = a_1\, \mathbf{i} + a_2\, \mathbf{j} + a_3\, \mathbf{k}$$

$$\mathbf{r} = r_1\, \mathbf{i} + r_2\, \mathbf{j} + r_3\, \mathbf{k} = x\, \mathbf{i} + y\, \mathbf{j} + z\, \mathbf{k}$$

expand $\nabla \times (\mathbf{a} \times \mathbf{r})$ formally as if it were the vector triple product of ∇, $\mathbf{a}$, and $\mathbf{r}$. Then

$$\nabla \times \mathbf{v} = \nabla \times \mathbf{v}_0 + (\nabla \cdot \mathbf{r})\, \mathbf{a} - (\nabla \cdot \mathbf{a})\, \mathbf{r}.$$

$\mathbf{v}_0$ is a constant vector. Hence the term $\nabla \times \mathbf{v}_0$ vanishes.

$$\nabla \cdot \mathbf{r} = \frac{\partial x}{\partial x} + \frac{\partial y}{\partial y} + \frac{\partial z}{\partial z} = 3.$$

As $\mathbf{a}$ is a constant vector it may be placed upon the other side of the differential operator, $\nabla \cdot \mathbf{a} = \mathbf{a} \cdot \nabla$.

$$\nabla\, \mathbf{r} = \left(a_1 \frac{\partial}{\partial x} + a_2 \frac{\partial}{\partial y} + a_3 \frac{\partial}{\partial z} \right) \mathbf{r} = a_1\, \mathbf{i} + a_2\, \mathbf{j} + a_3\, \mathbf{k} = \mathbf{a}.$$

Hence
$$\nabla \times \mathbf{v} = 3\, \mathbf{a} - \mathbf{a} = 2\, \mathbf{a}.$$

Therefore in the case of the motion of a rigid body the curl of the linear velocity at any point is equal to twice the angular velocity in magnitude and in direction.

$$\nabla \times \mathbf{v} = \text{curl}\, \mathbf{v} = 2\,\mathbf{a},$$

$$\mathbf{a} = \frac{1}{2}\, \nabla \times \mathbf{v} = \frac{1}{2}\, \text{curl}\, \mathbf{v}.$$

$$\mathbf{v} = \mathbf{v}_0 + \frac{1}{2}\, (\nabla \times \mathbf{v}) \times \mathbf{r} = \mathbf{v}_0 + \frac{1}{2}\, (\text{curl}\, \mathbf{v}) \times \mathbf{r}. \quad (34)$$

The expansion of $\nabla \times (\mathbf{a} \times \mathbf{r})$ formally may be avoided by multiplying $\mathbf{a} \times \mathbf{r}$ out and then applying the operator $\nabla \times$ to the result.

73.] It frequently happens, as in the case of the application just cited, that the operators $\nabla, \nabla \cdot, \nabla \times$, have to be applied to combinations of scalar functions, vector functions, or both. The following rules of operation will be found useful. Let u, v be scalar functions and $\mathbf{u}, \mathbf{v}$ vector functions of position in space. Then

$$\nabla\, (u + v) = \nabla u + \nabla v \quad (35)$$

$$\nabla \cdot (\mathbf{u} + \mathbf{v}) = \nabla \cdot \mathbf{u} + \nabla \cdot \mathbf{v} \quad (36)$$

$$\nabla \times (\mathbf{u} + \mathbf{v}) = \nabla \times \mathbf{u} + \nabla \times \mathbf{v} \quad (37)$$

$$\nabla\, (u\, v) = v\, \nabla u + u\, \nabla v \quad (38)$$

$$\nabla \cdot (u\, \mathbf{v}) = \nabla u \cdot \mathbf{v} + u\, \nabla \cdot \mathbf{v} \quad (39)$$

$$\nabla \times (u\, \mathbf{v}) = \nabla u \times \mathbf{v} + u\, \nabla \times \mathbf{v} \quad (40)$$

$$\nabla\, (\mathbf{u} \cdot \mathbf{v}) = \mathbf{v} \cdot \nabla \mathbf{u} + \mathbf{u} \cdot \nabla \mathbf{v} \quad (41)$$
$$+ \mathbf{v} \times (\nabla \times \mathbf{u}) + \mathbf{u} \times (\nabla \times \mathbf{v})^1$$

$$\nabla \cdot (\mathbf{u} \times \mathbf{v}) = \mathbf{v} \cdot \nabla \times \mathbf{u} - \mathbf{u} \cdot \nabla \times \mathbf{v} \quad (42)$$

$$\nabla \times (\mathbf{u} \times \mathbf{v}) = \mathbf{v} \cdot \nabla \mathbf{u} - \mathbf{v}\, \nabla \cdot \mathbf{u} - \mathbf{u} \cdot \nabla \mathbf{v} + \mathbf{u}\, \nabla \cdot \mathbf{v}.^1 \quad (43)$$

A word is necessary upon the matter of the interpretation of such expressions as

$$\nabla u\, v, \qquad \nabla u \cdot \mathbf{v}, \qquad \nabla u \times \mathbf{v}.$$

The rule followed in this book is that the operator ∇ applies to the nearest term only. That is,

[1] By Art. 69 the expressions $\mathbf{v} \cdot \nabla\, \mathbf{u}$ and $\mathbf{u} \cdot \nabla\, \mathbf{v}$ are to be interpreted as $(\mathbf{v} \cdot \nabla)\, \mathbf{u}$ and $(\mathbf{u} \cdot \nabla)\, \mathbf{v}$.

$$\nabla u v = (\nabla u)\, v$$

$$\nabla u \cdot \mathbf{v} = (\nabla u) \cdot \mathbf{v}$$

$$\nabla u \times \mathbf{v} = (\nabla u) \times \mathbf{v}.$$

If ∇ is to be applied to more than the one term which follows it, the terms to which it is applied are enclosed in a parenthesis as upon the left-hand side of the above equations.

The proofs of the formulæ may be given most naturally by expanding the expressions in terms of three assumed unit vectors **i, j, k.** The sign Σ of summation will be found convenient. By means of it the operators ∇, $\nabla\cdot$, $\triangle \times$ take the form

$$\nabla = \sum \mathbf{i} \frac{\partial}{\partial x},$$

$$\nabla \cdot = \sum \mathbf{i} \frac{\partial}{\partial x},$$

$$\nabla \times = \sum \mathbf{i} \times \frac{\partial}{\partial x}.$$

The summation extends over x, y, z.

To demonstrate $\quad \nabla \times (u\,\mathbf{v}) = \nabla u \times \mathbf{v} + u\, \nabla \times \mathbf{v}.$

$$\nabla \times (u\,\mathbf{v}) = \sum \mathbf{i} \times \frac{\partial}{\partial x}(u\,\mathbf{v}) = \sum \mathbf{i} \times \left(\frac{\partial u}{\partial x}\mathbf{v} + u\,\frac{\partial \mathbf{v}}{\partial x}\right),$$

$$\nabla \times (u\,\mathbf{v}) = \sum \mathbf{i} \times \left(\frac{\partial u}{\partial x}\,\mathbf{v}\right) + \sum \mathbf{i} \times \left(u\,\frac{\partial \mathbf{v}}{\partial x}\right)$$

$$= \sum \left(\mathbf{i}\,\frac{\partial u}{\partial x}\right) \times \mathbf{v} + \sum u\,\mathbf{i} \times \frac{\partial \mathbf{v}}{\partial x}.$$

Hence $\quad \nabla \times (u\,\mathbf{v}) = \nabla u \times \mathbf{v} + u \nabla \times \mathbf{v}.$

To demonstrate

$$\nabla (\mathbf{u} \cdot \mathbf{v}) = \mathbf{v} \cdot \nabla \mathbf{u} + \mathbf{u} \cdot \nabla \mathbf{v} + \mathbf{v} \times (\nabla \times \mathbf{u}) + \mathbf{u} \times (\nabla \times \mathbf{v}).$$

$$\nabla (\mathbf{u} \cdot \mathbf{v}) = \sum \mathbf{i} \frac{\partial}{\partial x} (\mathbf{u} \cdot \mathbf{v}) = \sum \mathbf{i} \left(\frac{\mathbf{u}}{\partial x} \cdot \mathbf{v} + \mathbf{u} \cdot \frac{\partial \mathbf{v}}{\partial x} \right)$$

$$\nabla (\mathbf{u} \cdot \mathbf{v}) = \sum \mathbf{i} \frac{\partial \mathbf{u}}{\partial x} \cdot \mathbf{v} + \sum \mathbf{i} \, \mathbf{u} \cdot \frac{\partial \mathbf{v}}{\partial x}$$

Now

$$\mathbf{v} \times (\nabla \times \mathbf{u}) = \mathbf{v} \times \sum \mathbf{i} \times \frac{\partial \mathbf{u}}{\partial x} = \sum \mathbf{v} \cdot \frac{\partial \mathbf{u}}{\partial x} \mathbf{i} - \sum \mathbf{v} \cdot \mathbf{i} \frac{\partial \mathbf{u}}{\partial x}.$$

$$\sum \mathbf{v} \cdot \frac{\partial \mathbf{u}}{\partial x} \mathbf{i} = \mathbf{v} \times (\nabla \times \mathbf{u}) + \sum \mathbf{v} \cdot \mathbf{i} \frac{\partial \mathbf{u}}{\partial x}$$

or

$$\sum \mathbf{v} \cdot \frac{\partial \mathbf{u}}{\partial x} \mathbf{i} = \mathbf{v} \times (\nabla \times \mathbf{u}) + \mathbf{v} \cdot \nabla \mathbf{u}.$$

In like manner $\sum \mathbf{u} \cdot \frac{\partial \mathbf{v}}{\partial x} \mathbf{i} = \mathbf{u} \times (\nabla \times \mathbf{v}) + \mathbf{u} \cdot \nabla \mathbf{v}.$

Hence

$$\nabla (\mathbf{u} \cdot \mathbf{v}) = \mathbf{v} \cdot \nabla \mathbf{u} + \mathbf{u} \cdot \nabla \mathbf{v} + \mathbf{v} \times (\nabla \times \mathbf{u}) + \mathbf{u} \times (\nabla \times \mathbf{v}).$$

The other formulæ are demonstrated in a similar manner.

74.] The notation[1]

$$\nabla (\mathbf{u} \cdot \mathbf{v})_{\mathbf{u}} \tag{44}$$

will be used to denote that in applying the operator ∇ to the product $(\mathbf{u} \cdot \mathbf{v})$, the quantity $\mathbf{u}$ is to be regarded as *constant.* That is, the operation ∇ is carried out only *partially* upon the product $(\mathbf{u} \cdot \mathbf{v})$. In general if ∇ is to be carried out partially upon any number of functions which occur after it in a parenthesis, those functions which are constant for the differentiations are written after the parenthesis as subscripts.

Let

$$\mathbf{u} = u_1 \mathbf{i} + u_2 \mathbf{j} + u_3 \mathbf{k},$$

$$\mathbf{v} = v_1 \mathbf{i} + v_2 \mathbf{j} + v_3 \mathbf{k}.$$

[1] This idea and notation of a partial ∇ so to speak may be avoided by means of the formula 41. But a certain amount of compactness and simplicity is lost thereby. The idea of $\nabla (\mathbf{u} \cdot \mathbf{v})_{\mathbf{u}}$ is surely no more complicated than $\mathbf{u} \cdot \nabla \mathbf{v}$ or $\mathbf{v} \times (\nabla \times \mathbf{u})$.

then $$\mathbf{u} \cdot \mathbf{v} = u_1 v_1 + u_2 v_2 + u_3 v_3$$

and $$\nabla (\mathbf{u} \cdot \mathbf{v}) = \sum \mathbf{i} \frac{\partial}{\partial x} (u_1 v_1 + u_2 v_2 + u_3 v_3).$$

$$\nabla (\mathbf{u} \cdot \mathbf{v}) = \sum \mathbf{i} \left(\frac{\partial u_1}{\partial x} v_1 + \frac{\partial u_2}{\partial x} v_2 + \frac{\partial u_3}{\partial x} v_3 \right)$$
$$+ \sum \mathbf{i} \left(u_1 \frac{\partial v_1}{\partial x} + u_2 \frac{\partial v_2}{\partial x} + u_3 \frac{\partial v_3}{\partial x} \right).$$

But $$\nabla (\mathbf{u} \cdot \mathbf{v})_{\mathbf{u}} = \sum \mathbf{i} \left(u_1 \frac{\partial v_1}{\partial x} + u_2 \frac{\partial v_2}{\partial x} + u_3 \frac{\partial v_3}{\partial x} \right)$$

and $$\nabla (\mathbf{u} \cdot \mathbf{v})_{\mathbf{v}} = \sum \mathbf{i} \left(v_1 \frac{\partial u_1}{\partial x} + v_2 \frac{\partial u_2}{\partial x} + v_3 \frac{\partial u_3}{\partial x} \right).$$

Hence $$\nabla (\mathbf{u} \cdot \mathbf{v}) = v_1 \nabla u_1 + v_2 \nabla u_2 + v_3 \nabla u_3$$
$$+ u_1 \nabla v_1 + u_2 \nabla v_2 + u_3 \nabla v_3.$$

But $$\nabla (\mathbf{u} \cdot \mathbf{v})_{\mathbf{u}} = u_1 \nabla v_1 + u_2 \nabla v_2 + u_3 \nabla v_3 \qquad (44)'$$

and $$\nabla (\mathbf{u} \cdot \mathbf{v})_{\mathbf{v}} = v_1 \nabla u_1 + v_2 \nabla u_2 + v_3 \nabla u_3.$$

Hence $$\nabla (\mathbf{u} \cdot \mathbf{v}) = \nabla (\mathbf{u} \cdot \mathbf{v})_{\mathbf{u}} + \nabla (\mathbf{u} \cdot \mathbf{v})_{\mathbf{v}}. \qquad (45)$$

This formula corresponds to the following one in the notation of differentials

$$d (\mathbf{u} \cdot \mathbf{v}) = d (\mathbf{u} \cdot \mathbf{v})_{\mathbf{u}} + d (\mathbf{u} \cdot \mathbf{v})_{\mathbf{v}}$$

or $$d (\mathbf{u} \cdot \mathbf{v}) = \mathbf{u} \cdot d \mathbf{v} + d \mathbf{u} \cdot \mathbf{v}.$$

The formulæ (35)–(43) given above (Art. 73) may be written in the following manner, as is obvious from analogy with the corresponding formulæ in differentials:

$$\nabla (u + v) = \nabla (u + v)_u + \nabla (u + v)_v \qquad (35)'$$

$$\nabla \cdot (\mathbf{u} + \mathbf{v}) = \nabla \cdot (\mathbf{u} + \mathbf{v})_{\mathbf{u}} + \nabla \cdot (\mathbf{u} + \mathbf{v})_{\mathbf{v}} \qquad (36)'$$

$$\nabla \times (\mathbf{u} + \mathbf{v}) = \nabla \times (\mathbf{u} + \mathbf{v})_{\mathbf{u}} + \nabla \times (\mathbf{u} + \mathbf{v})_{\mathbf{v}} \qquad (37)'$$

$$\nabla (u\, v) = \nabla (u\, v)_u + \nabla (u\, v)_v \qquad (38)'$$

$$\nabla \cdot (u\, \mathbf{v}) = \nabla \cdot (u\, \mathbf{v})_u + \nabla \cdot (u\, \mathbf{v})_{\mathbf{v}} \qquad (39)'$$

$$\nabla \times (u\, \mathbf{v}) = \nabla \times (u\, \mathbf{v})_u + \nabla \times (u\, \mathbf{v})_{\mathbf{v}} \qquad (40)'$$

$$\nabla (\mathbf{u} \cdot \mathbf{v}) = \nabla (\mathbf{u} \cdot \mathbf{v})_{\mathbf{u}} + \nabla (\mathbf{u} \cdot \mathbf{v})_{\mathbf{v}} \qquad (41)'$$

$$\nabla \cdot (\mathbf{u} \times \mathbf{v}) = \nabla \cdot (\mathbf{u} \times \mathbf{v})_{\mathbf{u}} + \nabla \cdot (\mathbf{u} \times \mathbf{v})_{\mathbf{v}} \qquad (42)'$$

$$\nabla \times (\mathbf{u} \times \mathbf{v}) = \nabla \times (\mathbf{u} \times \mathbf{v})_{\mathbf{u}} + \nabla \times (\mathbf{u} \times \mathbf{v})_{\mathbf{v}}. \qquad (43)'$$

This notation is particularly useful in the case of the scalar product $\mathbf{u} \cdot \mathbf{v}$ and for this reason it was introduced. In almost all other cases it can be done away without loss of simplicity. Take for instance (43)′. Expand $\nabla \times (\mathbf{u} \times \mathbf{v})_{\mathbf{u}}$ formally.

$$\nabla \times (\mathbf{u} \times \mathbf{v})_{\mathbf{u}} = (\nabla \cdot \mathbf{v})\, \mathbf{u} - (\nabla \cdot \mathbf{u})\, \mathbf{v},$$

where it must be understood that $\mathbf{u}$ is constant for the differentiations which occur in ∇. Then in the last term the factor $\mathbf{u}$ may be placed before the sign ∇. Hence

$$\nabla \times (\mathbf{u} \times \mathbf{v})_u = \mathbf{u}\, \nabla \cdot \mathbf{v} - \mathbf{u} \cdot \nabla\, \mathbf{v}.$$

In like manner $\nabla \times (\mathbf{u} \times \mathbf{v})_{\mathbf{v}} = \mathbf{v} \cdot \nabla\, \mathbf{u} - \mathbf{v}\, \nabla \cdot \mathbf{u}.$

Hence $\nabla \times (\mathbf{u} \times \mathbf{v}) = \mathbf{v} \cdot \nabla\, \mathbf{u} - \mathbf{v}\, \nabla \cdot \mathbf{u} - \mathbf{u} \cdot \nabla\, \mathbf{v} + \mathbf{u}\, \nabla \cdot \mathbf{v}.$

75.] There are a number of important relations in which the partial operation $\nabla\, (\mathbf{u} \cdot \mathbf{v})_{\mathbf{u}}$ figures.

$$\mathbf{u} \times (\nabla \times \mathbf{v}) = \nabla\, (\mathbf{u} \cdot \mathbf{v})_{\mathbf{u}} - \mathbf{u} \cdot \nabla\, \mathbf{v}, \qquad (46)$$

or
$$\nabla\, (\mathbf{u} \cdot \mathbf{v})_{\mathbf{u}} = \mathbf{u} \cdot \nabla\, \mathbf{v} + \mathbf{u} \times (\nabla \times \mathbf{v}), \qquad (46)'$$

or
$$\mathbf{u} \cdot \nabla\, \mathbf{v} = \nabla\, (\mathbf{u} \cdot \mathbf{v})_{\mathbf{u}} + (\nabla \times \mathbf{v}) \times \mathbf{u}. \qquad (46)''$$

The proof of this relation may be given by expanding in terms of **i, j, k**. A method of remembering the result easily is as follows. Expand the product

$$\mathbf{u} \times (\nabla \times \mathbf{v})$$

formally as if ∇, **u**, **v** were all real vectors. Then

$$\mathbf{u} \times (\nabla \times \mathbf{v}) = \mathbf{u} \cdot \mathbf{v} \nabla - \mathbf{u} \cdot \nabla \mathbf{v}.$$

The second term is capable of interpretation as it stands. The first term, however, is not. The operator ∇ has nothing upon which to operate. It therefore must be transposed so that it shall have $\mathbf{u} \cdot \mathbf{v}$ as an operand. But **u** being outside of the parenthesis in $\mathbf{u} \times (\nabla \times \mathbf{v})$ is constant for the differentiations. Hence

$$\mathbf{u} \cdot \mathbf{v} \nabla = \nabla (\mathbf{u} \cdot \mathbf{v})_{\mathbf{u}}$$

and $$\mathbf{u} \times (\nabla \times \mathbf{v}) = \nabla (\mathbf{u} \cdot \mathbf{v})_{\mathbf{u}} - \mathbf{u} \cdot \nabla \mathbf{v}. \tag{46}$$

If **u** be a unit vector, say **a**, the formula

$$\mathbf{a} \cdot \nabla \mathbf{v} = \nabla (\mathbf{a} \cdot \mathbf{v})_{\mathbf{a}} + (\nabla \times \mathbf{v}) \times \mathbf{a} \tag{47}$$

expresses the fact that the directional derivative $\mathbf{a} \cdot \nabla \mathbf{v}$ of a vector function **v** in the direction **a** is equal to the derivative of the projection of the vector **v** in that direction plus the vector product of the curl of **v** into the direction **a**.

Consider the values of **v** at two neighboring points.

$$\mathbf{v}(x, y, z)$$

and $$\mathbf{v}(x + dx,\ y + dy,\ z + dz)$$

$$d\mathbf{v} = \mathbf{v}(x + dx, y + dy, z + dz) - \mathbf{v}(x, y, z).$$

Let $$\mathbf{v} = v_1 \mathbf{i} + v_2 \mathbf{j} + v_3 \mathbf{k}$$

$$d\mathbf{v} = d v_1 \mathbf{i} + d v_2 \mathbf{j} + d v_3 \mathbf{k}.$$

But by (25)′ $$d v_1 = d\mathbf{r} \cdot \nabla v_1$$

$$d v_2 = d\mathbf{r} \cdot \nabla v_2$$

$$d v_3 = d\mathbf{r} \cdot \nabla v_3.$$

Hence $$d\mathbf{v} = d\mathbf{r} \cdot (\nabla v_1 \mathbf{i} + \nabla v_2 \mathbf{j} + \nabla v_3 \mathbf{k}).$$

Hence $$d\mathbf{v} = d\mathbf{r} \cdot \nabla \mathbf{v}.$$

By (46)″ $$d\mathbf{v} = \nabla (d\mathbf{r} \cdot \mathbf{v})_{d\mathbf{r}} + (\nabla \times \mathbf{v}) \times d\mathbf{r}. \tag{48}$$

Or if $\mathbf{v}_0$ denote the value of $\mathbf{v}$ at the point (x, y, z) and $\mathbf{v}$ the value at a neighboring point

$$\mathbf{v} = \mathbf{v}_0 + \nabla (d\,\mathbf{r} \cdot \mathbf{v})_{d\mathbf{r}} + (\nabla \times \mathbf{v}) \times d\,\mathbf{r}. \qquad (49)$$

This expression of $\mathbf{v}$ in terms of its value $\mathbf{v}_0$ at a given point, the *dels*, and the displacement $d\,\mathbf{r}$ is analogous to the expansion of a scalar functor of one variable by Taylor's theorem,

$$f(x) = f(x_0) + f'(x_0)\, d\,x.$$

The derivative of $(\mathbf{r} \cdot \mathbf{v})$ when $\mathbf{v}$ is constant is equal to $\mathbf{v}$.

That is
$$\nabla (\mathbf{r} \cdot \mathbf{v})_{\mathbf{v}} = \mathbf{v}.$$

For
$$\nabla (\mathbf{r} \cdot \mathbf{v})_{\mathbf{v}} = \mathbf{v} \cdot \nabla\, \mathbf{r} - (\nabla \times \mathbf{r}) \times \mathbf{v},$$

$$\mathbf{v} = v_1\, \mathbf{i} + v_2\, \mathbf{j} + v_3\, \mathbf{k},$$

$$\mathbf{v} \cdot \nabla = v_1 \frac{\partial}{\partial x} + v_2 \frac{\partial}{\partial y} + v_3 \frac{\partial}{\partial z},$$

$$\mathbf{r} = x\, \mathbf{i} + y\, \mathbf{j} + z\, \mathbf{k},$$

$$\mathbf{v} \cdot \nabla\, \mathbf{r} = v_1\, \mathbf{i} + v_2\, \mathbf{j} + v_3\, \mathbf{k} = \mathbf{v},$$

$$\nabla \times \mathbf{r} = 0.$$

Hence
$$\nabla (\mathbf{r} \cdot \mathbf{v})_{\mathbf{v}} = \mathbf{v}.$$

In like manner if instead of the finite vector $\mathbf{r}$, an infinitesimal vector $d\,\mathbf{r}$ be substituted, the result still is

$$\nabla (d\,\mathbf{r} \cdot \mathbf{v})_{\mathbf{v}} = \mathbf{v}.$$

By (49)
$$\mathbf{v} = \mathbf{v}_0 + \nabla (d\,\mathbf{r} \cdot \mathbf{v})_{d\mathbf{r}} + (\nabla \times \mathbf{v}) \times d\,\mathbf{r}$$

$$\nabla (d\,\mathbf{r} \cdot \mathbf{v}) = \nabla (d\,\mathbf{r} \cdot \mathbf{v})_{d\mathbf{r}} + \nabla (d\,\mathbf{r} \cdot \mathbf{v})_{\mathbf{v}}.$$

Hence
$$\nabla (d\,\mathbf{r} \cdot \mathbf{v})_{d\mathbf{r}} = \nabla (d\,\mathbf{r} \cdot \mathbf{v}) - \mathbf{v}.$$

Substituting:

$$\mathbf{v} = \frac{1}{2}\mathbf{v}_0 + \frac{1}{2} \nabla (d\,\mathbf{r} \cdot \mathbf{v}) + \frac{1}{2} (\nabla \times \mathbf{v}) \times d\,\mathbf{r}. \qquad (50)$$

This gives another form of (49) which is sometimes more convenient. It is also slightly more symmetrical.

* **76.**] Consider a moving fluid. Let $\mathbf{v}\,(x, y, z, t)$ be the velocity of the fluid at the point (x, y, z) at the time t. Surround a point (x_0, y_0, z_0) with a small sphere.

$$d\,\mathbf{r} \cdot d\,\mathbf{r} = c^2.$$

At each point of this sphere the velocity is

$$\mathbf{v} = \mathbf{v}_0 + d\,\mathbf{r} \cdot \nabla\,\mathbf{v}.$$

In the increment of time $\delta\,t$ the points of this sphere will have moved the distance

$$(\mathbf{v}_0 + d\,\mathbf{r} \cdot \nabla\,\mathbf{v})\,\delta\,t.$$

The point at the center will have moved the distance

$$\mathbf{v}_0\,\delta\,t.$$

The distance between the center and the points that were upon the sphere of radius $d\,\mathbf{r}$ at the commencement of the interval $\delta\,t$ has become at the end of that interval $\delta\,t$

$$d\,\mathbf{r}' = d\,\mathbf{r} + d\,\mathbf{r} \cdot \nabla\,\mathbf{v}\,\delta\,t.$$

To find the locus of the extremity of $d\,\mathbf{r}'$ it is necessary to eliminate $d\,\mathbf{r}$ from the equations

$$d\,\mathbf{r}' = d\,\mathbf{r} + d\,\mathbf{r} \cdot \nabla\,\mathbf{v}\,\delta\,t,$$

$$c^2 = d\,\mathbf{r} \cdot d\,\mathbf{r}.$$

The first equation may be solved for $d\,\mathbf{r}$ by the method of Art. 47, page 90, and the solution substituted into the second. The result will show that the infinitesimal sphere

$$d\,\mathbf{r} \cdot d\,\mathbf{r} = c^2$$

has been transformed into an ellipsoid by the motion of the fluid during the time $\delta\,t$.

A more definite account of the change that has taken place may be obtained by making use of equation (50)

$$\mathbf{v} = \frac{1}{2}\mathbf{v}_0 + \frac{1}{2}\nabla (d\,\mathbf{r} \cdot \mathbf{v}) + \frac{1}{2}(\nabla \times \mathbf{v}) \times d\,\mathbf{r},$$

$$\mathbf{v} = \mathbf{v}_0 + \frac{1}{2}[\nabla (d\,\mathbf{r} \cdot \mathbf{v}) - \mathbf{v}_0] + \frac{1}{2}(\nabla \times \mathbf{v}) \times d\,\mathbf{r};$$

or of the equation (49)

$$\mathbf{v} = \mathbf{v}_0 + \nabla (d\,\mathbf{r} \cdot \mathbf{v})_{d\mathbf{r}} + (\nabla \times \mathbf{v}) \times d\,\mathbf{r},$$

$$\mathbf{v} = \mathbf{v}_0 + \left[\nabla (d\,\mathbf{r} \cdot \mathbf{v})_{d\mathbf{r}} + \frac{1}{2}(\nabla \times \mathbf{v}) \times d\,\mathbf{r}\right] + \frac{1}{2}(\nabla \times \mathbf{v}) \times d\,\mathbf{r}.$$

The first term $\mathbf{v}_0$ in these equations expresses the fact that the infinitesimal sphere is moving as a whole with an instantaneous velocity equal to $\mathbf{v}_0$. This is the translational element of the motion. The last term

$$\frac{1}{2}(\nabla \times \mathbf{v}) \times d\,\mathbf{r} = \frac{1}{2}\,\text{curl}\,\mathbf{v} \times d\,\mathbf{r}$$

shows that the sphere is undergoing a rotation about an instantaneous axis in the direction of curl $\mathbf{v}$ and with an angular velocity equal in magnitude to one half the magnitude of curl $\mathbf{v}$. The middle term

$$\frac{1}{2}\nabla (d\,\mathbf{r} \cdot \mathbf{v}) - \mathbf{v}_0,$$

or

$$\nabla (d\,\mathbf{r} \cdot \mathbf{v})_{d\mathbf{r}} - \frac{1}{2}(\nabla \times \mathbf{v}) \times d\,\mathbf{r}$$

expresses the fact that the sphere is undergoing a deformation known as *homogeneous strain* by virtue of which it becomes ellipsoidal. For this term is equal to

$$dx\,\nabla v_1 + dy\,\nabla v_2 + dz\,\nabla v_3,$$

if v_1, v_2, v_3 be respectively the components of $\mathbf{v}$ in the directions $\mathbf{i}$, $\mathbf{j}$, $\mathbf{k}$. It is fairly obvious that at any given point (x_0, y_0, z_0) a set of three mutually perpendicular axes $\mathbf{i}$, $\mathbf{j}$, $\mathbf{k}$ may be chosen such that at that point ∇v_1, ∇v_2, ∇v_3 are respectively parallel to them. Then the expression above becomes simply

$$d x \frac{\partial v_1}{\partial x} \mathbf{i} + d y \frac{\partial v_2}{\partial y} \mathbf{j} + d z \frac{\partial v_3}{\partial z} \mathbf{k}.$$

The point whose coördinates referred to the center of the infinitesimal sphere are

$$d x, \quad d y, \quad d z$$

is therefore endowed with this velocity. In the time δt it will have moved to a new position

$$d x \left(1 + \frac{\partial v_1}{\partial x} \delta t\right), \quad d y \left(1 + \frac{\partial v_2}{\partial y} \delta t\right), \quad d z \left(1 + \frac{\partial v_3}{\partial z} \delta t\right).$$

The totality of the points upon the sphere

$$d \mathbf{r} \cdot d \mathbf{r} = d x^2 + d y^2 + d z^2 = c^2$$

goes over into the totality of points upon the ellipsoid of which the equation is

$$\frac{x^2}{\left(1 + \frac{\partial v_1}{\partial x} \delta t\right)^2} + \frac{y^2}{\left(1 + \frac{\partial v_2}{\partial y} \delta t\right)^2} + \frac{z^2}{\left(1 + \frac{\partial v_3}{\partial z} \delta t\right)^2} = c^2.$$

The statements made before (Art. 72) concerning the three types of motion which an infinitesimal sphere of fluid may possess have therefore now been demonstrated.

77.] The symbolic operator ∇ may be applied several times in succession. This will correspond in a general way to forming derivatives of an order higher than the first. The expressions found by thus repeating ∇ will all be independent of the axes because ∇ itself is. There are six of these dels of the second order.

Let $V\ (x, y, z)$ be a scalar function of position in space. The derivative ∇V is a vector function and hence has a curl and a divergence. Therefore

$$\nabla \cdot \nabla V, \qquad \nabla \times \nabla V$$

are the two derivatives of the second order which may be obtained from V.

$$\nabla \cdot \nabla V = \text{div } \nabla V \tag{51}$$

$$\nabla \times \nabla V = \text{curl } \nabla V. \tag{52}$$

The second expression $\nabla \times \nabla V$ *vanishes identically*. That is, *the derivative of any scalar function V possesses no curl.* This may be seen by expanding $\nabla \times \nabla V$ in terms of **i, j, k**. All the terms cancel out. Later (Art. 83) it will be shown conversely that if a vector function **W** possesses no curl, *i. e.* if

$$\nabla \times \mathbf{W} = \text{curl } \mathbf{W} \equiv 0, \text{ then } \mathbf{W} = \nabla V,$$

W is the derivative of some scalar function V.

The first expression $\nabla \cdot \nabla V$ when expanded in terms of **i, j, k** becomes

$$\nabla \cdot \nabla V = \frac{\partial^2 V}{\partial x^2} + \frac{\partial^2 V}{\partial y^2} + \frac{\partial^2 V}{\partial z^2}. \tag{51$'$}$$

Symbolically, $$\nabla \cdot \nabla = \frac{\partial^2}{\partial x^2} + \frac{\partial^2}{\partial y^2} + \frac{\partial^2}{\partial z^2}.$$

The operator $\nabla \cdot \nabla$ is therefore the well-known operator of Laplace. Laplace's Equation

$$\Delta V = \frac{\partial^2 V}{\partial x^2} + \frac{\partial^2 V}{\partial y^2} + \frac{\partial^2 V}{\partial z^2} = 0 \tag{53}$$

becomes in the notation here employed

$$\nabla \cdot \nabla V = 0. \tag{53$'$}$$

When applied to a scalar function V the operator $\nabla \cdot \nabla$ yields a scalar function which is, moreover, the divergence of the derivative.

Let T be the temperature in a body. Let c be the conductivity, ρ the density, and k the specific heat. The flow **f** is

$$\mathbf{f} = -c \nabla T.$$

The rate at which heat is leaving a point per unit volume per unit time is $\nabla \cdot \mathbf{f}$. The increment of temperature is

$$dT = -\frac{1}{\rho k} \nabla \cdot \mathbf{f}\, dt.$$

$$\frac{dT}{dt} = \frac{c}{\rho k} \nabla \cdot \nabla T.$$

This is Fourier's equation for the rate of change of temperature.

Let $\mathbf{V}$ be a vector function, and V_1, V_2, V_3 its three components. The operator $\nabla \cdot \nabla$ of Laplace may be applied to $\mathbf{V}$.

$$\nabla \cdot \nabla \mathbf{V} = \nabla \cdot \nabla V_1\, \mathbf{i} + \nabla \cdot \nabla V_2\, \mathbf{j} + \nabla \cdot \nabla V_3\, \mathbf{k} \quad (54)$$

If a vector function $\mathbf{V}$ satisfies Laplace's Equation, each of its three scalar components does. Other *dels* of the second order may be obtained by considering the divergence and curl of $\mathbf{V}$. The divergence $\nabla \cdot \mathbf{V}$ has a derivative

$$\nabla \nabla \cdot \mathbf{V} = \nabla \operatorname{div} \mathbf{V}. \quad (55)$$

The curl $\nabla \times \mathbf{V}$ has in turn a divergence and a curl,

$$\nabla \cdot \nabla \times \mathbf{V}, \quad \nabla \times \nabla \times \mathbf{V}.$$

and

$$\nabla \cdot \nabla \times \mathbf{V} = \operatorname{div} \operatorname{curl} \mathbf{V} \quad (56)$$

and

$$\nabla \times \nabla \times \mathbf{V} = \operatorname{curl} \operatorname{curl} \mathbf{V}. \quad (57)$$

Of these expressions $\nabla \cdot \nabla \times \mathbf{V}$ *vanishes identically*. That is, the *divergence of the curl of any vector is zero*. This may be seen by expanding $\nabla \cdot \nabla \times \mathbf{V}$ in terms of **i**, **j**, **k**. Later (Art. 83) it will be shown conversely that if the divergence of a vector function $\mathbf{W}$ vanishes identically, *i. e.* if

$$\nabla \cdot \mathbf{W} = \operatorname{div} \mathbf{W} \equiv 0, \text{ then } \mathbf{W} = \nabla \times \mathbf{V} = \operatorname{curl} \mathbf{V},$$

$\mathbf{W}$ is the curl of some vector function $\mathbf{V}$.

If the expression $\nabla \times (\nabla \times \mathbf{V})$ were expanded formally according to the law of the triple vector product,

$$\nabla \times (\nabla \times \mathbf{V}) = \nabla \cdot \mathbf{V} \nabla - \nabla \cdot \nabla \mathbf{V}.$$

The term $\nabla \cdot \mathbf{V} \nabla$ is meaningless until ∇ be transposed to the beginning so that it operates upon $\mathbf{V}$.

$$\nabla \times \nabla \times \mathbf{V} = \nabla \nabla \cdot \mathbf{V} - \nabla \cdot \nabla \mathbf{V}, \qquad (58)$$

or

$$\text{curl curl } \mathbf{V} = \nabla \text{ div } \mathbf{V} - \nabla \cdot \nabla \mathbf{V}. \qquad (58)'$$

This formula is very important. It expresses the curl of the curl of a vector in terms of the derivative of the divergence and the operator of Laplace. Should the vector function $\mathbf{V}$ satisfy Laplace's Equation,

$$\nabla \cdot \nabla \mathbf{V} = 0 \text{ and}$$

$$\text{curl curl } \mathbf{V} = \nabla \text{ div } \mathbf{V}.$$

Should the divergence of $\mathbf{V}$ be zero,

$$\text{curl curl } \mathbf{V} = - \nabla \cdot \nabla \mathbf{V}.$$

Should the curl of the curl of $\mathbf{V}$ vanish,

$$\nabla \text{ div } \mathbf{V} = \nabla \cdot \nabla \mathbf{V}.$$

To sum up. There are six of the *dels* of the second order.

$$\nabla \cdot \nabla V, \quad \nabla \times \nabla V,$$

$$\nabla \cdot \nabla \mathbf{V}, \quad \nabla \nabla \cdot \mathbf{V}, \quad \nabla \cdot \nabla \times \mathbf{V}, \quad \nabla \times \nabla \times \mathbf{V}.$$

Of these, two vanish identically.

$$\nabla \times \nabla V \equiv 0, \quad \nabla \cdot \nabla \times \mathbf{V} \equiv 0.$$

A third may be expressed in terms of two others.

$$\nabla \times \nabla \times \mathbf{V} = \nabla \nabla \cdot \mathbf{V} - \nabla \cdot \nabla \mathbf{V}. \qquad (58)$$

The operator $\nabla \cdot \nabla$ is equivalent to the operator of Laplace.

***78.**] The geometric interpretation of $\nabla \cdot \nabla u$ is interesting. It depends upon a geometric interpretation of the second derivative of a scalar function u of the one scalar variable x. Let u_i be the value of u at the point x_i. Let it be required to find the second derivative of u with respect to x at the point x_o. Let x_1 and x_2 be two points equidistant from x_o. That is, let

$$x_2 - x_o = x_o - x_1 = a.$$

Then

$$\frac{\dfrac{u_1 + u_2}{2} - u_o}{a^2}$$

is the ratio of the difference between the average of u at the points x_1 and x_2 and the value of u at x_o to the square of the distance of the points x_1, x_2 from x_o. That

$$\frac{1}{2}\frac{d^2 u}{d x^2} = \underset{a \doteq 0}{\text{Lim}} \frac{\dfrac{u_1 + u_2}{2} - u_o}{a^2}$$

is easily proved by Taylor's theorem.

Let u be a scalar function of position in space. Choose three mutually orthogonal lines **i**, **j**, **k** and evaluate the expressions

$$\frac{\partial^2 u}{\partial x^2}, \quad \frac{\partial^2 u}{\partial y^2}, \quad \frac{\partial^2 u}{\partial z^2}.$$

Let x_2 and x_1 be two points on the line **i** at a distance a from x_o; x_4 and x_3, two points on **j** at the same distance a from x_o; x_6 and x_5, two points on **k** at the same distance a from x_o.

$$\frac{1}{2}\frac{\partial^2 u}{\partial x^2} = \underset{a \doteq 0}{\text{Lim}} \frac{\dfrac{u_1 + u_2}{2} - u_o}{a^2}$$

$$\frac{1}{2}\frac{\partial^2 u}{\partial y^2} = \underset{a \doteq 0}{\text{Lim}} \frac{\dfrac{u_3 + u_4}{2} - u_o}{a^2}$$

$$\frac{1}{2}\frac{\partial^2 u}{\partial z^2} = \underset{a \doteq 0}{\text{Lim}} \frac{\dfrac{u_6 + u_5}{2} - u_o}{a^2}$$

Add:

$$\frac{1}{6}\left(\frac{\partial^2 u}{\partial x^2} + \frac{\partial^2 u}{\partial y^2} + \frac{\partial^2 u}{\partial z^2}\right) = \frac{1}{6}\nabla \cdot \nabla u$$

$$= \underset{a \doteq 0}{\text{Lim}} \left[\frac{\dfrac{u_1 + u_2 + u_3 + u_4 + u_5 + u_6}{6} - u_o}{a^2} \right].$$

As ∇ and $\nabla \cdot$ are independent of the particular axes chosen, this expression may be evaluated for a different set of axes, then for still a different one, etc. By adding together all these results

$$\frac{1}{6}\nabla \cdot \nabla u = \underset{a \doteq 0}{\text{Lim}} \frac{\dfrac{u_1 + u_2 + \cdots 6\,n \text{ terms}}{6\,n} - u_o}{a^2}.$$

Let n become infinite and at the same time let the different sets of axes point in every direction issuing from x_o. The fraction

$$\frac{u_1 + u_2 + \cdots 6\,n \text{ terms}}{6\,n}$$

then approaches the *average value* of u upon the surface of a sphere of radius a surrounding the point x_o. Denote this by u_a.

$$\frac{1}{6}\nabla \cdot \nabla u = \underset{a \doteq 0}{\text{Lim}} \frac{u_a - u_o}{a^2}.$$

$\nabla \cdot \nabla u$ is equal to six times the limit approached by the ratio of the *excess* of u on the surface of a sphere above the value at the center to the square of the radius of the sphere. The same reasoning holds in case u is a *vector* function.

If u be the temperature of a body $\nabla \cdot \nabla\ u$ (except for a constant factor which depends upon the material of the body) is equal to the rate of increase of temperature (Art. 77). If $\nabla \cdot \nabla u$ is positive the average temperature upon a small sphere is greater than the temperature at the center. The center of the sphere is growing warmer. In the case of a *steady* flow the temperature at the center must remain constant. Evidently therefore the condition for a steady flow is

$$\nabla \cdot \nabla u = 0.$$

That is, the temperature is a solution of Laplace's Equation.

Maxwell gave the name *concentration* to $-\nabla \cdot \nabla u$ whether u be a scalar or vector function. Consequently $\nabla \cdot \nabla u$ may be called the *dispersion* of the function u whether it be scalar or vector. The dispersion is proportional to the excess of the average value of the function on an infinitesimal surface above the value at the center. In case u is a vector function the average is a vector average. The additions in it are vector additions.

SUMMARY OF CHAPTER III

If a vector **r** is a function of a scalar t the derivative of **r** with respect to t is a vector quantity whose direction is that of the tangent to the curve described by the terminus of **r** and whose magnitude is equal to the rate of advance of that terminus along the curve per unit change of t. The derivatives of the components of a vector are the components of the derivatives.

$$\frac{d^n \mathbf{r}}{d t^n} = \frac{d^n r_1}{d t^n} \mathbf{i} + \frac{d^n r_2}{d t^n} \mathbf{j} + \frac{d^n r_3}{d t^n} \mathbf{k}. \qquad (2)'$$

A combination of vectors or of vectors and scalars may be differentiated just as in ordinary scalar analysis except that the differentiations must be performed *in situ*.

$$\frac{d}{d\,t}(\mathbf{a}\cdot\mathbf{b})=\frac{d\,\mathbf{a}}{d\,t}\cdot\mathbf{b}+\mathbf{a}\cdot\frac{d\,\mathbf{b}}{d\,t}, \tag{3}$$

$$\frac{d}{d\,t}(\mathbf{a}\times\mathbf{b})=\frac{d\,\mathbf{a}}{d\,t}\times\mathbf{b}+\mathbf{a}\times\frac{d\,\mathbf{b}}{d\,t}, \tag{4}$$

or

$$d\,(\mathbf{a}\cdot\mathbf{b})=d\,\mathbf{a}\cdot\mathbf{b}+\mathbf{a}\cdot d\,\mathbf{b}, \tag{3$'$}$$

$$d\,(\mathbf{a}\times\mathbf{b})=d\,\mathbf{a}\times\mathbf{b}+\mathbf{a}\times d\,\mathbf{b}, \tag{4$'$}$$

and so forth. The differential of a unit vector is perpendicular to that vector.

The derivative of a vector $\mathbf{r}$ with respect to the arc s of the curve which the terminus of the vector describes is the unit tangent to the curves directed toward that part of the curve along which s is supposed to increase.

$$\frac{d\,\mathbf{r}}{d\,s}=\mathbf{t}. \tag{8}$$

The derivative of $\mathbf{t}$ with respect to the arc s is a vector whose direction is normal to the curve on the concave side and whose magnitude is equal to the curvature of the curve.

$$\mathbf{C}=\frac{d\,\mathbf{t}}{d\,s}=\frac{d^2\,\mathbf{r}}{d\,s^2}. \tag{9}$$

The tortuosity of a curve in space is the derivative of the unit normal $\mathbf{n}$ to the osculating plane with respect to the arc s.

$$\mathbf{T}=\frac{d\,\mathbf{n}}{d\,s}=\frac{d}{d\,s}\left(\frac{d\,\mathbf{r}}{d\,s}\times\frac{d^2\,\mathbf{r}}{d\,s^2}\cdot\frac{1}{\sqrt{\mathbf{C}\cdot\mathbf{C}}}\right). \tag{11}$$

The magnitude of the tortuosity is

$$T=\frac{\left[\frac{d\,\mathbf{r}}{d\,s}\;\frac{d^2\,\mathbf{r}}{d\,s^2}\;\frac{d^3\,\mathbf{r}}{d\,s^3}\right]}{\frac{d^2\,\mathbf{r}}{d\,s^2}\cdot\frac{d^2\,\mathbf{r}}{d\,s^2}} \tag{13}$$

If $\mathbf{r}$ denote the position of a moving particle, t the time, $\mathbf{v}$ the velocity, $\mathbf{A}$ the acceleration,

$$\mathbf{v} = \frac{d\,\mathbf{r}}{d\,t} = \dot{\mathbf{r}} \qquad (15)$$

$$v = \frac{d\,s}{d\,t} = \dot{s} \qquad (16)$$

$$\mathbf{A} = \dot{\mathbf{v}} = \frac{d\,\mathbf{v}}{d\,t} = \frac{d^2\,\mathbf{r}}{d\,t^2} = \ddot{\mathbf{r}}. \qquad (18)$$

The acceleration may be broken up into two components of which one is parallel to the tangent and depends upon the rate of change of the scalar velocity v of the particle in its path, and of which the other is perpendicular to the tangent and depends upon the velocity of the particle and the curvature of the path.

$$\mathbf{A} = \ddot{s}\,\mathbf{t} + v^2\,\mathbf{C}. \qquad (19)$$

Applications to the hodograph, in particular motion in a circle, parabola, or under a central acceleration. Application to the proof of the theorem that the motion of a rigid body one point of which is fixed is an instantaneous rotation about an axis through the fixed point.

Integration with respect to a scalar is merely the inverse of differentiation. Application to finding the paths due to given accelerations.

The operator ∇ applied to a scalar function of position in space gives a vector whose direction is that of most rapid increase of that function and whose magnitude is equal to the rate of that increase per unit change of position in that direction

$$\nabla V = \mathbf{i}\,\frac{\partial V}{\partial x} + \mathbf{j}\,\frac{\partial V}{\partial y} + \mathbf{k}\,\frac{\partial V}{\partial z}, \qquad (21)$$

$$\nabla = \mathbf{i}\,\frac{\partial}{\partial x} + \mathbf{j}\,\frac{\partial}{\partial y} + \mathbf{k}\,\frac{\partial}{\partial z}. \qquad (22)$$

The operator ∇ is invariant of the axes **i, j, k.** It may be defined by the equation

$$\nabla V = \frac{dV}{dn}\mathbf{n}, \tag{24}$$

or

$$\nabla V \cdot d\mathbf{r} = dV. \tag{25)'$$

Computation of the derivative ∇V by two methods depending upon equations (21) and (25)′. Illustration of the occurrence of ∇ in mathematical physics.

∇ may be looked upon as a fictitious vector, a vector differentiator. It obeys the formal laws of vectors just in so far as the scalar differentiators of $\partial/\partial x$, $\partial/\partial y$, $\partial/\partial z$ obey the formal laws of scalar quantities

$$\mathbf{A} \cdot \nabla V = A_1 \frac{\partial V}{\partial x} + A_2 \frac{\partial V}{\partial y} + A_3 \frac{\partial V}{\partial z}. \tag{28}$$

If **a** be a unit vector $\mathbf{a} \cdot \nabla V$ is the directional derivative of V in the direction **a**.

$$\mathbf{a} \cdot \nabla V = (\mathbf{a} \cdot \nabla)\, V = \mathbf{a} \cdot (\nabla V). \tag{30}$$

If **V** is a vector function $\mathbf{a} \cdot \nabla \mathbf{V}$ is the directional derivative of that vector function in the direction **a**.

$$\nabla \cdot \mathbf{V} = \mathbf{i} \cdot \frac{\partial \mathbf{V}}{\partial x} + \mathbf{j} \cdot \frac{\partial \mathbf{V}}{\partial y} + \mathbf{k} \cdot \frac{\partial \mathbf{V}}{\partial z}, \tag{32)'$$

$$\nabla \times \mathbf{V} = \mathbf{i} \times \frac{\partial \mathbf{V}}{\partial x} + \mathbf{j} \times \frac{\partial \mathbf{V}}{\partial y} + \mathbf{k} \times \frac{\partial \mathbf{V}}{\partial z}, \tag{33)'$$

$$\nabla \cdot \mathbf{V} = \frac{\partial V_1}{\partial x} + \frac{\partial V_2}{\partial y} + \frac{\partial V_3}{\partial z}, \tag{32)''$$

$$\nabla \times \mathbf{V} = \mathbf{i}\left(\frac{\partial V_3}{\partial y} - \frac{\partial V_2}{\partial z}\right) + \mathbf{j}\left(\frac{\partial V_1}{\partial z} - \frac{\partial V_3}{\partial x}\right)$$

$$+ \mathbf{k}\left(\frac{\partial V_2}{\partial x} - \frac{\partial V_1}{\partial y}\right). \tag{33)''$$

Proof that $\nabla \cdot \mathbf{V}$ is the *divergence* of $\mathbf{V}$ and $\nabla \times \mathbf{V}$, the *curl* of $\mathbf{V}$.

$$\nabla \cdot \mathbf{V} = \text{div } \mathbf{V},$$

$$\nabla \times \mathbf{V} = \text{curl } \mathbf{V}.$$

$$\nabla (u + v) = \nabla u + \nabla v, \tag{35}$$

$$\nabla \cdot (\mathbf{u} + \mathbf{v}) = \nabla \cdot \mathbf{u} + \nabla \cdot \mathbf{v}, \tag{36}$$

$$\nabla \times (\mathbf{u} + \mathbf{v}) = \nabla \times \mathbf{u} + \nabla \times \mathbf{v}, \tag{37}$$

$$\nabla (u\, v) = v\, \nabla u + u\, \nabla v, \tag{38}$$

$$\nabla \cdot (u\, \mathbf{v}) = \nabla u \cdot \mathbf{v} + u\, \nabla \cdot \mathbf{v}, \tag{39}$$

$$\nabla \times (u\, \mathbf{v}) = \nabla u \times \mathbf{v} + u\, \nabla \times \mathbf{v}, \tag{40}$$

$$\nabla (\mathbf{u} \cdot \mathbf{v}) = \mathbf{v} \cdot \nabla \mathbf{u} + \mathbf{u} \cdot \nabla \mathbf{v} + \mathbf{v} \times (\nabla \times \mathbf{u}) + \mathbf{u} \times (\nabla \times \mathbf{v}), \tag{41}$$

$$\nabla \cdot (\mathbf{u} \times \mathbf{v}) = \mathbf{v} \;\; \nabla \times \mathbf{u} - \mathbf{u} \cdot \nabla \times \mathbf{v}, \tag{42}$$

$$\nabla \times (\mathbf{u} \times \mathbf{v}) = \mathbf{v} \cdot \nabla \mathbf{u} - \mathbf{v}\, \nabla \cdot \mathbf{u} - \mathbf{u} \cdot \nabla \mathbf{v} + \mathbf{u}\, \nabla \cdot \mathbf{v}. \tag{43}$$

Introduction of the partial *del*, $\nabla (\mathbf{u} \cdot \mathbf{v})_{\mathbf{u}}$, in which the differentiations are performed upon the hypothesis that $\mathbf{u}$ is constant.

$$\mathbf{u} \times (\nabla \times \mathbf{v}) = \nabla (\mathbf{u} \cdot \mathbf{v})_{\mathbf{u}} - \mathbf{u} \cdot \nabla \mathbf{v}. \tag{46}$$

If $\mathbf{a}$ be a unit vector the directional derivative

$$\mathbf{a} \cdot \nabla \mathbf{v} = \nabla (\mathbf{a} \cdot \mathbf{v})_{\mathbf{a}} + (\nabla \times \mathbf{v}) \times \mathbf{a}. \tag{47}$$

The expansion of any vector function $\mathbf{v}$ in the neighborhood of a point (x_0, y_0, z_0) at which it takes on the value of $\mathbf{v}_0$ is

$$\mathbf{v} = \mathbf{v}_0 + \nabla (d\,\mathbf{r} \cdot \mathbf{v})_{d\mathbf{r}} + (\nabla \times \mathbf{v}) \times d\,\mathbf{r}, \tag{49}$$

or $$\mathbf{v} = \tfrac{1}{2}\, \mathbf{v}_0 + \nabla (d\,\mathbf{r} \cdot \mathbf{v}) + \tfrac{1}{2} (\nabla \times \mathbf{v}) \times d\,\mathbf{r}. \tag{50}$$

Application to hydrodynamics.

The *dels* of the second order are six in number.

$$\nabla \times \nabla V = \text{curl}\, \nabla V = 0, \tag{52}$$

$$\nabla \cdot \nabla V \,\text{div}\, \nabla V = \frac{\partial^2 V}{\partial x^2} + \frac{\partial^2 V}{\partial y^2} + \frac{\partial^2 V}{\partial z^2}, \tag{51}$$

$\nabla \cdot \nabla$ is Laplace's operator. If $\nabla \cdot \nabla V = 0$, V satisfies Laplace's Equation. The operator may be applied to a vector.

$$\nabla \cdot \nabla \mathbf{V} = \frac{\partial^2 \mathbf{V}}{\partial x^2} + \frac{\partial^2 \mathbf{V}}{\partial y^2} + \frac{\partial^2 \mathbf{V}}{\partial z^2},$$

$$\nabla \nabla \cdot \mathbf{V} = \nabla \,\text{div}\, \mathbf{V}, \tag{55}$$

$$\nabla \cdot \nabla \times \mathbf{V} = \text{div curl}\, \mathbf{V} = 0, \tag{56}$$

$$\nabla \times \nabla \times \mathbf{V} = \text{curl curl}\, \mathbf{V} = \nabla \nabla \cdot \mathbf{V} - \nabla \cdot \nabla \mathbf{V}. \tag{58}$$

The geometric interpretation of $\nabla \cdot \nabla$ as giving the *dispersion* of a function.

Exercises on Chapter III

1. Given a particle moving in a plane curve. Let the plane be the $\mathbf{i}\,\mathbf{j}$-plane. Obtain the formulæ for the components of the velocity parallel and perpendicular to the radius vector $\mathbf{r}$. These are

$$\dot{r}\frac{\mathbf{r}}{r}, \quad \dot{\theta}\, \mathbf{k} \times \mathbf{r},$$

where θ is the angle the radius vector $\mathbf{r}$ makes with $\mathbf{i}$, and $\mathbf{k}$ is the normal to the plane.

2. Obtain the accelerations of the particle parallel and perpendicular to the radius vector. These are

$$(\ddot{r} - r\,\dot{\theta}^2)\frac{\mathbf{r}}{r}, \quad (r\,\ddot{\theta} + 2\,\dot{r}\,\dot{\theta})\,\mathbf{k} \times \frac{\mathbf{r}}{r}.$$

Express these formulæ in the usual manner in terms of x and y.

3. Obtain the accelerations of a moving particle parallel and perpendicular to the tangent to the path and reduce the results to the usual form.

4. If r, ϕ, θ be a system of polar coördinates in space, where r is the distance of a point from the origin, ϕ the meridianal angle, and θ the polar angle; obtain the expressions for the components of the velocity and acceleration along the radius vector, a meridian, and a parallel of latitude. Reduce these expressions to the ordinary form in terms of x, y, z.

5. Show by the direct method suggested in Art. 63 that the operator ∇ is independent of the axes.

6. By the second method given for computing ∇ find the derivative ∇ of a triple product $[\mathbf{a}\,\mathbf{b}\,\mathbf{c}]$ each term of which is a function of x, y, z in case

$$\mathbf{a} = (\mathbf{r} \cdot \mathbf{r})\, \mathbf{r}, \qquad \mathbf{b} = (\mathbf{r} \cdot \mathbf{a})\, \mathbf{e}, \qquad \mathbf{c} = \mathbf{r} \times \mathbf{f},$$

where $\mathbf{d}$, $\mathbf{e}$, $\mathbf{f}$ are constant vectors.

7. Compute $\nabla \cdot \nabla\, V$ when V is r^2, $\quad r$, $\quad \frac{1}{r}$, $\quad$ or $\frac{1}{r^2}$.

8. Compute $\nabla \cdot \nabla \mathbf{V}$, $\nabla \nabla \cdot \mathbf{V}$, and $\nabla \times \nabla \times \mathbf{V}$ when $\mathbf{V}$ is equal to $\mathbf{r}$ and when $\mathbf{V}$ is equal to $\frac{\mathbf{r}}{r^3}$, and show that in these cases the formula (58) holds.

9. Expand $\nabla \times \nabla V$ and $\nabla \cdot \nabla \times \mathbf{V}$ in terms of $\mathbf{i}, \mathbf{j}, \mathbf{k}$ and show that they vanish (Art. 77).

10. Show by expanding in terms of $\mathbf{i}$, $\mathbf{j}$, $\mathbf{k}$ that

$$\nabla \times \nabla \times \mathbf{V} = \nabla \nabla \cdot \mathbf{V} - \nabla \cdot \nabla \mathbf{V}.$$

11. Prove $\quad \mathbf{A} \cdot \nabla\, (\mathbf{V} \cdot \mathbf{W}) = \mathbf{V}\, \mathbf{A} \cdot \nabla \mathbf{W} + \mathbf{W}\, \mathbf{A} \cdot \nabla \mathbf{V},$

and

$$(\nabla \times \mathbf{V}) \times \mathbf{W} = \nabla \times (\mathbf{V} \times \mathbf{W})_{\mathbf{W}} + \mathbf{W}\, \nabla \cdot \mathbf{V} - \nabla\, (\mathbf{V} \cdot \mathbf{W})_{\mathbf{W}}.$$

CHAPTER IV

THE INTEGRAL CALCULUS OF VECTORS

79.] Let $\mathbf{W}(x, y, z)$ be a vector function of position in space. Let C be any curve in space, and $\mathbf{r}$ the radius vector drawn from some fixed origin to the points of the curve. Divide the curve into infinitesimal elements $d\,\mathbf{r}$. From the sum of the scalar product of these elements $d\,\mathbf{r}$ and the value of the function $\mathbf{W}$ at some point of the element —

thus
$$\Sigma\, \mathbf{W} \cdot d\,\mathbf{r}.$$

The limit of this sum when the elements $d\mathbf{r}$ become infinite in number, each approaching zero, is called the *line integral* of $\mathbf{W}$ along the curve C and is written

$$\int_C \mathbf{W} \cdot d\,\mathbf{r}.$$

If
$$\mathbf{W} = W_1\,\mathbf{i} + W_2\,\mathbf{j} + W_3\,\mathbf{k},$$

and
$$d\,\mathbf{r} = \mathbf{i}\,d\,x + \mathbf{j}\,d\,y + \mathbf{k}\,d\,z,$$

$$\int_C \mathbf{W} \cdot d\,\mathbf{r} = \int_C [W_1\,d\,x + W_2\,d\,y + W_3\,d\,z]. \qquad (1)$$

The definition of the line integral therefore coincides with the definition usually given. It is however necessary to specify in which direction the radius vector $\mathbf{r}$ is supposed to describe the curve during the integration. For the elements $d\,\mathbf{r}$ have opposite signs when the curve is described in opposite directions. If one method of description be denoted by C and the other by $-C$,

$$\int_{-C} \mathbf{W} \cdot d\,\mathbf{r} = -\int_{C} \mathbf{W} \cdot d\,\mathbf{r}.$$

In case the curve C is a closed curve bounding a portion of surface the curve will always be regarded as described in such a direction that the enclosed area appears positive (Art. 25).

If $\mathbf{f}$ denote the force which may be supposed to vary from point to point along the curve C, the work done by the force when its point of application is moved from the initial point $\mathbf{r}_0$ of the curve C to its final point $\mathbf{r}$ is the line integral

$$\int_{C} \mathbf{f} \cdot d\,\mathbf{r} = \int_{\mathbf{r}_0}^{\mathbf{r}} \mathbf{f} \cdot d\,\mathbf{r}.$$

Theorem: The line integral of the derivative ∇V of a scalar function $V(x, y, z)$ along any curve from the point $\mathbf{r}_0$ to the point $\mathbf{r}$ is equal to the difference between the values of the function $V(x, y, z)$ at the point $\mathbf{r}$ and at the point $\mathbf{r}_0$. That is,

$$\int_{\mathbf{r}_0}^{\mathbf{r}} \nabla V \cdot d\,\mathbf{r} = V(\mathbf{r}) - V(\mathbf{r}_0) = V(x, y, z) - V(x_0, y_0, z_0).$$

By definition $\qquad d\,\mathbf{r} \cdot \nabla V = d\,V$

$$\int_{\mathbf{r}_0}^{\mathbf{r}} d\,V = V(\mathbf{r}) - V(\mathbf{r}_0) = V(x, y, z) - V(x_0, y_0, z_0). \quad (2)$$

Theorem: The line integral of the derivative ∇V of a single valued scalar function of position V taken around a *closed* curve vanishes.

The fact that the integral is taken around a closed curve is denoted by writing a circle at the foot of the integral sign. To show

$$\int_{\circ} \nabla V \cdot d\,\mathbf{r} = 0. \quad (3)$$

The initial point $\mathbf{r}_0$ and the final point $\mathbf{r}$ coincide. Hence

$$V(\mathbf{r}) = V(\mathbf{r}_0).$$

Hence by (2)

$$\int_{\bigcirc} \nabla V \cdot d\mathbf{r} = 0.$$

Theorem: Conversely if the line integral of $\mathbf{W}$ about every closed curve vanishes, $\mathbf{W}$ is the derivative of some scalar function $V(x, y, z)$ of position in space.

Given

$$\int_{\bigcirc} \mathbf{W} \cdot d\mathbf{r} = 0.$$

To show

$$\mathbf{W} = \nabla V.$$

Let $\mathbf{r}_0$ be any fixed point in space and $\mathbf{r}$ a variable point. The line integral

$$\int_{\mathbf{r}_0}^{\mathbf{r}} \mathbf{W} \cdot d\mathbf{r}$$

is independent of the path of integration C. For let any two paths C and C' be drawn between $\mathbf{r}_0$ and $\mathbf{r}$. The curve which consists of the path C from $\mathbf{r}_0$ to $\mathbf{r}$ and the path $-C'$ from $\mathbf{r}$ to $\mathbf{r}_0$ is a closed curve. Hence by hypothesis

$$\int_C \mathbf{W} \cdot d\mathbf{r} + \int_{-C'} \mathbf{W} \cdot d\mathbf{r} = 0,$$

$$\int_{-C'} \mathbf{W} \cdot d\mathbf{r} = - \int_{C'} \mathbf{W} \cdot d\mathbf{r}.$$

Hence

$$\int_C \mathbf{W} \cdot d\mathbf{r} = \int_{C'} \mathbf{W} \cdot d\mathbf{r}.$$

Hence the value of the integral is independent of the path of integration and depends only upon the final point $\mathbf{r}$.

The value of the integral is therefore a scalar function of the position of the point $\mathbf{r}$ whose coördinates are x, y, z.

$$\int_{\mathbf{r}_0}^{\mathbf{r}} \mathbf{W} \cdot d\,\mathbf{r} = V\,(x, y, z).$$

Let the integral be taken between two points infinitely near together.

$$\mathbf{W} \cdot d\,\mathbf{r} = d\,V\,(x, y, z).$$

But by definition $$\nabla V \cdot d\,\mathbf{r} = d\,V.$$

Hence $$\mathbf{W} = \nabla V.$$

The theorem is therefore demonstrated.

80.] Let $\mathbf{f}$ be the force which acts upon a unit mass near the surface of the earth under the influence of gravity. Let a system of axes $\mathbf{i}$, $\mathbf{j}$, $\mathbf{k}$ be chosen so that $\mathbf{k}$ is vertical. Then

$$\mathbf{f} = -\,g\,\mathbf{k}.$$

The work done by the force when its point of application moves from the position $\mathbf{r}_0$ to the position $\mathbf{r}$ is

$$w = \int_{\mathbf{r}_0}^{\mathbf{r}} \mathbf{f} \cdot d\,\mathbf{r} = \int_{\mathbf{r}_0}^{\mathbf{r}} -\,g\;\mathbf{k} \cdot d\,\mathbf{r} = -\int_{\mathbf{r}_0}^{\mathbf{r}} g\,dz.$$

Hence $$w = -\,g\,(z - z_0) = g\,(z_0 - z).$$

The force $\mathbf{f}$ is said to be derivable from a *force-function* V when there exists a scalar function of position V such that the force is equal at each point of the derivative ∇V. Evidently if V is one force-function, another may be obtained by adding to V any arbitrary constant. In the above example the force-function is

$$V = w = g\,(z_0 - z).$$

Or more simply $$V = -\,g\,z.$$

The force is $$\mathbf{f} = \nabla V = -\,g\,\mathbf{k}.$$

The necessary and sufficient condition that a force-function $V(x, y, z)$ exist, is that the work done by the force when its point of application moves around a closed circuit be zero.

The work done by the force is

$$w = \int \mathbf{f} \cdot d\,\mathbf{r}.$$

If this integral vanishes when taken around every closed contour

$$\mathbf{f} = \nabla V = \nabla w.$$

And conversely if

$$\mathbf{f} = \nabla V$$

the integral vanishes. The force-function and the work done differ only by a constant.

$$V = w + \text{const.}$$

In case there is *friction* no force-function can exist. For the work done by friction when a particle is moved around in a closed circuit is never zero.

The force of attraction exerted by a fixed mass M upon a unit mass is directed toward the fixed mass and is proportional to the inverse square of the distance between the masses.

$$\mathbf{f} = -c \frac{M}{r^3} \mathbf{r}.$$

This is the law of universal gravitation as stated by Newton. It is easy to see that this force is derivable from a force-function V. Choose the origin of coördinates at the center of the attracting mass M. Then the work done is

$$w = -\int_{\mathbf{r}_0}^{\mathbf{r}} c \frac{M}{r^3} \mathbf{r} \cdot d\,\mathbf{r}.$$

But

$$\mathbf{r} \cdot d\,\mathbf{r} = r\,d\,r,$$

$$w = -c\,M \int_{r_0}^{r} \frac{d\,r}{r^2} = -c\,M \left\{ \frac{1}{r} - \frac{1}{r_0} \right\}.$$

By a proper choice of units the constant c may be made equal to unity. The force-function V may therefore be chosen as

$$V = -\frac{M}{r}.$$

If there had been several attracting bodies $M_1, M_2, M_3, \cdots$ the force-function would have been

$$V = -\left\{\frac{M_1}{r_1} + \frac{M_2}{r_2} + \frac{M_3}{r_3} + \cdots\right\}$$

where $r_1, r_2, r_3, \cdots$ are the distances of the attracted unit mass from the attracting masses $M_1, M_2, M_3 \cdots$

The law of the conservation of mechanical energy requires that the work done by the forces when a point is moved around a closed curve shall be zero. This is on the assumption that none of the mechanical energy has been converted into other forms of energy during the motion. The law of conservation of energy therefore requires the forces to be derivable from a force-function. Conversely if a force-function exists the work done by the forces when a point is carried around a closed curve is zero and consequently there is no loss of energy. A mechanical system for which a force-function exists is called a *conservative* system. From the example just cited above it is clear that bodies moving under the law of universal gravitation form a conservative system — at least so long as they do not collide.

81.] Let $\mathbf{W}(x, y, z)$ be any vector function of position in space. Let S be any surface. Divide this surface into infinitesimal elements. These elements may be regarded as plane and may be represented by infinitesimal vectors of which the direction is at each point the direction of the normal to the surface at that point and of which the magnitude is equal to the magnitude of the area of the infinitesimal element. Let this infinitesimal vector which represents the element of surface in magnitude and direction be denoted by $d\mathbf{a}$. Form the sum

$$\Sigma \mathbf{W} \cdot d\mathbf{a},$$

which is the sum of the scalar products of the value of **W** at each element of surface and the (vector) element of surface. The limit of this sum when the elements of surface approach zero is called the *surface integral* of **W** over the surface S, and is written

$$\iint_S \mathbf{W} \cdot d\mathbf{a}. \tag{4}$$

The value of the integral is scalar. If **W** and $d\mathbf{a}$ be expressed in terms of their three components parallel to **i**, **j**, **k**

$$\mathbf{W} = W_1 \mathbf{i} + W_2 \mathbf{j} + W_3 \mathbf{k},$$

$$d\mathbf{a} = (d\mathbf{a} \cdot \mathbf{i})\, \mathbf{i} + (d\mathbf{a} \cdot \mathbf{j})\, \mathbf{j} + (d\mathbf{a} \cdot \mathbf{k})\, \mathbf{k},$$

or

$$d\mathbf{a} = dy\, dz\, \mathbf{i} + dz\, dx\, \mathbf{j} + dx\, dy\, \mathbf{k},$$

$$\iint_S \mathbf{W} \cdot d\mathbf{a} = \iint [W_1\, dy\, dz + W_2\, dz\, dx + W_3\, dx\, dy]. \tag{5}$$

The surface integral therefore has been defined as is customary in ordinary analysis. It is however necessary to determine with the greatest care which normal to the surface $d\mathbf{a}$ is. That is, which *side* of the surface (so to speak) the integral is taken over. For the normals upon the two sides are the negatives of each other. Hence the surface integrals taken over the two sides will differ in sign. In case the surface be looked upon as bounding a portion of space $d\mathbf{a}$ is always considered to be the exterior normal.

If **f** denote the flux of any substance the surface integral

$$\iint_S \mathbf{f} \cdot d\mathbf{a}$$

gives the amount of that substance which is passing through the surface per unit time. It was seen before (Art. 71) that the rate at which matter was leaving a point per unit volume per unit time was $\nabla \cdot \mathbf{f}$. The total amount of matter which leaves a closed space bounded by a surface S per unit time is the ordinary triple integral

$$\iiint \nabla \cdot \mathbf{f} \, dv. \tag{6}$$

Hence the very important relation connecting a surface integral of a flux taken over a *closed* surface and the volume integral of the divergence of the flux taken over the space enclosed by the surface —

$$\iint_S \mathbf{f} \cdot d\,\mathbf{a} = \iiint \nabla \cdot \mathbf{f} \, dv. \tag{7}$$

Written out in the notation of the ordinary calculus this becomes

$$\iint [X \, dy \, dz + Y \, dz \, dx + Z \, dx \, dy]$$

$$= \iiint \left(\frac{\partial X}{\partial x} + \frac{\partial Y}{\partial y} + \frac{\partial Z}{\partial z} \right) dx \, dy \, dz \tag{8}$$

where X, Y, Z are the three components of the flux $\mathbf{f}$. The theorem is perhaps still more familiar when each of the three components is treated separately.

$$\iint X \, dx \, dy = \iiint \frac{\partial X}{\partial x} \, dx \, dy \, dz. \tag{8$'$}$$

This is known as *Gauss's Theorem.* It states that the surface integral (taken over a closed surface) of the product of a function X and the cosine of the angle which the exterior normal to that surface makes with the X-axis is equal to the volume integral of the partial derivative of that function with respect to x taken throughout the volume enclosed by that surface.

If the surface S be the surface bounding an infinitesimal sphere or cube

$$\iint_S \mathbf{f} \cdot d\,\mathbf{a} = \nabla \cdot \mathbf{f}\, d\,v$$

where $d\,v$ is the volume of that sphere or cube. Hence

$$\nabla \cdot \mathbf{f} = \frac{1}{d\,v} \iint_S \mathbf{f} \cdot d\,\mathbf{a}. \tag{9}$$

This equation may be taken as a *definition* of the divergence $\nabla \cdot \mathbf{f}$. The divergence of a vector function $\mathbf{f}$ is equal to the limit approached by the surface integral of $\mathbf{f}$ taken over a surface bounding an infinitesimal body divided by that volume when the volume approaches zero as its limit. That is

$$\nabla \cdot \mathbf{f} = \mathop{\mathrm{Lim}}_{d\,v \doteq 0} \frac{1}{d\,v} \iint_S \mathbf{f} \cdot d\,\mathbf{a}. \tag{10}$$

From this definition which is evidently independent of the axes all the properties of the divergence may be deduced. In order to make use of this definition it is necessary to develop at least the elements of the integral calculus of vectors before the differentiating operators can be treated. This definition of $\nabla \cdot \mathbf{f}$ consequently is interesting more from a theoretical than from a practical standpoint.

82.] *Theorem:* The surface integral of the curl of a vector function is equal to the line integral of that vector function taken around the closed curve bounding that surface.

$$\iint_S \nabla \times \mathbf{W} \cdot d\,\mathbf{a} = \int_{\mathrm{O}} \mathbf{W} \cdot d\,\mathbf{r}. \tag{11}$$

This is the celebrated theorem of *Stokes.* On account of its great importance in all branches of mathematical physics a number of different proofs will be given.

First Proof: Consider a small triangle *123* upon the surface S (Fig. 32). Let the value of **W** at the vertex *1* be $\mathbf{W}_0$. Then by (50), Chap. III., the value at any neighboring point is

$$\mathbf{W} = \frac{1}{2}\left\{ \mathbf{W}_0 + \nabla (\mathbf{W} \cdot \delta \mathbf{r}) + (\nabla \times \mathbf{W}) \times \delta \mathbf{r} \right\},$$

where the symbol $\delta \mathbf{r}$ has been introduced for the sake of distinguishing it from $d\,\mathbf{r}$ which is to be used as the element of integration. The integral of **W** taken around the triangle *123* is

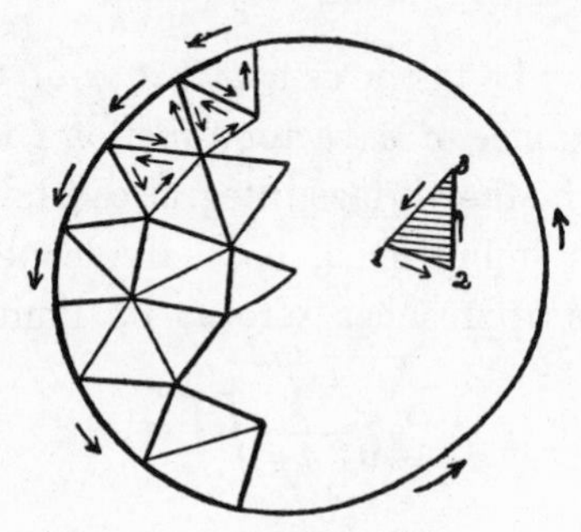

FIG. 32.

$$\int_\Delta \mathbf{W} \cdot d\,\mathbf{r} = \frac{1}{2} \int_\Delta \mathbf{W}_0 \cdot d\,\mathbf{r} + \frac{1}{2} \int_\Delta \nabla (\mathbf{W} \cdot \delta \mathbf{r}) \cdot d\,\mathbf{r}$$

$$+ \frac{1}{2} \int_\Delta (\nabla \times \mathbf{W}) \times \delta \mathbf{r} \cdot d\,\mathbf{r}.$$

The first term $\quad \frac{1}{2} \int_\Delta \mathbf{W}_0 \cdot d\,\mathbf{r} = \frac{1}{2} \mathbf{W}_0 \cdot \int_\Delta d\,\mathbf{r}$

vanishes because the integral of $d\,\mathbf{r}$ around a closed figure, in this case a small triangle, is zero. The second term

$$\frac{1}{2} \int_\Delta \nabla (\mathbf{W} \cdot \delta \mathbf{r}) \cdot d\,\mathbf{r}$$

vanishes by virtue of (3) page 180. Hence

$$\int_\Delta \mathbf{W} \cdot d\,\mathbf{r} = \tfrac{1}{2} \int_\Delta \nabla \times \mathbf{W} \times \delta\,\mathbf{r} \cdot d\,\mathbf{r}.$$

Interchange the dot and the cross in this triple product.

$$\int_\Delta \mathbf{W} \cdot d\,\mathbf{r} = \tfrac{1}{2} \int_\Delta \nabla \times \mathbf{W} \cdot \delta\,\mathbf{r} \times d\,\mathbf{r}.$$

When $d\,\mathbf{r}$ is equal to the side *12* of the triangle, $\delta\,\mathbf{r}$ is also equal to this side. Hence the product

$$\delta\,\mathbf{r} \times d\,\mathbf{r}$$

vanishes because $\delta\,\mathbf{r}$ and $d\,\mathbf{r}$ are collinear. In like manner when $d\,\mathbf{r}$ is the side *31*, $\delta\,\mathbf{r}$ is the same side *13*, but taken in the opposite direction. Hence the vector product vanishes. When $d\,\mathbf{r}$ is the side *23*, $\delta\,\mathbf{r}$ is a line drawn from the vertex *1* at which $\mathbf{W} = \mathbf{W}_0$ to this side *23*. Hence the product $\delta\,\mathbf{r} \times d\,\mathbf{r}$ is twice the area of the triangle. This area, moreover, is the positive area *123*. Hence

$$\tfrac{1}{2}\, \delta\,\mathbf{r} \times d\,\mathbf{r} = d\,\mathbf{a},$$

where $d\,\mathbf{a}$ denotes the positive area of the triangular element of surface. For the infinitesimal triangle therefore the relation

$$\int_\Delta \mathbf{W} \cdot d\,\mathbf{r} = \nabla \times \mathbf{W} \cdot d\,\mathbf{a}$$

holds.

Let the surface S be divided into elementary triangles. For convenience let the curve which bounds the surface be made up of the sides of these triangles. Perform the integration

$$\int_\Delta \mathbf{W} \cdot d\,\mathbf{r}$$

around each of these triangles and add the results together.

$$\sum_S \int_\Delta \mathbf{W} \cdot d\,\mathbf{r} = \sum_S \nabla \times \mathbf{W} \cdot d\,\mathbf{a}.$$

The second member $\sum_S \nabla \times \mathbf{W} \cdot d\,\mathbf{a}$

is the surface integral of the curl of **W**.

$$\sum_S \nabla \times \mathbf{W} \cdot d\,\mathbf{a} = \iint_S \nabla \times \mathbf{W} \cdot d\,\mathbf{a}.$$

In adding together the line integrals which occur in the first member it is necessary to notice that all the sides of the elementary triangles except those which lie along the bounding curve of the surface are traced twice in *opposite* directions. Hence all the terms in the sum

$$\sum_S \int_\Delta \mathbf{W} \cdot d\,\mathbf{r}$$

which arise from those sides of the triangles lying *within* the surface S cancel out, leaving in the sum only the terms which arise from those sides which make up the bounding curve of the surface. Hence the sum reduces to the line integral of **W** along the curve which bounds the surface S.

$$\sum_S \int_\Delta \mathbf{W} \cdot d\,\mathbf{r} = \int_\bigcirc \mathbf{W} \cdot d\,\mathbf{r}.$$

Hence
$$\iint_S \nabla \times \mathbf{W} \cdot d\,\mathbf{a} = \int_\bigcirc \mathbf{W} \cdot d\,\mathbf{r}. \tag{11}$$

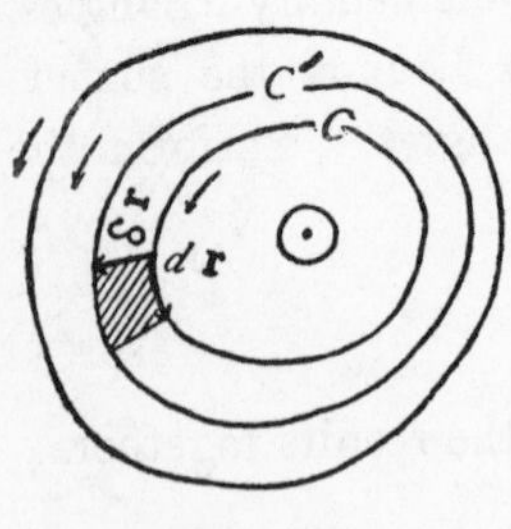

FIG. 33.

Second Proof: Let C be any closed contour drawn upon the surface S (Fig. 33). It will be assumed that C is continuous and does not cut itself. Let C' be another such contour near to C. Consider the variation δ which takes place in the line integral of **W** in passing from the contour C to the contour C'.

$$\delta \int \mathbf{W} \cdot d\mathbf{r} = \int_{C'} \mathbf{W} \cdot d\mathbf{r} - \int_{C} \mathbf{W} \cdot d\mathbf{r},$$

$$\delta \int \mathbf{W} \cdot d\mathbf{r} = \int \delta (\mathbf{W} \cdot d\mathbf{r}) = \int \mathbf{W} \cdot \delta\, d\mathbf{r} + \int \delta \mathbf{W} \cdot d\mathbf{r}.$$

But $$d(\mathbf{W} \cdot \delta \mathbf{r}) = d\mathbf{W} \cdot \delta \mathbf{r} + \mathbf{W} \cdot d\, \delta \mathbf{r}$$

and $$\delta\, d\mathbf{r} = d\, \delta \mathbf{r}.$$

Hence $$\int \mathbf{W} \cdot \delta\, d\mathbf{r} = \int \mathbf{W} \cdot d\, \delta \mathbf{r} = \int d(\mathbf{W} \cdot \delta \mathbf{r}) - \int d\mathbf{W} \cdot \delta \mathbf{r}.$$

The expression $d(\mathbf{W} \cdot \delta \mathbf{r})$ is by its form a perfect differential. The value of the integral of that expression will therefore be the difference between the values of $\mathbf{W} \cdot d\mathbf{r}$ at the end and at the beginning of the path of integration. In this case the integral is taken around the closed contour C. Hence

$$\int_C d(\mathbf{W} \cdot \delta \mathbf{r}) = 0.$$

Hence $$\int \mathbf{W} \cdot \delta\, d\mathbf{r} = -\int d\mathbf{W} \cdot \delta \mathbf{r},$$

and $$\delta \int \mathbf{W} \cdot d\mathbf{r} = \int \delta \mathbf{W} \cdot d\mathbf{r} - \int d\mathbf{W} \cdot \delta \mathbf{r},$$

$$\delta \int \mathbf{W} \cdot d\mathbf{r} = \int \left\{ \delta \mathbf{W} \cdot d\mathbf{r} - d\mathbf{W} \cdot \delta \mathbf{r} \right\}.$$

But $$d\mathbf{W} = \frac{\partial \mathbf{W}}{\partial x} dx + \frac{\partial \mathbf{W}}{\partial y} dy + \frac{\partial \mathbf{W}}{\partial z} dz,$$

or $$d\mathbf{W} = \frac{\partial \mathbf{W}}{\partial x} \mathbf{i} \cdot d\mathbf{r} + \frac{\partial \mathbf{W}}{\partial y} \mathbf{j} \cdot d\mathbf{r} + \frac{\partial \mathbf{W}}{\partial z} \mathbf{k} \cdot d\mathbf{r},$$

and $$\delta \mathbf{W} = \frac{\partial \mathbf{W}}{\partial x} \mathbf{i} \cdot \delta \mathbf{r} + \frac{\partial \mathbf{W}}{\partial y} \mathbf{j} \cdot \delta \mathbf{r} + \frac{\partial \mathbf{W}}{\partial z} \mathbf{k} \cdot \delta \mathbf{r}.$$

Substituting these values

$$\delta \int \mathbf{W} \cdot d\,\mathbf{r} = \int \left\{ \frac{\partial \mathbf{W}}{\partial x} \cdot d\,\mathbf{r}\ \ \mathbf{i} \cdot \delta\,\mathbf{r} - \frac{\partial \mathbf{W}}{\partial x} \cdot \delta\,\mathbf{r}\,\mathbf{i} \cdot d\mathbf{r} \right.$$

$$\left. + \text{ similar terms in } y \text{ and } z. \right\}.$$

But by (25) page 111

$$\left(\mathbf{i} \times \frac{\partial \mathbf{W}}{\partial x}\right) \cdot (\delta\,\mathbf{r} \times d\,\mathbf{r}) = \frac{\partial \mathbf{W}}{\partial x} \cdot d\,\mathbf{r}\ \ \mathbf{i} \cdot \delta\,\mathbf{r} - \frac{\partial \mathbf{W}}{\partial x} \cdot \delta\,\mathbf{r}\ \mathbf{i} \cdot d\,\mathbf{r}.$$

Hence
$$\delta \int \mathbf{W} \cdot d\,\mathbf{r} = \int \left\{ \mathbf{i} \times \frac{\partial \mathbf{W}}{\partial x} \cdot \delta\,\mathbf{r} \times d\,\mathbf{r} \right.$$

$$\left. + \text{ similar terms in } y \text{ and } z \right\}.$$

or
$$\delta \int \mathbf{W} \cdot d\,\mathbf{r} = \int \nabla \times \mathbf{W} \cdot \delta\,\mathbf{r} \times d\,\mathbf{r}.$$

In Fig. 33 it will be seen that $d\,\mathbf{r}$ is the element of arc along the curve C and $\delta\,\mathbf{r}$ is the distance from the curve C to the curve C'. Hence $\delta\,\mathbf{r} \times d\,\mathbf{r}$ is equal to the area of an elementary parallelogram included between C and C' upon the surface S. That is

$$\delta\,\mathbf{r} \times d\,\mathbf{r} = d\,\mathbf{a},$$

$$\delta \int \mathbf{W} \cdot d\,\mathbf{r} = \int \nabla \times \mathbf{W} \cdot d\,\mathbf{a}.$$

Let the curve C starting at a point O in S expand until it coincides with the contour bounding S. The line integral

$$\int \mathbf{W} \cdot d\,\mathbf{r}$$

will vary from the value 0 at the point O to the value

$$\int_{\circ} \mathbf{W} \cdot d\,\mathbf{r}$$

taken around the contour which bounds the surface S. This total variation of the integral will be equal to the sum of the variations δ

$$\sum \delta \int \mathbf{W} \cdot d\,\mathbf{r} = \sum \int \nabla \times \mathbf{W} \cdot d\,\mathbf{a}.$$

Or

$$\int_{\circ} \mathbf{W} \cdot d\,\mathbf{r} = \iint_{s} \nabla \times \mathbf{W} \cdot d\,\mathbf{a}. \qquad (11)$$

83.] Stokes's theorem that the surface integral of the curl of a vector function is equal to the line integral of the function taken along the closed curve which bounds the surface has been proved. The converse is also true. *If the surface integral of a vector function* **U** *is equal to the line integral of the function* **W** *taken around the curve bounding the surface and if this relation holds for all surfaces in space, then* **U** *is the curl of* **W**. That is

$$\text{if} \quad \iint_{s} \mathbf{U} \cdot d\,\mathbf{a} = \int_{\circ} \mathbf{W} \cdot d\,\mathbf{r}, \text{ then } \mathbf{U} = \nabla \times \mathbf{W}. \qquad (12)$$

Form the surface integral of the difference between **U** and $\nabla \times \mathbf{W}$.

$$\iint_{s} (\mathbf{U} - \nabla \times \mathbf{W}) \cdot d\,\mathbf{a} = \int_{\circ} \mathbf{W} \cdot d\,\mathbf{r} - \int_{\circ} \mathbf{W} \cdot d\,\mathbf{r} = 0,$$

or

$$\iint_{s} (\mathbf{U} - \nabla \times \mathbf{W}) \cdot d\,\mathbf{a} = 0.$$

Let the surface S over which the integration is performed be infinitesimal. The integral reduces to merely a single term

$$(\mathbf{U} - \nabla \times \mathbf{W}) \cdot d\,\mathbf{a} = 0.$$

As this equation holds for *any* element of surface $d\,\mathbf{a}$, the first factor vanishes. Hence

$$\mathbf{U} - \nabla \times \mathbf{W} = 0.$$

Hence

$$\mathbf{U} = \nabla \times \mathbf{W}.$$

The converse is therefore demonstrated.

A definition of $\nabla \times \mathbf{W}$ which is independent of the axes $\mathbf{i}, \mathbf{j}, \mathbf{k}$ may be obtained by applying Stokes's theorem to an infinitesimal plane area. Consider a point P. Pass a plane through P and draw in it, concentric with P, a small circle of area $d\,\mathbf{a}$.

$$\nabla \times \mathbf{W} \cdot d\,\mathbf{a} = \int_{\circ} \mathbf{W} \cdot d\,\mathbf{r}. \qquad (13)$$

When $d\,\mathbf{a}$ has the same direction as $\nabla \times \mathbf{W}$ the value of the line integral will be a maximum, for the cosine of the angle between $\nabla \times \mathbf{W}$ and $d\,\mathbf{a}$ will be equal to unity. For this value of $d\,\mathbf{a}$,

$$\nabla \times \mathbf{W} = \underset{d\,\mathbf{a} \doteq 0}{\mathrm{Lim}} \left[\frac{d\,\mathbf{a}}{d\,\mathbf{a} \cdot d\,\mathbf{a}} \int_{\circ} \mathbf{W} \cdot d\,\mathbf{r} \right]. \qquad (13)'$$

Hence the curl $\nabla \times \mathbf{W}$ of a vector function $\mathbf{W}$ has at each point of space the direction of the normal to that plane in which the line integral of $\mathbf{W}$ taken about a small circle concentric with the point in question is a maximum. The magnitude of the curl at the point is equal to the magnitude of that line integral of maximum value divided by the area of the circle about which it is taken. This definition like the one given in Art. 81 for the divergence is interesting more from theoretical than from practical considerations.

Stokes's theorem or rather its converse may be used to deduce Maxwell's equations of the electro-magnetic field in a simple manner. Let $\mathbf{E}$ be the electric force, $\mathbf{B}$ the magnetic induction, $\mathbf{H}$ the magnetic force, and $\mathbf{C}$ the flux of electricity per unit area per unit time (*i. e.* the current density).

It is a fact learned from experiment that the total electromotive force around a closed circuit is equal to the negative of the rate of change of total magnetic induction through the circuit. The total electromotive force is the line integral of the electric force taken around the circuit. That is

$$\int_{\circ} \mathbf{E} \cdot d\,\mathbf{r}.$$

The total magnetic induction through the circuit is the surface integral of the magnetic induction $\mathbf{B}$ taken over a surface bounded by the circuit. That is

$$\iint_s \mathbf{B} \cdot d\,\mathbf{a}.$$

Experiment therefore shows that

$$\int_\circ \mathbf{E} \cdot d\,\mathbf{r} = -\frac{d}{d\,t} \iint_s \mathbf{B} \cdot d\,\mathbf{a},$$

or

$$\int_\circ \mathbf{E} \cdot d\,\mathbf{r} = \iint_s -\dot{\mathbf{B}} \cdot d\,\mathbf{a}.$$

Hence by the converse of Stokes's theorem

$$\nabla \times \mathbf{E} = -\dot{\mathbf{B}}, \quad \text{curl}\, \mathbf{E} = -\dot{\mathbf{B}}.$$

It is also a fact of experiment that the work done in carrying a unit positive magnetic pole around a closed circuit is equal to 4π times the total electric flux through the circuit. The work done in carrying a unit pole around a circuit is the line integral of $\mathbf{H}$ around the circuit. That is

$$\int_\circ \mathbf{H} \cdot d\,\mathbf{r}.$$

The total flux of electricity through the circuit is the surface integral of $\mathbf{C}$ taken over a surface bounded by the circuit. That is

$$\iint_s \mathbf{C} \cdot d\,\mathbf{a}.$$

Experiment therefore teaches that

$$\int_\circ \mathbf{H} \cdot d\,\mathbf{r} = 4\pi \iint_s \mathbf{C} \cdot d\,\mathbf{a}.$$

By the converse of Stokes's theorem

$$\nabla \times \mathbf{H} = 4\,\pi\,\mathbf{C}.$$

With a proper interpretation of the current $\mathbf{C}$, as the displacement current in addition to the conduction current, an interpretation depending upon one of Maxwell's primary hypotheses, this relation and the preceding one are the fundamental equations of Maxwell's theory, in the form used by Heaviside and Hertz.

The theorems of Stokes and Gauss may be used to demonstrate the identities.

$$\nabla \cdot \nabla \times \mathbf{W} = 0, \qquad \text{div curl } \mathbf{W} = 0.$$

$$\nabla \times \nabla V = 0, \qquad \text{curl } \nabla V = 0.$$

According to Gauss's theorem

$$\iiint \nabla \cdot \nabla \times \mathbf{W}\, d\,v = \iint_s \nabla \times \mathbf{W} \cdot d\,\mathbf{a}.$$

According to Stokes's theorem

$$\iint_s \nabla \times \mathbf{W} \cdot d\,\mathbf{a} = \int_0 \mathbf{W} \cdot d\,\mathbf{r}.$$

Hence $$\iiint \nabla \cdot \nabla \times \mathbf{W}\, d\,v = \int_0 \mathbf{W} \cdot d\,\mathbf{r}.$$

Apply this to an infinitesimal sphere. The surface bounding the sphere is closed. Hence its bounding curve reduces to a point; and the integral around it, to zero.

$$\nabla \cdot \nabla \times \mathbf{W}\, d\,v = \int_0 \mathbf{W} \cdot d\,\mathbf{r} = 0,$$

$$\nabla \cdot \nabla \times \mathbf{W} = 0.$$

Again according to Stokes's theorem

$$\iint_s \nabla \times \nabla V \cdot d\mathbf{a} = \int_\circ \nabla V \cdot d\mathbf{r}.$$

Apply this to any infinitesimal portion of surface. The curve bounding this surface is closed. Hence the line integral of the derivative ∇V vanishes.

$$\nabla \times \nabla V \cdot d\mathbf{a} = 0.$$

As this equation holds for any $d\mathbf{a}$, it follows that

$$\nabla \times \nabla V = 0.$$

In a similar manner the converse theorems may be demonstrated. If the divergence $\nabla \cdot \mathbf{U}$ of a vector function $\mathbf{U}$ is everywhere zero, then $\mathbf{U}$ is the curl of some vector function $\mathbf{W}$.

$$\mathbf{U} = \nabla \times \mathbf{W}.$$

If the curl $\nabla \times \mathbf{U}$ of a vector function $\mathbf{U}$ is everywhere zero, then $\mathbf{U}$ is the derivative of some scalar function V,

$$\mathbf{U} = \nabla V.$$

84.] By making use of the three fundamental relations between the line, surface, and volume integrals, and the *dels*, viz.:

$$\int_{\mathbf{r}_0}^{\mathbf{r}} \nabla V \cdot d\mathbf{r} = V(\mathbf{r}) - V(\mathbf{r}_0), \qquad (2)$$

$$\iint_s \nabla \times \mathbf{W} \cdot d\mathbf{a} = \int_\circ \mathbf{W} \cdot d\mathbf{r}, \qquad (11)$$

$$\iiint \nabla \cdot \mathbf{W}\, dv = \iint_s \mathbf{W} \cdot d\mathbf{a}, \qquad (7)$$

it is possible to obtain a large number of formulæ for the transformation of integrals. These formulæ correspond to those connected with "integration by parts" in ordinary calculus. They are obtained by integrating both sides of the formulæ, page 161, for differentiating.

First $$\nabla (u\, v) = u \nabla v + v \nabla u.$$

$$\int \nabla (u\, v) \cdot d\mathbf{r} = \int u \nabla v \cdot d\mathbf{r} + \int v \nabla u \cdot d\mathbf{r}.$$

Hence $$\int u \nabla v \cdot d\mathbf{r} = \Big[u\, v\Big]_{\mathbf{r}_0}^{\mathbf{r}} - \int v \nabla u \cdot d\mathbf{r}. \qquad (14)$$

The expression $$\Big[u\, v\Big]_{\mathbf{r}_0}^{\mathbf{r}}$$

represents the difference between the value of $(u\, v)$ at $\mathbf{r}$, the end of the path, and the value at $\mathbf{r}_0$, the beginning of the path. If the path be closed

$$\int_{\circ} u \nabla v \cdot d\mathbf{r} = - \int_{\circ} v \nabla u \cdot d\mathbf{r}. \qquad (14)'$$

Second $$\nabla \times (u\, \mathbf{v}) = u \nabla \times \mathbf{v} + \nabla u \times \mathbf{v}.$$

$$\iint_s \nabla \times (u\, \mathbf{v}) \cdot d\,\mathbf{a} = \iint_s u \nabla \times \mathbf{v} \cdot d\,\mathbf{a} + \iint_s \nabla u \times \mathbf{v} \cdot d\,\mathbf{a}.$$

Hence

$$\iint_s \nabla u \times \mathbf{v} \cdot d\,\mathbf{a} = \int_{\circ} u\, \mathbf{v} \cdot d\,\mathbf{r} - \iint_s u \nabla \times \mathbf{v} \cdot d\,\mathbf{a}, \qquad (15)$$

or

$$\iint_s u \nabla \times \mathbf{v} \cdot d\,\mathbf{a} = \int_{\circ} u\, \mathbf{v} \cdot d\,\mathbf{r} - \iint_s \nabla u \times \mathbf{v} \cdot d\,\mathbf{a}, \qquad (15)'$$

Third $$\nabla \times (u \nabla v) = u \nabla \times \nabla v + \nabla u \times \nabla v.$$

But $$\nabla \times \nabla v = 0$$

Hence $$\nabla \times (u \nabla v) = \nabla u \times \nabla v,$$

$$\iint_S \nabla \times (u \nabla v) \cdot d\mathbf{a} = \iint_S \nabla u \times \nabla v \cdot d\mathbf{a}.$$

Hence

$$\iint_S \nabla u \times \nabla v \cdot d\mathbf{a} = \int_\circ u \nabla v \cdot d\mathbf{r} = -\int_\circ v \nabla u \cdot d\mathbf{r}, \quad (16)$$

Fourth $\quad \nabla \cdot (u\,\mathbf{v}) = u \nabla \cdot \mathbf{v} + \nabla u \cdot \mathbf{v}.$

$$\iiint \nabla \cdot (u\,\mathbf{v})\, dv = \iiint u \nabla \cdot \mathbf{v}\, dv + \iiint \nabla u \cdot \mathbf{v}\, dv.$$

Hence

$$\iiint u \nabla \cdot \mathbf{v}\, dv = \iint_S u\,\mathbf{v} \cdot d\mathbf{a} - \iiint \nabla u \cdot \mathbf{v}\, dv, \quad (17)$$

or

$$\iiint \nabla u \cdot \mathbf{v}\, dv = \iint_S u\,\mathbf{v} \cdot d\mathbf{a} - \iiint u \nabla \cdot \mathbf{v}\, dv, \quad (17)'$$

Fifth $\quad \nabla \cdot (\nabla u \times \mathbf{v}) = \nabla \times \nabla u \cdot \mathbf{v} - \nabla u \cdot \nabla \times \mathbf{v}.$

$$\nabla \cdot (\nabla u \times \mathbf{v}) = -\nabla u \cdot \nabla \times \mathbf{v},$$

$$\iiint \nabla \cdot (\nabla u \times \mathbf{v})\, dv = -\iiint \nabla u \cdot \nabla \times \mathbf{v}\, dv.$$

Hence $\iint_S \nabla u \times \mathbf{v} \cdot d\mathbf{a} = -\iiint \nabla u \cdot \nabla \times \mathbf{v}\, dv. \quad (18)$

In all these formulæ which contain a triple integral the surface S is the closed surface bounding the body throughout which the integration is performed.

Examples of integration by parts like those above can be multiplied almost without limit. Only one more will be given here. It is known as *Green's Theorem* and is perhaps the most important of all. If u and v are any two scalar functions of position,

$$\nabla \cdot (u \nabla v) = \nabla u \cdot \nabla v + u \nabla \cdot \nabla v,$$

$$\nabla \cdot (v \nabla u) = \nabla u \cdot \nabla v + v \nabla \cdot \nabla u,$$

$$\nabla u \cdot \nabla v = \nabla \cdot (u \nabla v) - u \nabla \cdot \nabla v = \nabla \cdot (v \nabla u) - v \nabla \cdot \nabla u,$$

$$\iiint \nabla u \cdot \nabla v \, dv = \iiint \nabla \cdot (u \nabla v) \, dv - \iiint u \nabla \cdot \nabla v \, dv,$$

$$= \iiint \nabla \cdot (v \nabla u) \, dv - \iiint v \nabla \cdot \nabla u \, dv.$$

Hence

$$\iiint \nabla u \cdot \nabla v \, dv = \iint u \nabla v \cdot d\mathbf{a} - \iiint u \nabla \cdot \nabla v \, dv,$$

$$= \iint v \nabla u \cdot d\mathbf{a} - \iiint v \nabla \cdot \nabla u \, dv. \quad (19)$$

By subtracting these equalities the formula (20)

$$\iiint (u \nabla \cdot \nabla v - v \nabla \cdot \nabla u) \, dv = \iint (u \nabla v - v \nabla u) \cdot d\mathbf{a}.$$

is obtained. By expanding the expression in terms of **i, j, k** the ordinary form of Green's theorem may be obtained. A further generalization due to Thomson (Lord Kelvin) is the following:

$$\iiint w \nabla u \cdot \nabla v \, dv = \iint u w \nabla v \cdot d\mathbf{a} - \iiint u \nabla \cdot [w \nabla v] \, dv,$$

$$= \iint v w \nabla u \cdot d\mathbf{a} - \iiint v \nabla \cdot [w \nabla u] \, dv, \quad (21)$$

where w is a third scalar function of position.

The element of volume dv has nothing to do with the scalar function v in these equations or in those that go before. The use of v in these two different senses can hardly give rise to any misunderstanding.

*** 85.**] In the preceding articles the scalar and vector functions which have been subject to treatment have been supposed to be continuous, single-valued, possessing derivatives of the first two orders at every point of space under consideration. When the functions are discontinuous or multiple-valued, or fail to possess derivatives of the first two orders in certain regions of space, some caution must be exercised in applying the results obtained.

Suppose for instance

$$V = \tan^{-1} \frac{y}{x},$$

$$\nabla V = - \frac{y}{x^2 + y^2} \mathbf{i} + \frac{x}{x^2 + y^2} \mathbf{j}.$$

The line integral

$$\int \nabla V \cdot d\mathbf{r} = \int \frac{x\,dy - y\,dx}{x^2 + y^2}.$$

Introducing polar coördinates

$$x = r \cos \theta,$$

$$y = r \sin \theta,$$

$$x\,dy - y\,dx = r\,d\theta,$$

$$\int \nabla \cdot d\mathbf{r} = \int \frac{d\theta}{r} = \frac{1}{r} \int d\theta.$$

Form the line integral from the point $(+1, 0)$ to the point $(-1, 0)$ along two different paths. Let one path be a semicircle lying above the X-axis; and the other, a semicircle lying below that axis. The value of the integral along the first path is

$$\frac{1}{r} \int_0^{\pi} d\theta = \pi;$$

along the second path, $\frac{1}{r} \int_0^{-\pi} d\theta = - \pi.$

From this it appears that the integral does not depend merely upon the limits of integration, but upon the path chosen, the value along one path being the negative of the value along the other. The integral around the circle which is a closed curve does not vanish, but is equal to $\pm 2\pi$.

It might seem therefore the results of Art. 79 were false and that consequently the entire bottom of the work which follows fell out. This however is not so. The difficulty is that the function

$$V = \tan^{-1}\frac{y}{x}$$

is not single-valued. At the point (1,1), for instance, the function V takes on not only the value

$$V = \tan^{-1} 1 = \frac{\pi}{4},$$

but a whole series of values

$$\frac{\pi}{4} + k\,\pi,$$

where k is any positive or negative integer. Furthermore at the origin, which was included between the two semicircular paths of integration, the function V becomes wholly indeterminate and fails to possess a derivative. It will be seen therefore that the origin is a peculiar or *singular* point of the function V. If the two paths of integration from $(+1, 0)$ to $(-1, 0)$ had not included the origin the values of the integral would not have differed. In other words the value of the integral around a closed curve which *does not include the origin* vanishes as it should.

Inasmuch as the origin appears to be the point which vitiates the results obtained, let it be considered as marked by an impassable barrier. Any closed curve C which does not contain the origin may be shrunk up or expanded at will; but a closed curve $\bar{C}$ which surrounds the origin cannot be so distorted as no longer to enclose that point without breaking its continuity. The curve C not surrounding the origin may shrink up to nothing without a break in its continuity; but $\overline{C}$ can only shrink down and fit closer and closer about the origin. It cannot be shrunk down to nothing. It must always remain encircling the origin. The curve C is said to be *reducible*; $\overline{C}$, *irreducible*. In case of the function V, then, it is true that the integral taken around any *reducible* circuit C vanishes; but the integral around any *irreducible* circuit $\overline{C}$ does not vanish.

Suppose next that V is any function whatsoever. Let all the points at which V fails to be continuous or to have continuous first partial derivatives be marked as impassable barriers. Then any circuit C which contains within it no such point may be shrunk up to nothing and is said to be *reducible*; but a circuit which contains one or more such points cannot be so shrunk up without breaking its continuity and it is said to be *irreducible*. The theorem may then be stated: *The line integral of the derivative ∇V of any function V vanishes around any reducible circuit C.* It may or may not vanish around an irreducible circuit. In case one irreducible circuit $\overline{C}$ may be distorted so as to coincide with another irreducible circuit $\overline{C}$ without passing through any of the singular points of V and without breaking its continuity, the two circuits are said to be *reconcilable* and the values of the line integral of ∇V about them are the same.

A region such that *any* closed curve C within it may be shrunk up to nothing without passing through any singular point of V and without breaking its continuity, that is, a region every closed curve in which is reducible, is said to be *acyclic*. All other regions are *cyclic*.

By means of a simple device any cyclic region may be rendered acyclic. Consider, for instance, the region (Fig. 34) enclosed between the surface of a cylinder and the surface of a cube which contains the cylinder and whose bases coincide with those of the cylinder. Such a region is realized in a room in which a column reaches from the floor to the ceiling. It is evident that this region is cyclic. A circuit which passes around the column is irreducible. It cannot be contracted to nothing without breaking its continuity. If now a diaphragm be inserted reaching from the surface of the cylinder or column to the surface of the cube the region thus formed bounded by the surface of the cylinder, the surface of the cube, and the *two sides of the diaphragm* is acyclic. Owing to the insertion of the diaphragm it is no longer possible to draw a circuit which shall pass completely around the cylinder — the diaphragm prevents it. Hence every closed circuit which may be drawn in the region is reducible and the region is acyclic.

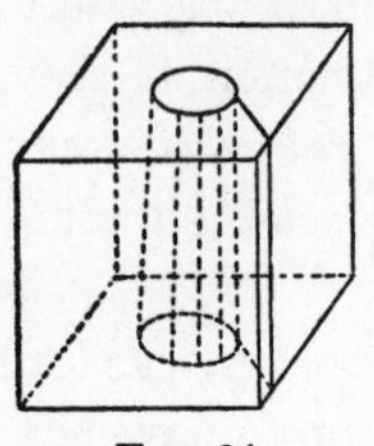

FIG. 34.

In like manner *any* region may be rendered acyclic by inserting a sufficient number of diaphragms. The bounding surfaces of the new region consist of the bounding surfaces of the given cyclic region and the *two* faces of each diaphragm.

In acyclic regions or regions rendered acyclic by the foregoing device all the results contained in Arts. 79 *et seq.* hold true. For cyclic regions they may or may not hold true. To enter further into these questions at this point is unnecessary. Indeed, even as much discussion as has been given them already may be superfluous. For they are questions which do not concern vector methods any more than the corresponding Cartesian ones. They belong properly to the subject of integration itself, rather than to the particular notation which may be employed in connection with it and which is the primary object of exposition here. In this respect these questions are similar to questions of rigor.

The Integrating Operators. The Potential

86.] Hitherto there have been considered line, surface, and volume integrals of functions both scalar and vector. There exist, however, certain special volume integrals which, owing to their intimate connection with the differentiating operators ∇, $\nabla \cdot$, $\nabla \times$, and owing to their especially frequent occurrence and great importance in physics, merit especial consideration. Suppose that

$$V(x_2, y_2, z_2)$$

is a scalar function of the position in space of the point

$$(x_2, y_2, z_2).$$

For the sake of definiteness V may be regarded as the density of matter at the point (x_2, y_2, z_2). In a homogeneous body V is constant. In those portions of space in which no matter exists V is identically zero. In non-homogeneous distributions of matter V varies from point to point; but at each point it has a definite value.

The vector

$$\mathbf{r}_2 = x_2\,\mathbf{i} + y_2\,\mathbf{j} + z_2\,\mathbf{k},$$

drawn from any assumed origin, may be used to designate the point (x_2, y_2, z_2). Let

$$(x_1, y_1, z_1)$$

be any other fixed point of space, represented by the vector

$$\mathbf{r}_1 = x_1\,\mathbf{i} + y_1\,\mathbf{j} + z_1\,\mathbf{k}$$

drawn from the same origin. Then

$$\mathbf{r}_2 - \mathbf{r}_1 = (x_2 - x_1)\,\mathbf{i} + (y_2 - y_1)\,\mathbf{j} + (z_2 - z_1)\,\mathbf{k}$$

is the vector drawn from the point (x_1, y_1, z_1) to the point (x_2, y_2, z_2). As this vector occurs a large number of times in the sections immediately following, it will be denoted by

$$\mathbf{r}_{12} = \mathbf{r}_2 - \mathbf{r}_1.$$

The length of $\mathbf{r}_{12}$ is then r_{12} and will be assumed to be positive.

$$r_{12} = \sqrt{\mathbf{r}_{12} \cdot \mathbf{r}_{12}} = \sqrt{(x_2 - x_1)^2 + (y_2 - y_1)^2 + (z_2 - z_1)^2}.$$

Consider the triple integral

$$I(x_1, y_1, z_1) = \iiint \frac{V(x_2, y_2, z_2)}{r_{12}} \, dx_2 \, dy_2 \, dz_2.$$

The integration is performed with respect to the variables x_2, y_2, z_2 — that is, with respect to the body of which V represents the density (Fig. 35). During the integration the point (x_1, y_1, z_1) remains fixed. The integral I has a definite value at each definite point (x_1, y_1, z_1). It is a function of that point. The interpretation of this integral I is easy, if the function V be regarded as the density of matter in space. The element of mass dm at (x_2, y_2, z_2) is

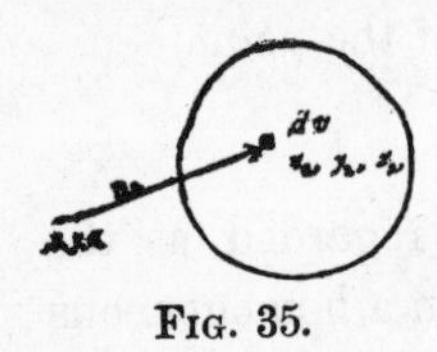

FIG. 35.

$$dm = V(x_2, y_2, z_2) \, dx_2 \, dy_2 \, dz_2 = V \, dv.$$

The integral I is therefore the sum of the elements of mass in a body, each divided by its distance from a fixed point (x_1, y_1, z_1).

$$I = \int \frac{dm}{r_{12}}.$$

This is what is termed the *potential* at the point (x_1, y_1, z_1) due to the body whose density is

$$V(x_2, y_2, z_2).$$

The limits of integration in the integral I may be looked at in either of two ways. In the first place they may be regarded as coincident with the limits of the body of which V is the density. This indeed might seem the most natural set of limits. On the other hand the integral I may be regarded as taken over *all space.* The value of the integral is the same in both cases. For when the limits are infinite the function V vanishes identically at every point (x_2, y_2, z_2) situated outside of the body and hence does not augment the value of the integral at all. It is found most convenient to consider the limits as infinite and the integral as extended over all space. This saves the trouble of writing in special limits for each particular case. The function V of itself then practically determines the limits owing to its vanishing identically at all points unoccupied by matter.

87.] The operation of finding the potential is of such frequent occurrence that a special symbol, *Pot*, is used for it.

$$\text{Pot}\, V = \iiint \frac{V(x_2, y_2, z_2)}{r_{12}}\, dx_2\, dy_2\, dz_2. \qquad (22)$$

The symbol is read "the potential of V." The potential, *Pot* V, is a function not of the variables x_2, y_2, z_2 with regard to which the integration is performed but of the point (x_1, y_1, z_1) which is fixed during the integration. These variables enter in the expression for r_{12}. The function V and *Pot* V therefore have different sets of variables.

It may be necessary to note that although V has hitherto been regarded as the density of matter in space, such an interpretation for V is entirely too restricted for convenience. Whenever it becomes necessary to form the integral

$$\iiint \frac{V(x_2, y_2, z_2)}{r_{12}}\, dx_2\, dy_2\, dz_2 = \iiint \frac{V_2}{r_{12}}\, dv_2. \qquad (22)'$$

of any scalar function V, no matter what V represents, that integral is called the potential of V. The reason for calling such an integral the potential even in cases in which it has no connection with physical potential is that it is formed according to the same formal law as the true potential and by virtue of that formation has certain simple rules of operation which other types of integrals do not possess.

Pursuant to this idea the potential of a *vector* function

$$\mathbf{W}\,(x_2, y_2, z_2)$$

may be written down.

$$\text{Pot}\,\mathbf{W} = \iiint \frac{\mathbf{W}\,(x_2, y_2, z_2)}{r_{12}}\, dx_2\, dy_2\, dz_2. \tag{23}$$

In this case the integral is the sum of vector quantities and is consequently itself a vector. Thus the potential of a vector function $\mathbf{W}$ is a vector function, just as the potential of a scalar function V was seen to be a scalar function of position in space. If $\mathbf{W}$ be resolved into its three components

$$\mathbf{W}\,(x_2, y_2, z_2) = \mathbf{i}\, X\,(x_2, y_2, z_2) + \mathbf{j}\, Y\,(x_2, y_2, z_2) + \mathbf{k}\, Z\,(x_2, y_2, z_2)$$

$$\text{Pot}\,\mathbf{W} = \mathbf{i}\,\text{Pot}\,X + \mathbf{j}\,\text{Pot}\,Y + \mathbf{k}\,\text{Pot}\,Z. \tag{24}$$

The potential of a vector function $\mathbf{W}$ is equal to the vector sum of the potentials of its three components X, Y, Z.

The potential of a scalar function V exists at a point $(x_1, y_1, z_1,)$ when and only when the integral

$$\text{Pot}\,V = \iiint \frac{V_2}{r_{12}}\, dv_2,$$

taken over all space converges to a definite value. If, for instance, V were everywhere constant in space the integral would become greater and greater without limit as the limits of integration were extended farther and farther out into space. Evidently therefore if the potential is to exist V must approach zero as its limit as the point (x_2, y_2, z_2) recedes indefinitely. A few important sufficient conditions for the convergence of the potential may be obtained by transforming to polar coördinates. Let

$$x = r \sin\theta \cos\phi,$$
$$y = r \sin\theta \sin\phi,$$
$$z = r \cos\theta,$$
$$dv = r^2 \sin\theta \, dr \, d\theta \, d\phi.$$

Let the point (x_1, y_1, z_1) which is fixed for the integration be chosen at the origin. Then

$$r_{12} = r$$

and the integral becomes

$$\iiint \frac{V_2}{r_{12}} dv_2 = \iiint \frac{V}{r} r^2 \sin\theta \, dr \, d\theta \, d\phi, \quad (22)$$

or simply $\quad \text{Pot}\, V = \iiint V r \sin\theta \, dr \, d\theta \, d\phi.$

If the function V *decrease so rapidly that the product*

$$V r^3$$

remains finite as r *increases indefinitely, then the integral converges as far as the distant regions of space are concerned.* For let

$$V r^3 < K$$

$$\left| \iiint_{r=R}^{r=\infty} V r \sin\theta \, dr \, d\theta \, d\phi \right| < \iiint_{r=R}^{r=\infty} \frac{K}{r^2} dr \, d\theta \, d\phi$$

$$\iiint_{r=R}^{r=\infty} \frac{K}{r^2} dr \, d\theta \, d\phi = 2\pi^2 \frac{K}{R}.$$

Hence the triple integral taken over all space outside of a sphere of radius R (where R is supposed to be a large quantity) is less than $2\pi^2 K / R$, and consequently converges as far as regions distant from the origin are concerned.

If the function V *remain finite or if it become infinite so weakly that the product*

$$V r$$

remains finite when r *approaches zero, then the integral converges as far as regions near to the origin are concerned.* For let

$$V r < K$$

$$\left| \int\int\int_{r=0}^{r=R} V r \sin \theta \, d r \; d \theta \; d \phi \right| < \int\int\int_{r=0}^{r=R} K d r \; d \theta \; d \phi.$$

$$\int\int\int_{r=0}^{r=R} K d r \; d \theta \; d \phi = 2 \pi^2 K R.$$

Hence the triple integral taken over all space inside a sphere of radius R (where R is now supposed to be a small quantity) is less than $2 \pi^2 K R$ and consequently converges as far as regions near to the origin which is the point (x_1, y_1, z_1) are concerned.

If at any point (x_2, y_2, z_2) *not coincident with the origin,* i. e. *the point* (x_1, y_1, z_1), *the function* V *becomes infinite so weakly that the product of the value of* V *at a point near to* (x_2, y_2, z_2) *by the square of the distance of that point from* (x_2, y_2, z_2) *remains finite as that distance approaches zero, then the integral converges as far as regions near to the point* (x_2, y_2, z_2) *are concerned.* The proof of this statement is like those given before. These three conditions for the convergence of the integral *Pot V* are sufficient. They are by no means necessary. The integral may converge when they do not hold. It is however indispensable to know whether or not an integral under discussion converges. Unless the tests given above show the convergence, more stringent ones must be resorted to. Such, however, will not be discussed here. They belong to the theory of integration in general rather than to the theory of the integrating operator *Pot.* The discussion of the convergence of the potential of a vector function **W** reduces at once to that of its *three* components which are scalar functions and may be treated as above.

88.] The potential is a function of the variables x_1, y_1, z_1 which are constant with respect to the integration. Let the value of the potential at the point (x_1, y_1, z_1) be denoted by

$$[\text{Pot}\, V]_{x_1, y_1, z_1}.$$

The first partial derivative of the potential with respect to x_1 is therefore

$$\frac{\partial\, \text{Pot}\, V}{\partial\, x_1} = \underset{\Delta x_1 \doteq 0}{\text{Lim}} \left\{ \frac{[\text{Pot}\, V]_{x_1 + \Delta x_1, y_1, z_1} - [\text{Pot}\, V]_{x_1, y_1, z_1}}{\Delta\, x_1} \right\} \quad (25)$$

The value of this limit may be determined by a simple device (Fig. 36). Consider the potential at the point

$$(x_1 + \Delta\, x_1, y_1, z_1)$$

due to a certain body T. This is the same as the potential at the point

$$(x_1, y_1, z_1)$$

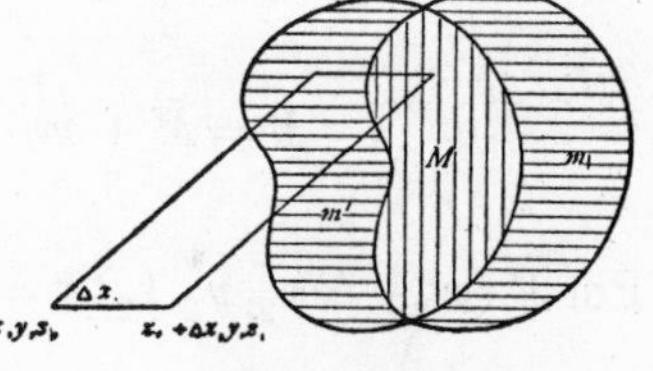

FIG. 36.

due to the same body T displaced in the negative direction by the amount $\Delta\, x_1$. For in finding the potential at a point P due to a body T the absolute positions in space of the body T and the point P are immaterial. It is only their positions *relative to each other* which determines the value of the potential. If both body and point be translated by the same amount in the same direction the value of the potential is unchanged. But now if T be displaced in the negative direction by the amount $\Delta\, x$, the value of V at each point of space is changed from

$$V(x_2, y_2, z_2) \text{ to } V(x_2 + \Delta\, x_2, y_2, z_2),$$

where $\Delta\, x_2 = \Delta\, x_1$.

Hence

$$[\text{Pot}\, V(x_2, y_2, z_2)]_{x_1+\Delta x_1,\, y_1,\, z_1} = [\text{Pot}\, V(x_2+\Delta x_2, y_2, z_2)]_{x_1, y_1, z_1}$$

Hence $$\underset{\Delta x_1 \doteq 0}{\text{Lim}} \left\{ \frac{[\text{Pot}\, V]_{x_1+\Delta x_1, y_1, z_1} - [\text{Pot}\, V]_{x_1, y_1, z_1}}{\Delta x_1} \right\} =$$

$$\underset{\Delta x_1 \doteq 0}{\text{Lim}} \left\{ \frac{[\text{Pot}\, V(x_2+\Delta x_2, y_2, z_2)]_{x_1, y_1, z_1} - [\text{Pot}\, V(x_2, y_2, z_2)]_{x_1, y_1, z_1}}{\Delta x_1} \right\}.$$

It will be found convenient to introduce the limits of integration. Let the portion of space originally filled by the body T be denoted by M; and let the portion filled by the body after its translation in the negative direction through the distance Δx_1 be denoted by M'. The regions M and M' overlap. Let the region common to both be $\overline{M}$; and let the remainder of M be m; the remainder of M', m'. Then

$$M = \overline{M} + m, \qquad M' = \overline{M} + m'.$$

$$\text{Pot}\, V(x_2+\Delta x_2, y_2, z_2)\dagger = \iiint_{M'} \frac{V(x_2+\Delta x_2, y_2, z_2)}{r_{12}}\, dv_2$$

$$= \iiint_{\overline{M}} \frac{V(x_2+\Delta x_2, y_2, z_2)}{r_{12}}\, dv_2 + \iiint_{m'} \frac{V(x_2+\Delta x_2, y_2, z_2)}{r_{12}}\, dv_2.$$

$$\text{Pot}\, V(x_2, y_2, z_2) = \iiint_{M} \frac{V(x_2, y_2, z_2)}{r_{12}}\, dv_2$$

$$= \iiint_{\overline{M}} \frac{V(x_2, y_2, z_2)}{r_{12}}\, dv_2 + \iiint_{m} \frac{V(x_2, y_2, z_2)}{r_{12}}\, dv_2.$$

Hence (25) becomes, when Δx_1 is replaced by its equal Δx_2,

† As all the following potentials are for the point x_1, y_1, z_1 the bracket and indices have been dropped.

$$\underset{\Delta x_2 \doteq 0}{\text{Lim}} \left\{ \frac{\iiint_{\bar{M}} \frac{(Vx_2 + \Delta x_2, y_2, z_2)}{r_{12}} dv_2 - \iiint_{\bar{M}} \frac{V(x_2, y_2, z_2)}{r_{12}} dv_2}{\Delta x_2 = 0} \right.$$

$$\left. + \iiint_{m'} \frac{V(x_2 + \Delta x_2, y_2, z_2)}{r_{12}} d v_2 - \iiint_m \frac{V(x_2, y_2, z_2)}{r_{12}} d v_2 \right\}$$

Or, $\underset{\Delta x_2 \doteq 0}{\text{Lim}} \iiint_{\bar{M}} \frac{V(x_2 + \Delta x_2, y_2, z_2) - V(x_2, y_2, z_2)}{r_{12} \Delta x_2} d v_2$

$$+ \underset{\Delta x_2 \doteq 0}{\text{Lim}} \iiint_{m'} \frac{V(x_2 + \Delta x_2, y_2, z_2)}{r_{12} \Delta x_2} d v_2$$

$$- \underset{\Delta x_2 \doteq 0}{\text{Lim}} \iiint_m \frac{V(x_2, y_2, z_2)}{r_{12} \Delta x_2} d v_2$$

$$\underset{\Delta x_2 \doteq 0}{\text{Lim}} \iiint_{\bar{M}} \frac{V(x_2 + \Delta x_2, y_2, z_2) - V(x_2, y_2, z_2)}{r_{12} \Delta x_2} d v_2$$

$$= \iiint_{\bar{M}} \underset{\Delta x_2 \doteq 0}{\text{Lim}} \left\{ \frac{V(x_2 + \Delta x_2, y_2, z_2) - V(x_2, y_2, z_2)}{r_{12} \Delta x_2} \right\} d v_2 \dagger$$

$$= \iiint_{\bar{M}} \frac{1}{r_{12}} \frac{\partial V(x_2, y_2, z_2)}{\partial x_2} d v_2 . \ddagger$$

when Δx_1 approaches zero as its limit the regions m and m', which are at no point thicker than Δx, approach zero; M' and $\bar{M}$ both approach M as a limit.

† There are cases in which this reversal of the order in which the two limits are taken gives incorrect results. This is a question of *double* limits and leads to the mazes of modern mathematical rigor.

‡ If the derivative of V is to exist at the surface bounding T the values of the function V must diminish continuously to zero upon the surface. If V changed suddenly from a finite value within the surface to a zero value outside the derivative $\partial V / \partial x_1$ would not exist and the triple integral would be meaningless. For the same reason V is supposed to be finite and continuous at every point within the region T.

Then if it be assumed that the region T is finite and that V vanishes upon the surface bounding T

$$\operatorname*{Lim}_{\Delta x_2 \doteq 0} \iiint_{m'} \frac{V(x_2 + \Delta x_2, y_2, z_2)}{r_{12}\, \Delta x_2} \, d v_2 = 0$$

$$\operatorname*{Lim}_{\Delta x_2 \doteq 0} \iiint_{m} \frac{V(x_2, y_2, z_2)}{r_{12}\, \Delta x_2} \, d v_2 = 0.$$

Consequently the expression for the derivative of the potential reduces to merely

$$\frac{\partial \operatorname{Pot} V}{\partial x_1} = \iiint_M \frac{1}{r_{12}} \frac{\partial V}{\partial x_2} \, d v_2 = \operatorname{Pot} \frac{\partial V}{\partial x_2}. \quad (26)$$

The partial derivative of the potential of a scalar function V *is equal to the potential of the partial derivative of* V.

The derivative ∇ *of the potential of* V *is equal to the potential of the derivative* ∇ V.

$$\nabla \operatorname{Pot} V = \operatorname{Pot} \nabla V. \quad (27)$$

This statement follows immediately from the former. As the ∇ upon the left-hand side applies to the set of variables x_1, y_1, z_1, it may be written ∇_1. In like manner the ∇ upon the right-hand side may be written ∇_2 to call attention to the fact that it applies to the variables x_2, y_2, z_2 of V.

Then $$\nabla_1 \operatorname{Pot} V = \operatorname{Pot} \nabla_2 V. \quad (27)'$$

To demonstrate this identity ∇ may be expanded in terms of **i, j, k.**

$$\mathbf{i} \frac{\partial \operatorname{Pot} V}{\partial x_1} + \mathbf{j} \frac{\partial \operatorname{Pot} V}{\partial y_1} + \mathbf{k} \frac{\partial \operatorname{Pot} V}{\partial z_1}$$

$$= \mathbf{i} \operatorname{Pot} \frac{\partial V}{\partial x_2} + \mathbf{j} \operatorname{Pot} \frac{\partial V}{\partial y_2} + \mathbf{k} \operatorname{Pot} \frac{\partial V}{\partial z_2}.$$

As **i**, **j**, **k** are constant vectors they may be placed under the sign of integration and the terms may be collected. Then by means of (26)

$$\nabla_1 \operatorname{Pot} V = \operatorname{Pot} \nabla_2 V.$$

The curl $\nabla \times$ *and divergence* $\nabla \cdot$ *of the potential of a vector function* **W** *are equal respectively to the potential of the curl and divergence of that function.*

$$\nabla_1 \times \operatorname{Pot} \mathbf{W} = \operatorname{Pot} \nabla_2 \times \mathbf{W},$$

or $$\text{curl Pot } \mathbf{W} = \text{Pot curl } \mathbf{W} \tag{28}$$

and $$\nabla_1 \cdot \operatorname{Pot} \mathbf{W} = \operatorname{Pot} \nabla_2 \cdot \mathbf{W},$$

or $$\text{div Pot } \mathbf{W} = \text{Pot div } \mathbf{W}. \tag{29}$$

These relations may be proved in a manner analogous to the above. It is even possible to go further and form the *dels* of higher order

$$\nabla \cdot \nabla \operatorname{Pot} V = \operatorname{Pot} \nabla \cdot \nabla V, \tag{30}$$

$$\nabla \cdot \nabla \operatorname{Pot} \mathbf{W} = \operatorname{Pot} \nabla \cdot \nabla \mathbf{W}, \tag{31}$$

$$\nabla \nabla \cdot \operatorname{Pot} \mathbf{W} = \operatorname{Pot} \nabla \nabla \cdot \mathbf{W}, \tag{32}$$

$$\nabla \times \nabla \times \operatorname{Pot} \mathbf{W} = \operatorname{Pot} \nabla \times \nabla \times \mathbf{W}. \tag{33}$$

The *dels* upon the left might have a subscript 1 attached to show that the differentiations are performed with respect to the variables x_1, y_1, z_1, and for a similar reason the *dels* upon the right might have been written with a subscript 2. The results of this article may be summed up as follows:

Theorem: *The differentiating operator* ∇ *and the integrating operator* Pot *are commutative.*

***89.**] In the foregoing work it has been assumed that the region T was finite and that the function V was everywhere finite and continuous inside of the region T and moreover decreased so as to approach zero continuously at the surface bounding that region. These restrictions are inconvenient and may be removed by making use of a surface integral. The derivative of the potential was obtained (page 213) in essentially the form

$$\frac{\partial \operatorname{Pot} V}{\partial x_1} = \iiint_M \frac{1}{r_{12}} \frac{\partial V}{\partial x_2} d v_2$$

$$+ \lim_{\Delta x_2 \doteq 0} \frac{1}{\Delta x_2} \iiint_{m'} \frac{V(x_2 + \Delta x_2, y_2, z_2)}{r_{12}} d v_2$$

$$- \lim_{\Delta x_2 \doteq 0} \frac{1}{\Delta x_2} \iiint_m \frac{V(x_2, y_2, z_2)}{r_{12}} d v_2.$$

Let $d\mathbf{a}$ be a directed element of the surface S bounding the region M. The element of volume $d v_2$ in the region m' is therefore equal to

$$d v_2 = \Delta x_2 \, \mathbf{i} \cdot d \mathbf{a}.$$

Hence $$\frac{1}{\Delta x_2} \iiint_{m'} \frac{V(x_2 + \Delta x_2, y_2, z_2)}{r_{12}} d v_2$$

$$= \iint \frac{V(x_2 + \Delta x_2, y_2, z_2)}{r_{12}} \mathbf{i} \cdot d \mathbf{a}.$$

The element of volume $d v_2$ in the region m is equal to

$$d v_2 = - \Delta x_2 \, \mathbf{i} \cdot d \mathbf{a}.$$

Hence $$- \frac{1}{\Delta x_2} \iiint_m \frac{V(x_2, y_2, z_2)}{r_{12}} d v_2$$

$$= \iint \frac{V(x_2, y_2, z_2)}{r_{12}} \mathbf{i} \cdot d \mathbf{a}.$$

Consequently

$$\frac{\partial \operatorname{Pot} V}{\partial x_1} = \iiint_M \frac{1}{r_{12}} \frac{\partial V}{\partial x_2} d v_2 + \iint_S \frac{V}{r_{12}} \mathbf{i} \cdot d \mathbf{a}. \qquad (34)$$

The volume integral is taken throughout the region M with the understanding that the value of the derivative of V at the surface S shall be equal to the limit of the value of that derivative when the surface is approached from the interior of M. This convention avoids the difficulty that arises in connection with the existence of the derivative at the surface S where V becomes discontinuous. The surface integral is taken over the surface S which bounds the region.

Suppose that the region M becomes infinite. By virtue of the conditions imposed upon V to insure the convergence of the potential

$$V r^3 < K.$$

Let the bounding surface S be a sphere of radius R, a quantity which is large.

$$\mathbf{i} \cdot d\,\mathbf{a} < r^2\, d\,\theta\, d\,\phi.$$

$$\left| \iint_S \frac{V}{r}\, \mathbf{i} \cdot d\,\mathbf{a} \right| < \iint_S \frac{K}{r^2}\, d\theta\, d\phi = 2\,\pi^2\, \frac{K}{R^2}.$$

The surface integral becomes smaller and smaller and approaches *zero* as its limit when the region M becomes infinite. Moreover the volume integral

$$\iiint_M \frac{1}{r_{12}} \frac{\partial V}{\partial x_2}\, d\,v_2,$$

remains finite as M becomes infinite. Consequently provided V is such a function that Pot V exists as far as the infinite regions of space are concerned, then the equation

$$\frac{\partial \operatorname{Pot} V}{\partial x_1} = \operatorname{Pot} \frac{\partial V}{\partial x_2}$$

holds as far as those regions of space are concerned.

Suppose that V ceases to be continuous or becomes infinite at a single point (x_1, y_1, z_1) within the region T. Surround this point with a small sphere of radius R. Let S denote the surface of this sphere and M all the region T not included within the sphere. Then

$$\frac{\partial \operatorname{Pot} V}{\partial x_1} = \iiint_M \frac{1}{r_{12}} \frac{\partial V}{\partial x_2} d v_2 + \iint_S \frac{V}{r_{12}} \mathbf{i} \cdot d \mathbf{a}.$$

By the conditions imposed upon V

$$V r < K$$

$$\left| \iint_S \frac{V}{r} \mathbf{i} \cdot d \mathbf{a} \right| < \iint_S K \, d\theta \; d\phi = 2 \pi^2 K.$$

Consequently when the sphere of radius R becomes smaller and smaller the surface integral may or may not become zero. Moreover the volume integral

$$\iiint_M \frac{1}{r_{12}} \frac{\partial V}{\partial x_2} d v_2$$

may or may not approach a limit when R becomes smaller and smaller. Hence the equation

$$\frac{\partial \operatorname{Pot} V}{\partial x_1} = \operatorname{Pot} \frac{\partial V}{\partial x_2}$$

has not always a definite meaning at a point of the region T at which V becomes infinite in such a manner that the product $V r$ remains finite.

If, however, V remains finite at the point in question so that the product $V r$ approaches zero, the constant K is zero and the surface integral becomes smaller and smaller as R approaches zero. Moreover the volume integral

$$\iiint_M \frac{1}{r_{12}} d v_2 \frac{\partial V}{\partial x_2}$$

approaches a definite limit as R becomes infinitesimal. Consequently the equation

$$\frac{\partial \operatorname{Pot} V}{\partial x_1} = \operatorname{Pot} \frac{\partial V}{\partial x_2}$$

holds in the neighborhood of all isolated points at which V remains finite even though it be discontinuous.

Suppose that V becomes infinite at some single point (x_2, y_2, z_2) not coincident with (x_1, y_1, z_1). According to the conditions laid upon V

$$V l^2 < K,$$

where l is the distance of the point (x_2, y_2, z_2) from a point near to it. Then the surface integral

$$\iint_s \frac{V}{r_{12}} \mathbf{i} \cdot d\,\mathbf{a}$$

need not become zero and consequently the equation

$$\frac{\partial \operatorname{Pot} V}{\partial x_1} = \operatorname{Pot} \frac{\partial V}{\partial x_2}$$

need not hold for any point (x_1, y_1, z_1) of the region. But if V becomes infinite at x_2, y_2, z_2 in such a manner that

$$V l < K,$$

then the surface integral will approach zero as its limit and the equation will hold.

Finally suppose the function V remains finite upon the surface S bounding the region T, but does not vanish there. In this case there exists a surface of discontinuities of V. Within this surface V is finite; without, it is zero. The surface integral

$$\iint_S \frac{V}{r_{12}} \mathbf{i} \cdot d\,\mathbf{a}$$

does not vanish in general. Hence the equation

$$\frac{\partial \operatorname{Pot} V}{\partial x_1} = \operatorname{Pot} \frac{\partial V}{\partial x_2}$$

cannot hold.

Similar reasoning may be applied to each of the three partial derivatives with respect to x_1, y_1, z_1. By combining the results it is seen that in general

$$\nabla_1 \operatorname{Pot} V = \operatorname{Pot} \nabla_2 V + \iint_S \frac{V}{r_{12}}\, d\mathbf{a}. \qquad (35)$$

Let V be any function in space, and let it be granted that Pot V exists. Surround each point of space at which V ceases to be finite by a small sphere. Let the surface of the sphere be denoted by S. Draw in space all those surfaces which are surfaces of discontinuity of V. Let these surfaces also be denoted by S. Then the formula (35) holds where the surface integral is taken over all the surfaces which have been designated by S. If the integral taken over all these surfaces vanishes when the radii of the spheres above mentioned become infinitesimal, then

$$\nabla_1 \operatorname{Pot} V = \operatorname{Pot} \nabla_2 V. \qquad (27)'$$

This formula

$$\nabla_1 \operatorname{Pot} V = \operatorname{Pot} \nabla_2 V.$$

will surely hold at a point (x_1, y_1, z_1) *if* V *remains always finite or becomes infinite at a point* (x_2, y_2, z_2) *so that the product* V 1 *remains finite, and if* V *possesses no surfaces of discontinuity, and if furthermore the product* $V r^3$ *remains finite as* r *becomes infinite.*[1] In other cases special tests must be applied to ascertain whether the formula (27)′ can be used or the more complicated one (35) must be resorted to.

[1] For extensions and modifications of this theorem, see exercises.

The relation (27) is so simple and so amenable to transformation that V will in general be assumed to be such a function that (27) holds. In cases in which V possesses a surface S of discontinuity it is frequently found convenient to consider V as replaced by another function $\bar{V}$ which has in general the same values as V but which instead of possessing a discontinuity at S merely changes very rapidly from one value to another as the point (x_2, y_2, z_2) passes from one side of S to the other. Such a device renders the potential of V simpler to treat analytically and probably conforms to actual physical states more closely than the more exact conception of a surface of discontinuity. This device practically amounts to including the surface integral in the symbol Pot ∇V.

In fact from the standpoint of pure mathematics it is better to state that where there exist surfaces at which the function V becomes discontinuous, the full value of *Pot* ∇V should always be understood as including the surface integral

$$\iint_s \frac{V}{r_{12}}\, d\mathbf{a}$$

in addition to the volume integral

$$\iiint \frac{\nabla V}{r_{12}}\, d v_2.$$

In like manner *Pot* $\nabla \cdot \mathbf{W}$, *Pot* $\nabla \times \mathbf{W}$, *New* $\nabla \cdot \mathbf{W}$ and other similar expressions to be met in the future must be regarded as consisting not only of a volume integral but of a surface integral in addition, whenever the vector function $\mathbf{W}$ possesses a surface of discontinuities.

It is precisely this convention in the interpretation of formulæ which permits such simple formulæ as (27) to hold in general, and which gives to the treatment of the integrating operators an elegance of treatment otherwise unobtainable.

The irregularities which may arise are thrown into the interpretation, not into the analytic appearance of the formulæ. This is the essence of Professor Gibbs's method of treatment.

90.] The first partial derivatives of the potential may also be obtained by differentiating under the sign of integration.[1]

$$\text{Pot}\, V = \iiint \frac{V(x_2, y_2, z_2)}{\sqrt{(x_2 - x_1)^2 + (y_2 - y_1)^2 + (z_2 - z_1)^2}}\, dx_2\, dy_2\, dz_2 \qquad (36)$$

$$\frac{\partial\, \text{Pot}\, V}{\partial x_1} = \iiint \frac{\partial}{\partial x_1} \left\{ \frac{V(x_2, y_2, z_2)}{\sqrt{[(x_2 - x_1)^2 + (y_2 - y_1)^2 + (z_2 - z_1)^2]}} \right\} dx_2 dy_2 dz_2$$

$$\frac{\partial\, \text{Pot}\, V}{\partial x_1} = \iiint \frac{(x_2 - x_1)\, V(x_2, y_2, z_2)}{\sqrt{[(x_2 - x_1)^2 + (y_2 - y_1)^2 + (z_2 - {}_2z)^2]^3}}\, dx_2\, dy_2\, dz_2 \qquad (37)$$

In like manner for a vector function **W**

$$\frac{\partial\, \text{Pot}\, \mathbf{W}}{\partial x_1} = \iiint \frac{(x_2 - x_1)\, \mathbf{W}(x_2, y_2, z_2)}{\sqrt{[(x_2 - x_1)^2 + (y_2 - y_1)^2 + (z_2 - z_1)^2]^3}}\, dx_2\, dy_2\, dz_2. \qquad (38)$$

Or
$$\frac{\partial\, \text{Pot}\, V}{\partial x_1} = \iiint \frac{(x_2 - x_1)\, V}{r^3{}_{12}}\, dv_2 \qquad (37)'$$

and
$$\frac{\partial\, \text{Pot}\, \mathbf{W}}{\partial x_1} = \iiint \frac{(x_2 - x_1)\, \mathbf{W}}{r^3{}_{12}}\, dv_2. \qquad (38)'$$

$$\nabla\, \text{Pot}\, V = \mathbf{i}\, \frac{\partial\, \text{Pot}\, V}{\partial x_1} + \mathbf{j}\, \frac{\partial\, \text{Pot}\, V}{\partial y_1} + \mathbf{k}\, \frac{\partial\, \text{Pot}\, V}{\partial z_1} =$$

$$\iiint \left\{ \frac{\mathbf{i}\,(x_2 - x_1)\, V}{r^3{}_{12}} + \frac{\mathbf{j}(y_2 - y_1)\, V}{r^3{}_{12}} + \frac{\mathbf{k}\,(z_2 - z_1)\, V}{r^3{}_{12}} \right\} dv_2.$$

But $\mathbf{i}\,(x_2 - x_1) + \mathbf{j}\,(y_2 - y_1) + \mathbf{k}\,(z_2 - z_1) = \mathbf{r}_{12}.$

[1] If an attempt were made to obtain the *second* partial derivatives in the same manner, it would be seen that the volume integrals no longer converged.

Hence $$\nabla \operatorname{Pot} V = \iiint \frac{\mathbf{r}_{12} V}{r^3_{12}} \, d v_2. \tag{39}$$

In like manner

$$\nabla \times \operatorname{Pot} \mathbf{W} = \iiint \frac{\mathbf{r}_{12} \times \mathbf{W}}{r^3_{12}} \, d v_2, \tag{40}$$

and $$\nabla \cdot \operatorname{Pot} \mathbf{W} = \iiint \frac{\mathbf{r}_{12} \cdot \mathbf{W}}{r^3_{12}} \, d v_2. \tag{41}$$

These three integrals obtained from the potential by the differentiating operators are of great importance in mathematical physics. Each has its own interpretation. Consequently although obtained so simply from the potential each is given a separate name. Moreover inasmuch as these integrals may exist even when the potential is divergent, they must be considered independent of it. They are to be looked upon as three new integrating operators defined each upon its own merits as the potential was defined.

Let, therefore,

$$\iiint \frac{\mathbf{r}_{12} \, V(x_2, y_2, z_2)}{r^3_{12}} \, d x_2 \, d y_2 \, d z_2 = \operatorname{New} V, \tag{42}$$

$$\iiint \frac{\mathbf{r}_{12} \times \mathbf{W}(x_2, y_2, z_2)}{r^3_{12}} \, d x_2 \, d y_2 \, d z_2 = \operatorname{Lap} \mathbf{W} \tag{43}$$

$$\iiint \frac{\mathbf{r}_{12} \cdot \mathbf{W}(x_2, y_2, z_2)}{r^3_{12}} \, d x_2 \, d y_2 \, d z_2 = \operatorname{Max} \mathbf{W}. \tag{44}$$

If the potential exists, then

$$\begin{aligned} \nabla \operatorname{Pot} V &= \operatorname{New} V \\ \nabla \times \operatorname{Pot} \mathbf{W} &= \operatorname{Lap} \mathbf{W} \\ \nabla \cdot \operatorname{Pot} \mathbf{W} &= \operatorname{Max} \mathbf{W}. \end{aligned} \tag{45}$$

The first is written *New V* and read "The Newtonian of V."

The reason for calling this integral the Newtonian is that if V represent the density of a body the integral gives the *force of attraction* at the point (x_1, y_1, z_1) due to the body. This will be proved later. The second is written *Lap* **W** and read "the Laplacian of **W**." This integral was used to a considerable extent by Laplace. It is of frequent occurrence in electricity and magnetism. If **W** represent the current **C** in space the Laplacian of **C** gives the *magnetic force* at the point (x_1, y_1, z_1) due to the current. The third is written *Max* **W** and read "the Maxwellian of **W**." This integral was used by Maxwell. It, too, occurs frequently in electricity and magnetism. For instance if **W** represent the intensity of magnetization **I**, the Maxwellian of **I** gives the *magnetic potential* at the point (x_1, y_1, z_1) due to the magnetization.

To show that the Newtonian gives the force of attraction according to the law of the inverse square of the distance. Let $d\,m_2$ be any element of mass situated at the point (x_2, y_2, z_2). The force at (x_1, y_1, z_1) due to $d\,m$ is equal to

$$\frac{d\,m_2}{r_{12}}$$

in magnitude and has the direction of the vector $\mathbf{r}_{12}$ from the point (x_1, y_1, z_1) to the point (x_2, y_2, z_2). Hence the force is

$$\frac{\mathbf{r}_{12}\,d\,m_2}{r^3_{12}}.$$

Integrating over the entire body, or over all space according to the convention here adopted, the total force is

$$\iiint \frac{\mathbf{r}_{12}\,d\,m_2}{r^3_{12}} = \iiint \frac{\mathbf{r}_{12}\,V}{r^3_{12}}\,d\,v_2 = \text{New}\ V,$$

where V denotes the density of matter.

The integral may be expanded in terms of **i, j, k,**

$$\text{New } V = \mathbf{i} \iiint \frac{(x_2 - x_1)\, V}{r^3_{12}}\, d\, v_2 + \mathbf{j} \iiint \frac{(y_2 - y_1)\, V}{r^3_{12}}\, d\, v_2$$

$$+ \mathbf{k} \iiint \frac{(z_2 - z_1)\, V}{r^3_{12}}\, d\, v_2.$$

The three components may be expressed in terms of the potential (if it exists) as

$$\iiint \frac{(x_2 - x_1)\, V}{r^3_{12}}\, d\, v_2 = \frac{\partial}{\partial\, x_1} \text{ Pot } V = \frac{\partial}{\partial\, x_1} \iiint \frac{V}{r_{12}}\, d\, v_2 \qquad (42)'$$

$$\iiint \frac{(y_2 - y_1)\, V}{r^3_{12}}\, d\, v_2 = \frac{\partial}{\partial\, y_1} \text{ Pot } V = \frac{\partial}{\partial\, y_1} \iiint \frac{V}{r_{12}}\, d\, v_2$$

$$\iiint \frac{(z_2 - z_1)\, V}{r^3_{12}}\, d\, v_2 = \frac{\partial}{\partial\, z_1} \text{ Pot } V = \frac{\partial}{\partial\, z_1} \iiint \frac{V}{r_{12}}\, d\, v_2.$$

It is in this form that the Newtonian is generally found in books.

To show that the Laplacian gives the magnetic force per unit positive pole at the point (x_1, y_1, z_1) due to a distribution $\mathbf{W}\,(x_2, y_2, z_2)$ of electric flux. The magnetic force at (x_1, y_1, z_1) due to an element of current $d\,\mathbf{C}_2$ is equal in magnitude to the magnitude $d\,C_2$ of that element of current divided by the square of the distance r_{12}; that is

$$\frac{d\, C_2}{r^2_{12}}.$$

The direction of the force is perpendicular both to the vector element of current $d\,\mathbf{C}_2$ and to the line $\mathbf{r}_{12}$ joining the points. The direction of the force is therefore the direction of the vector product of $\mathbf{r}_{12}$ and $d\,\mathbf{C}_2$. The force is therefore

$$\frac{\mathbf{r}_{12} \times d\, \mathbf{C}_2}{r^3_{12}}.$$

Integrating over all space, the total magnetic force acting at the point (x_1, y_1, z_1) upon a unit positive pole is

$$\iiint \frac{\mathbf{r}_{12} \times d\,\mathbf{C}_2}{r^3_{12}} = \iiint \frac{\mathbf{r}_{12} \times \mathbf{W}}{r^3_{12}}\, d\,v_2 = \text{Lap}\,\mathbf{W}.$$

This integral may be expanded in terms of **i, j, k.** Let

$$\mathbf{W}\,(x_2, y_2, z_2) = \mathbf{i}\,X(x_2, y_2, z_2) + \mathbf{j}\,Y(x_2, y_2, z_2) + \mathbf{k}\,Z(x_2, y_2, z_2).$$

$$\mathbf{r}_{12} = (x_2 - x_1)\,\mathbf{i} + (y_2 - y_1)\,\mathbf{j} + (z_2 - z_1)\,\mathbf{k}.$$

The **i, j, k** components of Lap **W** are respectively

$$\mathbf{i} \cdot \text{Lap}\,\mathbf{W} = \iiint \frac{(y_2 - y_1)\,Z - (z_2 - z_1)\,Y}{r^3_{12}}\, d\,v_2$$

$$\mathbf{j} \cdot \text{Lap}\,\mathbf{W} = \iiint \frac{(z_2 - z_1)\,X - (x_2 - x_1)\,Z}{r^3_{12}}\, d\,v_2 \qquad (43)'$$

$$\mathbf{k} \cdot \text{Lap}\,\mathbf{W} = \iiint \frac{(x_2 - x_1)\,Y - (y_2 - y_1)\,X}{r^3_{12}}\, d\,v_2$$

In terms of the potential (if one exists) this may be written

$$\mathbf{i} \cdot \text{Lap}\,\mathbf{W} = \frac{\partial\,\text{Pot}\,Z}{\partial\,y_1} - \frac{\partial\,\text{Pot}\,Y}{\partial\,z_1}$$

$$\mathbf{j} \cdot \text{Lap}\,\mathbf{W} = \frac{\partial\,\text{Pot}\,X}{\partial\,z_1} - \frac{\partial\,\text{Pot}\,Z}{\partial\,x_1}, \qquad (43)''$$

$$\mathbf{k} \cdot \text{Lap}\,\mathbf{W} = \frac{\partial\,\text{Pot}\,Y}{\partial\,x_1} - \frac{\partial\,\text{Pot}\,X}{\partial\,y_1}.$$

To show that if **I** be the intensity of magnetization at the point (x_2, y_2, z_2), that is, if **I** be a vector whose magnitude is equal to the magnetic moment per unit volume and whose direction is the direction of magnetization of the element $d\,v_2$ from south pole to north pole, then the Maxwellian of **I** is the magnetic potential due to the distribution of magnetization. The magnetic moment of the element of volume $d\,v_2$ is $\mathbf{I}\,d\,v_2$. The potential at (x_1, y_1, z_1) due to this element is equal to its magnetic moment divided by the square of the distance r_{12} and multiplied by the cosine of the angle between the direction of magnetization **I** and the vector $\mathbf{r}_{12}$. The potential is therefore

$$\frac{\mathbf{r}_{12} \cdot \mathbf{I}\,d\,v_2}{r^3_{12}}.$$

Integrating, the total magnetic potential is seen to be

$$\iiint \frac{\mathbf{r}_{12} \cdot \mathbf{I}}{r^3_{12}}\,d\,v_2 = \text{Max}\ \mathbf{I}.$$

This integral may also be written out in terms of x, y, z. Let

$$\mathbf{I}\,(x_2, y_2, z_2) = \mathbf{i}\,A\,(x_2, y_2, z_2) + \mathbf{j}\,B\,(x_2, y_2, z_2) + \mathbf{k}\,C\,(x_2, y_2, z_2)$$

$$\mathbf{r}_{12} \cdot \mathbf{I} = (x_2 - x_1)\,A + (y_2 - y_1)\,B + (z_2 - z_1)\,C.$$

If instead of x_1, y_1, z_1 the variables x, y, z; and instead of x_2, y_2, z_2 the variables ξ, η, ζ be used[1] the expression takes on the form given by Maxwell.

$$\text{Max}\ \mathbf{I} = \iiint \Big\{ A\,(\xi - x) + B\,(\eta - y) + C\,(\zeta - z) \Big\} \frac{1}{r^3}\,d\,v.$$

According to the notation employed for the Laplacian

$$\text{Max}\ \mathbf{W} = \iiint \frac{(x_2 - x_1)\,X + (y_2 - y_1)\,Y + (z_2 - z_1)\,Z}{r^3_{12}}\,d\,v_2. \tag{44}'$$

[1] Maxwell: Electricity and Magnetism, Vol. II. p. 9.

The Maxwellian of a vector function is a scalar quantity. It may be written in terms of the potential (if it exists) as

$$\text{Max}\,\mathbf{W} = \frac{\partial\,\text{Pot}\,X}{\partial\,x_1} + \frac{\partial\,\text{Pot}\,Y}{\partial\,y_1} + \frac{\partial\,\text{Pot}\,Z}{\partial\,z_1}. \qquad (44)''$$

This form of expression is much used in ordinary treatises upon mathematical physics.

The Newtonian, Laplacian, and Maxwellian, however, should not be associated indissolubly with the particular physical interpretations given to them above. They should be looked upon as integrating operators which may be applied, as the potential is, to any functions of position in space. The Newtonian is applied to a scalar function and yields a vector function. The Laplacian is applied to a vector function and yields a function of the same sort. The Maxwellian is applied to a vector function and yields a scalar function. Moreover, these integrals should not be looked upon as the derivatives of the potential. If the potential exists they are its derivatives. But they frequently exist when the potential fails to converge.

91.] Let V and $\mathbf{W}$ be such functions that their potentials exist and have in general definite values. Then by (27) and (29)

$$\nabla \cdot \nabla\,\text{Pot}\,V = \nabla \cdot \text{Pot}\,\nabla V = \text{Pot}\,\nabla \cdot \nabla V.$$

But by (45) $$\nabla\,\text{Pot}\,V = \text{New}\,V,$$

and $$\nabla \cdot \text{Pot}\,\nabla V = \text{Max}\,\nabla V.$$

Hence $$\nabla \cdot \nabla\,\text{Pot}\,V = \nabla \cdot \text{New}\,V = \text{Max}\,\nabla V$$

$$= \text{Pot}\,\nabla \cdot \nabla V \qquad (46)$$

By (27) and (29) $\nabla\nabla \cdot \text{Pot}\,\mathbf{W} = \nabla\,\text{Pot}\,\nabla \cdot \mathbf{W} = \text{Pot}\,\nabla\nabla \cdot \mathbf{W}.$

But by (45) $$\nabla \cdot \text{Pot}\,\mathbf{W} = \text{Max}\,\mathbf{W},$$

and by (45) $$\nabla\,\text{Pot}\,\nabla \cdot \mathbf{W} = \text{New}\,\nabla \cdot \mathbf{W}.$$

Hence $$\nabla \nabla \cdot \text{Pot}\, \mathbf{W} = \nabla\, \text{Max}\, \mathbf{W} = \text{New}\, \nabla\, \mathbf{W}$$
$$= \text{Pot}\, \nabla \nabla \cdot \mathbf{W} \qquad (47)$$

By (28) $$\nabla \times \nabla \times \text{Pot}\, \mathbf{W} = \nabla \times \text{Pot}\, \nabla \times \mathbf{W}$$
$$= \text{Pot}\, \nabla \times \nabla \times \mathbf{W}.$$

But by (45) $$\nabla \times \text{Pot}\, \mathbf{W} = \text{Lap}\, \mathbf{W},$$

and $$\nabla \times \text{Pot}\, \nabla \times \mathbf{W} = \text{Lap}\, \nabla \times \mathbf{W}.$$

Hence $$\nabla \times \nabla \times \text{Pot}\, \mathbf{W} = \nabla \times \text{Lap}\, \mathbf{W} = \text{Lap}\, \nabla \times \mathbf{W}$$
$$= \text{Pot}\, \nabla \times \nabla \times \mathbf{W}. \qquad (48)$$

By (56), Chap. III. $$\nabla \cdot \nabla \times \text{Pot}\, \mathbf{W} = 0,$$

or $$\nabla \cdot \text{Pot}\, \nabla \times \mathbf{W} = 0.$$

Hence $$\nabla \cdot \text{Lap}\, \mathbf{W} = \text{Max}\, \nabla \times \mathbf{W} = 0. \qquad (49)$$

And by (52), Chap. III. $$\nabla \times \nabla\, \text{Pot}\, V = 0,$$

or $$\nabla \times \text{Pot}\, \nabla V = 0.$$

Hence $$\nabla \times \text{New}\, V = \text{Lap}\, \nabla V = 0. \qquad (50)$$

And by (58), Chap. III. $$\nabla \times \nabla \times \mathbf{W} = \nabla \nabla \cdot \mathbf{W} - \nabla \cdot \nabla \mathbf{W},$$

$$\nabla \cdot \nabla \mathbf{W} = \nabla \nabla \cdot \mathbf{W} - \nabla \times \nabla \times \mathbf{W}.$$

Hence $$\nabla \cdot \nabla\, \text{Pot}\, \mathbf{W} = \text{New}\, \nabla \cdot \mathbf{W} - \text{Lap}\, \nabla \times \mathbf{W}, \qquad (51)$$

or $$\nabla \cdot \nabla\, \text{Pot}\, \mathbf{W} = \nabla\, \text{Max}\, \mathbf{W} - \nabla \times \text{Lap}\, \mathbf{W}.$$

These formulæ may be written out in terms of *curl* and *div* if desired. Thus

$$\text{div New}\, V = \text{Max}\, \nabla V, \qquad (46)'$$

$$\nabla\, \text{Max}\, \mathbf{W} = \text{New div}\, \mathbf{W} \qquad (47)'$$

$$\text{curl Lap}\, \mathbf{W} = \text{Lap curl}\, \mathbf{W} \qquad (48)'$$

$$\text{div Lap}\, \mathbf{W} = \text{Max curl}\, \mathbf{W} = 0 \qquad (49)'$$

$$\text{curl New}\, V = \text{Lap}\, \nabla V = 0 \qquad (50)'$$

$$\nabla \cdot \nabla\, \text{Pot}\, \mathbf{W} = \text{New div}\, \mathbf{W} - \text{Lap curl}\, \mathbf{W}. \qquad (51)'$$

Poisson's Equation

92.] *Let V be any function in space such that the potential*

$$\operatorname{Pot} V$$

has in general a definite value. Then

$$\nabla \cdot \nabla \operatorname{Pot} V = -4 \pi V, \tag{52}$$

or
$$\frac{\partial^2 \operatorname{Pot} V}{\partial x_1^{\,2}} + \frac{\partial^2 \operatorname{Pot} V}{\partial y_1^{\,2}} + \frac{\partial^2 \operatorname{Pot} V}{\partial z_1^{\,2}} = -4 \pi V.$$

This equation is known as Poisson's Equation.

The integral which has been defined as the *potential* is a solution of Poisson's Equation. The proof is as follows.

$$\operatorname{Pot} V = \iiint \frac{V}{r_{12}} \, d v_2.$$

$$\nabla_1 \operatorname{Pot} V = \operatorname{New} V = \iiint \frac{\mathbf{r}_{12} V}{r^3_{\;12}} \, d v_2 = \iiint \nabla_1 \frac{1}{r_{12}} \, V \, d v_2$$

$$\nabla_1 \cdot \nabla_1 \operatorname{Pot} V = \nabla_1 \cdot \operatorname{New} V = \operatorname{Max} \nabla_2 V = \iiint \frac{\mathbf{r}_{12} \cdot \nabla_2 V}{r^3_{\;12}} \, d v_2.$$

The subscripts *1* and *2* have been attached to designate clearly what are variables with respect to which the differentiations are performed.

$$\nabla_1 \cdot \nabla_1 \operatorname{Pot} V = \nabla_1 \cdot \operatorname{New} V = \iiint \nabla_1 \frac{1}{r_{12}} \cdot \nabla_2 V \, d v_2.$$

But
$$\nabla_1 \frac{1}{r_{12}} = - \nabla_2 \frac{1}{r_{12}}$$

and
$$\nabla_2 \cdot \left(V \, \nabla_2 \frac{1}{r_{12}} \right) = \nabla_2 \frac{1}{r_{12}} \cdot \nabla_2 V + V \, \nabla_2 \cdot \nabla_2 \frac{1}{r_{12}} \cdot$$

Hence $-\nabla_2 \frac{1}{r_{12}} \cdot \nabla_2 V = V \nabla_2 \cdot \nabla_2 \frac{1}{r_{12}} - \nabla_2 \cdot \left(V \nabla_2 \frac{1}{r_{12}} \right)$

or $\nabla_1 \frac{1}{r_{12}} \cdot \nabla_2 V = V \nabla_2 \cdot \nabla_2 \frac{1}{r_{12}} + \nabla_2 \cdot \left(V \nabla_1 \frac{1}{r_{12}} \right).$

Integrate:

$$\iiint \nabla_1 \frac{1}{r_{12}} \cdot \nabla_2 V \, d v_2 = \iiint V \nabla_2 \cdot \nabla_2 \frac{1}{r_{12}} \, d v_2$$
$$+ \iiint \nabla_2 \cdot \left(V \nabla_1 \frac{1}{r_{12}} \right) d v_2.$$

But $$\nabla_2 \cdot \nabla_2 \frac{1}{r_{12}} = 0.$$

That is to say $\frac{1}{r}$ satisfies Laplace's Equation. And by (8)

$$\iiint \nabla_2 \cdot \left(V \nabla_1 \cdot \frac{1}{r_{12}} \right) d v_2 = \iint_s V \nabla_1 \frac{1}{r_{12}} \cdot d \mathbf{a}.$$

Hence $$\nabla_1 \cdot \nabla_1 \operatorname{Pot} V = \iiint \nabla_1 \frac{1}{r_{12}} \cdot \nabla_2 V \, d v_2 \quad (53)$$

$$= \iint_s V \nabla_1 \frac{1}{r_{12}} \cdot d \mathbf{a}.$$

The surface integral is taken over the surface which bounds the region of integration of the volume integral. This is taken "over all space." Hence the surface integral must be taken over a sphere of radius R, a large quantity, and R must be allowed to increase without limit. At the point (x_1, y_1, z_1), however, the integrand of the surface integral becomes infinite owing to the presence of the term

$$\nabla_1 \frac{1}{r_{12}}.$$

Hence the surface S must include not only the surface of the sphere of radius R, but also the surface of a sphere of radius R', a small quantity, surrounding the point (x_1, y_1, z_1) and R' must be allowed to approach zero as its limit.

As it has been assumed that the potential of V exists, it is assumed that the conditions given (Art. 87) for the existence of the potential hold. That is

$$V r^3 < K, \text{ when } r \text{ is large}$$

$$V r < K, \text{ when } r \text{ is small.}$$

Introduce polar coördinates with the origin at the point (x_1, y_1, z_1). Then r_{12} becomes simply r

and
$$\nabla_1 \frac{1}{r_{12}} = - \nabla_2 \frac{1}{r_{12}} = \frac{\mathbf{r}}{r^3}.$$

Then for the large sphere of radius R

$$\nabla_1 \frac{1}{r_{12}} \cdot d\,\mathbf{a} = \frac{r}{r^3} r^2 \sin\theta \, d\theta \, d\phi.$$

Hence the surface integral over that sphere approaches zero as its limit. For

$$\left| \iint V \nabla_1 \frac{1}{r_{12}} \cdot d\,\mathbf{a} \right| < \iint \frac{K}{r^3} \, d\theta \, d\phi = \frac{2\pi}{R^3}.$$

Hence when R becomes infinite the surface integral over the large sphere approaches zero as its limit.

For the small sphere

$$\nabla_1 \frac{1}{r_{12}} \cdot d\,\mathbf{a} = - \frac{r}{r^3} r^2 \sin\theta \, d\theta \, d\phi.$$

Hence the integral over that sphere becomes

$$- \iint V \sin\theta \, d\theta \, d\phi.$$

Let V be supposed to be finite and continuous at the point (x_1, y_1, z_1) which has been selected as origin. Then for the surface integral V is practically constant and equal to its value

$$V(x_1, y_1, z_1)$$

at the point in question.

$$\iint \sin\theta \, d\theta \, d\phi = 4\pi.$$

Hence $$-\iint V \sin\theta \, d\theta \, d\phi \doteq -4\pi V$$

when the radius R' of the sphere of integration approaches zero as its limit. Hence

$$\iiint \nabla_1 \frac{1}{r_{12}} \cdot \nabla_2 V \, dv_2 = \iint_s V \nabla_1 \frac{1}{r_{12}} \cdot d\mathbf{a} = -4\pi V \quad (53)'$$

and $$\nabla \cdot \nabla \operatorname{Pot} V = -4\pi V. \quad (52)$$

In like manner if $\mathbf{W}$ is a vector function which has in general a definite potential, then that potential satisfies Poisson's Equation.

$$\nabla \cdot \nabla \operatorname{Pot} \mathbf{W} = -4\pi \mathbf{W}. \quad (52)'$$

The proof of this consists in resolving $\mathbf{W}$ into its three components. For each component the equation holds. Let

$$\mathbf{W} = X\mathbf{i} + Y\mathbf{j} + Z\mathbf{k},$$

$$\nabla \cdot \nabla \operatorname{Pot} X = -4\pi X,$$

$$\nabla \cdot \nabla \operatorname{Pot} Y = -4\pi Y,$$

$$\nabla \cdot \nabla \operatorname{Pot} Z = -4\pi Z.$$

Consequently

$$\nabla \cdot \nabla \operatorname{Pot} (X\mathbf{i} + Y\mathbf{j} + Z\mathbf{k}) = -4\pi (X\mathbf{i} + Y\mathbf{j} + Z\mathbf{k}).$$

Theorem: *If* V *and* **W** *are such functions of position in space that their potentials exist in general, then for all points at which* V *and* **W** *are finite and continuous those potentials satisfy Poisson's Equation,*

$$\nabla \cdot \nabla \text{ Pot } V = -4\pi V, \tag{52}$$

$$\nabla \cdot \nabla \text{ Pot } \mathbf{W} = -4\pi \mathbf{W}. \tag{52'}$$

The modifications in this theorem which are to be made at points at which V and **W** become *discontinuous* will not be taken up here.

93.] It was seen (46) Art. 91 that

$$\nabla \cdot \nabla \text{ Pot } V = \nabla \cdot \text{New } V = \text{Max } \nabla V.$$

Hence
$$\nabla \cdot \text{New } V = -4\pi V \tag{53}$$

or
$$\text{Max } \nabla V = -4\pi V.$$

In a similar manner it was seen (51) Art. 91 that

$$\begin{aligned} \nabla \cdot \nabla \text{ Pot } \mathbf{W} &= \nabla \text{ Max } \mathbf{W} - \nabla \times \text{Lap } \mathbf{W} \\ &= \text{New } \nabla \cdot \mathbf{W} - \text{Lap } \nabla \times \mathbf{W}. \end{aligned}$$

Hence
$$\nabla \text{ Max } \mathbf{W} - \nabla \times \text{Lap } \mathbf{W} = -4\pi \mathbf{W}, \tag{54}$$

or
$$\text{New } \nabla \cdot \mathbf{W} - \text{Lap } \nabla \times \mathbf{W} = -4\pi \mathbf{W}. \tag{54'}$$

By virtue of this equality **W** is divided into two parts.

$$\mathbf{W} = \frac{1}{4\pi} \text{Lap } \nabla \times \mathbf{W} - \frac{1}{4\pi} \text{New } \nabla \cdot \mathbf{W}. \tag{55}$$

Let
$$\mathbf{W} = \mathbf{W}_1 + \mathbf{W}_2,$$

where
$$\mathbf{W}_1 = \frac{1}{4\pi} \text{Lap } \nabla \times \mathbf{W} = \frac{1}{4\pi} \text{Lap curl } \mathbf{W} \tag{56}$$

and
$$\mathbf{W}_2 = -\frac{1}{4\pi} \text{New } \nabla \cdot \mathbf{W} = -\frac{1}{4\pi} \text{New div } \mathbf{W}. \tag{57}$$

Equation (55) states that any vector function $\mathbf{W}$ multiplied by 4π is equal to the difference of the Laplacian of its curl and the Newtonian of its divergence. Furthermore

$$\nabla \cdot \mathbf{W}_1 = \frac{1}{4\pi} \nabla \cdot \text{Lap} \nabla \times \mathbf{W} = \frac{1}{4\pi} \nabla \cdot \nabla \times \text{Lap} \, \mathbf{W}_1.$$

But the divergence of the curl of a vector function is zero.

Hence
$$\nabla \cdot \mathbf{W}_1 = \text{div} \, \mathbf{W}_1 = 0 \tag{58}$$

$$\nabla \times \mathbf{W}_2 = -\frac{1}{4\pi} \nabla \times \text{New} \nabla \cdot \mathbf{W}_2 = -\frac{1}{4\pi} \nabla \times \nabla \, \text{Max} \, \mathbf{W}_2.$$

But the curl of the derivative of a scalar function is zero.

Hence
$$\nabla \times \mathbf{W}_2 = \text{curl} \, \mathbf{W}_2 = 0. \tag{59}$$

Consequently any vector function $\mathbf{W}$ which has a potential may be divided into two parts of which one has no divergence and of which the other has no curl. This division of $\mathbf{W}$ into two such parts is unique.

In case a vector function has no potential but both its curl and divergence possess potentials, the vector function may be divided into three parts of which the first has no divergence; the second, no curl; the third, neither divergence nor curl.

Let
$$\mathbf{W} = \frac{1}{4\pi} \text{Lap} \nabla \times \mathbf{W} - \frac{1}{4\pi} \text{New} \nabla \cdot \mathbf{W} + \mathbf{W}_3. \tag{55$'$}$$

As before

$$\frac{1}{4\pi} \nabla \cdot \text{Lap} \nabla \times \mathbf{W} = \frac{1}{4\pi} \nabla \cdot \nabla \times \text{Pot} \nabla \times \mathbf{W} = 0$$

and
$$\frac{-1}{4\pi} \nabla \times \text{New} \nabla \cdot \mathbf{W} = \frac{-1}{4\pi} \nabla \times \nabla \, \text{Pot} \nabla \cdot \mathbf{W} = 0.$$

The divergence of the first part and the curl of the second part of $\mathbf{W}$ are therefore zero.

$$\frac{1}{4\pi}\nabla\times\text{Lap}\,\nabla\times\mathbf{W}=\frac{1}{4\pi}\nabla\times\nabla\times\text{Pot}\,\nabla\times\mathbf{W}$$

$$=\frac{1}{4\pi}\nabla\nabla\cdot\text{Pot}\,\nabla\times\mathbf{W}-\frac{1}{4\pi}\nabla\cdot\nabla\,\text{Pot}\,\nabla\times\mathbf{W}.$$

$$\frac{1}{4\pi}\nabla\nabla\cdot\text{Pot}\,\nabla\times\mathbf{W}=\frac{1}{4\pi}\nabla\,\text{Pot}\,\nabla\cdot\nabla\times\mathbf{W}=0,$$

for $$\nabla\cdot\nabla\times\mathbf{W}=0.$$

Hence $$\frac{-1}{4\pi}\nabla\cdot\nabla\,\text{Pot}\,\nabla\times\mathbf{W}=\nabla\times\mathbf{W}.$$

Hence $$\frac{1}{4\pi}\nabla\times\text{Lap}\,\nabla\times\mathbf{W}=\nabla\times\mathbf{W}=\nabla\times\mathbf{W}_1.$$

The curl of $\mathbf{W}$ is equal to the curl of the first part

$$\frac{1}{4\pi}\text{Lap}\,\nabla\times\mathbf{W}$$

into which $\mathbf{W}$ is divided. Hence as the second part has no curl, the third part can have none. Moreover

$$-\frac{1}{4\pi}\nabla\,\text{New}\,\nabla\cdot\mathbf{W}=\nabla\cdot\mathbf{W}=\nabla\cdot\mathbf{W}_1.$$

Thus the divergence of $\mathbf{W}$ is equal to the divergence of the second part

$$\frac{-1}{4\pi}\text{New}\,\nabla\cdot\mathbf{W}.$$

into which $\mathbf{W}$ is divided. Hence as the first part has no divergence the third can have none. Consequently the third part $\mathbf{W}_3$ has neither curl nor divergence. This proves the statement.

By means of Art. 96 it may be seen that any function $\mathbf{W}_3$ which possesses neither curl nor divergence, must either vanish throughout all space or must not become zero at infinity. In physics functions generally vanish at infinity. Hence functions which represent actual phenomena may be divided into two parts, of which one has no divergence and the other no curl.

94.] *Definition:* A vector function the divergence of which vanishes at every point of space is said to be *solenoidal.* A vector function the curl of which vanishes at every point of space is said to be *irrotational.*

In general a vector function is neither solenoidal nor irrotational. But it has been shown that any vector function which possesses a potential may be divided in one and only one way into two parts $\mathbf{W}_1$, $\mathbf{W}_2$ of which one is solenoidal and the other irrotational. The following theorems may be stated. They have all been proved in the foregoing sections.

With respect to a solenoidal *function* $\mathbf{W}_1$, *the operators*

$$\frac{1}{4\pi}\,\text{Lap} \quad \textit{and} \quad \nabla\times \quad \textit{or} \quad \text{curl}$$

are inverse operators. That is

$$\frac{1}{4\pi}\,\text{Lap}\,\nabla\times\mathbf{W}_1 = \nabla\times\frac{1}{4\pi}\,\text{Lap}\,\mathbf{W}_1 = \mathbf{W}_1. \tag{60}$$

Applied to an irrotational *function* $\mathbf{W}_2$ *either of these operators gives zero.* That is

$$\frac{1}{4\pi}\,\text{Lap}\,\mathbf{W}_2 = 0\,, \quad \nabla\times\mathbf{W}_2 = 0. \tag{61}$$

With respect to an irrotational *function* $\mathbf{W}_2$, *the operators*

$$\frac{1}{4\pi}\,\text{New} \quad \textit{and} \quad -\nabla\cdot \quad \textit{or} \quad -\text{div}$$

are inverse operators. That is

$$-\frac{1}{4\pi}\,\text{New}\,\nabla\cdot\mathbf{W}_2 = -\nabla\cdot\frac{1}{4\pi}\,\text{New}\,\mathbf{W}_2 = \mathbf{W}_2. \tag{62}$$

With respect to a scalar *function* V *the operators*

$$-\nabla\cdot \textit{ or } -\mathrm{div} \textit{ and } \frac{1}{4\pi}\,\mathrm{New},$$

and also

$$-\frac{1}{4\pi}\,\mathrm{Max} \textit{ and } \nabla$$

are inverse operators. That is

$$-\nabla\cdot\frac{1}{4\pi}\,\mathrm{New}\, V = V \tag{63}$$

and

$$-\frac{1}{4\pi}\,\mathrm{Max}\,\nabla V = V.$$

With respect to a solenoidal *function* $\mathbf{W}_1$ *the operators*

$$\frac{1}{4\pi}\,\mathrm{Pot} \textit{ and } \nabla\times\nabla\times \textit{ or } \mathrm{curl\ curl}$$

are inverse operators. That is

$$\frac{1}{4\pi}\,\mathrm{Pot}\,\nabla\times\nabla\times\mathbf{W}_1 = \nabla\times\nabla\times\frac{1}{4\pi}\,\mathrm{Pot}\,\mathbf{W}_1 = \mathbf{W}_1. \tag{64}$$

With respect to an irrotational *function* $\mathbf{W}_2$ *the operators*

$$\frac{1}{4\pi}\,\mathrm{Pot} \textit{ and } -\nabla\nabla\cdot$$

are inverse operators. That is

$$-\frac{1}{4\pi}\,\mathrm{Pot}\,\nabla\nabla\cdot\mathbf{W}_2 = -\nabla\nabla\cdot\frac{1}{4\pi}\,\mathrm{Pot}\,\mathbf{W}_2 = \mathbf{W}_2. \tag{65}$$

With respect to any scalar or vector *function* V, **W** *the operators*

$$\frac{1}{4\pi}\,\mathrm{Pot} \textit{ and } -\nabla\cdot\nabla$$

are inverse operators. That is

$$-\frac{1}{4\pi}\operatorname{Pot}\nabla\cdot\nabla V = -\nabla\cdot\nabla\frac{1}{4\pi}\operatorname{Pot} V = V$$

and $$-\frac{1}{4\pi}\operatorname{Pot}\nabla\cdot\nabla\mathbf{W} = -\nabla\cdot\nabla\frac{1}{4\pi}\operatorname{Pot}\mathbf{W} = \mathbf{W}. \quad (66)$$

With respect to a solenoidal *function* $\mathbf{W}_1$ *the differentiating operators of the second order*

$$-\nabla\cdot\nabla \textit{ and } \nabla\times\nabla\times$$

are equivalent

$$-\nabla\cdot\nabla\mathbf{W}_1 = \nabla\times\nabla\times\mathbf{W}_1. \quad (67)$$

With respect to an irrotational *function* $\mathbf{W}_2$ *the differentiating operators of the second order*

$$\nabla\cdot\nabla \textit{ and } \nabla\nabla\cdot$$

are equivalent. That is

$$\nabla\cdot\nabla\mathbf{W}_2 = \nabla\nabla\cdot\mathbf{W}_2. \quad (68)$$

By integrating the equations

$$4\pi V = -\nabla\cdot\operatorname{New} V$$

and $$4\pi\mathbf{W} = \nabla\times\operatorname{Lap}\mathbf{W} - \nabla\operatorname{Max}\mathbf{W}$$

by means of the potential integral *Pot*

$$4\pi\operatorname{Pot} V = -\operatorname{Pot}\nabla\cdot\operatorname{New} V = -\operatorname{Max}\operatorname{New} V \quad (69)$$

$$4\pi\operatorname{Pot}\mathbf{W} = \operatorname{Pot}\nabla\times\operatorname{Lap}\mathbf{W} - \operatorname{Pot}\nabla\operatorname{Max}\mathbf{W}$$

$$4\pi\operatorname{Pot}\mathbf{W} = \operatorname{Lap}\operatorname{Lap}\mathbf{W} - \operatorname{New}\operatorname{Max}\mathbf{W}. \quad (70)$$

Hence for scalar *functions and* irrotational *vector functions*

$$-\frac{1}{4\pi}\operatorname{New}\operatorname{Max}$$

is an operator which is equivalent to Pot. *For* solenoidal *vector functions the operator*

$$-\frac{1}{4\pi}\operatorname{Lap}\operatorname{Lap}$$

gives the potential. For any *vector function the first operator gives the potential of the* irrotational *part; the second, the potential of the* solenoidal *part.*

***95.**] There are a number of *double* volume integrals which are of such frequent occurrence in mathematical physics as to merit a passing mention, although the theory of them will not be developed to any considerable extent. These double integrals are all scalar quantities. They are not scalar functions of position in space. They have but a single value. The integrations in the expressions may be considered for convenience as extended over all space. The functions by vanishing identically outside of certain finite limits determine for all practical purposes the limits of integration in case they are finite.

Given two scalar functions U, V of position in space. The *mutual potential* or *potential product*, as it may be called, of the two functions is the sextuple integral

$$\text{Pot}\,(U,V)=\iiiint\!\!\iint \frac{U(x_1,y_1,z_1)\,V\,(x_2,y_2,z_2)}{r_{12}}\,dv_1\,dv_2. \tag{71}$$

One of the integrations may be performed

$$\text{Pot}\,(U,V)=\iiint U(x_1,y_1,z_1)\ \text{Pot}\,V\,dv_1$$

$$=\iiint V(x_2,y_2,z_2)\ \text{Pot}\,U\,dv_2. \tag{72}$$

In a similar manner the mutual potential or potential product of two vector functions $\mathbf{W}'$, $\mathbf{W}''$ is

$$\text{Pot}\,(\mathbf{W}',\mathbf{W}'')=\iiiint\!\!\iint \frac{\mathbf{W}'(x_1,y_1,z_1)\cdot\mathbf{W}''(x_2,y_2,z_2)}{r_{12}}\,dv_1\,dv_2. \tag{71$'$}$$

This is also a scalar quantity. One integration may be carried out

$$\text{Pot}\,(\mathbf{W}', \mathbf{W}'') = \iiint \mathbf{W}'(x_1, y_1, z_1) \cdot \text{Pot}\,\mathbf{W}''\, d\,v_1$$

$$= \iiint \mathbf{W}''(x_2, y_2, z_2) \cdot \text{Pot}\,\mathbf{W}'\, d\,v_2. \qquad (72)'$$

The *mutual Laplacian* or Laplacian product of two vector functions $\mathbf{W}'$, $\mathbf{W}''$ of position in space is the sextuple integral

$$\text{Lap}\,(\mathbf{W}', \mathbf{W}'')$$

$$= \iiiiiint \mathbf{W}'(x_1, y_1, z_1) \cdot \frac{\mathbf{r}_{12}}{r^3_{12}} \times \mathbf{W}''(x_2, y_2, z_2)\, d\,v_1\, d\,v_2. \qquad (73)$$

One integration may be performed.

$$\text{Lap}\,(\mathbf{W}', \mathbf{W}'') = \iiint \mathbf{W}''(x_2, y_2, z_2) \cdot \text{Lap}\,\mathbf{W}'\, d\,v_2$$

$$= \iiint \mathbf{W}'(x_1, y_1, z_1) \cdot \text{Lap}\,\mathbf{W}''\, d\,v_1. \qquad (74)$$

The *Newtonian product* of a scalar function V, and a vector function $\mathbf{W}$ of position in space is the sextuple integral

$$\text{New}\,(V, \mathbf{W}) = \iiiiiint \mathbf{W}(x_1, y_1, z_1) \cdot \frac{\mathbf{r}_{12}}{r^3_{12}}\, V(x_1, y_1, z_1)\, d\,v_1\, d\,v_2. \qquad (75)$$

By performing one integration

$$\text{New}\,(V, \mathbf{W}) = \iiint \mathbf{W}(x_2, y_2, z_2) \cdot \text{New}\,V\, d\,v_2. \qquad (76)$$

In like manner the *Maxwellian product* of a vector function $\mathbf{W}$ and a scalar function V of position in space is the integral

$$\text{Max}\,(\mathbf{W}, V) = \iiiiiint V(x_1, y_1, z_1)\, \frac{\mathbf{r}_{12}}{r^3_{12}} \cdot \mathbf{W}(x_2, y_2, z_2)\, d\,v_1\, d\,v_2. \qquad (77)$$

One integration yields

$$\text{Max}\,(\mathbf{W}, V) = \iiint V(x_1, y_1, z_1)\,\text{Max}\,\mathbf{W}\,d\,v_1 = -\,\text{New}\,(V, \mathbf{W}). \tag{78}$$

By (53) Art. 93.

$$4\,\pi\,U\,\text{Pot}\,V = -\,(\nabla \cdot \text{New}\,U)\,\text{Pot}\,V.$$

$$\nabla \cdot [\text{New}\,U\,\text{Pot}\,V] = (\nabla \cdot \text{New}\,U)\,\text{Pot}\,V + (\text{New}\,U) \cdot \nabla\,\text{Pot}\,V.$$

$$-\,(\nabla \cdot \text{New}\,U)\,\text{Pot}\,V = -\,\nabla \cdot [\text{New}\,U\,\text{Pot}\,V] + \text{New}\,U \cdot \text{New}\,V.$$

Integrate:

$$4\,\pi \iiint U\,\text{Pot}\,V\,d\,v = -\iiint \nabla \cdot [\text{New}\,U\,\text{Pot}\,V]\,d\,v$$

$$+ \iiint \text{New}\,U \cdot \text{New}\,V\,d\,v.$$

$$4\,\pi\,\text{Pot}\,(U, V) = \iiint \text{New}\,U \cdot \text{New}\,V\,d\,v$$

$$-\iint_S \text{Pot}\,V\,\text{New}\,U \cdot d\,\mathbf{a}. \tag{79}$$

The surface integral is to be taken over the entire surface S bounding the region of integration of the volume integral. As this region of integration is "all space," the surface S may be looked upon as the surface of a large sphere of radius R. If the functions U and V vanish identically for all points outside of certain finite limits, the surface integral must vanish. Hence

$$4\,\pi\,\text{Pot}\,(U, V) = \iiint \text{New}\,U \cdot \text{New}\,V\,d\,v. \tag{79$'$}$$

By (54) Art. 93,

$$4\,\pi\,\mathbf{W}'' \cdot \text{Pot}\,\mathbf{W}' = \nabla \times \text{Lap}\,\mathbf{W}'' \cdot \text{Pot}\,\mathbf{W}'$$

$$-\,\nabla\,\text{Max}\,\mathbf{W}'' \cdot \text{Pot}\,\mathbf{W}'.$$

But $\nabla \cdot [\text{Lap}\,\mathbf{W}'' \times \text{Pot}\,\mathbf{W}'] = \text{Pot}\,\mathbf{W}' \cdot \nabla \times \text{Lap}\,\mathbf{W}''$
$- \text{Lap}\,\mathbf{W}'' \cdot \nabla \times \text{Pot}\,\mathbf{W}',$

and $\nabla \cdot [\text{Max}\,\mathbf{W}''\ \text{Pot}\,\mathbf{W}'] = \text{Pot}\,\mathbf{W}' \cdot \nabla\ \text{Max}\,\mathbf{W}''$
$+ \text{Max}\,\mathbf{W}''\ \nabla \cdot \text{Pot}\,\mathbf{W}'.$

Hence $\nabla \times \text{Lap}\,\mathbf{W}'' \cdot \text{Pot}\,\mathbf{W}' = \nabla \cdot [\text{Lap}\,\mathbf{W}'' \times \text{Pot}\,\mathbf{W}']$
$+ \text{Lap}\,\mathbf{W}'' \cdot \text{Lap}\,\mathbf{W}',$

and $\nabla\ \text{Max}\,\mathbf{W}'' \cdot \text{Pot}\,\mathbf{W}' = \nabla \cdot [\text{Max}\,\mathbf{W}''\ \text{Pot}\,\mathbf{W}']$
$- \text{Max}\,\mathbf{W}''\ \text{Max}\,\mathbf{W}'.$

Hence substituting:

$$\begin{aligned} 4\,\pi\,\mathbf{W}'' \cdot \text{Pot}\,\mathbf{W}' = {} & \text{Lap}\,\mathbf{W}' \cdot \text{Lap}\,\mathbf{W}' + \text{Max}\,\mathbf{W}'\ \text{Max}\,\mathbf{W}'' \\ & + \nabla \cdot [\text{Lap}\,\mathbf{W}'' \times \text{Pot}\,\mathbf{W}'] \\ & - \nabla \cdot [\text{Max}\,\mathbf{W}''\ \text{Pot}\,\mathbf{W}']. \end{aligned}$$

Integrating:

$$\begin{aligned} 4\,\pi\,\text{Pot}\,(\mathbf{W}', \mathbf{W}'') = {} & \iiint \text{Lap}\,\mathbf{W}' \cdot \text{Lap}\,\mathbf{W}''\,dv \\ & + \iiint \text{Max}\,\mathbf{W}'\ \text{Max}\,\mathbf{W}''\,dv \quad (80) \\ & - \iint_s \text{Pot}\,\mathbf{W}' \times \text{Lap}\,\mathbf{W}''\,d\mathbf{a} - \iint_s \text{Max}\,\mathbf{W}''\ \text{Pot}\,\mathbf{W}' \cdot d\mathbf{a}. \end{aligned}$$

If now $\mathbf{W}'$ and $\mathbf{W}''$ exist only in finite space these surface integrals taken over a large sphere of radius R must vanish and then

$$\begin{aligned} 4\,\pi\,\text{Pot}\,(\mathbf{W}', \mathbf{W}'') = {} & \iiint \text{Lap}\,\mathbf{W}' \cdot \text{Lap}\,\mathbf{W}''\,dv \\ & + \iiint \text{Max}\,\mathbf{W}'\ \text{Max}\,\mathbf{W}''\,dv. \quad (80)' \end{aligned}$$

*** 96.**] There are a number of useful theorems of a function-theoretic nature which may perhaps be mentioned here owing to their intimate connection with the integral calculus of vectors. The proofs of them will in some instances be given and in some not. The theorems are often useful in practical applications of vector analysis to physics as well as in purely mathematical work.

Theorem: If $V(x, y, z)$ be a scalar function of position in space which possesses in general a definite derivative ∇V and if in any portion of space, finite or infinite but necessarily continuous, that derivative vanishes, then the function V is constant throughout that portion of space.

Given $$\nabla V = 0.$$

To show $$V = \text{const.}$$

Choose a fixed point (x_1, y_1, z_1) in the region. By (2) page 180

$$\int_{x_1, y_1, z_1}^{x, y, z} \nabla V \cdot d\mathbf{r} = V(x, y, z) - V(x_1, y_1, z_1).$$

But $$\int \nabla V \cdot d\mathbf{r} = \int 0 \cdot d\mathbf{r} = 0.$$

Hence $$V(x, y, z) = V(x_1, y_1, z_1) = \text{const.}$$

Theorem: If $V(x, y, z)$ be a scalar function of position in space which possesses in general a definite derivative ∇V; if the divergence of that derivative exists and is zero throughout any region of space,[1] finite or infinite but necessarily continuous; and if furthermore the derivative ∇V vanishes at every point of any finite volume or of any finite portion of surface in that region or bounding it, then the derivative vanishes throughout all that region and the function V reduces to a constant by the preceding theorem.

[1] The term *throughout any region of space* must be regarded as including the *boundaries* of the region as well as the region itself.

Given $\nabla \cdot \nabla V = 0$ for a region T,

and $\nabla V = 0$ for a finite portion of surface S.

To show $V = \text{const.}$

Since ∇V vanishes for the portion of surface S, V is certainly constant in S. Suppose that, upon one side of S and in the region T, V were not constant. The derivative ∇V upon this side of S has in the main the direction of the normal to the surface S. Consider a sphere which lies for the most part upon the outer side of S but which projects a little through the surface S. The surface integral of ∇V over the small portion of the sphere which projects through the surface S cannot be zero. For, as ∇V is in the main normal to S, it must be nearly parallel to the normal to the portion of spherical surface under consideration. Hence the terms

$$\nabla V \cdot d\mathbf{a}$$

in the surface integral all have the same sign and cannot cancel each other out. The surface integral of ∇V over that portion of S which is intercepted by the spherical surface vanishes because ∇V is zero. Consequently the surface integral of ∇V taken over the entire surface of the spherical segment which projects through S is not zero.

But $$\iint \nabla V \cdot d\mathbf{a} = \iiint \nabla \cdot \nabla V \, dv = 0.$$

Hence $$\iint \nabla V \cdot d\mathbf{a} = 0.$$

It therefore appears that the supposition that V is not constant upon one side of S leads to results which contradict the given relation $\nabla \cdot \nabla V = 0$. The supposition must therefore have been incorrect and V must be constant not only in S but in all portions of space near to S in the region T. By an extension of the reasoning V is seen to be constant throughout the entire region T.

Theorem: If $V\ (x, y, z)$ be a scalar function of position in space possessing in general a derivative ∇V and if throughout a certain region[1] T of space, finite or infinite, continuous or discontinuous, the divergence $\nabla \cdot \nabla V$ of that derivative exists and is zero, and if furthermore the function V possesses a constant value c in *all* the surfaces bounding the region and $V(x, y, z)$ approaches c as a limit when the point (x, y, z) recedes to infinity, then throughout the entire region T the function V has the same constant value c and the derivative ∇V vanishes.

The proof does not differ essentially from the one given in the case of the last theorem. The theorem may be generalized as follows:

Theorem: If $V(x, y, z)$ be any scalar function of position in space possessing in general a derivative ∇V; if $U\ (x, y, z)$ be any other scalar function of position which is either positive or negative throughout and upon the boundaries of a region T, finite or infinite, continuous or discontinuous; if the divergence $\nabla \cdot [U \nabla V]$ of the product of U and ∇V exists and is zero throughout and upon the boundaries of T and at infinity; and if furthermore V be constant and equal to c upon *all* the boundaries of T and at infinity; then the function V is constant throughout the entire region T and is equal to c.

Theorem: If $V\ (x, y, z)$ be any scalar function of position in space possessing in general a derivative ∇V; if throughout any region T of space, finite or infinite, continuous or discontinuous, the divergence $\nabla \cdot \nabla V$ of this derivative exists and is zero; and if in *all* the bounding surfaces of the region T the normal component of the derivative ∇V vanishes and at infinite distances in T (if such there be) the product

[1] The region includes its boundaries.

$r^2 \partial V / \partial r$ vanishes, where r denotes the distance measured from any fixed origin; then throughout the entire region T the derivative ∇V vanishes and in each continuous portion of T V is constant, although for different continuous portions this constant may not be the same.

This theorem may be generalized as the preceding one was by the substitution of the relation $\nabla \cdot (U \nabla V) = 0$ for $\nabla \cdot \nabla V = 0$ and $U r^2 \partial V / \partial r = 0$ for $r^2 \partial V / \partial r = 0$.

As corollaries of the foregoing theorems the following statements may be made. The language is not so precise as in the theorems themselves, but will perhaps be understood when they are borne in mind.

If $\nabla U = \nabla V$, then U and V differ at most by a constant.

If $\nabla \cdot \nabla U = \nabla \cdot \nabla V$ and if $\nabla U = \nabla V$ in any finite portion of surface S, then $\nabla U = \nabla V$ at all points and U differs from V only by a constant at most.

If $\nabla \cdot \nabla U = \nabla \cdot \nabla V$ and if $U = V$ in all the bounding surfaces of the region and at infinity (if the region extend thereto), then at all points U and V arc equal.

If $\nabla \cdot \nabla U = \nabla \cdot \nabla V$ and if in all the bounding surfaces of the region the normal components of ∇U and ∇V are equal and if at infinite distances $r^2 (\partial U / \partial r - \partial V / \partial r)$ is zero, then ∇U and ∇V are equal at all points of the region and U differs from V only by a constant.

Theorem: If $\mathbf{W}'$ and $\mathbf{W}''$ are two vector functions of position in space which in general possess curls and divergences; if for any region T, finite or infinite but necessarily continuous, the curl of $\mathbf{W}'$ is equal to the curl of $\mathbf{W}''$ and the divergence of $\mathbf{W}'$ is equal to the divergence of $\mathbf{W}''$; and if moreover the two functions $\mathbf{W}'$ and $\mathbf{W}''$ are equal to each other at every point of any finite volume in T or of any finite surface in T or bounding it; then $\mathbf{W}'$ is equal to $\mathbf{W}''$ at every point of the region T.

Since $\nabla \times \mathbf{W}' = \nabla \times \mathbf{W}''$, $\nabla \times (\mathbf{W}' - \mathbf{W}'') = 0$. A vector function whose curl vanishes is equal to the derivative[1] of a scalar function V (page 197). Let $\nabla V = \mathbf{W}' - \mathbf{W}''$. Then $\nabla \cdot \nabla V = 0$ owing to the equality of the divergences. The theorem therefore becomes a corollary of a preceding one.

Theorem: If $\mathbf{W}'$ and $\mathbf{W}''$ are two vector functions of position which in general possess definite curls and divergences; if throughout any *aperiphractic*[2] region T, finite but not necessarily continuous, the curl of $\mathbf{W}'$ is equal to the curl of $\mathbf{W}''$ and the divergence of $\mathbf{W}'$ is equal to the divergence of $\mathbf{W}''$; and if furthermore in *all* the bounding surfaces of the region T the *tangential* components $\mathbf{W}'$ and $\mathbf{W}''$ are equal; then $\mathbf{W}'$ is equal to $\mathbf{W}''$ throughout the aperiphractic region T.

Theorem: If $\mathbf{W}'$ and $\mathbf{W}''$ are two vector functions of position in space which in general possess definite curls and divergences; if throughout any *acyclic* region T, finite but not necessarily continuous, the curl of $\mathbf{W}'$ is equal to the curl $\mathbf{W}''$ and the divergence of $\mathbf{W}'$ is equal to the divergence of $\mathbf{W}''$; and if in *all* the bounding surfaces of the region T the *normal* components of $\mathbf{W}'$ and $\mathbf{W}''$ are equal; then the functions $\mathbf{W}'$ and $\mathbf{W}''$ are equal throughout the region acyclic T.

The proofs of these two theorems are carried out by means of the device suggested before.

Theorem: If $\mathbf{W}'$ and $\mathbf{W}''$ are two vector functions such that $\nabla \cdot \nabla \mathbf{W}'$ and $\nabla \cdot \nabla \mathbf{W}''$ have in general definite values in a certain region T, finite or infinite, continuous or discontinuous; and if in *all* the bounding surfaces of the region and at infinity the functions $\mathbf{W}'$ and $\mathbf{W}''$ are equal; then $\mathbf{W}'$ is equal to $\mathbf{W}''$ throughout the entire region T.

The proof is given by treating separately the three components of $\mathbf{W}'$ and $\mathbf{W}''$.

[1] The region T may have to be made acyclic by the insertion of diaphragms.

[2] A region which encloses within itself another region is said to be *periphractic.* If it encloses no region it is *aperiphractic.*

SUMMARY OF CHAPTER IV

The line integral of a vector function $\mathbf{W}$ along a curve C is defined as

$$\int_C \mathbf{W} \cdot d\mathbf{r} = \int_C [W_1\, dx + W_2\, dy + W_3\, dz]. \qquad (1)$$

The line integral of the derivative ∇V of a scalar function V along a curve C from $\mathbf{r}_0$ to $\mathbf{r}$ is equal to the difference between the values of V at the points $\mathbf{r}$ and $\mathbf{r}_0$ and hence the line integral taken around a closed curve is zero; and conversely if the line integral of a vector function $\mathbf{W}$ taken around any closed curve vanishes, then $\mathbf{W}$ is the derivative ∇V of some scalar function V.

$$\int_{\mathbf{r}_0}^{\mathbf{r}_1} \nabla V \cdot d\mathbf{r} = V(\mathbf{r}) - V(\mathbf{r}_0) \qquad (2)$$

$$\int_{\circ} \nabla V \cdot d\mathbf{r} = 0 \qquad (3)$$

and if $\displaystyle\int_{\circ} \mathbf{W} \cdot d\mathbf{r} = 0$, then $\mathbf{W} = \nabla V$.

Illustration of the theorem by application to mechanics.

The surface integral of a vector function $\mathbf{W}$ over a surface S is defined as

$$\iint_S \mathbf{W} \cdot d\mathbf{a} = \iint_S [W_1\, dy\, dz + W_2\, dz\, dx + W_3\, dx\, dy].$$

Gauss's Theorem: The surface integral of a vector function taken over a closed surface is equal to the volume integral of the divergence of that function taken throughout the volume enclosed by that surface

$$\iiint \nabla \cdot \mathbf{W}\, dv = \iint_S \mathbf{W} \cdot d\mathbf{a}, \qquad (7)$$

or
$$\iiint\left(\frac{\partial X}{\partial x}+\frac{\partial Y}{\partial y}+\frac{\partial Z}{\partial z}\right)dv$$
$$=\iint_s [X\,dy\,dz+Y\,dz\,dx+Z\,dx\,dy], \qquad (8)$$

if X, Y, Z be the three components of the vector function $\mathbf{W}$.

Stokes's Theorem: The surface integral of the curl of a vector function taken over any surface is equal to the line integral of the function taken around the line bounding the surface. And conversely if the surface integral of a vector function $\mathbf{U}$ taken over any surface is equal to the line integral of a function $\mathbf{W}$ taken around the boundary, then $\mathbf{U}$ is the curl of $\mathbf{W}$.

$$\iint_s \nabla\times\mathbf{W}\cdot d\,\mathbf{a}=\int_\circ \mathbf{W}\cdot d\,\mathbf{r}, \qquad (11)$$

and if $\iint_s \mathbf{U}\cdot d\,\mathbf{a}=\int_\circ \mathbf{W}\cdot d\,\mathbf{r}$, then $\mathbf{U}=\nabla\times\mathbf{W}$. (12)

Application of the theorem of Stokes to deducing the equations of the electro-magnetic field from two experimental facts due to Faraday. Application of the theorems of Stokes and Gauss to the proof that the divergence of the curl of a vector function is zero and the curl of the derivative of a scalar function is zero.

Formulæ analogous to integration by parts

$$\int u\,\nabla v\cdot d\,\mathbf{r}=[u\,v]_{\mathbf{r}_0}^{\mathbf{r}}-\int v\,\nabla u\cdot d\,\mathbf{r}, \qquad (14)$$

$$\iint_s \nabla u\times\mathbf{v}\cdot d\,\mathbf{a}=\int_\circ u\,\mathbf{v}\cdot d\,\mathbf{r}-\iint_s u\,\nabla\times\mathbf{v}\cdot d\,\mathbf{a}, \qquad (15)$$

$$\iint_s \nabla u\times\nabla v\cdot d\mathbf{a}=\int_\circ u\,\nabla v\cdot d\,\mathbf{r}=-\int_\circ v\,\nabla u\cdot d\,\mathbf{r}, \qquad (16)$$

$$\iiint u \nabla \cdot \mathbf{v}\, dv = \iint_s u \mathbf{v} \cdot d\mathbf{a} - \iiint \nabla u \cdot \mathbf{v}\, dv, \quad (17)$$

$$\iint_s \nabla u \times \mathbf{v} \cdot d\mathbf{a} = -\iiint \nabla u \cdot \nabla \times \mathbf{v}\, dv. \quad (18)$$

Green's Theorem:

$$\iiint \nabla u \cdot \nabla v\, dv = \iint_s u \nabla v \cdot d\mathbf{a} - \iiint u \nabla \cdot \nabla v\, dv$$

$$= \iint_s v \nabla u \cdot d\mathbf{a} - \iiint v \nabla \cdot \nabla u\, dv, \quad (19)$$

$$\iiint (u \nabla \cdot \nabla v - v \nabla \cdot \nabla u)\, dv = \iint_s (u \nabla v - v \nabla u) \cdot d\mathbf{a}. \quad (20)$$

Kelvin's generalization:

$$\iiint w \nabla u \cdot \nabla v\, dv = \iint_s u w \nabla v \cdot d\mathbf{a} - \iiint u \cdot \nabla [w \nabla v]\, dv$$

$$= \iint_s v w \nabla u \cdot d\mathbf{a} - \iiint v \nabla \cdot [w \nabla u]\, dv. \quad (21)$$

The integrating operator known as the *potential* is defined by the equation

$$\operatorname{Pot} V = \iiint \frac{V(x_2, y_2, z_2)}{r_{12}}\, dx_2\, dy_2\, dz_2. \quad (22)$$

$$\operatorname{Pot} \mathbf{W} = \iiint \frac{\mathbf{W}(x_2, y_2, z_2)}{r_{12}}\, dx_2\, dy_2\, dz_2. \quad (23)$$

$$\nabla \operatorname{Pot} V = \operatorname{Pot} \nabla V, \quad (27)$$

$$\nabla \times \operatorname{Pot} \mathbf{W} = \operatorname{Pot} \nabla \times \mathbf{W}, \quad (28)$$

$$\nabla \cdot \operatorname{Pot} \mathbf{W} = \operatorname{Pot} \nabla \cdot \mathbf{W}, \quad (29)$$

$$\nabla \cdot \nabla \operatorname{Pot} V = \operatorname{Pot} \nabla \cdot \nabla V, \quad (30)$$

$$\nabla \cdot \nabla \operatorname{Pot} \mathbf{W} = \operatorname{Pot} \nabla \cdot \nabla \mathbf{W}, \tag{31}$$

$$\nabla \nabla \cdot \operatorname{Pot} \mathbf{W} = \operatorname{Pot} \nabla \nabla \cdot \mathbf{W}, \tag{32}$$

$$\nabla \times \nabla \times \operatorname{Pot} \mathbf{W} = \operatorname{Pot} \nabla \times \nabla \times \mathbf{W}. \tag{33}$$

The integrating operator Pot and the differentiating operator ∇ are commutative.

The three additional integrating operators known as the *Newtonian*, the *Laplacian*, and the *Maxwellian*.

$$\operatorname{New} V = \iiint \frac{\mathbf{r}_{12}\, V(x_2, y_2, z_2)}{r^3_{12}}\, dx_2\, dy_2\, dz_2. \tag{42}$$

$$\operatorname{Lap} \mathbf{W} = \iiint \frac{\mathbf{r}_{12} \times \mathbf{W}\,(x_2, y_2, z_2)}{r^3_{12}}\, dx_2\, dy_2\, dz_2. \tag{43}$$

$$\operatorname{Max} \mathbf{W} = \iiint \frac{\mathbf{r}_{12} \cdot \mathbf{W}\,(x_2, y_2, z_2)}{r^3_{12}}\, dx_2\, dy_2\, dz_2. \tag{44}$$

If the potential exists these integrals are related to it as follows:

$$\nabla \operatorname{Pot} V = \operatorname{New} V,$$
$$\nabla \times \operatorname{Pot} \mathbf{W} = \operatorname{Lap} \mathbf{W}, \tag{45}$$
$$\nabla \cdot \operatorname{Pot} \mathbf{W} = \operatorname{Max} \mathbf{W}.$$

The interpretation of the physical meaning of the Newtonian on the assumption that V is the density of an attracting body, of the Laplacian on the assumption that $\mathbf{W}$ is electric flux, of the Maxwellian on the assumption that $\mathbf{W}$ is the intensity of magnetization. The expression of these integrals or their components in terms of x, y, z; formulæ (42)′, (43)′, (44)′ and (42)″, (43)″, (44)″.

$$\nabla \cdot \operatorname{New} V = \operatorname{Max} \nabla V, \tag{46}$$

$$\nabla \operatorname{Max} \mathbf{W} = \operatorname{New} \nabla \cdot \mathbf{W}, \tag{47}$$

$$\nabla \times \operatorname{Lap} \mathbf{W} = \operatorname{Lap} \nabla \times \mathbf{W}, \tag{48}$$

$$\nabla \cdot \text{Lap}\ \mathbf{W} = \text{Max}\ \nabla \times \mathbf{W} = 0, \tag{49}$$

$$\nabla \times \text{New}\ V = \text{Lap}\ \nabla V = 0, \tag{50}$$

$$\begin{aligned} \nabla \cdot \nabla\ \text{Pot}\ \mathbf{W} &= \text{New}\ \nabla \cdot \mathbf{W} - \text{Lap}\ \nabla \times \mathbf{W} \\ &= \nabla\ \text{Max}\ \mathbf{W} - \nabla \times \text{Lap}\ \mathbf{W}. \end{aligned} \tag{51}$$

The potential is a solution of Poisson's Equation. That is,

$$\nabla \cdot \nabla\ \text{Pot}\ V = -4\pi V, \tag{52}$$

and

$$\nabla \cdot \nabla\ \text{Pot}\ V = -4\pi \mathbf{W}. \tag{52'}$$

$$V = \frac{-1}{4\pi} \nabla \cdot \text{New}\ V, \tag{53}$$

$$\mathbf{W} = \frac{1}{4\pi} \text{Lap}\ \nabla \times \mathbf{W} - \frac{1}{4\pi} \text{New}\ \nabla \cdot \mathbf{W}. \tag{55}$$

Hence $\mathbf{W}$ is divided into two parts of which one is solenoidal and the other irrotational, provided the potential exists. In case the potential does not exist a third term $\mathbf{W}_3$ must be added of which both the divergence and the curl vanish. A list of theorems which follow immediately from equations (52), (52)′, (53), (55) and which state that certain integrating operators are inverse to certain differentiating operators. Let V be a scalar function, $\mathbf{W}_1$, a solenoidal vector function, and $\mathbf{W}_2$ an irrotational vector function. Then

$$\frac{1}{4\pi} \text{Lap}\ \nabla \times \mathbf{W}_1 = \nabla \times \frac{1}{4\pi} \text{Lap}\ \mathbf{W}_1 = \mathbf{W}_1. \tag{60}$$

$$\frac{1}{4\pi} \text{Lap}\ \mathbf{W}_2 = 0, \quad \nabla \times \mathbf{W}_2 = 0 \tag{61}$$

$$-\frac{1}{4\pi} \text{New}\ \nabla \cdot \mathbf{W}_2 = -\nabla \cdot \frac{1}{4\pi} \text{New}\ \mathbf{W}_2 = \mathbf{W}_2. \tag{62}$$

$$\begin{cases} -\nabla \cdot \dfrac{1}{4\pi} \text{ New } V = V \\ -\dfrac{1}{4\pi} \text{ Max } \nabla V = V. \end{cases} \tag{63}$$

$$\frac{1}{4\pi} \text{ Pot } \nabla \times \nabla \times \mathbf{W}_1 = \nabla \times \nabla \times \text{ Pot } \mathbf{W}_1 = \mathbf{W}_1 \tag{64}$$

$$-\frac{1}{4\pi} \text{ Pot } \nabla \nabla \cdot \mathbf{W}_2 = - \nabla \cdot \nabla \frac{1}{4\pi} \text{ Pot } \mathbf{W}_2 = \mathbf{W}_2. \tag{65}$$

$$\begin{cases} -\dfrac{1}{4\pi} \text{ Pot } \nabla \cdot \nabla V = - \nabla \cdot \nabla \dfrac{1}{4\pi} \text{ Pot } V = V \\ -\dfrac{1}{4\pi} \text{ Pot } \nabla \cdot \nabla \mathbf{W} = - \nabla \cdot \nabla \dfrac{1}{4\pi} \text{ Pot } \mathbf{W} = \mathbf{W}. \end{cases} \tag{66}$$

$$- \nabla \cdot \nabla \mathbf{W}_1 = \nabla \times \nabla \times \mathbf{W}_1 \tag{67}$$

$$\nabla \cdot \nabla \mathbf{W}_2 = \nabla \nabla \cdot \mathbf{W}_2 \tag{68}$$

$$4\pi \text{ Pot } V = - \text{ Max New } V \tag{69}$$

$$4\pi \text{ Pot } \mathbf{W} = \text{Lap Lap } \mathbf{W} - \text{New Max } \mathbf{W}. \tag{70}$$

Mutual potentials Newtonians, Laplacians, and Maxwellians may be formed. They are sextuple integrals. The integrations cannot all be performed immediately ; but the first three may be. Formulæ (71) to (80) inclusive deal with these integrals. The chapter closes with the enunciation of a number of theorems of a function-theoretic nature. By means of these theorems certain facts concerning functions may be inferred from the conditions that they satisfy Laplace's equation and have certain boundary conditions.

Among the exercises number 6 is worthy of especial attention. *The work done in the text has for the most part assumed that the potential exists. But many of the formulæ connecting Newtonians, Laplacians, and Maxwellians hold when the potential does not exist.* These are taken up in Exercise 6 referred to.

Exercises on Chapter IV

1.[1] If V is a scalar function of position in space the line integral

$$\int_C V\, d\,\mathbf{r}$$

is a vector quantity. Show that

$$\int_\circ V\, d\,\mathbf{r} = - \iint_s \nabla V \times d\,\mathbf{a} = \iint_s d\,\mathbf{a} \times \nabla V.$$

That is; the line integral of a scalar function around a closed curve is equal to the *skew* surface integral of the derivative of the function taken over any surface spanned into the contour of the curve. Show further that if V is constant the integral around any closed curve is zero and conversely if the integral around any closed curve is zero the function V is constant.

Hint: Instead of treating the integral as it stands multiply it (with a dot) by an arbitrary constant unit vector and thus reduce it to the line integral of a vector function.

2. If $\mathbf{W}$ is a vector function the line integral

$$\mathbf{H} = \int_C \mathbf{W} \times d\,\mathbf{r}$$

is a vector quantity. It may be called the *skew* line integral of the function $\mathbf{W}$. If $\mathbf{c}$ is any constant vector, show that if the integral be taken around a closed curve

$$\mathbf{H} \cdot \mathbf{c} = \iint_s (\mathbf{c}\, \nabla \cdot \mathbf{W} - \mathbf{c} \cdot \nabla\, \mathbf{W}) \cdot d\,\mathbf{a} = \mathbf{c} \cdot \int_\circ \mathbf{W} \times d\,\mathbf{r},$$

[1] The first four exercises are taken from Föppl's Einführung in die Maxwell'sche Theorie der Electricität where they are worked out.

and $$\mathbf{H}\cdot\mathbf{c}=\mathbf{c}\cdot\left\{\iint_s \nabla\cdot\mathbf{W}\,d\,\mathbf{a}-\iint_s \nabla\,(\mathbf{W}\cdot d\,\mathbf{a})\right\}$$

$$+\iint_s \mathbf{W}\cdot[\mathbf{c}\cdot\nabla\,(d\,\mathbf{a})].$$

In case the integral is taken over a plane curve and the surface S is the portion of plane included by the curve

$$\mathbf{H}=\iint_s [\,\nabla\cdot\mathbf{W}\,d\,\mathbf{a}-\nabla\,(\mathbf{W}\cdot d\,\mathbf{a})].$$

Show that the integral taken over a plane curve vanishes when **W** is constant and conversely if the integral over any plane curve vanishes **W** must be constant.

3. The surface integral of a scalar function V is

$$\mathbf{S}=\iint_s V\,d\,\mathbf{a}.$$

This is a vector quantity. Show that the surface integral of V taken over any closed surface is equal to the volume integral of ∇V taken throughout the volume bounded by that surface. That is

$$\iint_s V\,d\,\mathbf{a}=-\iiint \nabla V\,d\,v.$$

Hence conclude that the surface integral over a closed surface vanishes if V be constant and conversely if the surface integral over any closed surface vanishes the function V must be constant.

4. If **W** be a vector function, the surface integral

$$\mathbf{T}=\iint_s d\,\mathbf{a}\times\mathbf{W}$$

may be called the *skew* surface integral. It is a vector quantity. Show that the skew surface integral of a vector function taken over a closed surface is equal to the volume integral of the vector function taken throughout the volume bounded by the surface. That is

$$\iint_s d\mathbf{a} \times \mathbf{W} = \iiint \nabla \times \mathbf{W}\, dv.$$

Hence conclude that the skew surface integral taken over any surface in space vanishes when and only when $\mathbf{W}$ is an irrotational function. That is, when and only when the line integral of $\mathbf{W}$ for every closed circuit vanishes.

5. Obtain some formulæ for these integrals which are analogous to integrating by parts.

6. The work in the text assumes for the most part that the potentials of V and $\mathbf{W}$ exist. Many of the relations, however, may be demonstrated without that assumption. Assume that the Newtonian, the Laplacian, the Maxwellian exist. For simplicity in writing let

$$p_{12} = \frac{1}{r_{12}}, \qquad \nabla_1 p_{12} = \frac{\mathbf{r}_{12}}{r^3_{12}}$$

Then

$$\text{New } V = \iiint \nabla_1 p_{12} V(x_2, y_2, z_2)\, d v_2, \tag{81}$$

$$\text{Lap } \mathbf{W} = \iiint \nabla_1 p_{12} \times \mathbf{W}\,(x_2, y_2, z_2)\, d v_2, \tag{82}$$

$$\text{Max } \mathbf{W} = \iiint \nabla_1 p_{12} \cdot \mathbf{W}\,(x_2, y_2, z_2)\, d v_2, \tag{83}$$

$$\nabla_1 p_{12} = - \nabla_2 p_{12} \tag{84}$$

$$\nabla_2 (p_{12} V) = \nabla_2 p_{12}\, V + p_{12} \nabla_2 V,$$

$$\iiint \nabla_2 p_{12}\, V\, d v_2 = \iiint \nabla_2 (p_{12}\, V)\, d v_2 - \iiint p_{12} \nabla_2\, V\, d v_2.$$

By exercise (3) $\iiint \nabla_2 (p_{12} V) \, d v_2 = \iint p_{12} V \, d \mathbf{a}.$

It can be shown that if V is such a function that New V exists, then this surface integral taken over a large sphere of radius R and a small sphere of radius R' approaches zero when R becomes indefinitely great; and R', indefinitely small. Hence

$$\iiint \nabla_2 p_{12} V \, d v_2 = - \iiint p_{12} \nabla_2 V \, d v_2,$$

or

$$\text{New } V = \text{Pot } \nabla V. \tag{85}$$

Prove in a similar manner that

$$\text{Lap } \mathbf{W} = \text{Pot } \nabla \times \mathbf{W}, \tag{86}$$

$$\text{Max } \mathbf{W} = \text{Pot } \nabla \cdot \mathbf{W}. \tag{87}$$

By means of (85), (86), (87) it is possible to prove that

$$\nabla \times \text{Lap } \mathbf{W} = \text{Lap } \nabla \times \mathbf{W},$$

$$\nabla \cdot \text{New } V = \text{Max } \nabla V,$$

$$\nabla \text{ Max } \mathbf{W} = \text{New } \nabla \cdot \mathbf{W}.$$

Then prove

$$\nabla \times \text{Lap } \mathbf{W} = \iiint p_{12} \nabla \nabla \cdot \mathbf{W} \, d v_2 - \iiint p_{12} \nabla \cdot \nabla \mathbf{W} \, d v_2$$

and

$$\nabla \text{ Max } \mathbf{W} = \iiint p_{12} \nabla \nabla \cdot \mathbf{W} \, d v_2.$$

Hence $\nabla \times \text{Lap } \mathbf{W} - \nabla \text{ Max } \mathbf{W} = - \iiint p_{12} \nabla \cdot \nabla \mathbf{W} \, d v_2.$

Hence

$$\nabla \times \text{Lap } \mathbf{W} - \nabla \text{ Max } \mathbf{W} = 4 \pi \mathbf{W}. \tag{88}$$

7. An integral used by Helmholtz is

$$H(V) = \iiint r_{12} V \, d v_2,$$

or if **W** be a vector function

$$H(\mathbf{W}) = \iiint r_{12}\, \mathbf{W}\, dv_2. \tag{90}$$

Show that the integral converges if V diminishes so rapidly that

$$V r^5 < K$$

when r becomes indefinitely great.

$$\nabla H(V) = H(\nabla V) = \text{New}\,(r^2\, V), \tag{91}$$

$$\nabla \cdot H(\mathbf{W}) = H(\nabla \cdot \mathbf{W}) = \text{Max}\,(r^2\, \mathbf{W}), \tag{92}$$

$$\nabla \times H(\mathbf{W}) = H(\nabla \times \mathbf{W}) = \text{Lap}\,(r^2\, \mathbf{W}), \tag{93}$$

$$\nabla \cdot \nabla H(V) = H(\nabla \cdot \nabla V) = \text{Max}\,(r^2 \nabla V) = 2 \text{ Pot } V \tag{94}$$

$$\nabla \cdot \nabla\, H(\mathbf{W}) = H(\nabla \cdot \nabla\, \mathbf{W}) = 2 \text{ Pot } \mathbf{W}. \tag{95}$$

$$H(V) = -\frac{1}{2\,\pi} \text{ Pot Pot } V. \tag{96}$$

$$H(\mathbf{W}) = -\frac{1}{2\,\pi} \text{ Pot Pot } \mathbf{W}. \tag{97}$$

$$-2\,\mathbf{W} = \nabla \times \nabla \times H(\mathbf{W}) + \nabla \nabla \cdot H(\mathbf{W}). \tag{98}$$

8. Give a proof of Gauss's Theorem which does not depend upon the physical interpretation of a function as the flux of a fluid. The reasoning is similar to that employed in Art. 51 and in the first proof of Stokes's Theorem.

9. Show that the division of **W** into two parts, page 235, is unique.

10. Treat, in a manner analogous to that upon page 220, the case in which V has curves of discontinuities.

CHAPTER V

LINEAR VECTOR FUNCTIONS

97.] AFTER the definitions of products had been laid down and applied, two paths of advance were open. One was differential and integral calculus; the other, higher algebra in the sense of the theory of linear homogeneous substitutions. The treatment of the first of these topics led to new ideas and new symbols — to the derivative, divergence, curl, scalar and vector potential, that is, to ∇, $\nabla\cdot$, $\nabla\times$, and *Pot* with the auxiliaries, the Newtonian, the Laplacian, and the Maxwellian. The treatment of the second topic will likewise introduce novelty both in concept and in notation — the linear vector function, the dyad, and the dyadic with their appropriate symbolization.

The simplest example of a linear vector function is the product of a scalar constant and a vector. The vector $\mathbf{r}'$

$$\mathbf{r}' = c\,\mathbf{r} \qquad (1)$$

is a linear function of $\mathbf{r}$. A more general linear function may be obtained by considering the components of $\mathbf{r}$ individually. Let $\mathbf{i}, \mathbf{j}, \mathbf{k}$ be a system of axes. The components of $\mathbf{r}$ are

$$\mathbf{i}\cdot\mathbf{r}, \quad \mathbf{j}\cdot\mathbf{r}, \quad \mathbf{k}\cdot\mathbf{r}.$$

Let each of these be multiplied by a scalar constant which may be different for the different components.

$$c_1\,\mathbf{i}\cdot\mathbf{r}, \quad c_2\,\mathbf{j}\cdot\mathbf{r}, \quad c_3\,\mathbf{k}\cdot\mathbf{r}.$$

Take these as the components of a new vector $\mathbf{r}'$

$$\mathbf{r}' = \mathbf{i}\ (c_1\,\mathbf{i} \cdot \mathbf{r}) + \mathbf{j}\ (c_2\,\mathbf{j} \cdot \mathbf{r}) + \mathbf{k}\ (c_3\,\mathbf{k} \cdot \mathbf{r}). \qquad (2)$$

The vector $\mathbf{r}'$ is then a linear function of $\mathbf{r}$. Its components are always equal to the corresponding components of $\mathbf{r}$ each multiplied by a definite scalar constant.

Such a linear function has numerous applications in geometry and physics. If, for instance, $\mathbf{i}$, $\mathbf{j}$, $\mathbf{k}$ be the axes of a homogeneous strain and c_1, c_2, c_3, the elongations along these axes, a point

$$\mathbf{r} = \mathbf{i}\,x + \mathbf{j}\,y + \mathbf{k}\,z$$

becomes
$$\mathbf{r}' = \mathbf{i}\,c_1\,x + \mathbf{j}\,c_2\,y + \mathbf{k}\,c_3\,z,$$

or
$$\mathbf{r}' = \mathbf{i}\,c_1\,\mathbf{i} \cdot \mathbf{r} + \mathbf{j}\,c_2\,\mathbf{j} \cdot \mathbf{r} + \mathbf{k}\,c_3\,\mathbf{k} \cdot \mathbf{r}.$$

This sort of linear function occurs in the theory of elasticity and in hydrodynamics. In the theory of electricity and magnetism, the electric force $\mathbf{E}$ is a linear function of the electric displacement $\mathbf{D}$ in a dielectric. For isotropic bodies the function becomes merely a constant

$$\mathbf{E} = k\,\mathbf{D}.$$

But in case the body be non-isotropic, the components of the force along the different axes will be multiplied by different constants k_1, k_2, k_3. Thus

$$\mathbf{E} = \mathbf{i}\,k_1\,\mathbf{i} \cdot \mathbf{D} + \mathbf{j}\,k_2\,\mathbf{j} \cdot \mathbf{D} + \mathbf{k}\,k_3\,\mathbf{k} \cdot \mathbf{D}.$$

The linear vector function is indispensable in dealing with the phenomena of electricity, magnetism, and optics in non-isotropic bodies.

98.] It is possible to define a linear vector function, as has been done above, by means of the components of a vector. The most general definition would be

Definition: A vector $\mathbf{r}'$ is said to be a linear vector function of another vector $\mathbf{r}$ when the components of $\mathbf{r}'$ along three non-coplanar vectors are expressible linearly with scalar coefficients in terms of the components of $\mathbf{r}$ along those same vectors.

If $$\mathbf{r} = x\,\mathbf{a} + y\,\mathbf{b} + z\,\mathbf{c}, \text{ where } [\mathbf{a}\,\mathbf{b}\,\mathbf{c}] \neq 0,$$

and $$\mathbf{r}' = x'\,\mathbf{a} + y'\,\mathbf{b} + z'\,\mathbf{c},$$

and if
$$x' = a_1 x + b_1 y + c_1 z,$$
$$y' = a_2 x + b_2 y + c_2 z, \qquad (3)$$
$$z' = a_3 x + b_3 y + c_3 z,$$

then $\mathbf{r}'$ is a linear function of $\mathbf{r}$. (The constants a_1, b_1, c_1, etc., have no connection with the components of $\mathbf{a}$, $\mathbf{b}$, $\mathbf{c}$ parallel to $\mathbf{i}$, $\mathbf{j}$, $\mathbf{k}$.) Another definition however is found to be more convenient and from it the foregoing may be deduced.

Definition: A continuous vector function of a vector is said to be a *linear* vector function when the function of the sum of any two vectors is the sum of the functions of those vectors. That is, the function f is linear if

$$f(\mathbf{r}_1 + \mathbf{r}_2) = f(\mathbf{r}_1) + f(\mathbf{r}_2). \qquad (4)$$

Theorem: If a be any positive or negative scalar and if f be a linear function, then the function of a times $\mathbf{r}$ is a times the function of $\mathbf{r}$.

$$f(a\,\mathbf{r}) = a\,f(\mathbf{r}). \qquad (5)$$

And hence

$$f(a_1\mathbf{r}_1 + a_2\mathbf{r}_2 + a_3\mathbf{r}_3 + \cdots)$$
$$= a_1 f(\mathbf{r}_1) + a_2 f(\mathbf{r}_2) + a_3 f(\mathbf{r}_3) + \cdots \qquad (5)$$

The proof of this theorem which appears more or less obvious is a trifle long. It depends upon making repeated use of relation (4).

$$f(\mathbf{r} + \mathbf{r}) = f(\mathbf{r}) + f(\mathbf{r}) = 2 f(\mathbf{r}).$$

Hence $$f(2\mathbf{r}) = 2 f(\mathbf{r}).$$

In like manner $$f(n\mathbf{r}) = n f(\mathbf{r})$$

where n is any positive integer.

Let m be any other positive integer. Then by the relation just obtained

$$f(\mathbf{r}) = f\left(m \frac{\mathbf{r}}{m}\right) = m f\left(\frac{\mathbf{r}}{m}\right)$$

and $$f\left(\frac{\mathbf{r}}{m}\right) = \frac{1}{m} f(\mathbf{r}).$$

Hence $$f\left(n \frac{\mathbf{r}}{m}\right) = f\left(\frac{n}{m} \mathbf{r}\right) = \frac{n}{m} f(\mathbf{r}).$$

That is, equation (5) has been proved in case the constant a is a rational positive number.

To show the relation for negative numbers note that

$$f(0) = f(0 + 0) = 2 f(0).$$

Hence $$f(0) = 0.$$

But $$f(0) = f(\mathbf{r} - \mathbf{r}) = f(\mathbf{r} + (-\mathbf{r})) = f(\mathbf{r}) + f(-\mathbf{r}).$$

Hence $$f(\mathbf{r}) = - f(-\mathbf{r}).$$

To prove (5) for incommensurable values of the constant a, it becomes necessary to make use of the continuity of the function f. That is

$$\underset{x = a}{\text{Lim}} f(x\mathbf{r}) = f\left(\underset{x \doteq a}{\text{Lim}} (x\mathbf{r})\right).$$

Let x approach the incommensurable number a by passing through a suite of commensurable values. Then

$$f(x\mathbf{r}) = x f(\mathbf{r}).$$

Hence $$\underset{x = a}{\text{Lim}} f(x\mathbf{r}) = a f(\mathbf{r})$$

$$\underset{x \doteq a}{\text{Lim}} (x\,\mathbf{r}) = a\,\mathbf{r}.$$

Hence $$f(a\,\mathbf{r}) = a\,f(\mathbf{r})$$

which proves the theorem.

Theorem: A linear vector function $f(\mathbf{r})$ is entirely determined when its values for three non-coplanar vectors $\mathbf{a}, \mathbf{b}, \mathbf{c}$ are known.

Let
$$\mathbf{l} = f(\mathbf{a}),$$
$$\mathbf{m} = f(\mathbf{b}),$$
$$\mathbf{n} = f(\mathbf{c}).$$

Since $\mathbf{r}$ is any vector whatsoever, it may be expressed as

$$\mathbf{r} = x\,\mathbf{a} + y\,\mathbf{b} + z\,\mathbf{c}.$$

Hence $$f(\mathbf{r}) = x\,\mathbf{l} + y\,\mathbf{m} + z\,\mathbf{n}.$$

99.] In Art. 97 a particular case of a linear function was expressed as

$$\mathbf{r}' = \mathbf{i}\,c_1\,\mathbf{i}\cdot\mathbf{r} + \mathbf{j}\,c_2\,\mathbf{j}\cdot\mathbf{r} + \mathbf{k}\,c_3\,\mathbf{k}\cdot\mathbf{r}.$$

For the sake of brevity and to save repeating the vector $\mathbf{r}$ which occurs in each of these terms in the same way this may be written in the symbolic form

$$\mathbf{r}' = (\mathbf{i}\,c_1\,\mathbf{i} + \mathbf{j}\,c_2\,\mathbf{j} + \mathbf{k}\,c_3\,\mathbf{k})\cdot\mathbf{r}.$$

In like manner if $\mathbf{a}_1, \mathbf{a}_2, \mathbf{a}_3 \cdots$ be any given vectors, and $\mathbf{b}_1, \mathbf{b}_2, \mathbf{b}_3, \cdots$ another set equal in number, the expression

$$\mathbf{r}' = \mathbf{a}_1\,\mathbf{b}_1\cdot\mathbf{r} + \mathbf{a}_2\,\mathbf{b}_2\cdot\mathbf{r} + \mathbf{a}_3\,\mathbf{b}_3\cdot\mathbf{r} + \cdots \qquad (6)$$

is a linear vector function of $\mathbf{r}$; for owing to the distributive character of the scalar product this function of $\mathbf{r}$ satisfies relation (4). For the sake of brevity $\mathbf{r}'$ may be written symbolically in the form

$$\mathbf{r}' = (\mathbf{a}_1\,\mathbf{b}_1 + \mathbf{a}_2\,\mathbf{b}_2 + \mathbf{a}_3\,\mathbf{b}_3 + \cdots)\cdot\mathbf{r}. \qquad (6)'$$

No particular physical or geometrical significance is to be attributed at present to the expression

$$(\mathbf{a}_1\,\mathbf{b}_1 + \mathbf{a}_2\,\mathbf{b}_2 + \mathbf{a}_3\,\mathbf{b}_3 + \cdots). \qquad (7)$$

It should be regarded as an operator or symbol which converts the vector $\mathbf{r}$ into the vector $\mathbf{r}'$ and which merely affords a convenient and quick way of writing the relation (6).

Definition: An expression $\mathbf{a}\,\mathbf{b}$ formed by the juxtaposition of two vectors without the intervention of a dot or a cross is called a *dyad.* The symbolic sum of two dyads is called a dyadic binomial; of three, a dyadic trinomial; of any number, a dyadic polynomial. For the sake of brevity dyadic binomials, trinomials, and polynomials will be called simply *dyadics.* The first vector in a dyad is called the *antecedent;* and the second vector, the *consequent.* The antecedents of a dyadic are the vectors which are the antecedents of the individual dyads of which the dyadic is composed. In like manner the consequents of a dyadic are the consequents of the individual dyads. Thus in the dyadic (7) $\mathbf{a}_1, \mathbf{a}_2, \mathbf{a}_3 \cdots$ are the antecedents and $\mathbf{b}_1, \mathbf{b}_2, \mathbf{b}_3 \cdots$ the consequents.

Dyadics will be represented symbolically by the capital Greek letters. When only one dyadic is present the letter Φ will generally be used. In case several are under consideration other Greek capitals will be employed also. With this notation (7) becomes

$$\Phi = \mathbf{a}_1\,\mathbf{b}_1 + \mathbf{a}_2\,\mathbf{b}_2 + \mathbf{a}_3\,\mathbf{b}_3 + \cdots, \qquad (7)'$$

and (6)′ may now be written briefly in the form

$$\mathbf{r}' = \Phi \cdot \mathbf{r}. \qquad (8)$$

By definition $\quad \Phi \cdot \mathbf{r} = \mathbf{a}_1\,\mathbf{b}_1 \cdot \mathbf{r} + \mathbf{a}_2\,\mathbf{b}_2 \cdot \mathbf{r} + \mathbf{a}_3\,\mathbf{b}_3 \cdot \mathbf{r} + \cdots$

The symbol $\Phi \cdot \mathbf{r}$ is read Φ *dot* $\mathbf{r}$. It is called the direct product of Φ into $\mathbf{r}$ because the consequents $\mathbf{b}_1, \mathbf{b}_2, \mathbf{b}_3 \cdots$ are multiplied into $\mathbf{r}$ by direct or scalar multiplication. The order of the factors Φ and $\mathbf{r}$ is important. The direct product of $\mathbf{r}$ into Φ is

$$\begin{aligned} \mathbf{r} \cdot \Phi &= \mathbf{r} \cdot (\mathbf{a}_1 \mathbf{b}_1 + \mathbf{a}_2 \mathbf{b}_2 + \mathbf{a}_3 \mathbf{b}_3 + \cdots) \\ &= \mathbf{r} \cdot \mathbf{a}_1 \mathbf{b}_1 + \mathbf{r} \cdot \mathbf{a}_2 \mathbf{b}_2 + \mathbf{r} \cdot \mathbf{a}_3 \mathbf{b}_3 + \cdots \end{aligned} \quad (9)$$

Evidently the vectors $\Phi \cdot \mathbf{r}$ and $\mathbf{r} \cdot \Phi$ are in general different.

Definition: When the dyadic Φ is multiplied into $\mathbf{r}$ as $\Phi \cdot \mathbf{r}$, Φ is said to be a *prefactor* to $\mathbf{r}$. When $\mathbf{r}$ is multiplied in Φ as $\mathbf{r} \cdot \Phi$, Φ is said to be a *postfactor* to $\mathbf{r}$.

A dyadic Φ used either as a prefactor or as a postfactor to a vector $\mathbf{r}$ determines a linear vector function of $\mathbf{r}$. The two linear vector functions thus obtained are in general different from one another. They are called *conjugate* linear vector functions. The two dyadics

$$\Phi = \mathbf{a}_1 \mathbf{b}_1 + \mathbf{a}_2 \mathbf{b}_2 + \mathbf{a}_3 \mathbf{b}_3 + \cdots$$

and

$$\Psi = \mathbf{b}_1 \mathbf{a}_1 + \mathbf{b}_2 \mathbf{a}_2 + \mathbf{b}_3 \mathbf{a}_3 + \cdots,$$

each of which may be obtained from the other by interchanging the antecedents and consequents, are called *conjugate dyadics.* The fact that one dyadic is the conjugate of another is denoted by affixing a subscript C to either.

Thus $\qquad \Psi = \Phi_C \qquad \Phi = \Psi_C.$

Theorem: A dyadic used as a postfactor gives the same result as its conjugate used as a prefactor. That is

$$\mathbf{r} \cdot \Phi = \Phi_C \cdot \mathbf{r}. \quad (9)$$

100.] *Definition:* Any two dyadics Φ and Ψ are said to be *equal*

when $\qquad \Phi \cdot \mathbf{r} = \Psi \cdot \mathbf{r} \qquad$ for all values of $\mathbf{r}$,

or when $\qquad \mathbf{r} \cdot \Phi = \mathbf{r} \cdot \Psi \qquad$ for all values of $\mathbf{r}$, $\qquad$ (10)

or when $\qquad \mathbf{s} \cdot \Phi \cdot \mathbf{r} = \mathbf{s} \cdot \Psi \cdot \mathbf{r} \qquad$ for all values of $\mathbf{s}$ and $\mathbf{r}$.

The third relation is equivalent to the first. For, if the vectors $\Phi \cdot \mathbf{r}$ and $\Psi \cdot \mathbf{r}$ are equal, the scalar products of any vector $\mathbf{s}$ into them must be equal. And conversely if the scalar product of any and *every* vector $\mathbf{s}$ into the vectors $\Phi \cdot \mathbf{r}$ and $\Psi \cdot \mathbf{r}$ are equal, then those vectors must be equal. In like manner it may be shown that the third relation is equivalent to the second. Hence all three are equivalent.

Theorem: A dyadic Φ is completely determined when the values

$$\Phi \cdot \mathbf{a}, \quad \Phi \cdot \mathbf{b}, \quad \Phi \cdot \mathbf{c},$$

where $\mathbf{a}, \mathbf{b}, \mathbf{c}$ are any three non-coplanar vectors, are known.

This follows immediately from the fact that a dyadic defines a linear vector function. If

$$\mathbf{r} = x\,\mathbf{a} + y\,\mathbf{b} + z\,\mathbf{c},$$

$$\Phi \cdot \mathbf{r} = \Phi \cdot (x\,\mathbf{a} + y\,\mathbf{b} + z\,\mathbf{c}) = x\,\Phi \cdot \mathbf{a} + y\,\Phi \cdot \mathbf{b} + z\,\Phi \cdot \mathbf{c},$$

consequently two dyadics Φ and Ψ are equal provided equations (10) hold for *three* non-coplanar vectors $\mathbf{r}$ and *three* non-coplanar vectors $\mathbf{s}$.

Theorem: Any linear vector function f may be represented by a dyadic Φ to be used as a prefactor and by a dyadic Ψ, which is the conjugate of Φ, to be used as a postfactor.

The linear vector function is completely determined when its values for three non-coplanar vectors (say $\mathbf{i}, \mathbf{j}, \mathbf{k}$) are known (page 264). Let

$$f(\mathbf{i}) = \mathbf{a}, \quad f(\mathbf{j}) = \mathbf{b}, \quad f(\mathbf{k}) = \mathbf{c}.$$

Then the linear function f is equivalent to the dyadic Φ given by

$$\Phi = \mathbf{a}\,\mathbf{i} + \mathbf{b}\,\mathbf{j} + \mathbf{c}\,\mathbf{k},$$

to be used as a postfactor; and to the dyadic Ψ

$$\Psi = \Phi_C = \mathbf{i}\,\mathbf{a} + \mathbf{j}\,\mathbf{b} + \mathbf{k}\,\mathbf{c},$$

to be used as a prefactor.

$$f(\mathbf{r}) = \Phi \cdot \mathbf{r} = \mathbf{r} \cdot \Phi_C.$$

The study of linear vector functions therefore is identical with the study of dyadics.

Definition: A dyad $\mathbf{a}\,\mathbf{b}$ is said to be multiplied by a scalar a when the antecedent or the consequent is multiplied by that scalar, or when a is distributed in any manner between the antecedent and the consequent. If $a = a' a''$

$$a\,(\mathbf{a}\,\mathbf{b}) = (a\,\mathbf{a})\,\mathbf{b} = \mathbf{a}\,(a\,\mathbf{b}) = (a'\,\mathbf{a})\,(a''\,\mathbf{b}).$$

A dyadic Φ is said to be multiplied by the scalar a when each of its dyads is multiplied by that scalar. The product is written

$$a\,\Phi \quad \text{or} \quad \Phi\,a.$$

The dyadic $a\,\Phi$ applied to a vector $\mathbf{r}$ either as a prefactor or as a postfactor yields a vector equal to a times the vector obtained by applying Φ to $\mathbf{r}$ — that is

$$(a\,\Phi) \cdot \mathbf{r} = a\,(\Phi \cdot \mathbf{r}).$$

Theorem: The combination of vectors in a dyad is distributive. That is

$$(\mathbf{a} + \mathbf{b})\,\mathbf{c} = \mathbf{a}\,\mathbf{c} + \mathbf{b}\,\mathbf{c}$$

and

$$\mathbf{a}\,(\mathbf{b} + \mathbf{c}) = \mathbf{a}\,\mathbf{b} + \mathbf{a}\,\mathbf{c}. \tag{11}$$

This follows immediately from the definition of equality of dyadics (10). For

$$[(\mathbf{a} + \mathbf{b})\,\mathbf{c}] \cdot \mathbf{r} = (\mathbf{a} + \mathbf{b})\,\mathbf{c} \cdot \mathbf{r} = \mathbf{a}\,\mathbf{c} \cdot \mathbf{r} + \mathbf{b}\,\mathbf{c} \cdot \mathbf{r} = (\mathbf{a}\,\mathbf{c} + \mathbf{b}\,\mathbf{c}) \cdot \mathbf{r}$$

and

$$[\mathbf{a}\,(\mathbf{b} + \mathbf{c})] \cdot \mathbf{r} = \mathbf{a}\,(\mathbf{b} + \mathbf{c}) \cdot \mathbf{r} = \mathbf{a}\,\mathbf{b} \cdot \mathbf{r} + \mathbf{a}\,\mathbf{c} \cdot \mathbf{r} = (\mathbf{a}\,\mathbf{b} + \mathbf{a}\,\mathbf{c}) \cdot \mathbf{r}.$$

Hence it follows that a dyad which consists of two factors, each of which is the sum of a number of vectors, may be multiplied out according to the law of ordinary algebra — except that *the order of the factors in the dyads must be maintained.*

$$(\mathbf{a}+\mathbf{b}+\mathbf{c}+\cdots)(\mathbf{l}+\mathbf{m}+\mathbf{n}+\cdots)=\mathbf{a}\,\mathbf{l}+\mathbf{a}\,\mathbf{m}+\mathbf{a}\,\mathbf{n}+\cdots$$
$$+\mathbf{b}\,\mathbf{l}+\mathbf{b}\,\mathbf{m}+\mathbf{b}\,\mathbf{n}+\cdots \qquad (11)'$$
$$+\mathbf{c}\,\mathbf{l}+\mathbf{c}\,\mathbf{m}+\mathbf{c}\,\mathbf{n}+\cdots$$
$$+\;.\;\;.\;\;.\;\;.\;\;.\;\;.\;\;.$$

The dyad therefore appears as a *product* of the two vectors of which it is composed, inasmuch as it obeys the characteristic law of products — the distributive law. This is a justification for writing a dyad with the antecedent and consequent in juxtaposition as is customary in the case of products in ordinary algebra.

The Nonion Form of a Dyadic

101.] From the three unit vectors **i, j, k** nine dyads may be obtained by combining two at a time. These are

$$\begin{array}{lll} \mathbf{i}\,\mathbf{i}, & \mathbf{i}\,\mathbf{j}, & \mathbf{i}\,\mathbf{k}, \\ \mathbf{j}\,\mathbf{i}, & \mathbf{j}\,\mathbf{j}, & \mathbf{j}\,\mathbf{k}, \\ \mathbf{k}\,\mathbf{i}, & \mathbf{k}\,\mathbf{j}, & \mathbf{k}\,\mathbf{k}. \end{array} \qquad (12)$$

If all the antecedents and consequents in a dyadic Φ be expressed in terms of **i, j, k**, and if the resulting expression be simplified by performing the multiplications according to the distributive law (11)′ and if the terms be collected, the dyadic Φ may be reduced to the sum of nine dyads each of which is a scalar multiple of one of the nine fundamental dyads given above.

$$\begin{aligned} \Phi = {} & a_{11}\,\mathbf{i}\,\mathbf{i} + a_{12}\,\mathbf{i}\,\mathbf{j} + a_{13}\,\mathbf{i}\,\mathbf{k} \\ & + a_{21}\,\mathbf{j}\,\mathbf{i} + a_{22}\,\mathbf{j}\,\mathbf{j} + a_{23}\,\mathbf{k}\,\mathbf{k} \qquad (13) \\ & + a_{31}\,\mathbf{k}\,\mathbf{i} + a_{32}\,\mathbf{k}\,\mathbf{j} + a_{33}\,\mathbf{k}\,\mathbf{k}. \end{aligned}$$

This is called the *nonion* form of Φ.

Theorem: The necessary and sufficient condition that two dyadics Φ and Ψ be equal is that, when expressed in nonion form, the scalar coefficients of the corresponding dyads be equal.

If the coefficients be equal, then obviously

$$\Phi \cdot \mathbf{r} = \Psi \cdot \mathbf{r}$$

for any value of **r** and the dyadics by (10) must be equal. Conversely, if the dyadics Φ and Ψ are equal, then by (10)

$$\mathbf{s} \cdot \Phi \cdot \mathbf{r} = \mathbf{s} \cdot \Psi \cdot \mathbf{r}$$

for all values of **s** and **r**. Let **s** and **r** each take on the values **i**, **j**, **k**. Then (14)

$$\mathbf{i} \cdot \Phi \cdot \mathbf{i} = \mathbf{i} \cdot \Psi \cdot \mathbf{i}, \quad \mathbf{i} \cdot \Phi \cdot \mathbf{j} = \mathbf{i} \cdot \Psi \cdot \mathbf{j}, \quad \mathbf{i} \cdot \Phi \cdot \mathbf{k} = \mathbf{i} \cdot \Psi \cdot \mathbf{k}$$

$$\mathbf{j} \cdot \Phi \cdot \mathbf{i} = \mathbf{j} \cdot \Psi \cdot \mathbf{i}, \quad \mathbf{j} \cdot \Phi \cdot \mathbf{j} = \mathbf{j} \cdot \Psi \cdot \mathbf{j}, \quad \mathbf{j} \cdot \Phi \cdot \mathbf{k} = \mathbf{j} \cdot \Psi \cdot \mathbf{k}$$

$$\mathbf{k} \cdot \Phi \cdot \mathbf{i} = \mathbf{k} \cdot \Psi \cdot \mathbf{i}, \quad \mathbf{k} \cdot \Phi \cdot \mathbf{j} = \mathbf{k} \cdot \Psi \cdot \mathbf{j}, \quad \mathbf{k} \cdot \Phi \cdot \mathbf{k} = \mathbf{k} \cdot \Psi \cdot \mathbf{k}.$$

But these quantities are precisely the nine coefficients in the expansion of the dyadics Φ and Ψ. Hence the corresponding coefficients are equal and the theorem is proved.[1] This analytic statement of the equality of two dyadics can sometimes be used to greater advantage than the more fundamental definition (10) based upon the conception of the dyadic as defining a linear vector function.

Theorem: A dyadic Φ may be expressed as the sum of nine dyads of which the antecedents are any three given non-coplanar vectors, **a**, **b**, **c** and the consequents any three given non-coplanar vectors **l**, **m**, **n**.

Every antecedent may be expressed in terms of **a**, **b**, **c**; and every consequent, in terms of **l**, **m**, **n**. The dyadic may then be reduced to the form

$$\begin{aligned} \Phi = {} & a_{11}\, \mathbf{a}\,\mathbf{l} + a_{12}\, \mathbf{a}\,\mathbf{m} + a_{13}\, \mathbf{a}\,\mathbf{n} \\ & + a_{21}\, \mathbf{b}\,\mathbf{l} + a_{22}\, \mathbf{b}\,\mathbf{m} + a_{23}\, \mathbf{b}\,\mathbf{n} \\ & + a_{31}\, \mathbf{b}\,\mathbf{l} + a_{32}\, \mathbf{c}\,\mathbf{m} + a_{33}\, \mathbf{c}\,\mathbf{n}. \end{aligned} \qquad (15)$$

[1] As a corollary of the theorem it is evident that the nine dyads (12) are independent. None of them may be expressed linearly in terms of the others.

This expression of Φ is more general than that given in (13). It reduces to that expression when each set of vectors **a**, **b**, **c** and **l**, **m**, **n** coincides with **i**, **j**, **k**.

Theorem: Any dyadic Φ may be reduced to the sum of *three* dyads of which either the antecedents or the consequents, but not both, may be arbitrarily chosen provided they be non-coplanar.

Let it be required to express Φ as the sum of three dyads of which **a**, **b**, **c** are the antecedents. Let **l**, **m**, **n** be any other three non-coplanar vectors. Φ may then be expressed as in (15). Hence

$$\Phi = \mathbf{a}\,(a_{11}\,\mathbf{l} + a_{12}\,\mathbf{m} + a_{13}\,\mathbf{n}) + \mathbf{b}\,(a_{21}\,\mathbf{l} + a_{22}\,\mathbf{m} + a_{23}\,\mathbf{n})$$
$$+ \mathbf{c}\,(a_{31}\,\mathbf{l} + a_{32}\,\mathbf{m} + a_{32}\,\mathbf{n}),$$

or

$$\Phi = \mathbf{a}\,\mathbf{A} + \mathbf{b}\,\mathbf{B} + \mathbf{c}\,\mathbf{C}. \qquad (16)$$

In like manner if it be required to express Φ as the sum of three dyads of which the three non-coplanar vectors **l**, **m**, **n** are the consequents

$$\Phi = \mathbf{L}\,\mathbf{l} + \mathbf{M}\,\mathbf{m} + \mathbf{N}\,\mathbf{n}, \qquad (16)'$$

where

$$\mathbf{L} = a_{11}\,\mathbf{a} + a_{21}\,\mathbf{b} + a_{31}\,\mathbf{c},$$
$$\mathbf{M} = a_{12}\,\mathbf{a} + a_{22}\,\mathbf{b} + a_{32}\,\mathbf{c},$$
$$\mathbf{N} = a_{13}\,\mathbf{a} + a_{23}\,\mathbf{b} + a_{33}\,\mathbf{c}.$$

The expressions (15), (16), (16)′ for Φ are *unique.* Two equal dyadics which have the same three non-coplanar antecedents, **a**, **b**, **c**, have the same consequents **A**, **B**, **C** — these however need not be non-coplanar. And two equal dyadics which have the same three non-coplanar consequents **l**, **m**, **n**, have the same three antecedents.

102.] *Definition:* The symbolic product formed by the juxtaposition of two vectors **a**, **b** without the intervention of a dot or a cross is called the *indeterminate product* of the two vectors **a** and **b**.

The reason for the term indeterminate is this. The two products $\mathbf{a} \cdot \mathbf{b}$ and $\mathbf{a} \times \mathbf{b}$ have definite meanings. One is a certain scalar, the other a certain vector. On the other hand the product $\mathbf{a}\,\mathbf{b}$ is neither vector nor scalar — it is purely symbolic and acquires a determinate physical meaning only when used as an operator. The product $\mathbf{a}\,\mathbf{b}$ does not obey the commutative law. It does however obey the distributive law (11) and the associative law as far as scalar multiplication is concerned (Art. 100).

Theorem : The indeterminate product $\mathbf{a}\,\mathbf{b}$ of two vectors is the most general product in which scalar multiplication is associative.

The most general product conceivable ought to have the property that when the product is known the two factors are also known. Certainly no product could be more general. Inasmuch as scalar multiplication is to be associative, that is

$$a\,(\mathbf{a}\,\mathbf{b}) = (a\,\mathbf{a})\,\mathbf{b} = \mathbf{a}\,(a\,\mathbf{b}) = (a'\,\mathbf{a})\,(a''\,\mathbf{b}),$$

it will be impossible to completely determine the vectors $\mathbf{a}$ and $\mathbf{b}$ when their product $\mathbf{a}\,\mathbf{b}$ is given. Any scalar factor may be transferred from one vector to the other. Apart from this possible transference of a scalar factor, the vectors composing the product are known when the product is known. In other words —

Theorem : If the two indeterminate products $\mathbf{a}\,\mathbf{b}$ and $\mathbf{a}'\,\mathbf{b}'$ are equal, the vectors $\mathbf{a}$ and $\mathbf{a}'$, $\mathbf{b}$ and $\mathbf{b}'$ must be collinear and the product of the lengths of $\mathbf{a}$ and $\mathbf{b}$ (taking into account the positive or negative sign according as $\mathbf{a}$ and $\mathbf{b}$ have respectively equal or opposite directions to $\mathbf{a}'$ and $\mathbf{b}'$) is equal to the product of the lengths of $\mathbf{a}'$ and $\mathbf{b}'$.

Let
$$\mathbf{a} = a_1\,\mathbf{i} + a_2\,\mathbf{j} + a_3\,\mathbf{k},$$

$$\mathbf{b} = b_1\,\mathbf{i} + b_2\,\mathbf{j} + b_3\,\mathbf{k},$$

$$\mathbf{a}' = a_1' \, \mathbf{i} + a_2' \, \mathbf{j} + a_3' \, \mathbf{k},$$

$$\mathbf{b}' = b_1' \, \mathbf{i} + b_2' \, \mathbf{j} + b_3' \, \mathbf{k}.$$

Then
$$\begin{aligned} \mathbf{a}\,\mathbf{b} = & \; a_1 b_1 \; \mathbf{i}\mathbf{i} + a_1 b_2 \; \mathbf{i}\mathbf{j} + a_1 b_3 \, \mathbf{i}\mathbf{k} \\ & + a_2 b_1 \; \mathbf{j}\mathbf{i} + a_2 b_2 \; \mathbf{j}\mathbf{j} + a_2 b_3 \, \mathbf{j}\mathbf{k} \\ & + a_3 b_1 \; \mathbf{k}\mathbf{j} + a_3 b_2 \; \mathbf{k}\mathbf{j} + a_3 b_3 \, \mathbf{k}\mathbf{k}. \end{aligned}$$

and
$$\begin{aligned} \mathbf{a}'\mathbf{b}' = & \; a_1' b_1' \; \mathbf{i}\mathbf{i} + a_1' b_2' \; \mathbf{i}\mathbf{j} + a_1' b_3' \; \mathbf{i}\mathbf{k} \\ & + a_2' b_1' \; \mathbf{j}\mathbf{i} + a_2' b_2' \; \mathbf{j}\mathbf{j} + a_2 b_3 \, \mathbf{j}\mathbf{k} \\ & + a_3' b_1' \; \mathbf{k}\mathbf{i} + a_3' b_2' \; \mathbf{k}\mathbf{j} + a_3' b_3' \; \mathbf{k}\mathbf{k}. \end{aligned}$$

Since $\mathbf{a}\,\mathbf{b} = \mathbf{a}'\mathbf{b}'$ corresponding coefficients are equal. Hence

$$a_1 : a_2 : a_3 = a_1' : a_2' : a_3',$$

which shows that the vectors $\mathbf{a}$ and $\mathbf{a}'$ are collinear.

And
$$b_1 : b_2 : b_3 = b_1' : b_2' : b_3',$$

which shows that the vectors $\mathbf{b}$ and $\mathbf{b}'$ are collinear.

But
$$a_1 \, b_1 = a_1' \, b_1'.$$

This shows that the product of the lengths (including sign) are equal and the theorem is proved.

The proof may be carried out geometrically as follows. Since $\mathbf{a}\,\mathbf{b}$ is equal to $\mathbf{a}'\mathbf{b}'$

$$\mathbf{a}\,\mathbf{b} \cdot \mathbf{r} = \mathbf{a}'\,\mathbf{b}' \cdot \mathbf{r}$$

for all values of $\mathbf{r}$. Let $\mathbf{r}$ be perpendicular to $\mathbf{b}$. Then $\mathbf{b} \cdot \mathbf{r}$ vanishes and consequently $\mathbf{b}' \cdot \mathbf{r}$ also vanishes. This is true for any vector $\mathbf{r}$ in the plane perpendicular to $\mathbf{b}$. Hence $\mathbf{b}$ and $\mathbf{b}'$ are perpendicular to the same plane and are collinear. In like manner by using $\mathbf{a}\,\mathbf{b}$ as a postfactor $\mathbf{a}$ and $\mathbf{a}'$ are seen to be parallel. Also

$$\mathbf{a}\,\mathbf{b} \cdot \mathbf{b} = \mathbf{a}'\,\mathbf{b}' \cdot \mathbf{b},$$

which shows that the products of the lengths are the same.

The indeterminate product $\mathbf{a}\,\mathbf{b}$ imposes *five* conditions upon the vectors $\mathbf{a}$ and $\mathbf{b}$. The directions of $\mathbf{a}$ and $\mathbf{b}$ are fixed and likewise the product of their lengths. The scalar product $\mathbf{a} \cdot \mathbf{b}$, being a scalar quantity, imposes only one condition upon $\mathbf{a}$ and $\mathbf{b}$. The vector product $\mathbf{a} \times \mathbf{b}$, being a vector quantity, imposes three conditions. The normal to the plane of $\mathbf{a}$ and $\mathbf{b}$ is fixed and also the area of the parallelogram of which they are the side. The nine indeterminate products (12) of $\mathbf{i}, \mathbf{j}, \mathbf{k}$ into themselves are independent. The nine scalar products are not independent. Only two of them are different.

$$\mathbf{i} \cdot \mathbf{i} = \mathbf{j} \cdot \mathbf{j} = \mathbf{k} \cdot \mathbf{k} = 1,$$

and $$\mathbf{i} \cdot \mathbf{j} = \mathbf{j} \cdot \mathbf{i} = \mathbf{j} \cdot \mathbf{k} = \mathbf{k} \cdot \mathbf{j} = \mathbf{k} \cdot \mathbf{i} = \mathbf{i} \cdot \mathbf{k} = 0.$$

The nine vector products are not independent either; for

$$\mathbf{i} \times \mathbf{i} = \mathbf{j} \times \mathbf{j} = \mathbf{k} \times \mathbf{k} = 0,$$

and $$\mathbf{i} \times \mathbf{j} = -\mathbf{j} \times \mathbf{i}, \quad \mathbf{j} \times \mathbf{k} = -\mathbf{k} \times \mathbf{j}, \quad \mathbf{k} \times \mathbf{i} = -\mathbf{i} \times \mathbf{k}.$$

The two products $\mathbf{a} \cdot \mathbf{b}$ and $\mathbf{a} \times \mathbf{b}$ obtained respectively from the indeterminate product by inserting a dot and a cross between the factors are functions of the indeterminate product. That is to say, when $\mathbf{a}\,\mathbf{b}$ is given, $\mathbf{a} \cdot \mathbf{b}$ and $\mathbf{a} \times \mathbf{b}$ are determined. For these products depend solely upon the directions of $\mathbf{a}$ and $\mathbf{b}$ and upon the product of the length of $\mathbf{a}$ and $\mathbf{b}$, all of which are known when $\mathbf{a}\,\mathbf{b}$ is known. That is

if $$\mathbf{a}\,\mathbf{b} = \mathbf{a}'\mathbf{b}', \quad \mathbf{a} \cdot \mathbf{b} = \mathbf{a}' \cdot \mathbf{b}' \text{ and } \mathbf{a} \times \mathbf{b} = \mathbf{a}' \times \mathbf{b}'. \tag{17}$$

It does not hold conversely that if $\mathbf{a} \cdot \mathbf{b}$ and $\mathbf{a} \times \mathbf{b}$ are known $\mathbf{a}\,\mathbf{b}$ is fixed; for taken together $\mathbf{a} \cdot \mathbf{b}$ and $\mathbf{a} \times \mathbf{b}$ impose upon the vectors only four conditions, whereas $\mathbf{a}\,\mathbf{b}$ imposes five. Hence $\mathbf{a}\,\mathbf{b}$ appears not only as the most general product but as the most fundamental product. The others are merely functions of it. Their functional nature is brought out clearly by the notation of the dot and the cross.

Definition: A scalar known as *the scalar of* Φ may be obtained by inserting a dot between the antecedent and consequent of each dyad in a dyadic. This scalar will be denoted by a subscript S attached to Φ.[1]

If
$$\Phi = \mathbf{a}_1 \mathbf{b}_1 + \mathbf{a}_2 \mathbf{b}_2 + \mathbf{a}_3 \mathbf{b}_3 + \cdots$$
$$\Phi_S = \mathbf{a}_1 \cdot \mathbf{b}_1 + \mathbf{a}_2 \cdot \mathbf{b}_2 + \mathbf{a}_3 \cdot \mathbf{b}_3 + \cdots \tag{18}$$

In like manner a vector known as *the vector of* Φ may be obtained by inserting a cross between the antecedent and consequent of each dyad in Φ. This vector will be denoted by attaching a subscript cross to Φ.

$$\Phi_\times = \mathbf{a}_1 \times \mathbf{b}_1 + \mathbf{a}_2 \times \mathbf{b}_2 + \mathbf{a}_3 \times \mathbf{b}_3 + \cdots \tag{19}$$

If Φ be expanded in nonion form in terms of **i, j, k,**

$$\Phi_S = a_{11} + a_{22} + a_{33}, \tag{20}$$
$$\Phi_\times = (a_{23} - a_{32})\, \mathbf{i} + (a_{31} - a_{13})\, \mathbf{j} + (a_{12} - a_{21})\, \mathbf{k}. \tag{21}$$

Or
$$\Phi_S = \mathbf{i} \cdot \Phi \cdot \mathbf{i} + \mathbf{j} \cdot \Phi \cdot \mathbf{j} + \mathbf{k} \cdot \Phi \cdot \mathbf{k}, \tag{20$'$}$$
$$\Phi_\times = (\mathbf{j} \cdot \Phi \cdot \mathbf{k} - \mathbf{k} \cdot \Phi \cdot \mathbf{j})\, \mathbf{i} + (\mathbf{k} \cdot \Phi \cdot \mathbf{i} - \mathbf{i} \cdot \Phi \cdot \mathbf{k})\, \mathbf{j} + (\mathbf{i} \cdot \Phi \cdot \mathbf{j} - \mathbf{j} \cdot \Phi \cdot \mathbf{i})\, \mathbf{k}. \tag{21$'$}$$

In equations (20) and (21) the scalar and vector of Φ are expressed in terms of the coefficients of Φ when expanded in the nonion form. Hence if Φ and Ψ are two equal dyadics, the scalar of Φ is equal to the scalar of Ψ and the vector of Φ is equal to the vector of Ψ.

If
$$\Phi = \Psi, \quad \Phi_S = \Psi_S \quad \text{and} \quad \Phi_\times = \Psi_\times. \tag{22}$$

From this it appears that Φ_S and $\Phi_\times$ are functions of Φ uniquely determined when Φ is given. They may sometimes be obtained more conveniently from (20) and (21) than from (18) and (19), and sometimes not.

[1] A subscript dot might be used for the scalar of Φ if it were sufficiently distinct and free from liability to misinterpretation.

Products of Dyadics

103.] In giving the definitions and proving the theorems concerning products of dyadics, the dyad is made the underlying principle. What is true for the dyad is true for the dyadic in general owing to the fact that dyads and dyadics obey the distributive law of multiplication.

Definition: The *direct product* of the *dyad* $\mathbf{a\,b}$ into the *dyad* $\mathbf{c\,d}$ is written

$$(\mathbf{a\,b}) \cdot (\mathbf{c\,d})$$

and is by definition equal to the dyad $(\mathbf{b} \cdot \mathbf{c})\, \mathbf{a\,d}$.

$$(\mathbf{a\,b}) \cdot (\mathbf{c\,d}) = \mathbf{a}\,(\mathbf{b} \cdot \mathbf{c})\, \mathbf{d} = \mathbf{b} \cdot \mathbf{c} \;\; \mathbf{a\,d}.^{1} \qquad (23)$$

That is, the antecedent of the first and the consequent of the second dyad are taken for the antecedent and consequent respectively of the product and the whole is multiplied by the scalar product of the consequent of the first and the antecedent of the second.

Thus the two vectors which stand together in the product

$$(\mathbf{a\,b}) \cdot (\mathbf{c\,d})$$

are multiplied as they stand. The other two are left to form a new dyad. The direct product of two *dyadics* may be defined as the formal expansion (according to the distributive law) of the product into a sum of products of dyads. Thus if

$$\Phi = (\mathbf{a}_1\, \mathbf{b}_1 + \mathbf{a}_2\, \mathbf{b}_2 + \mathbf{a}_3\, \mathbf{b}_3 + \cdots)$$

and

$$\Psi = (\mathbf{c}_1\, \mathbf{d}_1 + \mathbf{c}_2\, \mathbf{d}_2 + \mathbf{c}_3\, \mathbf{d}_3 + \cdots)$$

$$\Phi \cdot \Psi = (\mathbf{a}_1\, \mathbf{b}_1 + \mathbf{a}_2\, \mathbf{b}_2 + \mathbf{a}_3\, \mathbf{b}_3 + \cdots) \cdot (\mathbf{c}_1\, \mathbf{d}_1 + \mathbf{c}_2\, \mathbf{d}_2 + \mathbf{c}_3\, \mathbf{d}_3 + \cdots)$$

$$\begin{aligned} &= \mathbf{a}_1\, \mathbf{b}_1 \cdot \mathbf{c}_1\, \mathbf{d}_1 + \mathbf{a}_1\, \mathbf{b}_1 \cdot \mathbf{c}_2\, \mathbf{d}_2 + \mathbf{a}_1\, \mathbf{b}_1 \cdot \mathbf{c}_3\, \mathbf{d}_3 + \cdots \\ &+ \mathbf{a}_2\, \mathbf{b}_2 \cdot \mathbf{c}_1\, \mathbf{d}_1 + \mathbf{a}_2\, \mathbf{b}_2 \cdot \mathbf{c}_2\, \mathbf{d}_2 + \mathbf{a}_2\, \mathbf{b}_2 \cdot \mathbf{c}_3\, \mathbf{d}_3 + \cdots \qquad (23)' \\ &+ \mathbf{a}_3\, \mathbf{b}_3 \cdot \mathbf{c}_1\, \mathbf{d}_1 + \mathbf{a}_3\, \mathbf{b}_3 \cdot \mathbf{c}_2\, \mathbf{d}_2 + \mathbf{a}_3\, \mathbf{b}_3 \cdot \mathbf{c}_3\, \mathbf{d}_3 + \cdots \\ &+ \cdot \; \cdot \; \cdot \; \cdot \; \cdot \; \cdot \; \cdot \; \cdot \; \cdot \; \cdot \; \cdot \; \cdot \; \cdot \; \cdot \; \cdot \end{aligned}$$

[1] The parentheses may be omitted in each of these three expressions.

$$\begin{aligned}\Phi \cdot \Psi = \mathbf{b}_1 \cdot \mathbf{c}_1\ \mathbf{a}_1 \mathbf{d}_1 + \mathbf{b}_1 \cdot \mathbf{c}_2\ \mathbf{a}_1 \mathbf{d}_2 + \mathbf{b}_1 \cdot \mathbf{c}_3\ \mathbf{a}_1 \mathbf{d}_3 + \cdots \\ + \mathbf{b}_2 \cdot \mathbf{c}_1\ \mathbf{a}_2 \mathbf{d}_1 + \mathbf{b}_2 \cdot \mathbf{c}_2\ \mathbf{a}_2 \mathbf{d}_2 + \mathbf{b}_2 \cdot \mathbf{c}_3\ \mathbf{a}_2 \mathbf{d}_3 + \cdots \\ + \mathbf{b}_3 \cdot \mathbf{c}_1\ \mathbf{a}_3 \mathbf{d}_1 + \mathbf{b}_3 \cdot \mathbf{c}_2\ \mathbf{a}_3 \mathbf{d}_2 + \mathbf{b}_3 \cdot \mathbf{c}_3\ \mathbf{a}_3 \mathbf{d}_3 + \cdots \\ + \ . \ . \ . \ . \ . \ . \ . \ . \ . \ . \ . \ . \ . \ . \ . \ . \quad (23)''\end{aligned}$$

The product of two dyadics Φ and Ψ is a dyadic $\Phi \cdot \Psi$.

Theorem: The product $\Phi \cdot \Psi$ of two dyadics Φ and Ψ when regarded as an operator to be used as a prefactor is equivalent to the operator Ψ followed by the operator Φ.

Let $$\Omega = \Phi \cdot \Psi.$$

To show $$\Omega \cdot \mathbf{r} = \Phi \cdot (\Psi \cdot \mathbf{r}),$$

or $$(\Phi \cdot \Psi) \cdot \mathbf{r} = \Phi \cdot (\Psi \cdot \mathbf{r}). \quad (24)$$

Let $\mathbf{a}\,\mathbf{b}$ be any dyad of Φ and $\mathbf{c}\,\mathbf{d}$ any dyad of Ψ.

$$(\mathbf{a}\,\mathbf{b} \cdot \mathbf{c}\,\mathbf{d}) \cdot \mathbf{r} = \mathbf{b} \cdot \mathbf{c}\,(\mathbf{a}\,\mathbf{d} \cdot \mathbf{r}) = (\mathbf{b} \cdot \mathbf{c})\ (\mathbf{d} \cdot \mathbf{r})\,\mathbf{a},$$

$$\mathbf{a}\,\mathbf{b} \cdot (\mathbf{c}\,\mathbf{d} \cdot \mathbf{r}) = \mathbf{a}\,\mathbf{b} \cdot \mathbf{c}\,(\mathbf{d} \cdot \mathbf{r}) = (\mathbf{b} \cdot \mathbf{c})\ (\mathbf{d} \cdot \mathbf{r})\,\mathbf{a},$$

Hence $$(\mathbf{a}\,\mathbf{b} \cdot \mathbf{c}\,\mathbf{d}) \cdot \mathbf{r} = \mathbf{a}\,\mathbf{b} \cdot (\mathbf{c}\,\mathbf{d} \cdot \mathbf{r}).$$

The theorem is true for dyads. Consequently by virtue of the distributive law it holds true for dyadics in general.

If $\mathbf{r}$ denote the position vector drawn from an assumed origin to a point P in space, $\mathbf{r}' = \Psi \cdot \mathbf{r}$ will be the position vector of another point P', and $\mathbf{r}'' = \Phi \cdot (\Psi \cdot \mathbf{r})$ will be the position vector of a third point P''. That is to say, Ψ defines a transformation of space such that the points P go over into the points P'. Φ defines a transformation of space such that the points P' go over into the points P''. Hence Ψ followed by Φ carries P into P''. The single operation $\Phi \cdot \Psi$ also carries P into P''.

Theorem: Direct multiplication of dyadics obeys the distributive law. That is

$$\Phi \cdot (\Psi + \Psi') = \Phi \cdot \Psi + \Phi \cdot \Psi'$$

and $$(\Phi' + \Phi) \cdot \Psi = \Phi' \cdot \Psi + \Phi \cdot \Psi. \tag{25}$$

Hence in general the product

$$(\Phi + \Phi' + \Phi'' + \cdots) \cdot (\Psi + \Psi' + \Psi'' + \cdots)$$

may be expanded formally according to the distributive law.

Theorem: The product of three dyadics Φ, Ψ, Ω is associative. That is

$$(\Phi \cdot \Psi) \cdot \Omega = \Phi \cdot (\Psi \cdot \Omega) \tag{26}$$

and consequently either product may be written without parentheses, as

$$\Phi \cdot \Psi \cdot \Omega. \tag{26$'$}$$

The proof consists in the demonstration of the theorem for three dyads $\mathbf{a\,b}$, $\mathbf{c\,d}$, $\mathbf{e\,f}$ taken respectively from the three dyadics Φ, Ψ, Ω.

$$(\mathbf{a\,b} \cdot \mathbf{c\,d}) \cdot \mathbf{e\,f} = (\mathbf{b} \cdot \mathbf{c})\, \mathbf{a\,d} \cdot \mathbf{e\,f} = (\mathbf{b} \cdot \mathbf{c})\, (\mathbf{d} \cdot \mathbf{e})\, \mathbf{a\,f},$$

$$\mathbf{a\,b} \cdot (\mathbf{c\,d} \cdot \mathbf{e\,f}) = (\mathbf{d} \cdot \mathbf{e})\, \mathbf{a\,b} \cdot \mathbf{c\,f} = (\mathbf{d} \cdot \mathbf{e})\, (\mathbf{b} \cdot \mathbf{c})\, \mathbf{a\,f}.$$

The proof may also be given by considering Φ, Ψ, and Ω as operators

$$\{(\Phi \cdot \Psi) \cdot \Omega\} \cdot \mathbf{r} = (\Phi \cdot \Psi) \cdot (\Omega \cdot \mathbf{r}).$$

Let $$\Omega \cdot \mathbf{r} = \mathbf{r}'$$

$$\{(\Phi \cdot \Psi) \cdot \Omega\} \cdot \mathbf{r} = (\Phi \cdot \Psi) \cdot \mathbf{r}' = \Phi \cdot (\Psi \cdot \mathbf{r}')$$

Let $$\Psi \cdot \mathbf{r}' = \mathbf{r}'',$$

$$\{(\Phi \cdot \Psi) \cdot \Omega\} \cdot \mathbf{r} = \Phi \cdot \mathbf{r}'' = \mathbf{r}'''.$$

Again $$\{\Phi \cdot (\Psi \cdot \Omega)\} \cdot \mathbf{r} = \Phi \cdot [(\Psi \cdot \Omega) \cdot \mathbf{r}].$$

$$(\Psi \cdot \Omega) \cdot \mathbf{r} = \Psi \cdot (\Omega \cdot \mathbf{r}) = \Psi \cdot \mathbf{r}' = \mathbf{r}''.$$

$$\{\Phi \cdot (\Psi \cdot \Omega)\} \cdot \mathbf{r} = \Phi \cdot [\Psi \cdot \mathbf{r}'] = \Phi \cdot \mathbf{r}'' = \mathbf{r}'''$$

Hence $$\{(\Phi \cdot \Psi) \cdot \Omega\} \cdot \mathbf{r} = \{\Phi \cdot (\Psi \cdot \Omega)\} \cdot \mathbf{r}$$

for all values of $\mathbf{r}$. Consequently

$$(\Phi \cdot \Psi) \cdot \Omega = \Phi \cdot (\Psi \cdot \Omega).$$

The theorem may be extended by mathematical induction to the case of any number of dyadics. The direct product of any number of dyadics is associative. Parentheses may be inserted or omitted at pleasure without altering the result.

It was shown above (24) that

$$(\Phi \cdot \Psi) \cdot \mathbf{r} = \Phi \cdot (\Psi \cdot \mathbf{r}) = \Phi \cdot \Psi \cdot \mathbf{r}. \qquad (24)'$$

Hence the product of two dyadics and a vector is associative. The theorem is true in case the vector precedes the dyadics and also when the number of dyadics is greater than two. But the theorem is *untrue* when the vector occurs between the dyadics. The product of a dyadic, a vector, and another dyadic is not associative.

$$(\Phi \cdot \mathbf{r}) \cdot \Psi \neq \Phi \cdot (\mathbf{r} \cdot \Psi). \qquad (27)$$

Let $\mathbf{a}\,\mathbf{b}$ be a dyad of Φ, and $\mathbf{c}\,\mathbf{d}$ a dyad of Ψ.

$$(\mathbf{a}\,\mathbf{b} \cdot \mathbf{r}) \cdot \mathbf{c}\,\mathbf{d} = \mathbf{b} \cdot \mathbf{r}\,(\mathbf{a} \cdot \mathbf{c}\,\mathbf{d}) = (\mathbf{b} \cdot \mathbf{r})\,(\mathbf{a} \cdot \mathbf{c})\,\mathbf{d},$$

$$\mathbf{a}\,\mathbf{b} \cdot (\mathbf{r} \cdot \mathbf{c}\,\mathbf{d}) = \mathbf{a}\,\mathbf{b} \cdot \mathbf{d}\,(\mathbf{r} \cdot \mathbf{c}) = \mathbf{b} \cdot \mathbf{d}\,(\mathbf{r} \cdot \mathbf{c})\,\mathbf{a}$$

Hence $\qquad (\mathbf{a}\,\mathbf{b} \cdot \mathbf{r}) \cdot \mathbf{c}\,\mathbf{d} \neq \mathbf{a}\,\mathbf{b} \cdot (\mathbf{r} \cdot \mathbf{c}\,\mathbf{d}).$

The results of this article may be summed up as follows:

Theorem: The direct product of any number of dyadics or of any number of dyadics with a vector factor at either end or at both ends obeys the distributive and associative laws of multiplication — parentheses may be inserted or omitted at pleasure. But the direct product of any number of dyadics with a vector factor at some other position than at either end is not associative — parentheses are necessary to give the expression a definite meaning.

Later it will be seen that by making use of the conjugate dyadics a vector factor which occurs between other dyadics may be placed at the end and hence the product may be made to assume a form in which it is associative.

104.] *Definition:* The *skew* products of a dyad $\mathbf{a}\mathbf{b}$ into a vector $\mathbf{r}$ and of a vector $\mathbf{r}$ into a dyad $\mathbf{a}\mathbf{b}$ are defined respectively by the equations

$$\begin{aligned}(\mathbf{a}\,\mathbf{b}) \times \mathbf{r} &= \mathbf{a}\,(\mathbf{b} \times \mathbf{r}),\\ \mathbf{r} \times (\mathbf{a}\,\mathbf{b}) &= (\mathbf{r} \times \mathbf{a})\,\mathbf{b}.\end{aligned} \tag{28}$$

The skew product of a dyad and a vector at either end is a dyad. The obvious extension to dyadics is

$$\begin{aligned}\Phi \times \mathbf{r} &= (\mathbf{a}_1\,\mathbf{b}_1 + \mathbf{a}_2\,\mathbf{b}_2 + \mathbf{a}_3\,\mathbf{b}_3 + \cdots) \times \mathbf{r}\\ &= \mathbf{a}_1\,\mathbf{b}_1 \times \mathbf{r} + \mathbf{a}_2\,\mathbf{b}_2 \times \mathbf{r} + \mathbf{a}_3\,\mathbf{b}_3 \times \mathbf{r} + \cdots\\ \mathbf{r} \times \Phi &= \mathbf{r} \times (\mathbf{a}_1\,\mathbf{b}_1 + \mathbf{a}_2\,\mathbf{b}_2 + \mathbf{a}_3\,\mathbf{b}_3 + \cdots)\\ &= \mathbf{r} \times \mathbf{a}_1\,\mathbf{b}_1 + \mathbf{r} \times \mathbf{a}_2\,\mathbf{b}_2 + \mathbf{r} \times \mathbf{a}_3\,\mathbf{b}_3 + \cdots\end{aligned} \tag{28$'$}$$

Theorem: The direct product of any number of dyadics multiplied at either end or at both ends by a vector whether the multiplication be performed with a cross or a dot is associative. But in case the vector occurs at any other position than the end the product is not associative. That is,

$$\begin{aligned}(\mathbf{r} \times \Phi) \cdot \Psi &= \mathbf{r} \times (\Phi \cdot \Psi) = \mathbf{r} \times \Phi \cdot \Psi,\\ (\Phi \cdot \Psi) \times \mathbf{r} &= \Phi \cdot (\Psi \times \mathbf{r}) = \Phi \cdot \Psi \times \mathbf{r},\\ (\mathbf{r} \times \Phi) \cdot \mathbf{s} &= \mathbf{r} \times (\Phi \cdot \mathbf{s}) = \mathbf{r} \times \Phi \cdot \mathbf{s},\\ \mathbf{r} \cdot (\Phi \times \mathbf{s}) &= (\mathbf{r} \cdot \Phi) \times \mathbf{s} = \mathbf{r} \cdot \Phi \times \mathbf{s},\\ \mathbf{r} \times (\Phi \times \mathbf{s}) &= (\mathbf{r} \times \Phi) \times \mathbf{s} = \mathbf{r} \times \Phi \times \mathbf{s},\end{aligned} \tag{29}$$

but

$$\Psi \cdot (\mathbf{r} \times \Phi) \neq (\Psi \cdot \mathbf{r}) \times \Psi.$$

Furthermore the expressions

$$\mathbf{s} \cdot \mathbf{r} \times \Phi \text{ and } \Phi \times \mathbf{r} \cdot \mathbf{s}$$

can have no other meaning than

$$\begin{aligned}\mathbf{s} \cdot \mathbf{r} \times \Phi &= \mathbf{s} \cdot (\mathbf{r} \times \Phi),\\ \Phi \times \mathbf{r} \cdot \mathbf{s} &= (\Phi \times \mathbf{r}) \cdot \mathbf{s},\end{aligned} \tag{30}$$

since the product of a dyadic Φ with a cross into a scalar $\mathbf{s} \cdot \mathbf{r}$ is meaningless. Moreover since the dot and the cross may be interchanged in the scalar triple product of three vectors it appears that

$$\mathbf{s} \cdot \mathbf{r} \times \Phi = (\mathbf{s} \times \mathbf{r}) \cdot \Phi,$$

$$\Phi \times \mathbf{r} \cdot \mathbf{s} = \Phi \cdot (\mathbf{r} \times \mathbf{s}), \tag{31}$$

and

$$\Phi \cdot (\mathbf{r} \times \Psi) = (\Phi \times \mathbf{r}) \cdot \Psi.$$

The parentheses in the following expressions cannot be omitted without incurring ambiguity.

$$\Phi \cdot (\mathbf{r} \times \mathbf{s}) \neq (\Phi \cdot \mathbf{r}) \times \mathbf{s},$$

$$(\mathbf{s} \times \mathbf{r}) \cdot \Phi \neq \mathbf{s} \times (\mathbf{r} \cdot \Phi), \tag{31$'$}$$

$$(\Phi \cdot \mathbf{r}) \times \Psi \neq \Phi \times (\mathbf{r} \cdot \Psi).$$

The formal skew product of two dyads $\mathbf{a}\,\mathbf{b}$ and $\mathbf{c}\,\mathbf{d}$ would be

$$(\mathbf{a}\,\mathbf{b}) \times (\mathbf{c}\,\mathbf{d}) = \mathbf{a}\,(\mathbf{b} \times \mathbf{c})\,\mathbf{d}.$$

In this expression three vectors $\mathbf{a}$, $\mathbf{b} \times \mathbf{c}$, $\mathbf{d}$ are placed side by side with no sign of multiplication uniting them. Such an expression

$$\mathbf{r\,s\,t} \tag{32}$$

is called a *triad;* and a sum of such expressions, a *triadic.* The theory of triadics is intimately connected with the theory of linear *dyadic* functions of a vector, just as the theory of dyadics is connected with the theory of linear *vector* functions of a vector. In a similar manner by going a step higher *tetrads* and *tetradics* may be formed, and finally *polyads* and *polyadics.* But the theory of these higher combinations of vectors will not be taken up in this book. The dyadic furnishes about as great a generality as is ever called for in practical applications of vector methods.

Degrees of Nullity of Dyadics

105.] It was shown (Art. 101) that a dyadic could always be reduced to a sum of three terms at most, and this reduction can be accomplished in only one way when the antecedents or the consequents are specified. In particular cases it may be possible to reduce the dyadic further to a sum of two terms or to a single term or to zero. Thus let

$$\Phi = \mathbf{a\,l} + \mathbf{b\,m} + \mathbf{c\,n}.$$

If **l**, **m**, **n** are coplanar one of the three may be expressed in terms of the other two as

$$\mathbf{l} = x\,\mathbf{m} + y\,\mathbf{n}.$$

Then
$$\Phi = \mathbf{a}\,x\,\mathbf{m} + \mathbf{a}\,y\,\mathbf{n} + \mathbf{b\,m} + \mathbf{c\,n},$$

$$\Phi = (\mathbf{a}\,x + \mathbf{b})\,\mathbf{m} + (\mathbf{a}\,y + \mathbf{c})\,\mathbf{n}.$$

The dyadic has been reduced to two terms. If **l**, **m**, **n** were all collinear the dyadic would reduce to a single term and if they all vanished the dyadic would vanish.

Theorem: If a dyadic Φ be expressed as the sum of three terms

$$\Phi = \mathbf{a\,l} + \mathbf{b\,m} + \mathbf{c\,n}$$

of which the antecedents **a**, **b**, **c** are known to be non-coplanar, then the dyadic may be reduced to the sum of two dyads when and only when the consequents are coplanar.

The proof of the first part of the theorem has just been given. To prove the second part suppose that the dyadic could be reduced to a sum of two terms

$$\Phi = \mathbf{d\,p} + \mathbf{e\,q}$$

and that the consequents **l**, **m**, **n** of Φ were non-coplanar. This supposition leads to a contradiction. For let **l**′, **m**′, **n**′ be the system reciprocal to **l**, **m**, **n**. That is,

$$\mathbf{l}' = \frac{\mathbf{m} \times \mathbf{n}}{[\mathbf{l\,m\,n}]}, \quad \mathbf{m}' = \frac{\mathbf{n} \times \mathbf{l}}{[\mathbf{l\,m\,n}]}, \quad \mathbf{n}' = \frac{\mathbf{l} \times \mathbf{m}}{[\mathbf{l\,m\,n}]}.$$

The vectors $\mathbf{l}'$, $\mathbf{m}'$, $\mathbf{n}'$ exist and are non-coplanar because $\mathbf{l}$, $\mathbf{m}$, $\mathbf{n}$ have been assumed to be non-coplanar. Any vector $\mathbf{r}$ may be expressed in terms of them as

$$\mathbf{r} = x\,\mathbf{l}' + y\,\mathbf{m}' + z\,\mathbf{n}',$$

$$\Phi \cdot \mathbf{r} = (\mathbf{a}\,\mathbf{l} + \mathbf{b}\,\mathbf{m} + \mathbf{c}\,\mathbf{n}) \cdot (x\,\mathbf{l}' + y\,\mathbf{m}' + z\,\mathbf{n}').$$

But $$\mathbf{l} \cdot \mathbf{l}' = \mathbf{m} \cdot \mathbf{m}' = \mathbf{n} \cdot \mathbf{n}' = 1,$$

and $\mathbf{l} \cdot \mathbf{m}' = \mathbf{l}' \cdot \mathbf{m} = \mathbf{m} \cdot \mathbf{n}' = \mathbf{m}' \cdot \mathbf{n} = \mathbf{n} \cdot \mathbf{l}' = \mathbf{n}' \cdot \mathbf{l} = 0.$

Hence $$\Phi \cdot \mathbf{r} = x\,\mathbf{a} + y\,\mathbf{b} + z\,\mathbf{c}.$$

By giving to $\mathbf{r}$ a suitable value the vector $\Phi \cdot \mathbf{r}$ may be made equal to any vector in space.

But $$\Phi \cdot \mathbf{r} = (\mathbf{d}\,\mathbf{p} + \mathbf{e}\,\mathbf{q}) \cdot \mathbf{r} = \mathbf{d}\,(\mathbf{p} \cdot \mathbf{r}) + \mathbf{e}\,(\mathbf{q} \cdot \mathbf{r}).$$

This shows that $\Phi \cdot \mathbf{r}$ must be coplanar with $\mathbf{d}$ and $\mathbf{e}$. Hence $\Phi \cdot \mathbf{r}$ can take on only those vector values which lie in the plane of $\mathbf{d}$ and $\mathbf{e}$. Thus the assumption that $\mathbf{l}$, $\mathbf{m}$, $\mathbf{n}$ are non-coplanar leads to a contradiction. Hence $\mathbf{l}$, $\mathbf{m}$, $\mathbf{n}$ must be coplanar and the theorem is proved.

Theorem: If a dyadic Φ be expressed as the sum of three terms

$$\Phi = \mathbf{a}\,\mathbf{l} + \mathbf{b}\,\mathbf{m} + \mathbf{c}\,\mathbf{n},$$

of which the antecedents $\mathbf{a}$, $\mathbf{b}$, $\mathbf{c}$ are known to be non-coplanar, the dyadic Φ can be reduced to a single dyad when and only when the consequents $\mathbf{l}$, $\mathbf{m}$, $\mathbf{n}$ are collinear.

The proof of the first part was given above. To prove the second part suppose Φ could be expressed as

$$\Phi = \mathbf{d}\,\mathbf{p}.$$

Let $$\Psi = \Phi \times \mathbf{p} = \mathbf{d}\,\mathbf{p} \times \mathbf{p} = \mathbf{d}\,0 = 0,$$

$$\Psi = \mathbf{a}\,\mathbf{l} \times \mathbf{p} + \mathbf{b}\,\mathbf{m} \times \mathbf{p} + \mathbf{c}\,\mathbf{n} \times \mathbf{p}.$$

From the second equation it is evident that Ψ used as a postfactor for any vector

$$\mathbf{r} = x\,\mathbf{a}' + y\,\mathbf{b}' + z\,\mathbf{c}',$$

where $\mathbf{a}'$, $\mathbf{b}'$, $\mathbf{c}'$ is the reciprocal system to $\mathbf{a}$, $\mathbf{b}$, $\mathbf{c}$ gives

$$\mathbf{r} \cdot \Psi = x\,\mathbf{l} \times \mathbf{p} + y\,\mathbf{m} \times \mathbf{p} + z\,\mathbf{n} \times \mathbf{p}.$$

From the first expression

$$\mathbf{r} \cdot 0 = 0.$$

Hence $$x\,\mathbf{l} \times \mathbf{p} + y\,\mathbf{m} \times \mathbf{p} + z\,\mathbf{n} \times \mathbf{p}$$

must be zero for every value of $\mathbf{r}$, that is, for every value of x, y, z. Hence

$$\mathbf{l} \times \mathbf{p} = 0, \qquad \mathbf{m} \times \mathbf{p} = 0, \qquad \mathbf{n} \times \mathbf{p} = 0.$$

Hence $\mathbf{l}$, $\mathbf{m}$, and $\mathbf{n}$ are all parallel to $\mathbf{p}$ and the theorem has been demonstrated.

If the three consequents $\mathbf{l}$, $\mathbf{m}$, $\mathbf{n}$ had been known to be non-coplanar instead of the three antecedents, the statement of the theorems would have to be altered by interchanging the words *antecedent* and *consequent* throughout. There is a further theorem dealing with the case in which both antecedents and consequents of Φ are coplanar. Then Φ is reducible to the sum of two dyads.

106.] *Definition:* A dyadic which cannot be reduced to the sum of fewer than three dyads is said to be *complete.* A dyadic which may be reduced to the sum of two dyads, but cannot be reduced to a single dyad is said to be *planar.* In case the plane of the antecedents and the plane of the consequents coincide when the dyadic is expressed as the sum of two dyads, the dyadic is said to be *uniplanar.* A dyadic which may be reduced to a single dyad is said to be *linear.* In case the antecedent and consequent of that dyad are collinear, the dyadic is said to be *unilinear*. If a dyadic may be so expressed that all of its terms vanish the dyadic is said to be zero. In this case the nine coefficients of the dyadic as expressed in nonion form must vanish.

The properties of complete, planar, uniplanar, linear, and unilinear dyadics when regarded as operators are as follows. Let

$$\mathbf{s} = \Phi \cdot \mathbf{r} \text{ and } \mathbf{t} = \mathbf{r} \cdot \Phi.$$

If Φ is complete **s** *and* **t** *may be made to take on any desired value by giving* **r** *a suitable value.*

$$\Phi = \mathbf{a}\,\mathbf{l} + \mathbf{b}\,\mathbf{m} + \mathbf{c}\,\mathbf{n}.$$

As Φ is complete **l**, **m**, **n** are non-coplanar and hence have a reciprocal system $\mathbf{l}'$, $\mathbf{m}'$, $\mathbf{n}'$.

$$\mathbf{s} = \Phi \cdot (x\,\mathbf{l}' + y\,\mathbf{m}' + z\,\mathbf{n}') = x\,\mathbf{a} + y\,\mathbf{b} + z\,\mathbf{c}.$$

In like manner **a**, **b**, **c** possess a system of reciprocals $\mathbf{a}'$, $\mathbf{b}'$, $\mathbf{c}'$.

$$\mathbf{t} = (x\,\mathbf{a}' + y\,\mathbf{b}' + z\,\mathbf{c}') \cdot \Phi = x\,\mathbf{l} + y\,\mathbf{m} + z\,\mathbf{n}.$$

A complete dyadic Φ applied to a vector **r** cannot give zero unless the vector **r** itself is zero.

If Φ is planar the vector **s** *may take on any value in the plane of the antecedents and* **t** *any value in the plane of the consequents of Φ; but no values out of those planes.* The dyadic Φ when used as a prefactor reduces every vector **r** in space to a vector in the plane of the antecedents. In particular any vector **r** perpendicular to the plane of the consequents of Φ is reduced to zero. The dyadic Φ used as a postfactor reduces every vector **r** in space to a vector in the plane of the consequents of Φ. In particular a vector perpendicular to the plane of the antecedents of Φ is reduced to zero. In case the dyadic is uniplanar the same statements hold.

If Φ is linear the vector **s** *may take on any value collinear with the antecedent of Φ and* **t** *any value collinear with the con-*

sequent of Φ; but no other values. The dyadic Φ used as a prefactor reduces any vector **r** to the line of the antecedent of Φ. In particular any vectors perpendicular to the consequent of Φ are reduced to zero. The dyadic Φ used as a postfactor reduces any vector **r** to the line of the consequent of Φ. In particular any vectors perpendicular to the antecedent of Φ are thus reduced to zero.

If Φ is a zero dyadic the vectors **s** *and* **t** *are both zero no matter what the value of* **r** *may be.*

Definition: A *planar* dyadic is said to possess *one degree of nullity.* A *linear* dyadic is said to possess *two degrees of nullity.* A zero dyadic is said to possess *three degrees of nullity* or *complete nullity.*

107.] *Theorem:* The direct product of two complete dyadics is complete; of a complete dyadic and a planar dyadic, planar; of a complete dyadic and a linear dyadic, linear.

Theorem: The product of two planar dyadics is planar except when the plane of the consequent of the first dyadic in the product is perpendicular to the plane of the antecedent of the second dyadic. In this case the product reduces to a linear dyadic—and only in this case.

Let
$$\Phi = \mathbf{a}_1 \mathbf{b}_1 + \mathbf{a}_2 \mathbf{b}_2,$$
$$\Psi = \mathbf{c}_1 \mathbf{d}_1 + \mathbf{c}_2 \mathbf{d}_2,$$
$$\Omega = \Phi \cdot \Psi.$$

The vector $\mathbf{s} = \Psi \cdot \mathbf{r}$ takes on all values in the plane of $\mathbf{c}_1$ and $\mathbf{c}_2$
$$\mathbf{s} = x\,\mathbf{c}_1 + y\,\mathbf{c}_2.$$

The vector $\mathbf{s}' = \Phi \cdot \mathbf{s}$ takes on the values

$$\mathbf{s}' = \Phi \cdot \mathbf{s} = x\,(\mathbf{b}_1 \cdot \mathbf{c}_1)\,\mathbf{a}_1 + y\,(\mathbf{b}_1 \cdot \mathbf{c}_2)\,\mathbf{a}_1$$
$$+ x\,(\mathbf{b}_2 \cdot \mathbf{c}_1)\,\mathbf{a}_2 + y\,(\mathbf{b}_2 \cdot \mathbf{c})\,\mathbf{a}_2,$$
$$\mathbf{s}' = \{x\,(\mathbf{b}_1 \cdot \mathbf{c}_1) + y\,(\mathbf{b}_1 \cdot \mathbf{c}_2)\}\,\mathbf{a}_1 + \{x\,(\mathbf{b}_2 \cdot \mathbf{c}_1) + y\,(\mathbf{b}_2 \cdot \mathbf{c}_2)\}\,\mathbf{a}_2.$$

Let $$\mathbf{s}' = x' \mathbf{a}_1 + y' \mathbf{a}_2,$$
where $$x' = x (\mathbf{b}_1 \cdot \mathbf{c}_1) + y (\mathbf{b}_1 \cdot \mathbf{c}_2),$$
and $$y' = x (\mathbf{b}_2 \cdot \mathbf{c}_1) + y (\mathbf{b}_2 \cdot \mathbf{c}_2).$$
These equations may always be solved for x and y when any desired values x' and y' are given — that is, when $\mathbf{s}'$ has any desired value in the plane of $\mathbf{a}_1$ and $\mathbf{a}_2$ — unless the determinant

$$\begin{vmatrix} \mathbf{b}_1 \cdot \mathbf{c}_1 & \mathbf{b}_1 \cdot \mathbf{c}_2 \\ \mathbf{b}_2 \cdot \mathbf{c}_1 & \mathbf{b}_2 \cdot \mathbf{c}_2 \end{vmatrix} = 0.$$

But by (25), Chap. II., this is merely the product

$$(\mathbf{b}_1 \times \mathbf{b}_2) \cdot (\mathbf{c}_1 \times \mathbf{c}_2) = 0.$$

The vector $\mathbf{b}_1 \times \mathbf{b}_2$ is perpendicular to the plane of the consequents of Φ; and $\mathbf{c}_1 \times \mathbf{c}_2$, to the plane of the antecedents of Ψ. Their scalar product vanishes when and only when the vectors are perpendicular — that is, when the planes are perpendicular. Consequently $\mathbf{s}'$ may take on any value in the plane of $\mathbf{a}_1$ and $\mathbf{a}_2$ and $\Phi \cdot \Psi$ is therefore a planar dyadic unless the planes of $\mathbf{b}_1$ and $\mathbf{b}_2$, $\mathbf{c}_1$ and $\mathbf{c}_2$ are perpendicular. If however $\mathbf{b}_1$ and $\mathbf{b}_2$, $\mathbf{c}_1$ and $\mathbf{c}_2$ are perpendicular $\mathbf{s}'$ can take on only values in a certain line of the plane of $\mathbf{a}_1$ and $\mathbf{a}_2$, and hence $\Phi \cdot \Psi$ is linear. The theorem is therefore proved.

Theorem : The product of two linear dyadics is linear except when the consequent of the first factor is perpendicular to the antecedent of the second. In this case the product is zero — and only in this case.

Theorem : The product of a planar dyadic into a linear is linear except when the plane of the consequents of the planar dyadic is perpendicular to the antecedent of the linear dyadic. In this case the product is zero — and only in this case.

Theorem : The product of a linear dyadic into a planar dyadic is linear except when the consequent of the linear dyadic is perpendicular to the plane of the antecedents of the planar dyadic. In this case the product is zero — and only in this case.

It is immediately evident that in the cases mentioned the products do reduce to zero. It is not quite so apparent that they can reduce to zero in only those cases. The proofs are similar to the one given above in the case of two planar dyadics. They are left to the reader. The proof of the first theorem stated, page 286, is also left to the reader.

The Idemfactor;[1] *Reciprocals and Conjugates of Dyadics*

108.] *Definition:* If a dyadic applied as a prefactor or as a postfactor to any vector always yields that vector the dyadic is said to be an *idemfactor*. That is

if $\quad \boldsymbol{\Phi} \cdot \mathbf{r} = \mathbf{r}$ for all values of $\mathbf{r}$,

or if $\quad \mathbf{r} \cdot \boldsymbol{\Phi} = \mathbf{r}$ for all values of $\mathbf{r}$,

then $\boldsymbol{\Phi}$ is an idemfactor. The capital I is used as the symbol for an idemfactor. The idemfactor is a *complete* dyadic. For there can be no direction in which $\mathrm{I} \cdot \mathbf{r}$ vanishes.

Theorem: When expressed in nonion form the idemfactor is

$$\mathrm{I} = \mathbf{i}\,\mathbf{i} + \mathbf{j}\,\mathbf{j} + \mathbf{k}\,\mathbf{k}. \tag{33}$$

Hence all idemfactors are equal.

To prove that the idemfactor takes the form (33) it is merely necessary to apply the idemfactor I to the vectors **i**, **j**, **k** respectively. Let

$$\begin{aligned} \mathrm{I} = a_{11}\, \mathbf{i}\,\mathbf{i} &+ a_{12}\, \mathbf{i}\,\mathbf{j} + a_{13}\, \mathbf{i}\,\mathbf{k} \\ + a_{21}\, \mathbf{j}\,\mathbf{i} &+ a_{22}\, \mathbf{j}\,\mathbf{j} + a_{23}\, \mathbf{j}\,\mathbf{k} \\ + a_{31}\, \mathbf{k}\,\mathbf{i} &+ a_{32}\, \mathbf{k}\,\mathbf{j} + a_{33}\, \mathbf{k}\,\mathbf{k}. \end{aligned}$$

[1] In the theory of dyadics the idemfactor I plays a rôle analogous to unity in ordinary algebra. The notation is intended to suggest this analogy.

$$\mathbf{I} \cdot \mathbf{i} = a_{11}\,\mathbf{i} + a_{21}\,\mathbf{j} + a_{31}\,\mathbf{k}.$$

If
$$\mathbf{I} \cdot \mathbf{i} = \mathbf{i},$$

$$a_{11} = 1 \text{ and } a_{21} = a_{31} = 0.$$

In like manner it may be shown that all the coefficients vanish except a_{11}, a_{22}, a_{33} all of which are unity. Hence

$$\mathbf{I} = \mathbf{i}\,\mathbf{i} + \mathbf{j}\,\mathbf{j} + \mathbf{k}\,\mathbf{k}. \tag{33}$$

Theorem: The direct product of any dyadic and the idemfactor is that dyadic. That is,

$$\Phi \cdot \mathbf{I} = \Phi \text{ and } \mathbf{I} \cdot \Phi = \Phi.$$

For
$$(\Phi \cdot \mathbf{I}) \cdot \mathbf{r} = \Phi \cdot (\mathbf{I} \cdot \mathbf{r}) = \Phi \cdot \mathbf{r},$$

no matter what the value of $\mathbf{r}$ may be. Hence, page 266,

$$\Phi \cdot \mathbf{I} = \Phi.$$

In like manner it may be shown that $\mathbf{I} \cdot \Phi = \Phi$.

Theorem: If $\mathbf{a}'$, $\mathbf{b}'$, $\mathbf{c}'$ and $\mathbf{a}$, $\mathbf{b}$, $\mathbf{c}$ be two reciprocal systems of vectors the expressions

$$\mathbf{I} = \mathbf{a}\,\mathbf{a}' + \mathbf{b}\,\mathbf{b}' + \mathbf{c}\,\mathbf{c}', \tag{34}$$

$$\mathbf{I} = \mathbf{a}'\,\mathbf{a} + \mathbf{b}'\,\mathbf{b} + \mathbf{c}'\,\mathbf{c}$$

are idemfactors.

For by (30) and (31) Chap. II.,

$$\mathbf{r} = \mathbf{r} \cdot \mathbf{a}\,\mathbf{a}' + \mathbf{r} \cdot \mathbf{b}\,\mathbf{b}' + \mathbf{r} \cdot \mathbf{c}\,\mathbf{c}',$$

and
$$\mathbf{r} = \mathbf{r} \cdot \mathbf{a}'\,\mathbf{a} + \mathbf{r} \cdot \mathbf{b}'\,\mathbf{b} + \mathbf{r} \cdot \mathbf{c}'\,\mathbf{c}.$$

Hence the expressions must be idemfactors by definition.

Theorem: Conversely if the expression

$$\Phi = \mathbf{a}\,\mathbf{l} + \mathbf{b}\,\mathbf{m} + \mathbf{c}\,\mathbf{n}$$

is an idemfactor $\mathbf{l}$, $\mathbf{m}$, $\mathbf{n}$ must be the reciprocal system of $\mathbf{a}$, $\mathbf{b}$, $\mathbf{c}$.

In the first place since Φ is the idemfactor, it is a complete dyadic. Hence the antecedents **a**, **b**, **c** are non-coplanar and possess a set of reciprocals **a**$'$, **b**$'$, **c**$'$. Let

$$\mathbf{r} = x\,\mathbf{a}' + y\,\mathbf{b}' + z\,\mathbf{c}'.$$

By hypothesis $\mathbf{r} \cdot \Phi = \mathbf{r}.$

Then $\mathbf{r} \cdot \Phi = x\,\mathbf{l} + y\,\mathbf{m} + z\,\mathbf{n} = x\,\mathbf{a}' + y\,\mathbf{b}' + z\,\mathbf{c}$

for all values of **r**, that is, for all values of x, y, z. Hence the corresponding coefficients must be equal. That is,

$$\mathbf{l} = \mathbf{a}', \quad \mathbf{m} = \mathbf{b}', \quad \mathbf{n} = \mathbf{c}'.$$

Theorem: If Φ and Ψ be any two dyadics, and if the product $\Phi \cdot \Psi$ is equal to the idemfactor;[1] then the product $\Psi \cdot \Phi$, when the factors are taken in the reversed order, is also equal to the idemfactor.

Let $\Phi \cdot \Psi = \mathrm{I}.$

To show $\Psi \cdot \Phi = \mathrm{I}.$

$$\mathbf{r} \cdot (\Phi \cdot \Psi) = \mathbf{r} \cdot \mathrm{I} = \mathbf{r},$$

$$\mathbf{r} \cdot (\Phi \cdot \Psi) \cdot \Phi = \mathbf{r} \cdot \Phi,$$

$$\mathbf{r} \cdot (\Phi \cdot \Psi) \cdot \Phi = (\mathbf{r} \cdot \Phi) \cdot (\Psi \cdot \Phi) = \mathbf{r} \cdot \Phi.$$

This relation holds for all values of **r**. As Φ is complete $\mathbf{r} \cdot \Phi$ must take on all desired values. Hence by definition

$$\Psi \cdot \Phi = \mathrm{I}.$$

If the product of two dyadics is an idemfactor, that product may be taken in either order.

109.] *Definition:* When two dyadics are so related that their product is equal to the idemfactor, they are said to be

[1] This necessitates both the dyadics Φ and Ψ to be complete. For the product of two incomplete dyadics is incomplete and hence could not be equal to the idemfactor.

reciprocals.[1] The notation used for reciprocals in ordinary algebra is employed to denote reciprocal dyadics. That is,

$$\text{if} \qquad \Phi \cdot \Psi = \mathrm{I}, \quad \Phi = \Psi^{-1} = \frac{\mathrm{I}}{\Psi} \text{ and } \Psi = \Phi^{-1} = \frac{\mathrm{I}}{\Phi} \cdot \qquad (35)$$

Theorem: Reciprocals of the same or equal dyadics are equal.

Let Φ and Ψ be two given equal dyadics, Φ^{-1} and Ψ^{-1} their reciprocals as defined above. By hypothesis

$$\Phi = \Psi,$$

$$\Phi \cdot \Phi^{-1} = \mathrm{I},$$

and
$$\Psi \cdot \Psi^{-1} = \mathrm{I}.$$

To show
$$\Phi^{-1} = \Psi^{-1}.$$

$$\Phi \cdot \Phi^{-1} = \mathrm{I} = \Psi \cdot \Psi^{-1}.$$

As
$$\Phi = \Psi, \quad \Phi \cdot \Phi^{-1} = \Phi \cdot \Psi^{-1},$$

$$\Phi^{-1} \cdot \Phi \cdot \Phi^{-1} = \Phi^{-1} \cdot \Phi \cdot \Psi^{-1},$$

$$\Phi^{-1} \cdot \Phi = \mathrm{I},$$

$$\mathrm{I} \cdot \Phi^{-1} = \Phi^{-1} = \mathrm{I} \cdot \Psi^{-1} = \Psi^{-1}.$$

Hence
$$\Phi^{-1} = \Psi^{-1}.$$

The reciprocal of Φ is the dyadic whose antecedents are the reciprocal system to the consequents of Φ and whose consequents are the reciprocal system to the antecedents of Φ.

If a complete dyadic Φ be written in the form

$$\Phi = \mathbf{a\,l} + \mathbf{b\,m} + \mathbf{c\,n},$$

its reciprocal is
$$\Phi^{-1} = \mathbf{l'\,a'} + \mathbf{m'\,b'} + \mathbf{n'\,c'}. \qquad (36)$$

For $(\mathbf{a\,l} + \mathbf{b\,m} + \mathbf{c\,n}) \cdot (\mathbf{l'\,a'} + \mathbf{n'\,b'} + \mathbf{n'\,c'}) = \mathbf{a\,a'} + \mathbf{b\,b'} + \mathbf{c\,c'}.$

Theorem: If the direct products of a complete dyadic Φ into two dyadics Ψ and Ω are equal as dyadics then Ψ and Ω

[1] An incomplete dyadic has no (finite) reciprocal.

are equal. If the product of a dyadic Φ into two vectors $\mathbf{r}$ and $\mathbf{s}$ (whether the multiplication be performed with a dot or a cross) are equal, then the vectors $\mathbf{r}$ and $\mathbf{s}$ are equal. That is,

$$\text{if} \qquad \Phi \cdot \Psi = \Phi \cdot \Omega, \qquad \text{then } \Psi = \Omega,$$

$$\text{and if} \qquad \Phi \cdot \mathbf{r} = \Phi \cdot \mathbf{s}, \qquad \text{then } \mathbf{r} = \mathbf{s}, \tag{37}$$

$$\text{and if} \qquad \Phi \times \mathbf{r} = \Phi \times \mathbf{s}, \qquad \text{then } \mathbf{r} = \mathbf{s}.$$

This may be seen by multiplying each of the equations through by the reciprocal of Φ,

$$\Phi^{-1} \cdot \Phi \cdot \Psi = \Psi = \Phi^{-1} \cdot \Phi \cdot \Omega = \Omega,$$

$$\Phi^{-1} \cdot \Phi \cdot \mathbf{r} = \mathbf{r} = \Phi^{-1} \cdot \Phi \cdot \mathbf{s} = \mathbf{s},$$

$$\Phi^{-1} \cdot \Phi \times \mathbf{r} = \mathrm{I} \times \mathbf{r} = \Phi^{-1} \cdot \Phi \times \mathbf{s} = \mathrm{I} \times \mathbf{s}.$$

To reduce the last equation proceed as follows. Let $\mathbf{t}$ be any vector,

$$\mathbf{t} \cdot \mathrm{I} \times \mathbf{r} = \mathbf{t} \cdot \mathrm{I} \times \mathbf{s},$$

$$\mathbf{t} \cdot \mathrm{I} = \mathbf{t}.$$

$$\text{Hence} \qquad \mathbf{t} \times \mathbf{r} = \mathbf{t} \times \mathbf{s}.$$

As $\mathbf{t}$ is any vector, $\mathbf{r}$ is equal to $\mathbf{s}$.

Equations (37) give what is equivalent to the law of cancelation for complete dyadics. Complete dyadics may be canceled from either end of an expression just as if they were scalar quantities. The cancelation of an incomplete dyadic is not admissible. It corresponds to the cancelation of a zero factor in ordinary algebra.

110.] *Theorem:* The reciprocal of the product of any number of dyadics is equal to the product of the reciprocals taken in the *opposite* order.

It will be sufficient to give the proof for the case in which the product consists of two dyadics. To show

$$(\Phi \cdot \Psi)^{-1} = \Psi^{-1} \cdot \Phi^{-1},$$

$$\Phi \cdot \Psi \cdot \Psi^{-1} \cdot \Phi^{-1} = \Phi \cdot (\Psi \cdot \Psi^{-1}) \cdot \Phi^{-1} = \Phi \cdot \Phi^{-1} = \mathrm{I}.$$

Hence $$(\Phi \cdot \Psi) \cdot (\Psi^{-1} \cdot \Phi^{-1}) = \mathrm{I}.$$

Hence $\Phi \cdot \Psi$ and $\Psi^{-1} \cdot \Phi^{-1}$ must be reciprocals. That is,

$$(\Phi \cdot \Psi)^{-1} = \Psi^{-1} \cdot \Phi^{-1}.$$

The proof for any number of dyadics may be given in the same manner or obtained by mathematical induction.

Definition: The products of a dyadic Φ, taken any number of times, by itself are called *powers* of Φ and are denoted in the customary manner.

$$\Phi \cdot \Phi = \Phi^2,$$

$$\Phi \cdot \Phi \cdot \Phi = \Phi \cdot \Phi^2 = \Phi^3,$$

and so forth.

Theorem: The reciprocal of a power of Φ is the power of the reciprocal of Φ.

$$(\Phi^n)^{-1} = (\Phi^{-1})^n = \Phi^{-n}. \quad (37)$$

The proof follows immediately as a corollary of the preceding theorem. The symbol Φ^{-n} may be interpreted as the nth power of the reciprocal of Φ or as the reciprocal of the nth power of Φ.

If Φ be interpreted as an operator determining a transformation of space, the positive powers of Φ correspond to repetitions of the transformation. The negative powers of Φ correspond to the inverse transformations. The idemfactor corresponds to the identical transformation—that is, no transformation at all. The fractional and irrational powers of Φ will not be defined. They are seldom used and are not single-valued. For instance the idemfactor I has the two square roots $\pm$ I. But in addition to these it has a doubly infinite system of square roots of the form

$$\Phi = -\mathbf{i}\,\mathbf{i} + \mathbf{j}\,\mathbf{j} + \mathbf{k}\,\mathbf{k}.$$

Geometrically the transformation

$$\mathbf{r}' = \Phi \cdot \mathbf{r}$$

is a reflection of space in the **j k**-plane. This transformation replaces each figure by a symmetrical figure, symmetrically situated upon the opposite side of the **j k**-plane. The transformation is sometimes called *perversion*. The idemfactor has also a doubly infinite system of square roots of the form

$$\Psi = \mathbf{i}\,\mathbf{i} - \mathbf{j}\mathbf{j} - \mathbf{k}\,\mathbf{k}.$$

Geometrically the transformation

$$\mathbf{r}' = \Psi \cdot \mathbf{r}$$

is a reflection in the **i**-axis. This transformation replaces each figure by its equal rotated about the **i**-axis through an angle of 180°. The idemfactor thus possesses not only two square roots; but in addition two doubly infinite systems of square roots; and it will be seen (Art. 129) that these are by no means all.

111.] The *conjugate* of a dyadic has been defined (Art. 99) as the dyadic obtained by interchanging the antecedents and consequents of a given dyadic and the notation of a subscript C has been employed. The equation

$$\mathbf{r} \cdot \Phi = \Phi_C \cdot \mathbf{r} \tag{9}$$

has been demonstrated. The following theorems concerning conjugates are useful.

Theorem: The conjugate of the sum or difference of two dyadics is equal to the sum or difference of the conjugates,

$$(\Phi \pm \Psi)_C = \Phi_C \pm \Psi_C.$$

Theorem: The conjugate of a product of dyadics is equal to the product of the conjugates taken in the *opposite* order.

It will be sufficient to demonstrate the theorem in case the product contains two factors. To show

$$(\Phi \cdot \Psi)_C = \Psi_C \cdot \Phi_C, \tag{40}$$

$$(\Phi \cdot \Psi)_C \cdot \mathbf{r} = \mathbf{r} \cdot (\Phi \cdot \Psi) = (\mathbf{r} \cdot \Phi) \cdot \Psi,$$

$$\mathbf{r} \cdot \Phi = \Phi_C \cdot \mathbf{r},$$

$$(\mathbf{r} \cdot \Phi) \cdot \Psi = \Psi_C \cdot (\mathbf{r} \cdot \Phi) = \Psi_C \cdot \Phi_C \cdot \mathbf{r}.$$

Hence
$$(\Phi \cdot \Psi)_C = \Psi_C \cdot \Phi_C.$$

Theorem: The conjugate of the power of a dyadic is the power of the conjugate of the dyadic.

$$(\Phi^n)_C = (\Phi_C)^n = \Phi^n_C. \tag{41}$$

This is a corollary of the foregoing theorem. The expression Φ^n_C may be interpreted in either of two equal ways.

Theorem: The conjugate of the reciprocal of a dyadic is equal to the reciprocal of the conjugate of the dyadic.

$$(\Phi^{-1})_C = (\Phi_C)^{-1} = \Phi^{-1}_C. \tag{42}$$

For
$$(\Phi^{-1})_C \cdot \Phi_C = (\Phi \cdot \Phi^{-1})_C = \mathrm{I}_C = \mathrm{I}.$$

The idemfactor is its own conjugate as may be seen from the nonion form.

$$\mathrm{I} = \mathbf{i}\,\mathbf{i} + \mathbf{j}\,\mathbf{j} + \mathbf{k}\,\mathbf{k}$$

$$(\Phi_C)^{-1} \cdot \Phi_C = \mathrm{I}.$$

Hence
$$(\Phi_C)^{-1} \cdot \Phi_C = (\Phi^{-1})_C \cdot \Phi_C.$$

Hence
$$(\Phi_C)^{-1} = (\Phi^{-1})_C.$$

The expression Φ_C^{-1} may therefore be interpreted in either of two equivalent ways — as the reciprocal of the conjugate or as the conjugate of the reciprocal.

Definition: If a dyadic is equal to its conjugate, it is said to be *self-conjugate*. If it is equal to the negative of its conjugate, it is said to be *anti-self-conjugate.* For *self*-conjugate dyadics.

$$\mathbf{r} \cdot \Phi = \Phi \cdot \mathbf{r}, \quad \Phi = \Phi_C.$$

For anti-self-conjugate dyadics

$$\mathbf{r} \cdot \Phi = - \Phi \cdot \mathbf{r}, \quad \Phi = - \Phi_C.$$

Theorem : Any dyadic may be divided in one and only one way into two parts of which one is self-conjugate and the other anti-self-conjugate.

For $$\Phi = \frac{1}{2}(\Phi + \Phi_C) + \frac{1}{2}(\Phi - \Phi_C). \tag{43}$$

But $$(\Phi + \Phi_C)_C = \Phi_C + \Phi_{CC} = \Phi_C + \Phi,$$

and $$(\Phi - \Phi_C)_C = \Phi_C - \Phi_{CC} = \Phi_C - \Phi.$$

Hence the part $\frac{1}{2}(\Phi + \Phi_C)$ is self-conjugate; and the part $\frac{1}{2}(\Phi - \Phi_C)$, anti-self-conjugate. Thus the division has been accomplished in one way. Let

$$\frac{1}{2}(\Phi + \Phi_C) = \Phi'$$

and $$\frac{1}{2}(\Phi - \Phi_C) = \Phi''.$$

$$\Phi = \Phi' + \Phi''.$$

Suppose it were possible to decompose Φ in another way into a self-conjugate and an anti-self-conjugate part. Let then

$$\Phi = (\Phi' + \Omega) + (\Phi'' - \Omega).$$

Where $$(\Phi' + \Omega) = (\Phi' + \Omega)_C = \Phi'_C + \Omega_C = \Phi' + \Omega_C.$$

Hence if $(\Phi' + \Omega)$ is self-conjugate, Ω is self-conjugate.

$$-(\Phi'' - \Omega) = (\Phi'' - \Omega)_C = \Phi''_C - \Omega_C = - \Phi'' - \Omega_C.$$

Hence if $(\Phi'' - \Omega)$ is anti-self-conjugate Ω is anti-self-conjugate.

$$\Omega = \Omega_C, \quad \Omega = - \Omega_C.$$

Any dyadic which is both self-conjugate and anti-self-conjugate is equal to its negative and consequently vanishes. Hence Ω is zero and the division of Φ into two parts is unique.

Anti-self-conjugate Dyadics. The Vector Product

112.] In case Φ is any dyadic the expression

$$\Phi'' = \frac{1}{2}(\Phi - \Phi_c)$$

gives the anti-self-conjugate part of Φ. If Φ should be entirely anti-self-conjugate Φ is equal to Φ''. Let therefore Φ'' be any anti-self-conjugate dyadic,

$$\Phi'' = \frac{1}{2}(\Phi - \Phi_c).$$

Suppose $$\Phi = \mathbf{a}\,\mathbf{l} + \mathbf{b}\,\mathbf{m} + \mathbf{c}\,\mathbf{n},$$

$$\Phi - \Phi_c = \mathbf{a}\mathbf{l} - \mathbf{l}\mathbf{a} + \mathbf{b}\mathbf{m} - \mathbf{m}\mathbf{b} + \mathbf{c}\mathbf{n} - \mathbf{n}\mathbf{c},$$

$$\Phi'' \cdot \mathbf{r} = \mathbf{a}\mathbf{l}\cdot\mathbf{r} - \mathbf{l}\mathbf{a}\cdot\mathbf{r} + \mathbf{b}\mathbf{m}\cdot\mathbf{r} - \mathbf{m}\mathbf{b}\cdot\mathbf{r} + \mathbf{c}\mathbf{n}\cdot\mathbf{r} - \mathbf{n}\mathbf{c}\cdot\mathbf{r}.$$

But $$\mathbf{a}\mathbf{l}\cdot\mathbf{r} - \mathbf{l}\mathbf{a}\cdot\mathbf{r} = -(\mathbf{a}\times\mathbf{l})\times\mathbf{r},$$

$$\mathbf{b}\mathbf{m}\cdot\mathbf{r} - \mathbf{m}\mathbf{b}\cdot\mathbf{r} = -(\mathbf{b}\times\mathbf{m})\times\mathbf{r},$$

$$\mathbf{c}\mathbf{n}\cdot\mathbf{r} - \mathbf{n}\mathbf{c}\cdot\mathbf{r} = -(\mathbf{c}\times\mathbf{n})\times\mathbf{r}.$$

Hence $$\Phi''\cdot\mathbf{r} = -\frac{1}{2}(\mathbf{a}\times\mathbf{l} + \mathbf{b}\times\mathbf{m} + \mathbf{c}\times\mathbf{n})\times\mathbf{r}.$$

But by definition $$\Phi_\times = \mathbf{a}\times\mathbf{l} + \mathbf{b}\times\mathbf{m} + \mathbf{c}\times\mathbf{n}.$$

Hence $$\Phi''\cdot\mathbf{r} = -\frac{1}{2}\Phi_\times\times\mathbf{r},$$

$$\mathbf{r}\cdot\Phi'' = \Phi''_c\cdot\mathbf{r} = -\Phi''\cdot\mathbf{r} = \frac{1}{2}\Phi_\times\times\mathbf{r} = -\frac{1}{2}\mathbf{r}\times\Phi_\times.$$

The results may be stated in a theorem as follows.

Theorem : The direct product of any anti-self-conjugate dyadic and the vector **r** is equal to the vector product of minus one half the vector of that dyadic and the vector **r**.

$$\frac{1}{2}(\Phi - \Phi_C) \cdot \mathbf{r} = -\frac{1}{2}\Phi_\times \times \mathbf{r},$$

$$\frac{1}{2}\mathbf{r} \cdot (\Phi - \Phi_C) = -\frac{1}{2}\mathbf{r} \times \Phi_\times. \qquad (44)$$

Theorem: Any anti-self-conjugate dyadic Φ'' possesses one degree of nullity. It is a uniplanar dyadic the plane of whose consequents and antecedents is perpendicular to $\Phi_\times''$, the vector of Φ.

This theorem follows as a corollary from equations (44).

Theorem: Any dyadic Φ may be broken up into two parts of which one is self-conjugate and the other equivalent to minus one half the vector of Φ used in cross multiplication.

$$\Phi \cdot \mathbf{r} = \Phi' \cdot \mathbf{r} - \frac{1}{2}\Phi_\times \times \mathbf{r},$$

or symbolically $$\Phi \cdot = \Phi' \cdot - \frac{1}{2}\Phi_\times \times. \qquad (45)$$

113.] Any vector $\mathbf{c}$ used in vector multiplication defines a linear vector function. For

$$\mathbf{c} \times (\mathbf{r} + \mathbf{s}) = \mathbf{c} \times \mathbf{r} + \mathbf{c} \times \mathbf{s}.$$

Hence it must be possible to represent the operator $\mathbf{c}\times$ as a dyadic. This dyadic will be uniplanar with plane of its antecedents and consequents perpendicular to $\mathbf{c}$, so that it will reduce all vectors parallel to $\mathbf{c}$ to zero. The dyadic may be found as follows

$$\mathbf{c} \times \mathbf{r} = \mathrm{I} \cdot \mathbf{c} \times \mathbf{r} = \mathrm{I} \times \mathbf{c} \cdot \mathbf{r} = (\mathrm{I} \times \mathbf{c}) \cdot \mathbf{r}.$$

By (31) $$\mathrm{I} \cdot (\mathbf{c} \times \mathrm{I}) = (\mathrm{I} \times \mathbf{c}) \cdot \mathrm{I},$$

$$(\mathrm{I} \times \mathbf{c}) \cdot \mathbf{r} = \{(\mathrm{I} \times \mathbf{c}) \cdot \mathrm{I}\} \cdot \mathbf{r} = \{\mathrm{I} \cdot (\mathbf{c} \times \mathrm{I})\} \cdot \mathbf{r}$$

$$= \mathrm{I} \cdot (\mathbf{c} \times \mathrm{I}) \cdot \mathbf{r} = (\mathbf{c} \times \mathrm{I}) \cdot \mathbf{r}.$$

Hence $$\mathbf{c} \times \mathbf{r} = (\mathrm{I} \times \mathbf{c}) \cdot \mathbf{r} = (\mathbf{c} \times \mathrm{I}) \cdot \mathbf{r},$$

and $$\mathbf{r} \times \mathbf{c} = \mathbf{r} \cdot (\mathrm{I} \times \mathbf{c}) = \mathbf{r} \cdot (\mathbf{c} \times \mathrm{I}). \qquad (46)$$

This may be stated in words.

Theorem: The vector **c** used in vector multiplication with a vector **r** is equal to the dyadic $\mathrm{I} \times \mathbf{c}$ or $\mathbf{c} \times \mathrm{I}$ used in direct multiplication with **r**. If **c** precedes **r** the dyadics are to be used as prefactors; if **c** follows **r**, as postfactors. The dyadics $\mathrm{I} \times \mathbf{c}$ and $\mathbf{c} \times \mathrm{I}$ are anti-self-conjugate.

In case the vector **c** is a unit vector the application of the operator $\mathbf{c} \times$ to any vector **r** in a plane perpendicular to **c** is equivalent to turning **r** through a positive right angle about the axis **c**. The dyadic $\mathbf{c} \times \mathrm{I}$ or $\mathrm{I} \times \mathbf{c}$ where **c** is a unit vector therefore turns any vector **r** perpendicular to **c** through a right angle about the line **c** as an axis. If **r** were a vector lying out of a plane perpendicular to **c** the effect of the dyadic $\mathrm{I} \times \mathbf{c}$ or $\mathbf{c} \times \mathrm{I}$ would be to annihilate that component of **r** which is parallel to **c** and turn that component of **r** which is perpendicular to **c** through a right angle about **c** as axis.

If the dyadic be applied twice the vectors perpendicular to **r** are rotated through two right angles. They are reversed in direction. If it be applied three times they are turned through three right angles. Applying the operator $\mathrm{I} \times \mathbf{c}$ or $\mathbf{c} \times \mathrm{I}$ four times brings a vector perpendicular to **c** back to its original position. The powers of the dyadic are therefore

$$
\begin{aligned}
(\mathrm{I} \times \mathbf{c})^2 &= (\mathbf{c} \times \mathrm{I})^2 = -(\mathrm{I} - \mathbf{c}\,\mathbf{c}), \\
(\mathrm{I} \times \mathbf{c})^3 &= (\mathbf{c} \times \mathrm{I})^3 = -\mathrm{I} \times \mathbf{c} = -\mathbf{c} \times \mathrm{I}, \\
(\mathrm{I} \times \mathbf{c})^4 &= (\mathbf{c} \times \mathrm{I})^4 = \mathrm{I} - \mathbf{c}\,\mathbf{c}, \\
(\mathrm{I} \times \mathbf{c})^5 &= (\mathbf{c} \times \mathrm{I})^5 = \mathrm{I} \times \mathbf{c} = \mathbf{c} \times \mathrm{I}.
\end{aligned}
\qquad (47)
$$

It thus appears that the dyadic $\mathrm{I} \times \mathbf{c}$ or $\mathbf{c} \times \mathrm{I}$ obeys the same law as far as its powers are concerned as the scalar imaginary $\sqrt{-1}$ in algebra.

The dyadic $\mathrm{I} \times \mathbf{c}$ or $\mathbf{c} \times \mathrm{I}$ is a quadrantal versor only for vectors perpendicular to **c**. For vectors parallel to **c** it acts as an annihilator. To avoid this effect and obtain a true quadrantal versor for all vectors $\mathbf{r}$ in space it is merely necessary to add the dyad $\mathbf{c}\,\mathbf{c}$ to the dyadic $\mathrm{I} \times \mathbf{c}$ or $\mathbf{c} \times \mathrm{I}$.

If
$$X = \mathrm{I} \times \mathbf{c} + \mathbf{c}\mathbf{c} = \mathbf{c} \times \mathrm{I} + \mathbf{c}\mathbf{c},$$
$$X^2 = -\,\mathrm{I},$$
$$X^3 = -\,X, \tag{48}$$
$$X^4 = \mathrm{I},$$
$$X^5 = X.$$

The dyadic X therefore appears as a fourth root of the idemfactor. The quadrantal versor X is analogous to the imaginary $\sqrt{-1}$ of a scalar algebra. The dyadic X is complete and consists of two parts of which $\mathrm{I} \times \mathbf{c}$ is anti-self-conjugate; and $\mathbf{c}\,\mathbf{c}$, self-conjugate.

114.] If $\mathbf{i}, \mathbf{j}, \mathbf{k}$ are three perpendicular unit vectors

$$\mathrm{I} \times \mathbf{i} = \mathbf{i} \times \mathrm{I} = \mathbf{k}\mathbf{j} - \mathbf{j}\mathbf{k},$$
$$\mathrm{I} \times \mathbf{j} = \mathbf{j} \times \mathrm{I} = \mathbf{i}\mathbf{k} - \mathbf{k}\mathbf{i}, \tag{49}$$
$$\mathrm{I} \times \mathbf{k} = \mathbf{k} \times \mathrm{I} = \mathbf{j}\mathbf{i} - \mathbf{i}\mathbf{j},$$

as may be seen by multiplying the idemfactor

$$\mathrm{I} = \mathbf{i}\,\mathbf{i} + \mathbf{j}\,\mathbf{j} + \mathbf{k}\,\mathbf{k}$$

into $\mathbf{i}$, $\mathbf{j}$, and $\mathbf{k}$ successively. These expressions represent quadrantal versors about the axis $\mathbf{i}, \mathbf{j}, \mathbf{k}$ respectively combined with annihilators along those axes. They are equivalent, when used in direct multiplication, to $\mathbf{i}\,\times$, $\mathbf{j}\times$, $\mathbf{k}\,\times$ respectively,

$$(\mathrm{I} \times \mathbf{k})^2 = (\mathbf{k} \times \mathrm{I})^2 = -\,(\mathbf{i}\,\mathbf{i} + \mathbf{j}\,\mathbf{j}),$$
$$(\mathrm{I} \times \mathbf{k})^3 = (\mathbf{k} \times \mathrm{I})^3 = -\,(\mathbf{j}\,\mathbf{i} - \mathbf{i}\mathbf{j}),$$
$$(\mathrm{I} \times \mathbf{k})^4 = (\mathbf{k} \times \mathrm{I})^4 = \mathbf{i}\mathbf{i} + \mathbf{j}\mathbf{j},$$

The expression $(\mathrm{I} \times \mathbf{k})^4$ is an idemfactor for the plane of $\mathbf{i}$ and $\mathbf{j}$, but an annihilator for the direction $\mathbf{k}$. In a similar manner the dyad $\mathbf{k}\,\mathbf{k}$ is an idemfactor for the direction $\mathbf{k}$, but an annihilator for the plane perpendicular to $\mathbf{k}$. These partial idemfactors are frequently useful.

If $\mathbf{a}, \mathbf{b}, \mathbf{c}$ are any three vectors and $\mathbf{a}', \mathbf{b}', \mathbf{c}'$ the reciprocal system,

$$\mathbf{a}\,\mathbf{a}' + \mathbf{b}\,\mathbf{b}'$$

used as a prefactor is an idemfactor for all vectors in the plane of $\mathbf{a}$ and $\mathbf{b}$, but an annihilator for vectors in the direction $\mathbf{c}$. Used as a postfactor it is an idemfactor for all vectors, in the plane of $\mathbf{a}'$ and $\mathbf{b}'$, but an annihilator for vectors in the direction $\mathbf{c}'$. In like manner the expression

$$\mathbf{c}\,\mathbf{c}'$$

used as a prefactor is an idemfactor for vectors in the direction $\mathbf{c}$, but for vectors in the plane of $\mathbf{a}$ and $\mathbf{b}$ it is an annihilator. Used as a postfactor it is an idemfactor for vectors in the direction $\mathbf{c}'$, but an annihilator for vectors in the plane of $\mathbf{a}'$ and $\mathbf{b}'$, that is, for vectors perpendicular of $\mathbf{c}$.

If $\mathbf{a}$ and $\mathbf{b}$ are any two vectors

$$(\mathbf{a} \times \mathbf{b}) \times \mathrm{I} = \mathrm{I} \times (\mathbf{a} \times \mathbf{b}) = \mathbf{b}\,\mathbf{a} - \mathbf{a}\,\mathbf{b}. \qquad (50)$$

For

$$\{(\mathbf{a} \times \mathbf{b}) \times \mathrm{I}\} \cdot \mathbf{r} = (\mathbf{a} \times \mathbf{b}) \times \mathbf{r} = \mathbf{b}\,\mathbf{a} \cdot \mathbf{r} - \mathbf{a}\,\mathbf{b} \cdot \mathbf{r} = (\mathbf{b}\,\mathbf{a} - \mathbf{a}\,\mathbf{b}) \cdot \mathbf{r}.$$

The vector $\mathbf{a} \times \mathbf{b}$ in cross multiplication is therefore equal to the dyadic $(\mathbf{b}\,\mathbf{a} - \mathbf{a}\,\mathbf{b})$ in direct multiplication. If the vector is used as a prefactor the dyadic must be so used.

$$(\mathbf{a} \times \mathbf{b}) \times \mathbf{r} = (\mathbf{b}\,\mathbf{a} - \mathbf{a}\,\mathbf{b}) \cdot \mathbf{r},$$

$$\mathbf{r} \times (\mathbf{a} \times \mathbf{b}) = \mathbf{r} \cdot (\mathbf{b}\,\mathbf{a} - \mathbf{a}\,\mathbf{b}). \qquad (51)$$

This is a symmetrical and easy form in which to remember the formula for expanding a triple vector product.

Reduction of Dyadics to Normal Form

115.] Let Φ be any complete dyadic and let $\mathbf{r}$ be a unit vector. Then the vector $\mathbf{r}'$

$$\mathbf{r}' = \Phi \cdot \mathbf{r}$$

is a linear function of $\mathbf{r}$. When $\mathbf{r}$ takes on all values consistent with its being a unit vector — that is, when the terminus of $\mathbf{r}$ describes the surface of a unit sphere, — the vector $\mathbf{r}'$ varies continuously and its terminus describes a surface. This surface is closed. It is in fact an ellipsoid.[1]

Theorem: It is always possible to reduce a complete dyadic to a sum of three terms of which the antecedents among themselves and the consequents among themselves are mutually perpendicular. This is called the *normal form* of Φ.

$$\Phi = a\,\mathbf{i}'\,\mathbf{i} + b\,\mathbf{j}'\,\mathbf{j} + c\,\mathbf{k}'\,\mathbf{k}.$$

To demonstrate the theorem consider the surface described by

$$\mathbf{r}' = \Phi \cdot \mathbf{r}.$$

As this is a closed surface there must be some direction of $\mathbf{r}$ which makes $\mathbf{r}'$ a maximum or at any rate gives $\mathbf{r}'$ as great a value as it is possible for $\mathbf{r}'$ to take on. Let this direction of $\mathbf{r}$ be called $\mathbf{i}$, and let the corresponding direction of $\mathbf{r}'$ — the direction in which $\mathbf{r}'$ takes on a value at least as great as any — be called $\mathbf{a}$. Consider next all the values of $\mathbf{r}$ which lie in a plane perpendicular to $\mathbf{i}$. The corresponding values of $\mathbf{r}'$ lie in a plane owing to a fact that $\Phi \cdot \mathbf{r}$ is a linear vector

[1] This may be proved as follows:

$$\mathbf{r}' = \Phi \cdot \mathbf{r} \quad \mathbf{r} = \Phi^{-1} \cdot \mathbf{r}' = \mathbf{r} \cdot \Phi_c{}^{-1}.$$

Hence
$$\mathbf{r} \cdot \mathbf{r} = 1 = \mathbf{r}' \cdot (\Phi_c{}^{-1} \cdot \Phi^{-1}) \cdot \mathbf{r}' = \mathbf{r}' \cdot \Psi \cdot \mathbf{r}'.$$

By expressing Ψ in nonion form, the equation $\mathbf{r}' \cdot \Psi \cdot \mathbf{r}' = 1$ is seen to be of the second degree. Hence $\mathbf{r}'$ describes a quadric surface. The only closed quadric surface is the ellipsoid.

function. Of these values of $\mathbf{r}'$ one must be at least as great as any other. Call this $\mathbf{b}$ and let the corresponding direction of $\mathbf{r}$ be called $\mathbf{j}$. Finally choose $\mathbf{k}$ perpendicular to $\mathbf{i}$ and $\mathbf{j}$ upon the positive side of plane of $\mathbf{i}$ and $\mathbf{j}$. Let $\mathbf{c}$ be the value of $\mathbf{r}'$ which corresponds to $\mathbf{r} = \mathbf{k}$. Since the dyadic Φ changes $\mathbf{i}, \mathbf{j}, \mathbf{k}$ into $\mathbf{a}, \mathbf{b}, \mathbf{c}$ it may be expressed in the form

$$\Phi = \mathbf{a}\,\mathbf{i} + \mathbf{b}\,\mathbf{j} + \mathbf{c}\,\mathbf{k}.$$

It remains to show that the vectors $\mathbf{a}, \mathbf{b}, \mathbf{c}$ as determined above are mutually perpendicular.

$$\mathbf{r}' = (\mathbf{a}\,\mathbf{i} + \mathbf{b}\,\mathbf{j} + \mathbf{c}\,\mathbf{k}) \cdot \mathbf{r},$$

$$d\mathbf{r}' = (\mathbf{a}\,\mathbf{i} + \mathbf{b}\,\mathbf{j} + \mathbf{c}\,\mathbf{k}) \cdot d\mathbf{r},$$

$$\mathbf{r}' \cdot d\mathbf{r}' = \mathbf{r}' \cdot \mathbf{a}\,\mathbf{i} \cdot d\mathbf{r} + \mathbf{r}' \cdot \mathbf{b}\,\mathbf{j} \cdot d\mathbf{r} + \mathbf{r}' \cdot \mathbf{c}\,\mathbf{k} \cdot d\mathbf{r}.$$

When $\mathbf{r}$ is parallel to $\mathbf{i}$, $\mathbf{r}'$ is a maximum and hence must be perpendicular to $d\mathbf{r}'$. Since $\mathbf{r}$ is a unit vector $d\mathbf{r}$ is always perpendicular to $\mathbf{r}$. Hence when $\mathbf{r}$ is parallel to $\mathbf{i}$

$$\mathbf{r}' \cdot \mathbf{b} \;\; \mathbf{j} \cdot d\mathbf{r} + \mathbf{r}' \cdot \mathbf{c} \;\; \mathbf{k} \cdot d\mathbf{r} = 0.$$

If further $d\mathbf{r}$ is perpendicular to $\mathbf{j}$, $\mathbf{r}' \cdot \mathbf{c}$ vanishes, and if $d\mathbf{r}$ is perpendicular to $\mathbf{k}$, $\mathbf{r}' \cdot \mathbf{b}$ vanishes. Hence when $\mathbf{r}$ is parallel to $\mathbf{i}$, $\mathbf{r}'$ is perpendicular to both $\mathbf{b}$ and $\mathbf{c}$. But when $\mathbf{r}$ is parallel to $\mathbf{i}$, $\mathbf{r}'$ is parallel to $\mathbf{a}$. Hence $\mathbf{a}$ is perpendicular to $\mathbf{b}$ and $\mathbf{c}$. Consider next the plane of $\mathbf{j}$ and $\mathbf{k}$ and the plane of $\mathbf{b}$ and $\mathbf{c}$. Let $\mathbf{r}$ be any vector in the plane of $\mathbf{j}$ and $\mathbf{k}$.

$$\mathbf{r}' = (\mathbf{b}\,\mathbf{j} + \mathbf{c}\,\mathbf{k}) \cdot \mathbf{r},$$

$$d\mathbf{r}' = (\mathbf{b}\,\mathbf{j} + \mathbf{c}\,\mathbf{k}) \cdot d\mathbf{r},$$

$$\mathbf{r}' \cdot d\mathbf{r}' = \mathbf{r}' \cdot \mathbf{b} \;\; \mathbf{j} \cdot \mathbf{dr} + \mathbf{r}' \cdot \mathbf{c} \;\; \mathbf{k} \cdot \mathbf{dr}.$$

When $\mathbf{r}$ takes the value $\mathbf{j}$, $\mathbf{r}'$ is a maximum in this plane and hence is perpendicular to $d\mathbf{r}'$. Since $\mathbf{r}$ is a unit vector it is perpendicular to $d\mathbf{r}$. Hence when $\mathbf{r}$ is parallel to $\mathbf{j}$, $d\mathbf{r}$ is perpendicular to $\mathbf{j}$, and

$$\mathbf{r}' \cdot d\mathbf{r}' = 0 = \mathbf{r}' \cdot \mathbf{c} \ \ \mathbf{k} \cdot d\mathbf{r}.$$

Hence $\mathbf{r}' \cdot \mathbf{c}$ is zero. But when $\mathbf{r}$ is parallel to $\mathbf{j}$, $\mathbf{r}'$ takes the value $\mathbf{b}$. Consequently $\mathbf{b}$ is perpendicular to $\mathbf{c}$.

It has therefore been shown that $\mathbf{a}$ is perpendicular to $\mathbf{b}$ and $\mathbf{c}$, and that $\mathbf{b}$ is perpendicular to $\mathbf{c}$. Consequently the three antecedents of Φ are mutually perpendicular. They may be denoted by $\mathbf{i}'$, $\mathbf{j}'$, $\mathbf{k}'$. Then the dyadic Φ takes the form

$$\Phi = a\,\mathbf{i}'\mathbf{i} + b\,\mathbf{j}'\mathbf{j} + c\,\mathbf{k}'\mathbf{k}, \tag{52}$$

where a, b, c are scalar constants positive or negative.

116.] *Theorem:* The complete dyadic Φ may always be reduced to a sum of three dyads whose antecedents and whose consequents form a right-handed rectangular system of unit vectors and whose scalar coefficients are either all positive or all negative.

$$\Phi = \pm\,(a\,\mathbf{i}'\mathbf{i} + b\,\mathbf{j}'\mathbf{j} + c\,\mathbf{k}'\mathbf{k}). \tag{53}$$

The proof of the theorem depends upon the statements made on page 20 that if one or three vectors of a right-handed system be reversed the resulting system is left-handed, but if two be reversed the system remains right-handed. If then *one* of the coefficients in (52) is negative, the directions of the other two axes may be reversed. Then all the coefficients are negative. If two of the coefficients in (52) are negative, the directions of the two vectors to which they belong may be reversed and then the coefficients in Φ are all positive. Hence in any case the reduction to the form in which all the coefficients are positive or all are negative has been performed.

As a limiting case between that in which the coefficients are all positive and that in which they are all negative comes the case in which one of them is zero. The dyadic then takes the form

$$\Phi = a\,\mathbf{i}'\mathbf{i} + b\,\mathbf{j}'\mathbf{j} \qquad (54)$$

and is planar. The coefficients a and b may always be taken positive. By a proof similar to the one given above it is possible to show that any planar dyadic may be reduced to this form. The vectors $\mathbf{i}'$ and $\mathbf{j}'$ are perpendicular, and the vectors $\mathbf{i}$ and $\mathbf{j}$ are likewise perpendicular.

It might be added that in case the three coefficients a, b, c in the reduction (53) are all different the reduction can be performed in only one way. If two of the coefficients (say a and b) are equal the reduction may be accomplished in an infinite number of ways in which the third vector $\mathbf{k}'$ is always the same, but the two vectors $\mathbf{i}'$, $\mathbf{j}'$ to which the equal coefficients belong may be any two vectors in the plane perpendicular to $\mathbf{k}$. In all these reductions the three scalar coefficients will have the same values as in any one of them. If the three coefficients a, b, c are all equal when Φ is reduced to the normal form (53), the reduction may be accomplished in a doubly infinite number of ways. The three vectors $\mathbf{i}'$, $\mathbf{j}'$, $\mathbf{k}'$ may be any right-handed rectangular system in space. In all of these reductions the three scalar coefficients are the same as in any one of them. These statements will not be proved. They correspond to the fact that the ellipsoid which is the locus of the terminus of $\mathbf{r}'$ may have three different principal axes or it may be an ellipsoid of revolution, or finally a sphere.

Theorem: Any self-conjugate dyadic may be expressed in the form

$$\Phi = a\,\mathbf{i}\mathbf{i} + b\,\mathbf{j}\mathbf{j} + c\,\mathbf{k}\mathbf{k} \qquad (55)$$

where a, b, and c are scalars, positive or negative.

Let

$$\Phi = a\,\mathbf{i}'\mathbf{i} + b\,\mathbf{j}'\mathbf{j} + c\,\mathbf{k}'\mathbf{k}, \qquad (52)$$

$$\Phi_C = a\,\mathbf{i}\mathbf{i}' + b\,\mathbf{j}\mathbf{j}' + c\,\mathbf{k}\mathbf{k}',$$

$$\Phi \cdot \Phi_C = a^2 \mathbf{i}'\mathbf{i}' + b^2 \mathbf{j}'\mathbf{j}' + c^2 \mathbf{k}'\mathbf{k}',$$

$$\Phi_C \cdot \Phi = a^2 \mathbf{i}\mathbf{i} + b^2 \mathbf{j}\mathbf{j} + c^2 \mathbf{k}\mathbf{k}.$$

Since
$$\Phi = \Phi_C,$$

$$\Phi \cdot \Phi_C = \Phi_C \cdot \Phi = \Phi^2.$$

$$\mathrm{I} = \mathbf{i}\mathbf{i} + \mathbf{j}\mathbf{j} + \mathbf{k}\mathbf{k} = \mathbf{i}'\mathbf{i}' + \mathbf{j}'\mathbf{j} + \mathbf{k}'\mathbf{k}',$$

$$\Phi^2 - a^2 \mathrm{I} = (b^2 - a^2)\, \mathbf{j}'\mathbf{j}' + (c^2 - a^2)\, \mathbf{k}'\mathbf{k}',$$

$$(\Phi^2 - a^2 \mathrm{I}) \cdot \mathbf{i}' = 0$$

$$\Phi^2 - a^2 \mathrm{I} = (b^2 - a^2)\, \mathbf{j}\mathbf{j} + (c^2 - a^2)\, \mathbf{k}\mathbf{k},$$

$$(\Phi^2 - a^2 \mathrm{I}) \cdot \mathbf{i} = 0.$$

If **i** and **i**$'$ were not parallel $(\Phi^2 - a^2\, \mathrm{I})$ would annihilate two vectors **i** and **i**$'$ and hence every vector in their plane. $(\Phi^2 - a^2\, \mathrm{I})$ would therefore possess two degrees of nullity and be linear. But it is apparent that if a, b, c are different this dyadic is not linear. It is planar. Hence **i** and **i**$'$ must be parallel. In like manner it may be shown that **j** and **j**$'$, **k** and **k**$'$ are parallel. The dyadic Φ therefore takes the form

$$\Phi = a\, \mathbf{i}\mathbf{i} + b\, \mathbf{j}\mathbf{j} + c\, \mathbf{k}\mathbf{k}$$

where a, b, c are positive or negative scalar constants.

Double Multiplication [1]

117.] *Definition:* The *double dot* product of two dyads is the *scalar* quantity obtained by multiplying the scalar product of the antecedents by the scalar product of the consequents. The product is denoted by inserting two dots between the dyads.

$$\mathbf{a}\mathbf{b} : \mathbf{c}\mathbf{d} = \mathbf{a} \cdot \mathbf{c}\ \mathbf{b} \cdot \mathbf{d}. \tag{56}$$

This product evidently obeys the commutative law

$$\mathbf{a}\mathbf{b} : \mathbf{c}\mathbf{d} = \mathbf{c}\mathbf{d} : \mathbf{a}\mathbf{b},$$

[1] The researches of Professor Gibbs upon *Double Multiplication* are here printed for the first time.

and the distributive law both with regard to the dyads and with regard to the vectors in the dyads. The double dot product of two dyadics is obtained by multiplying the product out formally according to the distributive law into the sum of a number of double dot products of dyads.

If $$\Phi = \mathbf{a}_1\mathbf{b}_1 + \mathbf{a}_2\mathbf{b}_2 + \mathbf{a}_3\mathbf{b}_3 + \cdots$$

and $$\Psi = \mathbf{c}_1\mathbf{d}_1 + \mathbf{c}_2\mathbf{d}_2 + \mathbf{c}_3\mathbf{d}_3 + \cdots$$

$$\begin{aligned}\Phi : \Psi = {} & (\mathbf{a}_1\mathbf{b}_1 + \mathbf{a}_2\mathbf{b}_2 + \mathbf{a}_3\mathbf{b}_3 + \cdots) : (\mathbf{c}_1\mathbf{d}_1 + \mathbf{c}_2\mathbf{d}_2 \\ & \qquad + \mathbf{c}_3\mathbf{d}_3 + \cdots) \\ = {} & \mathbf{a}_1\mathbf{b}_1 : \mathbf{c}_1\mathbf{d}_1 + \mathbf{a}_1\mathbf{b}_1 : \mathbf{c}_2\mathbf{d}_2 + \mathbf{a}_1\mathbf{b}_1 : \mathbf{c}_3\mathbf{d}_3 + \cdots \\ & + \mathbf{a}_2\mathbf{b}_2 : \mathbf{c}_1\mathbf{d}_1 + \mathbf{a}_2\mathbf{b}_2 : \mathbf{c}_2\mathbf{d}_2 + \mathbf{a}_2\mathbf{b}_2 : \mathbf{c}_3\mathbf{d}_3 + \cdots \qquad (56)' \\ & + \mathbf{a}_3\mathbf{b}_3 : \mathbf{c}_1\mathbf{d}_1 + \mathbf{a}_3\mathbf{b}_3 : \mathbf{c}_2\mathbf{d}_2 + \mathbf{a}_3\mathbf{b}_3 : \mathbf{c}_3\mathbf{d}_3 + \cdots \\ & + \cdot \;\cdot \;\cdot \;\cdot \;\cdot \;\cdot \;\cdot \;\cdot \;\cdot \;\cdot \;\cdot \;\cdot \;\cdot \;\cdot \;\cdot \;\cdot\end{aligned}$$

$$\begin{aligned}\Phi : \Psi = {} & \mathbf{a}_1\cdot\mathbf{c}_1 \;\; \mathbf{b}_1\cdot\mathbf{d}_1 + \mathbf{a}_1\cdot\mathbf{c}_2 \;\; \mathbf{b}_1\cdot\mathbf{d}_2 + \mathbf{a}_1\cdot\mathbf{c}_3 \;\; \mathbf{b}_1\cdot\mathbf{d}_3 + \cdots \\ & + \mathbf{a}_2\cdot\mathbf{c}_1 \;\; \mathbf{b}_2\cdot\mathbf{d}_1 + \mathbf{d}_2\cdot\mathbf{c}_2 \;\; \mathbf{b}_2\cdot\mathbf{d}_2 + \mathbf{a}_2\cdot\mathbf{c}_3 \;\; \mathbf{b}_2\cdot\mathbf{d}_3 + \cdots \\ & + \mathbf{a}_3\cdot\mathbf{c}_1 \;\; \mathbf{b}_3\cdot\mathbf{d}_1 + \mathbf{a}_3\cdot\mathbf{c}_2 \;\; \mathbf{b}_3\cdot\mathbf{d}_2 + \mathbf{a}_3\cdot\mathbf{c}_3 \;\; \mathbf{b}_3\cdot\mathbf{d}_3 + \cdots \\ & + \cdot \;\cdot \;\cdot \;\cdot \;\cdot \;\cdot \;\cdot \;\cdot \;\cdot \;\cdot \;\cdot \;\cdot \;\cdot \;\cdot \;\cdot \;\cdot \qquad (56)''\end{aligned}$$

Definition: The *double cross* product of two dyads is the *dyad* of which the antecedent is the vector product of the antecedents of the two dyads and of which the consequent is the vector product of the consequent of the two dyads. The product is denoted by inserting two crosses between the dyads

$$\mathbf{a}\mathbf{b} \overset{\times}{\times} \mathbf{c}\mathbf{d} = \mathbf{a}\times\mathbf{c} \;\; \mathbf{b}\times\mathbf{d}. \qquad (57)$$

This product also evidently obeys the commutative law

$$\mathbf{a}\mathbf{b} \overset{\times}{\times} \mathbf{c}\mathbf{d} = \mathbf{c}\mathbf{d} \overset{\times}{\times} \mathbf{a}\mathbf{b},$$

and the distributive law both with regard to the dyads and with regard to the vectors of which the dyads are composed. The double cross product of two dyadics is therefore defined as the formal expansion of the product according to the distributive law into a sum of double cross products of dyads.

If $$\Phi = \mathbf{a}_1 \mathbf{b}_1 + \mathbf{a}_2 \mathbf{b}_2 + \mathbf{a}_3 \mathbf{b}_3 + \cdots$$

and $$\Psi = \mathbf{c}_1 \mathbf{d}_1 + \mathbf{c}_2 \mathbf{d}_2 + \mathbf{c}_3 \mathbf{d}_3 + \cdots$$

$$\begin{aligned}\Phi \overset{\times}{\times} \Psi = (\mathbf{a}_1 \mathbf{b}_1 + \mathbf{a}_2 \mathbf{b}_2 + \mathbf{a}_3 \mathbf{b}_3 + \cdots) \overset{\times}{\times} (\mathbf{c}_1 \mathbf{d}_1 + \mathbf{c}_2 \mathbf{d}_2 \\ + \mathbf{c}_3 \mathbf{d}_3 + \cdots)\end{aligned}$$

$$\begin{aligned}&= \mathbf{a}_1 \mathbf{b}_1 \overset{\times}{\times} \mathbf{c}_1 \mathbf{d}_1 + \mathbf{a}_1 \mathbf{b}_1 \overset{\times}{\times} \mathbf{c}_2 \mathbf{d}_2 + \mathbf{a}_1 \mathbf{b}_1 \overset{\times}{\times} \mathbf{c}_3 \mathbf{d}_3 + \cdots \\ &+ \mathbf{a}_2 \mathbf{b}_2 \overset{\times}{\times} \mathbf{c}_1 \mathbf{d}_1 + \mathbf{a}_2 \mathbf{b}_2 \overset{\times}{\times} \mathbf{c}_2 \mathbf{d}_2 + \mathbf{a}_2 \mathbf{b}_2 \overset{\times}{\times} \mathbf{c}_3 \mathbf{d}_3 + \cdots \qquad (57)' \\ &+ \mathbf{a}_3 \mathbf{b}_3 \overset{\times}{\times} \mathbf{c}_1 \mathbf{d}_1 + \mathbf{a}_3 \mathbf{b}_3 \overset{\times}{\times} \mathbf{c}_2 \mathbf{d}_2 + \mathbf{a}_3 \mathbf{b}_3 \overset{\times}{\times} \mathbf{c}_3 \mathbf{d}_3 + \cdots \\ &+ \cdot \cdot \cdot \cdot \cdot \cdot \cdot \cdot \cdot \cdot \cdot \cdot \cdot \cdot \cdot \cdot \cdot\end{aligned}$$

$$\begin{aligned}\Phi \overset{\times}{\times} \Psi = \mathbf{a}_1 \times \mathbf{c}_1 \;\; \mathbf{b}_1 \times \mathbf{d}_1 + \mathbf{a}_1 \times \mathbf{c}_2 \;\; \mathbf{b}_1 \times \mathbf{d}_2 + \mathbf{a}_1 \times \mathbf{c}_3 \;\; \mathbf{b}_1 \times \mathbf{d}_3 + \cdots \\ + \mathbf{a}_2 \times \mathbf{c}_1 \;\; \mathbf{b}_2 \times \mathbf{d}_1 + \mathbf{a}_2 \times \mathbf{c}_2 \;\; \mathbf{b}_2 \times \mathbf{d}_2 + \mathbf{a}_2 \times \mathbf{c}_3 \;\; \mathbf{b}_2 \times \mathbf{d}_3 + \cdots \\ + \mathbf{a}_3 \times \mathbf{c}_3 \;\; \mathbf{b}_3 \times \mathbf{d}_1 + \mathbf{a}_3 \times \mathbf{c}_2 \;\; \mathbf{b}_3 \times \mathbf{d}_1 + \mathbf{a}_3 \times \mathbf{c}_3 \;\; \mathbf{b}_3 \times \mathbf{d}_3 + \cdots \\ + \cdot \cdot \cdot \cdot \cdot \cdot \cdot \cdot \cdot \cdot \cdot \cdot \cdot \cdot \cdot \qquad (57)''\end{aligned}$$

Theorem: The double dot and double cross products of two dyadics obey the commutative and distributive laws of multiplication. But the double products of more than two dyadics (whenever they have any meaning) do not obey the associative law.

$$\begin{aligned}\Phi : \Psi &= \Psi : \Phi \\ \Phi \overset{\times}{\times} \Psi &= \Psi \overset{\times}{\times} \Phi \qquad (58) \\ (\Phi \overset{\times}{\times} \Psi) \overset{\times}{\times} \Omega &\neq \Phi \overset{\times}{\times} (\Psi \overset{\times}{\times} \Omega).\end{aligned}$$

The theorem is sufficiently evident without demonstration.

Theorem: The double dot product of two fundamental dyads is equal to unity or to zero according as the two dyads are equal or different.

$$\mathbf{i\,j : i\,j} = \mathbf{i \cdot i}\ \ \mathbf{j \cdot j} = 1$$

$$\mathbf{i\,j : k\,i} = \mathbf{i \cdot k}\ \ \mathbf{j \cdot i} = 0.$$

Theorem: The double cross product of two fundamental dyads (12) is equal to zero if either the antecedents or the consequents are equal. But if neither antecedents nor consequents are equal the product is equal to one of the fundamental dyads taken with a positive or a negative sign. That is

$$\mathbf{i\,j} \overset{\times}{\times} \mathbf{i\,k} = \mathbf{i \times i}\ \ \mathbf{j \times k} = 0$$

$$\mathbf{i\,j} \overset{\times}{\times} \mathbf{k\,i} = \mathbf{i \times k}\ \ \mathbf{j \times i} = + \mathbf{j\,k}.$$

There exists a scalar triple product of three dyads in which the multiplications are double. Let Φ, Ψ, Ω be any three dyadics. The expression

$$\Phi \overset{\times}{\times} \Psi : \Omega$$

is a scalar quantity. The multiplication with the double cross must be performed first. This product is entirely independent of the order in which the factors are arranged or the position of the dot and crosses. Let $\mathbf{a\,b}$, $\mathbf{c\,d}$, and $\mathbf{e\,f}$ be three dyads,

$$\mathbf{a\,b} \overset{\times}{\times} \mathbf{c\,d} : \mathbf{e\,f} = [\mathbf{a\,c\,e}]\ [\mathbf{b\,d\,f}]. \qquad (59)$$

That is, the product of three dyads united by a double cross and a double dot is equal to the product of the scalar triple product of the three antecedents by the scalar triple product of the three consequents. From this the statement made above follows. For if the dots and crosses be interchanged or if the order of the factors be permuted cyclicly the two scalar triple products are not altered. If the cyclic order of the factors is reversed each scalar triple product changes sign. Their product therefore is not altered.

118.] A dyadic Φ may be multiplied by itself with double cross. Let

$$\Phi = \mathbf{a}\mathbf{l} + \mathbf{b}\mathbf{m} + \mathbf{c}\mathbf{n}$$

$$\begin{aligned}\Phi \overset{\times}{\times} \Phi &= (\mathbf{a}\mathbf{l} + \mathbf{b}\mathbf{m} + \mathbf{c}\mathbf{n}) \overset{\times}{\times} (\mathbf{a}\mathbf{l} + \mathbf{b}\mathbf{m} + \mathbf{c}\mathbf{n}) \\ &= \mathbf{a}\times\mathbf{a}\ \ \mathbf{l}\times\mathbf{l} + \mathbf{a}\times\mathbf{b}\ \ \mathbf{l}\times\mathbf{m} + \mathbf{a}\times\mathbf{c}\ \ \mathbf{l}\times\mathbf{n} \\ &+ \mathbf{b}\times\mathbf{a}\ \ \mathbf{m}\times\mathbf{l} + \mathbf{b}\times\mathbf{b}\ \ \mathbf{m}\times\mathbf{m} + \mathbf{b}\times\mathbf{c}\ \ \mathbf{m}\times\mathbf{n} \\ &+ \mathbf{c}\times\mathbf{a}\ \ \mathbf{n}\times\mathbf{l} + \mathbf{c}\times\mathbf{b}\ \ \mathbf{n}\times\mathbf{m} + \mathbf{c}\times\mathbf{c}\ \ \mathbf{n}\times\mathbf{n}.\end{aligned}$$

The products in the main diagonal vanish. The others are equal in pairs. Hence

$$\Phi \overset{\times}{\times} \Phi = 2(\mathbf{b}\times\mathbf{c}\ \ \mathbf{m}\times\mathbf{n} + \mathbf{c}\times\mathbf{a}\ \ \mathbf{n}\times\mathbf{l} + \mathbf{a}\times\mathbf{b}\ \ \mathbf{l}\times\mathbf{m}). \quad (60)$$

If **a**, **b**, **c** and **l**, **m**, **n** are non-coplanar this may be written

$$\Phi \overset{\times}{\times} \Phi = \frac{2}{[\mathbf{a}\mathbf{b}\mathbf{c}]\ [\mathbf{l}\mathbf{m}\mathbf{n}]} (\mathbf{a}'\mathbf{l}' + \mathbf{b}'\mathbf{m}' + \mathbf{c}'\mathbf{n}'). \quad (60)'$$

The product $\Phi \overset{\times}{\times} \Phi$ is a species of power of Φ. It may be regarded as a square of Φ. The notation Φ_2 will be employed to represent this product after the scalar factor 2 has been stricken out.

$$\Phi_2 = \frac{\Phi \overset{\times}{\times} \Phi}{2} = (\mathbf{b}\times\mathbf{c}\ \ \mathbf{m}\times\mathbf{n} + \mathbf{c}\times\mathbf{a}\ \ \mathbf{n}\times\mathbf{l} + \mathbf{a}\times\mathbf{b}\ \ \mathbf{l}\times\mathbf{m}) \quad (61)$$

The triple product of a dyadic Φ expressed as the sum of three dyads with itself twice repeated is

$$\Phi \overset{\times}{\times} \Phi : \Phi = 2\ \Phi_2 : \Phi$$

$$\begin{aligned}\Phi_2 : \Phi = (\mathbf{b}\times\mathbf{c}\ \ \mathbf{m}\times\mathbf{n} + \mathbf{c}\times\mathbf{a}\ \ \mathbf{n}\times\mathbf{l} + \mathbf{a}\times\mathbf{b}\ \ \mathbf{l}\times\mathbf{m}) \\ : (\mathbf{a}\mathbf{l} + \mathbf{b}\mathbf{m} + \mathbf{c}\mathbf{n}).\end{aligned}$$

In expanding this product every term in which a letter is repeated vanishes. For a scalar triple product of three vectors two of which are equal is zero. Hence the product reduces to three terms only

$$\Phi_2 : \Phi = [\mathbf{b\,c\,a}]\,[\mathbf{m\,n\,l}] + [\mathbf{c\,a\,b}]\,[\mathbf{n\,l\,m}] + [\mathbf{a\,b\,c}]\,[\mathbf{l\,m\,n}]$$

or
$$\Phi_2 : \Phi = 3\,[\mathbf{a\,b\,c}]\,[\mathbf{l\,m\,n}]$$

$$\Phi \overset{\times}{\times} \Phi : \Phi = 6\,[\mathbf{a\,b\,c}]\,[\mathbf{l\,m\,n}].$$

The triple product of a dyadic by itself twice repeated is equal to six times the scalar triple product of its antecedents multiplied by the scalar triple product of its consequents. The product is a species of cube. It will be denoted by Φ_3 after the scalar factor 6 has been stricken out.

$$\Phi_3 = \frac{\Phi \overset{\times}{\times} \Phi : \Phi}{6} = [\mathbf{a\,b\,c}]\,[\mathbf{l\,m\,n}]. \qquad (62)$$

119.] If Φ_2 be called the *second* of Φ; and Φ_3, the *third* of Φ, the following theorems may be stated concerning the seconds and thirds of conjugates, reciprocals, and products.

Theorem: The second of the conjugate of a dyadic is equal to the conjugate of the second of that dyadic. The third of the conjugate is equal to the third of the dyadic.

$$(\Phi_2)_C = (\Phi_C)_2$$
$$\Phi_3 = (\Phi_C)_3. \qquad (63)$$

Theorem: The second and third of the reciprocal of a dyadic are equal respectively to the reciprocals of the second and third.

$$(\Phi^{-1})_2 = (\Phi_2)^{-1} = \Phi_2^{-1}$$
$$(\Phi^{-1})_3 = (\Phi_3)^{-1} = \Phi_3^{-1} \qquad (64)$$

Let
$$\Phi = \mathbf{a\,l} + \mathbf{b\,m} + \mathbf{c\,n}$$

$$\Phi^{-1} = \mathbf{l}'\,\mathbf{a}' + \mathbf{m}'\,\mathbf{b}' + \mathbf{n}'\,\mathbf{c}' \qquad (36)$$

$$\Phi_2 = \frac{\mathbf{a}'\,\mathbf{l}' + \mathbf{b}'\,\mathbf{m}' + \mathbf{c}'\,\mathbf{n}'}{[\mathbf{a\,b\,c}]\,[\mathbf{l\,m\,n}]} \qquad (60)'$$

$$(\Phi_2)^{-1} = [\mathbf{a\,b\,c}]\,[\mathbf{l\,m\,n}]\,(\mathbf{la} + \mathbf{mb} + \mathbf{nc})$$

$$(\Phi^{-1})_2 = \frac{\mathbf{la} + \mathbf{mb} + \mathbf{nc}}{[\mathbf{a'\,b'\,c'}]\,[\mathbf{l'\,m'\,n'}]}.$$

But $\quad [\mathbf{a'\,b'\,c'}]\,[\mathbf{a\,b\,c}] = 1$ and $[\mathbf{l'\,m'\,n'}]\,[\mathbf{l\,m\,n}] = 1.$

Hence
$$(\Phi_2)^{-1} = (\Phi^{-1})_2 = \Phi_2^{-1}.$$

$$\Phi_3 = [\mathbf{a\,b\,c}]\,[\mathbf{l\,m\,n}],$$

$$(\Phi_3)^{-1} = \frac{1}{[\mathbf{a\,b\,c}]\,[\mathbf{l\,m\,n}]},$$

$$(\Phi^{-1})_3 = [\mathbf{a'\,b'\,c'}]\,[\mathbf{l'\,m'\,n'}].$$

Hence
$$(\Phi_3)^{-1} = (\Phi^{-1})_3 = \Phi_3^{-1}.$$

Theorem: The second and third of a product are equal respectively to the product of the seconds and the product of the thirds.

$$\begin{aligned} (\Phi \cdot \Psi)_2 &= \Phi_2 \cdot \Psi_2 \\ (\Phi \cdot \Psi)_3 &= \Phi_3\, \Psi_3. \end{aligned} \qquad (65)$$

Choose any three non-coplanar vectors $\mathbf{l}, \mathbf{m}, \mathbf{n}$ as consequents of Φ and let $\mathbf{l}', \mathbf{m}', \mathbf{n}'$ be the antecedents of Ψ.

$$\Phi = \mathbf{a\,l} + \mathbf{b\,m} + \mathbf{c\,n},$$

$$\Psi = \mathbf{l'\,d} + \mathbf{m'\,e} + \mathbf{n'\,f},$$

$$\Phi \cdot \Psi = \mathbf{a\,d} + \mathbf{b\,e} + \mathbf{c\,f},$$

$$(\Phi \cdot \Psi)_2 = \mathbf{b} \times \mathbf{c} \;\; \mathbf{e} \times \mathbf{f} + \mathbf{c} \times \mathbf{a} \;\; \mathbf{f} \times \mathbf{d} + \mathbf{a} \times \mathbf{b} \;\; \mathbf{d} \times \mathbf{e},$$

$$\Phi_2 = \mathbf{b} \times \mathbf{c} \;\; \mathbf{m} \times \mathbf{n} + \mathbf{c} \times \mathbf{a} \;\; \mathbf{n} \times \mathbf{l} + \mathbf{a} \times \mathbf{b} \;\; \mathbf{l} \times \mathbf{m},$$

$$\Psi_2 = \mathbf{m}' \times \mathbf{n}' \;\; \mathbf{e} \times \mathbf{f} + \mathbf{n}' \times \mathbf{l}' \;\; \mathbf{f} \times \mathbf{d} + \mathbf{l}' \times \mathbf{m}' \;\; \mathbf{d} \times \mathbf{e}.$$

Hence $\Phi_2 \cdot \Psi_2 = \mathbf{b} \times \mathbf{c} \;\; \mathbf{e} \times \mathbf{f} + \mathbf{c} \times \mathbf{a} \;\; \mathbf{f} \times \mathbf{d} + \mathbf{a} \times \mathbf{b} \;\; \mathbf{d} \times \mathbf{e}.$

Hence
$$(\Phi \cdot \Psi)_2 = \Phi_2 \cdot \Psi_2 \cdot$$

$$(\Phi \cdot \Psi)_3 = [\mathbf{a\,b\,c}]\,[\mathbf{d\,e\,f}]$$

$$\Phi_3 = [\mathbf{a\,b\,c}]\,[\mathbf{l\,m\,n}],$$

$$\Psi_3 = [\mathbf{l'\,m'\,n'}]\,[\mathbf{d\,e\,f}].$$

Hence $$\Phi_3\,\Psi_3 = [\mathbf{a\,b\,c}]\,[\mathbf{d\,e\,f}].$$

Hence $$(\Phi \cdot \Psi)_3 = \Phi_3\,\Psi_3.$$

Theorem: The second and third of a power of a dyadic are equal respectively to the powers of the second and third of the dyadic.

$$\begin{aligned}(\Phi^n)_2 &= (\Phi_2)^n = \Phi_2{}^n \\ (\Phi^n)_3 &= (\Phi_3)^n = \Phi_3{}^n.\end{aligned} \qquad (66)$$

Theorem: The second of the idemfactor is the idemfactor. The third of the idemfactor is unity.

$$\begin{aligned}\mathrm{I}_2 &= \mathrm{I} \\ \mathrm{I}_3 &= 1.\end{aligned} \qquad (67)$$

Theorem: The product of the second and conjugate of a dyadic is equal to the product of the third and the idemfactor.

$$\Phi_2 \cdot \Phi_C = \Phi_3\,\mathrm{I}, \qquad (68)$$

$$\Phi_2 = \mathbf{b \times c}\ \ \mathbf{m \times n} + \mathbf{c \times a}\ \ \mathbf{n \times l} + \mathbf{a \times b}\ \ \mathbf{l \times m},$$

$$\Phi_C = \mathbf{l\,a} + \mathbf{m\,b} + \mathbf{n\,c},$$

$$\Phi_2 \cdot \Phi_C = [\mathbf{l\,m\,n}]\,(\mathbf{b \times c}\ \ \mathbf{a} + \mathbf{c \times a}\ \ \mathbf{b} + \mathbf{a \times b}\ \ \mathbf{c}).$$

The antecedents **a**, **b**, **c** of the dyadic Φ may be assumed to be non-coplanar. Then

$$\begin{aligned}(\mathbf{b \times c}\ \ \mathbf{a} + \mathbf{c \times a}\ \ \mathbf{b} + \mathbf{a \times b}\ \ \mathbf{c}) &= [\mathbf{a\,b\,c}]\,(\mathbf{a'\,a} + \mathbf{b'\,b} + \mathbf{c'\,c}) \\ &= [\mathbf{a\,b\,c}]\,\mathrm{I}.\end{aligned}$$

Hence $$\Phi_2 \cdot \Phi_3 = \Phi_3\,\mathrm{I}.$$

120.] Let a dyadic Φ be given. Let it be reduced to the sum of three dyads of which the three antecedents are non-coplanar.

$$\Phi = \mathbf{a}\,\mathbf{l} + \mathbf{b}\,\mathbf{m} + \mathbf{c}\,\mathbf{n},$$

$$\Phi_2 = \mathbf{b} \times \mathbf{c}\;\; \mathbf{m} \times \mathbf{n} + \mathbf{c} \times \mathbf{a}\;\; \mathbf{n} \times \mathbf{l} + \mathbf{a} \times \mathbf{b}\;\; \mathbf{l} \times \mathbf{m},$$

$$\Phi_3 = [\mathbf{a}\,\mathbf{b}\,\mathbf{c}]\;[\mathbf{l}\,\mathbf{m}\,\mathbf{n}].$$

Theorem: The necessary and sufficient condition that a dyadic Φ be complete is that the third of Φ be different from zero.

For it was shown (Art. 106) that both the antecedents and the consequents of a complete dyadic are non-coplanar. Hence the two scalar triple products which occur in Φ_3 cannot vanish.

Theorem: The necessary and sufficient condition that a dyadic Φ be planar is that the third of Φ shall vanish but the second of Φ shall not vanish.

It was shown (Art. 106) that if a dyadic Φ be planar its consequents **l**, **m**, **n** must be planar and conversely if the consequents be coplanar the dyadic is planar. Hence for a planar dyadic Φ_3 must vanish. But Φ_2 cannot vanish. Since **a**, **b**, **c** have been assumed non-coplanar, the vectors $\mathbf{b} \times \mathbf{c}$, $\mathbf{c} \times \mathbf{a}$, $\mathbf{a} \times \mathbf{b}$ are non-coplanar. Hence if Φ_2 vanishes each of the vectors $\mathbf{m} \times \mathbf{n}$, $\mathbf{n} \times \mathbf{l}$, $\mathbf{l} \times \mathbf{m}$ vanishes — that is, **l**, **m**, **n** are collinear. But this is impossible since the dyadic Φ is planar and not linear.

Theorem: The necessary and sufficient condition that a non-vanishing dyadic be linear is that the second of Φ, and consequently the third of Φ, vanishes.

For if Φ be linear the consequents **l**, **m**, **n**, are collinear. Hence their vector products vanish and the consequents of Φ_2 vanish. If conversely Φ_2 vanishes, each of its consequents must be zero and hence these consequents of Φ are collinear.

The vanishing of the third, unaccompanied by the vanishing of the second of a dyadic, implies one degree of nullity. The vanishing of the second implies two degrees of nullity.

The vanishing of the dyadic itself is complete nullity. The results may be put in tabular form.

$$\Phi_3 \neq 0, \quad \Phi \text{ is complete.}$$
$$\Phi_3 = 0, \quad \Phi_2 \neq 0, \quad \Phi \text{ is planar.} \qquad (69)$$
$$\Phi_3 = 0, \quad \Phi_2 = 0, \quad \Phi \neq 0, \quad \Phi \text{ is linear.}$$

It follows immediately that the third of any anti-self-conjugate dyadic vanishes; but the second does not. For any such dyadic is planar but cannot be linear.

Nonion Form. Determinants.[1] *Invariants of a Dyadic*

121.] If Φ be expressed in nonion form

$$\begin{aligned}\Phi = a_{11}\,\mathbf{i\,i} + a_{12}\,\mathbf{i\,j} + a_{13}\,\mathbf{i\,k} \\ + a_{21}\,\mathbf{j\,i} + a_{22}\,\mathbf{j\,j} + a_{23}\,\mathbf{j\,k} \\ + a_{31}\,\mathbf{k\,i} + a_{32}\,\mathbf{k\,j} + a_{33}\,\mathbf{k\,k}.\end{aligned} \qquad (13)$$

The conjugate of Φ has the same scalar coefficients as Φ, but they are arranged symmetrically with respect to the main diagonal. Thus

$$\begin{aligned}\Phi_C = a_{11}\,\mathbf{i\,i} + a_{21}\,\mathbf{i\,j} + a_{31}\,\mathbf{i\,k}, \\ + a_{12}\,\mathbf{j\,i} + a_{22}\,\mathbf{j\,j} + a_{32}\,\mathbf{j\,k}, \\ + a_{13}\,\mathbf{k\,i} + a_{23}\,\mathbf{j\,k} + a_{33}\,\mathbf{k\,k}.\end{aligned} \qquad (70)$$

The second of Φ may be computed. Take, for instance, one term. Let it be required to find the coefficient of $\mathbf{i\,j}$ in Φ_2. What terms in Φ can yield a double cross product equal to $\mathbf{i\,j}$? The vector product of the antecedents must be $\mathbf{i}$ and the vector product of the consequents must be $\mathbf{j}$. Hence the antecedents must be $\mathbf{j}$ and $\mathbf{k}$; and the consequents, $\mathbf{k}$ and $\mathbf{i}$. These terms are

$$a_{21}\,\mathbf{j\,i} \overset{\times}{\times} a_{33}\,\mathbf{k\,k} = -\,a_{21}\,a_{33}\,\mathbf{i\,j}$$
$$a_{31}\,\mathbf{k\,i} \overset{\times}{\times} a_{23}\,\mathbf{j\,k} = a_{31}\,a_{23}\,\mathbf{i\,j}.$$

[1] The results hold only for determinants of the third order. The extension to determinants of higher orders is through Multiple Algebra.

Hence the term in $\mathbf{i\,j}$ in Φ_2 is

$$(a_{31}\,a_{23} - a_{21}\,a_{33})\,\mathbf{i\,j}.$$

This is the first minor of a_{12} in the determinant

$$\begin{vmatrix} a_{11} & a_{12} & a_{13} \\ a_{21} & a_{22} & a_{23} \\ a_{31} & a_{32} & a_{33} \end{vmatrix}.$$

This minor is taken with the negative sign. That is, the coefficient of $\mathbf{i\,j}$ in Φ_2 is what is termed the *cofactor* of the coefficient of $\mathbf{i\,j}$ in the determinant. The cofactor is merely the first minor taken with the positive or negative sign according as the sum of the subscripts of the term whose first minor is under consideration is even or odd. The coefficient of any dyad in Φ_2 is easily seen to be the cofactor of the corresponding term in Φ. The cofactors are denoted generally by large letters.

$$A_{11} = \begin{vmatrix} a_{22} & a_{23} \\ a_{32} & a_{33} \end{vmatrix} \text{ is the cofactor of } a_{11}.$$

$$A_{12} = -\begin{vmatrix} a_{21} & a_{23} \\ a_{31} & a_{33} \end{vmatrix} \text{ is the cofactor of } a_{12}.$$

$$A_{32} = -\begin{vmatrix} a_{11} & a_{12} \\ a_{21} & a_{23} \end{vmatrix} \text{ is the cofactor of } a_{32}.$$

With this notation the second of Φ becomes

$$\begin{aligned} \Phi_2 = A_{11}\,\mathbf{i\,i} + A_{12}\,\mathbf{i\,j} + A_{13}\,\mathbf{i\,k} \\ + A_{21}\,\mathbf{j\,i} + A_{22}\,\mathbf{j\,j} + A_{23}\,\mathbf{k\,k} \\ + A_{31}\,\mathbf{k\,i} + A_{32}\,\mathbf{k\,j} + A_{33}\,\mathbf{k\,k}. \end{aligned} \qquad (71)$$

The value of the third of Φ may be obtained by writing Φ as the sum of three dyads

$$\Phi = (a_{11}\,\mathbf{i} + a_{21}\,\mathbf{j} + a_{31}\,\mathbf{k})\,\mathbf{i} + (a_{12}\,\mathbf{i} + a_{22}\,\mathbf{j} + a_{32}\,\mathbf{k})\,\mathbf{j} \\ + (a_{13}\,\mathbf{i} + a_{23}\,\mathbf{j} + a_{33}\,\mathbf{k})\,\mathbf{k}$$

$$\Phi_3 = [(a_{11}\,\mathbf{i} + a_{21}\,\mathbf{j} + a_{31}\,\mathbf{k}) \quad (a_{21}\,\mathbf{i} + a_{22}\,\mathbf{j} + a_{33}\,\mathbf{k})$$

$$(a_{13}\,\mathbf{i} + a_{23}\,\mathbf{j} + a_{33}\,\mathbf{k})]\ [\mathbf{i}\ \mathbf{j}\ \mathbf{k}]$$

This is easily seen to be equal to the determinant

$$\Phi_3 = \begin{vmatrix} a_{11} & a_{12} & a_{13} \\ a_{21} & a_{22} & a_{23} \\ a_{31} & a_{32} & a_{33} \end{vmatrix} \tag{72}$$

For this reason Φ_3 is frequently called the determinant of Φ and is written

$$\Phi_3 = |\,\Phi\,| \tag{72'}$$

The idea of the determinant is very natural when Φ is regarded as expressed in nonion form. On the other hand unless Φ be expressed in that form the conception of Φ_3, the third of Φ, is more natural.

The reciprocal of a dyadic in nonion form may be found most easily by making use of the identity

$$\Phi_2 \cdot \Phi_C = \Phi_3\,\mathbf{I} \tag{68}$$

or

$$\Phi_C^{-1} = \frac{1}{\Phi_3}\,\Phi_2$$

or

$$\Phi^{-1} = \frac{1}{\Phi_3}\,\Phi_{2C}.$$

Hence

$$\Phi^{-1} = \frac{\left\{\begin{matrix} A_{11}\,\mathbf{i}\mathbf{i} + A_{21}\,\mathbf{i}\mathbf{j} + A_{31}\,\mathbf{i}\mathbf{k} \\ + A_{12}\,\mathbf{j}\mathbf{i} + A_{22}\,\mathbf{j}\mathbf{j} + A_{32}\,\mathbf{j}\mathbf{k} \\ + A_{13}\,\mathbf{k}\mathbf{i} + A_{23}\,\mathbf{k}\mathbf{j} + A_{33}\,\mathbf{k}\mathbf{k} \end{matrix}\right\}}{\begin{vmatrix} a_{11} & a_{12} & a_{13} \\ a_{21} & a_{22} & a_{23} \\ a_{31} & a_{32} & a_{33} \end{vmatrix}} \tag{73}$$

If the determinant be denoted by D

$$\Phi^{-1} = \frac{A_{11}}{D}\,\mathbf{i\,i} + \frac{A_{21}}{D}\,\mathbf{i\,j} + \frac{A_{31}}{D}\,\mathbf{i\,k}$$

$$+ \frac{A_{12}}{D}\,\mathbf{j\,i} + \frac{A_{22}}{D}\,\mathbf{j\,j} + \frac{A_{32}}{D}\,\mathbf{j\,k} \qquad (73)'$$

$$+ \frac{A_{13}}{D}\,\mathbf{k\,i} + \frac{A_{23}}{D}\,\mathbf{k\,j} + \frac{A_{33}}{D}\,\mathbf{k\,k}.$$

If Ψ is a second dyadic given in nonion form as

$$\Psi = b_{11}\,\mathbf{i\,i} + b_{12}\,\mathbf{i\,j} + b_{13}\,\mathbf{i\,k},$$

$$+ b_{21}\,\mathbf{j\,i} + b_{22}\,\mathbf{j\,j} + b_{23}\,\mathbf{j\,k},$$

$$+ b_{31}\,\mathbf{k\,i} + b_{32}\,\mathbf{k\,j} + b_{33}\,\mathbf{k\,k},$$

the product $\Phi \cdot \Psi$ of the two dyadics may readily be found by actually performing the multiplication

$$\Phi \cdot \Psi = (a_{11}\,b_{11} + a_{12}\,b_{21} + a_{13}\,b_{31})\,\mathbf{i\,i} + (a_{11}\,b_{12} + a_{12}\,b_{22}$$

$$+ a_{13}\,b_{32})\,\mathbf{i\,j} + (a_{11}\,b_{13} + a_{12}\,b_{23} + a_{13}\,b_{33})\,\mathbf{i\,k}$$

$$+ (a_{21}\,b_{11} + a_{22}\,b_{21} + a_{23}\,b_{31})\,\mathbf{j\,i} + (a_{21}\,b_{12} + a_{22}\,b_{22}$$

$$+ a_{23}\,b_{32})\,\mathbf{j\,j} + (a_{21}\,b_{13} + a_{22}\,b_{23} + a_{23}\,b_{33})\,\mathbf{j\,k}$$

$$+ (a_{31}\,b_{11} + a_{32}\,b_{21} + a_{33}\,b_{31})\,\mathbf{k\,i} + (a_{31}\,b_{12} + a_{32}\,b_{22}$$

$$+ a_{33}\,b_{32})\,\mathbf{k\,j} + (a_{31}\,b_{12} + a_{32}\,b_{23} + a_{33}\,b_{33})\,\mathbf{k\,k}.$$

$$\Phi : \Psi = a_{11}\,b_{11} + a_{12}\,b_{12} + a_{13}\,b_{13}$$

$$+ a_{21}\,b_{21} + a_{22}\,b_{22} + a_{23}\,b_{23} \qquad (75)$$

$$+ a_{31}\,b_{31} + a_{32}\,b_{32} + a_{33}\,b_{33}.$$

Since the third or determinant of a product is equal to the product of the determinants, the law of multiplication of determinants follows from (65) and (74).

$$\begin{vmatrix} a_{11} & a_{12} & a_{13} \\ a_{21} & a_{22} & a_{23} \\ a_{31} & a_{32} & a_{33} \end{vmatrix} \begin{vmatrix} b_{11} & b_{12} & b_{13} \\ b_{21} & b_{22} & b_{23} \\ b_{31} & b_{32} & b_{33} \end{vmatrix} = \begin{vmatrix} a_{11} b_{11} + a_{12} b_{21} + a_{13} b_{31} & a_{11} b_{12} + a_{12} b_{22} + a_{13} b_{32} & a_{11} b_{13} + a_{12} b_{23} + a_{13} b_{33}, \\ a_{21} b_{11} + a_{22} b_{21} + a_{23} b_{31} & a_{21} b_{12} + a_{22} b_{22} + a_{23} b_{32} & a_{21} b_{13} + a_{22} b_{23} + a_{23} b_{33}, \\ a_{31} b_{11} + a_{32} b_{21} + a_{33} b_{31} & a_{31} b_{12} + a_{32} b_{22} + a_{33} b_{32} & a_{31} b_{13} + a_{32} b_{23} + a_{33} b_{33}. \end{vmatrix} \quad (76)$$

The rule may be stated in words. To multiply two determinants form the determinant of which the element in the mth row and nth column is the sum of the products of the elements in the mth row of the first determinant and nth column of the second.

If
$$\Phi = \mathbf{a\,l} + \mathbf{b\,m} + \mathbf{c\,n},$$

$$\Phi_2 = \mathbf{b \times c} \;\; \mathbf{m \times n} + \mathbf{c \times a} \;\; \mathbf{n \times l} + \mathbf{a \times b} \;\; \mathbf{l \times m}.$$

Then

$$|\Phi_2| = (\Phi_2)_3 = [\mathbf{b \times c} \;\; \mathbf{c \times a} \;\; \mathbf{a \times b}] \;\; [\mathbf{m \times n} \;\; \mathbf{n \times l} \;\; \mathbf{l \times m}]$$

Hence $\quad |\Phi_2| = (\Phi_2)_3 = [\mathbf{a\,b\,c}]^2 \, [\mathbf{l\,m\,n}]^2 = \Phi_3^2.$

Hence

$$|\Phi_2| = \begin{vmatrix} A_{11} & A_{12} & A_{13} \\ A_{21} & A_{22} & A_{23} \\ A_{31} & A_{32} & A_{33} \end{vmatrix} = \begin{vmatrix} a_{11} & a_{12} & a_{13} \\ a_{21} & a_{22} & a_{23} \\ a_{31} & a_{32} & a_{33} \end{vmatrix}^2 \quad (77)$$

The determinant of the cofactors of a given determinant of the third order is equal to the square of the given determinant.

122.] A dyadic Φ has three scalar invariants — that is three scalar quantities which are independent of the form in which Φ is expressed. These are

$$\Phi_S, \quad (\Phi_2)_S, \quad \Phi_3,$$

the scalar of Φ, the scalar of the second of Φ, and the third or determinant of Φ. If Φ be expressed in nonion form these quantities are

$$\Phi_S = a_{11} + a_{22} + a_{23}$$

$$(\Phi_2)_S = A_{11} + A_{22} + A_{33} \tag{78}$$

$$\Phi_3 = \begin{vmatrix} a_{11} & a_{12} & a_{13} \\ a_{21} & a_{22} & a_{23} \\ a_{31} & a_{32} & a_{33} \end{vmatrix}$$

No matter in terms of what right-handed rectangular system of these unit vectors Φ may be expressed these quantities are the same. The scalar of Φ is the sum of the three coefficients in the main diagonal. The scalar of the second of Φ is the sum of the first minors or cofactors of the terms in the main diagonal. The third of Φ is the determinant of the coefficients. These three invariants are by far the most important that a dyadic Φ possesses.

Theorem: Any dyadic satisfies a cubic equation of which the three invariants Φ_S, Φ_{2S}, Φ_3 are the coefficients.

By (68) $\qquad (\Phi - x\,\mathrm{I})_2 \cdot (\Phi - x\,\mathrm{I})_C = (\Phi - x\,\mathrm{I})_3$

$$(\Phi - x\,\mathrm{I})_3 = \begin{vmatrix} a_{11} - x & a_{12} & a_{13} \\ a_{21} & a_{22} - x & a_{23} \\ a_{31} & a_{32} & a_{33} - x \end{vmatrix}$$

Hence $\qquad (\Phi - x\,\mathrm{I})_3 = \Phi_3 - x\,\Phi_{2S} + x^2\,\Phi_S - x^3$

as may be seen by actually performing the expansion.

$$(\Phi - x\,\mathrm{I})_2 \cdot (\Phi - x\,\mathrm{I})_C = \Phi_3 - x\,\Phi_{2S} + x^2\,\Phi_S - x^3.$$

This equation is an identity holding for *all* values of the scalar x. It therefore holds, if in place of the scalar x, the dyadic Φ which depends upon nine scalars be substituted. That is

$$(\Phi - \Phi \cdot \mathrm{I})_2 \cdot (\Phi - \Phi \cdot \mathrm{I})_C = \mathrm{I}\,\Phi_3 - \Phi\,\Phi_{2S} + \Phi^2\,\Phi_S - \Phi^3.$$

But the terms upon the left are identically zero. Hence

$$\Phi^3 - \Phi_S\,\Phi^2 + \Phi_{2S}\,\Phi - \Phi_3\,\mathrm{I} = 0. \tag{79}$$

This equation may be called the Hamilton-Cayley equation. Hamilton showed that a quaternion satisfied an equation analogous to this one and Cayley gave the generalization to matrices. A matrix of the *n*th order satisfies an algebraic equation of the *n*th degree. The analogy between the theory of dyadics and the theory of matrices is very close. In fact, a dyadic may be regarded as a matrix of the third order and conversely a matrix of the third order may be looked upon as a dyadic. The addition and multiplication of matrices and dyadics are then performed according to the same laws. A generalization of the idea of a dyadic to spaces of higher dimensions than the third leads to Multiple Algebra and the theory of matrices of orders higher than the third.

Summary of Chapter V

A vector $\mathbf{r}'$ is said to be a linear function of a vector $\mathbf{r}$ when the components of $\mathbf{r}'$ are linear homogeneous functions of the components of $\mathbf{r}$. Or a function of $\mathbf{r}$ is said to be a linear vector function of $\mathbf{r}$ when the function of the sum of two vectors is the sum of the functions of those vectors.

$$\mathbf{f}(\mathbf{r}_1 + \mathbf{r}_2) = \mathbf{f}(\mathbf{r}_1) + \mathbf{f}(\mathbf{r}_2). \qquad (4)$$

These two ideas of a linear vector function are equivalent.

A sum of a number of symbolic products of two vectors, which are obtained by placing the vectors in juxtaposition without intervention of a dot or cross and which are called dyads, is called a dyadic and is represented by a Greek capital. A dyadic determines a linear vector function of a vector by direct multiplication with that vector

$$\Phi = \mathbf{a}_1 \mathbf{b}_1 + \mathbf{a}_2 \mathbf{b}_2 + \mathbf{a}_3 \mathbf{b}_3 + \cdots \qquad (7)$$

$$\Phi \cdot \mathbf{r} = \mathbf{a}_1 \mathbf{b}_1 \cdot \mathbf{r} + \mathbf{a}_2 \mathbf{b}_2 \cdot \mathbf{r} + \mathbf{a}_3 \mathbf{b}_3 \cdot \mathbf{r} + \cdots \qquad (8)$$

Two dyadics are equal when they are equal as operators upon all vectors or upon three non-coplanar vectors. That is, when

$$\Phi \cdot \mathbf{r} = \Psi \cdot \mathbf{r} \text{ for all values or for three non-coplanar values of } \mathbf{r}, \tag{10}$$

or $$\mathbf{r} \cdot \Phi = \mathbf{r} \cdot \Psi \text{ for all values or for three non-coplanar values of } \mathbf{r},$$

or $$\mathbf{s} \cdot \Phi \cdot \mathbf{r} = \mathbf{s} \cdot \Psi \cdot \mathbf{r} \text{ for all values or for three non-coplanar values of } \mathbf{r} \text{ and } \mathbf{s}.$$

Any linear vector function may be represented by a dyadic.

Dyads obey the distributive law of multiplication with regard to the two vectors composing the dyad

$$\begin{aligned}(\mathbf{a}+\mathbf{b}+\mathbf{c}+\cdots)(\mathbf{l}+\mathbf{m}+\mathbf{n}+\cdots) = \mathbf{a}\,\mathbf{l} &+ \mathbf{a}\,\mathbf{m} + \mathbf{a}\,\mathbf{n} + \cdots \\ &+ \mathbf{b}\,\mathbf{l} + \mathbf{b}\,\mathbf{m} + \mathbf{b}\,\mathbf{n} + \cdots \\ &+ \mathbf{c}\,\mathbf{l} + \mathbf{c}\,\mathbf{m} + \mathbf{c}\,\mathbf{n} + \cdots \\ &+ \quad \cdot \quad \cdot \quad \cdot \quad \cdot \quad \cdot \quad \cdot \end{aligned} \tag{11$'$}$$

Multiplication by a scalar is associative. In virtue of these two laws a dyadic may be expanded into a sum of nine terms by means of the fundamental dyads,

$$\begin{matrix} \mathbf{i}\,\mathbf{i}, & \mathbf{i}\,\mathbf{j}, & \mathbf{i}\,\mathbf{k}, \\ \mathbf{j}\,\mathbf{i}, & \mathbf{j}\,\mathbf{j}, & \mathbf{j}\,\mathbf{k}, \\ \mathbf{k}\,\mathbf{i}, & \mathbf{k}\,\mathbf{j}, & \mathbf{k}\,\mathbf{k}, \end{matrix} \tag{12}$$

as $$\begin{aligned}\Phi &= a_{11}\,\mathbf{i}\,\mathbf{i} + a_{12}\,\mathbf{i}\mathbf{j} + a_{13}\,\mathbf{i}\,\mathbf{k}, \\ &= a_{21}\,\mathbf{j}\,\mathbf{i} + a_{22}\,\mathbf{j}\mathbf{j} + a_{23}\,\mathbf{j}\,\mathbf{k}, \\ &= a_{31}\,\mathbf{k}\,\mathbf{i} + a_{32}\,\mathbf{k}\mathbf{j} + a_{33}\,\mathbf{k}\,\mathbf{k}.\end{aligned} \tag{13}$$

If two dyadics are equal the corresponding coefficients in their expansions into nonion form are equal and conversely.

Any dyadic may be expressed as the sum of three dyads of which the antecedents or the consequents are any three given non-coplanar vectors. This expression of the dyadic is unique.

The symbolic product **a b** known as a dyad is the most general product of two vectors in which multiplication by a scalar is associative. It is called the indeterminate product. The product imposes five conditions upon the vectors **a** and **b**. Their directions and the product of their lengths are determined by the product. The scalar and vector products are functions of the indeterminate product. A scalar and a vector may be obtained from any dyadic by inserting a dot and a cross between the vectors in each dyad. This scalar and vector are functions of the dyadic.

$$\Phi_S = \mathbf{a}_1 \cdot \mathbf{b}_1 + \mathbf{a}_2 \cdot \mathbf{b}_2 + \mathbf{a}_3 \cdot \mathbf{b}_3 + \cdots \qquad (18)$$

$$\Phi_\times = \mathbf{a}_1 \times \mathbf{b}_1 + \mathbf{a}_2 \times \mathbf{b}_2 + \mathbf{a}_3 \times \mathbf{b}_3 + \cdots \qquad (19)$$

$$\Phi_S = \mathbf{i} \cdot \Phi \cdot \mathbf{i} + \mathbf{j} \cdot \Phi \cdot \mathbf{j} + \mathbf{k} \cdot \Phi \cdot \mathbf{k} \qquad (20)$$

$$= a_{11} + a_{22} + a_{33},$$

$$\Phi_\times = (\mathbf{j} \cdot \Phi \cdot \mathbf{k} - \mathbf{k} \cdot \Phi \cdot \mathbf{j})\, \mathbf{i} + (\mathbf{k} \cdot \Phi \cdot \mathbf{i} - \mathbf{i} \cdot \Phi \cdot \mathbf{k})\, \mathbf{j}$$

$$+ (\mathbf{i} \cdot \Phi \cdot \mathbf{j} - \mathbf{j} \cdot \Phi \cdot \mathbf{i})\, \mathbf{k} \qquad (21)$$

$$= (a_{23} - a_{32})\, \mathbf{i} + (a_{31} - a_{13})\, \mathbf{j} + (a_{12} - a_{21})\, \mathbf{k}.$$

The direct product of two dyads is the dyad whose antecedent and consequent are respectively the antecedent of the first dyad and the consequent of the second multiplied by the scalar product of the consequent of the first dyad and the antecedent of the second.

$$(\mathbf{a}\,\mathbf{b}) \cdot (\mathbf{c}\,\mathbf{d}) = (\mathbf{b} \cdot \mathbf{c})\, \mathbf{a}\,\mathbf{b}. \qquad (23)$$

The direct product of two dyadics is the formal expansion, according to the distributive law, of the product into the sum of products of dyads. Direct multiplication of dyadics or of dyadics and a vector at either end or at both ends obeys the distributive and associative laws of multiplication. Consequently such expressions as

$$\Phi \cdot \Psi \cdot \mathbf{r}, \quad \mathbf{s} \cdot \Phi \cdot \Psi, \quad \mathbf{s} \cdot \Phi \cdot \Psi \cdot \mathbf{r}, \quad \Phi \cdot \Psi \cdot \Omega \qquad (24)\text{–}(26)$$

may be written without parentheses; for parentheses may be inserted at pleasure without altering the value of the product. In case the vector occurs at other positions than at the end the product is no longer associative.

The skew product of a dyad and a vector may be defined by the equation

$$(\mathbf{a}\,\mathbf{b}) \times \mathbf{r} = \mathbf{a}\ \mathbf{b} \times \mathbf{r},$$

$$\mathbf{r} \times (\mathbf{a}\,\mathbf{b}) = \mathbf{r} \times \mathbf{a}\ \ \mathbf{b}. \qquad (28)$$

The skew product of a dyadic and a vector is equal to the formal expansion of that product into a sum of products of dyads and that vector. The statement made concerning the associative law for direct products holds when the vector is connected with the dyadics in skew multiplication. The expressions

$$\mathbf{r} \times \Phi \cdot \Psi, \quad \Phi \cdot \Psi \times \mathbf{r}, \quad \mathbf{r} \times \Phi \cdot \mathbf{s}, \quad \mathbf{r} \cdot \Phi \times \mathbf{s}, \quad \mathbf{r} \times \Phi \times \mathbf{s} \qquad (29)$$

may be written without parentheses and parentheses may be inserted at pleasure without altering the value of the product. Moreover

$$\mathbf{s} \cdot (\mathbf{r} \times \Phi) = (\mathbf{s} \times \mathbf{r}) \cdot \Phi, \quad (\Phi \times \mathbf{r}) \cdot \mathbf{s} = \Phi \cdot (\mathbf{r} \times \mathbf{s}),$$

$$\Phi \cdot (\mathbf{r} \times \Psi) = (\Phi \times \mathbf{r}) \cdot \Psi. \qquad (31)'$$

But the parentheses cannot be omitted.

The necessary and sufficient condition that a dyadic may be reduced to the sum of two dyads or to a single dyad or to zero is that, when expressed as the sum of three dyads of which the antecedents (or consequents) are known to be non-coplanar, the consequents (or antecedents) shall be respectively coplanar or collinear or zero. A complete dyadic is one which cannot be reduced to a sum of fewer than three dyads. A planar dyadic is one which can be reduced to a sum of just two dyads. A linear dyadic is one which can be reduced to a single dyad.

A complete dyadic possesses no degree of nullity. There is no direction in space for which it is an annihilator. A planar dyadic possesses one degree of nullity. There is one direction in space for which it is an annihilator when used as a prefactor and one when used as a postfactor. A linear dyadic possesses two degrees of nullity. There are two independent directions in space for which it is an annihilator when used as a prefactor and two directions when used as a postfactor. A zero dyadic possesses three degrees of nullity or complete nullity. It annihilates every vector in space.

The products of a complete dyadic and a complete, planar, or linear dyadic are respectively complete, planar, or linear. The products of a planar dyadic with a planar or linear dyadic are respectively planar or linear, except in certain cases where relations of perpendicularity between the consequents of the first dyadic and the antecedents of the second introduce one more degree of nullity into the product. The product of a linear dyadic by a linear dyadic is in general linear; but in case the consequent of the first is perpendicular to the antecedent of the second the product vanishes. The product of any dyadic by a zero dyadic is zero.

A dyadic which when applied to any vector in space reproduces that vector is called an idemfactor. All idemfactors are equal and reducible to the form

$$\mathbf{I} = \mathbf{i}\,\mathbf{i} + \mathbf{j}\,\mathbf{j} + \mathbf{k}\,\mathbf{k}. \qquad (33)$$

Or

$$\mathbf{I} = \mathbf{a}\,\mathbf{a}' + \mathbf{b}\,\mathbf{b}' + \mathbf{c}\,\mathbf{c}'. \qquad (34)$$

The product of any dyadic and an idemfactor is that dyadic.

If the product of two complete dyadics is equal to the idemfactor the dyadics are commutative and either is called the reciprocal of the other. A complete dyadic may be canceled from either end of a product of dyadics and vectors as in ordinary algebra; for the cancelation is equivalent to multiplication by the reciprocal of that dyadic. Incomplete dyadics possess no reciprocals. They correspond to zero in ordinary algebra. The reciprocal of a product is equal to the product of the reciprocals taken in inverse order.

$$(\Phi \cdot \Psi)^{-1} = \Psi^{-1} \cdot \Phi^{-1}. \tag{38}$$

The conjugate of a dyadic is the dyadic obtained by interchanging the order of the antecedents and consequents. The conjugate of a product is equal to the product of the conjugates taken in the opposite order.

$$(\Phi \cdot \Psi)_C = \Psi_C \cdot \Phi_C. \tag{40}$$

The conjugate of the reciprocal is equal to the reciprocal of the conjugate. A dyadic may be divided in one and only one way into the sum of two parts of which one is self-conjugate and the other anti-self-conjugate.

$$\Phi = \tfrac{1}{2}(\Phi + \Phi_C) + \tfrac{1}{2}(\Phi - \Phi_C). \tag{43}$$

Any anti-self-conjugate dyadic or the anti-self-conjugate part of any dyadic, used in direct multiplication, is equivalent to minus one-half the vector of that dyadic used in skew multiplication.

$$\tfrac{1}{2}(\Phi - \Phi_C) \cdot \mathbf{r} = -\tfrac{1}{2}\Phi_\times \times \mathbf{r},$$

$$\tfrac{1}{2}\mathbf{r} \cdot (\Phi - \Phi_C) = -\tfrac{1}{2}\mathbf{r} \times \Phi_\times. \tag{44}$$

A dyadic of the form $\mathbf{c} \times \mathrm{I}$ or $\mathrm{I} \times \mathbf{c}$ is anti-self-conjugate and used in direct multiplication is equivalent to the vector $\mathbf{c}$ used in skew multiplication.

Also
$$\mathbf{c} \times \mathbf{r} = (\mathrm{I} \times \mathbf{c}) \cdot \mathbf{r} = (\mathbf{c} \times \mathrm{I}) \cdot \mathbf{r}, \tag{46}$$
$$\mathbf{c} \times \Phi = (\mathrm{I} \times \mathbf{c}) \cdot \Phi = (\mathbf{c} \times \mathrm{I}) \cdot \Phi.$$

The dyadic $\mathbf{c} \times \mathrm{I}$ or $\mathrm{I} \times \mathbf{c}$, where $\mathbf{c}$ is a unit vector is a quadrantal versor for vectors perpendicular to $\mathbf{c}$ and an annihilator for vectors parallel to $\mathbf{c}$. The dyadic $\mathrm{I} \times \mathbf{c} + \mathbf{c}\,\mathbf{c}$ is a true quadrantal versor for all vectors. The powers of these dyadics behave like the powers of the imaginary unit $\sqrt{-1}$, as may be seen from the geometric interpretation. Applied to the unit vectors $\mathbf{i}, \mathbf{j}, \mathbf{k}$

$$\mathrm{I} \times \mathbf{i} = \mathbf{i} \times \mathrm{I} = \mathbf{k}\,\mathbf{j} - \mathbf{j}\,\mathbf{k}, \text{ etc.} \tag{49}$$

The vector $\mathbf{a} \times \mathbf{b}$ in skew multiplication is equivalent to $(\mathbf{a} \times \mathbf{b}) \times \mathrm{I}$ in direct multiplication.

$$(\mathbf{a} \times \mathbf{b}) \times \mathrm{I} = \mathrm{I} \times (\mathbf{a} \times \mathbf{b}) = \mathbf{b}\,\mathbf{a} - \mathbf{a}\,\mathbf{b} \tag{50}$$
$$(\mathbf{a} \times \mathbf{b}) \times \mathbf{r} = (\mathbf{b}\,\mathbf{a} - \mathbf{a}\,\mathbf{b}) \cdot \mathbf{r}$$
$$\mathbf{r} \times (\mathbf{a} \times \mathbf{b}) = \mathbf{r} \cdot (\mathbf{b}\,\mathbf{a} - \mathbf{a}\,\mathbf{b}). \tag{51}$$

A complete dyadic may be reduced to a sum of three dyads of which the antecedents among themselves and the consequents among themselves each form a right-handed rectangular system of three unit vectors and of which the scalar coefficients are all positive or all negative.

$$\Phi = \pm\,(a\,\mathbf{i}'\mathbf{i} + b\,\mathbf{j}'\mathbf{j} + c\,\mathbf{k}'\mathbf{k}). \tag{53}$$

This is called the normal form of the dyadic. An incomplete dyadic may be reduced to this form but one or more of the coefficients are zero. The reduction is unique in case the constants a, b, c are different. In case they are not different the reduction may be accomplished in more than one way. Any self-conjugate dyadic may be reduced to the normal form

$$\Phi = a\,\mathbf{i}\,\mathbf{i} + b\,\mathbf{j}\,\mathbf{j} + c\,\mathbf{k}\,\mathbf{k}, \tag{55}$$

in which the constants a, b, c are not necessarily positive.

The double dot and double cross multiplication of dyads is defined by the equations

$$\mathbf{a}\,\mathbf{b} : \mathbf{c}\,\mathbf{d} = \mathbf{a} \cdot \mathbf{c} \quad \mathbf{b} \cdot \mathbf{d}, \tag{56}$$

$$\mathbf{a}\,\mathbf{b} \overset{\times}{\times} \mathbf{c}\,\mathbf{d} = \mathbf{a} \times \mathbf{c} \quad \mathbf{b} \times \mathbf{d}. \tag{57}$$

The double dot and double cross multiplication of dyadics is obtained by expanding the product formally, according to the distributive law, into a sum of products of dyads. The double dot and double cross multiplication of dyadics is commutative but not associative.

One-half the double cross product of a dyadic Φ by itself is called the second of Φ. If

$$\Phi = \mathbf{a}\,\mathbf{l} + \mathbf{b}\,\mathbf{m} + \mathbf{c}\,\mathbf{n},$$

$$\Phi_2 = \tfrac{1}{2}\,\Phi \overset{\times}{\times} \Phi = \mathbf{b} \times \mathbf{c} \quad \mathbf{m} \times \mathbf{n} + \mathbf{c} \times \mathbf{a} \quad \mathbf{n} \times \mathbf{l} + \mathbf{a} \times \mathbf{b} \quad \mathbf{l} \times \mathbf{m}. \tag{61}$$

One-third of the double dot product of the second of Φ and Φ is called the third of Φ and is equal to the product of the scalar triple product of the antecedents of Φ and the scalar triple product of the consequent of Φ.

$$\Phi_3 = \tfrac{1}{6}\,\Phi \overset{\times}{\times} \Phi : \Phi = [\mathbf{a}\,\mathbf{b}\,\mathbf{c}]\,[\mathbf{l}\,\mathbf{m}\,\mathbf{n}]. \tag{62}$$

The second of the conjugate is the conjugate of the second. The third of the conjugate is equal to the third of the original dyadic. The second and third of the reciprocal are the reciprocals of the second and third of the second and third of a dyadic. The second and third of a product are the products of the seconds and thirds.

$$(\Phi_C)_2 = (\Phi_2)_C,$$

$$(\Phi_C)_3 = \Phi_3, \tag{63}$$

$$(\Phi^{-1})_2 = (\Phi_2)^{-1},$$

$$(\Phi^{-1})_3 = (\Phi_3)^{-1}. \tag{64}$$

$$(\Phi \cdot \Psi)_2 = \Phi_2 \cdot \Psi_2 \tag{65}$$

$$(\Phi \cdot \Psi)_3 = \Phi_3\, \Psi_3.$$

The product of the second and conjugate of a dyadic is equal to the product of the third and the idemfactor.

$$\Phi_2 \cdot \Phi_C = \Phi_3 \, \mathrm{I} \qquad (68)$$

The conditions for the various degrees of nullity may be expressed in terms of the second and third of Φ.

$$\begin{aligned} \Phi_3 \neq 0, \quad & \Phi \text{ is complete} \\ \Phi_3 = 0,\ \Phi_2 \neq 0, \quad & \Phi \text{ is planar} \\ \Phi_3 = 0,\ \Phi_2 = 0,\ \Phi \neq 0, \quad & \Phi \text{ is linear.} \end{aligned} \qquad (69)$$

The closing sections of the chapter contain the expressions (70)–(78) of a number of the results in nonion form and the deduction therefrom of a number of theorems concerning determinants. They also contain the cubic equation which is satisfied by a dyadic Φ.

$$\Phi^3 - \Phi_S \, \Phi^2 + \Phi_{2S} \, \Phi^3 + \Phi_3 \, \mathrm{I} = 0. \qquad (79)$$

This is called the Hamilton-Cayley equation. The coefficients Φ_S, Φ_{2S}, and Φ_3 are the three fundamental scalar invariants of Φ.

Exercises on Chapter V

1. Show that the two definitions given in Art. 98 for a linear vector function are equivalent.

2. Show that the reduction of a dyadic as in (15) can be accomplished in only one way if **a**, **b**, **c**, **l**, **m**, **n**, are given.

3. Show $\qquad (\Phi \times \mathbf{a})_C = -\mathbf{a} \times \Phi_C.$

4. Show that if $\Phi \times \mathbf{r} = \Psi \times \mathbf{r}$ for any value of **r** different from zero, then Φ must equal Ψ — unless both Φ and Ψ are linear and the line of their consequents is parallel to **r**.

5. Show that if $\Phi \cdot \mathbf{r} = 0$ for any three non-coplanar values of **r**, then $\Phi = 0$.

6. Prove the statements made in Art. 106 and the converse of the statements.

7. Show that if Ω is complete and if $\Phi \cdot \Omega = \Psi \cdot \Omega$, then Φ and Ψ are equal. Give the proof by means of theory developed prior to Art. 109.

8. *Definition:* Two dyadics such that $\Phi \cdot \Psi = \Psi \cdot \Phi$ – that is to say, two dyadics that are commutative — are said to be *homologous*. Show that if any number of dyadics are homogeneous to one another, any other dyadics which may be obtained from them by addition, subtraction, and direct multiplication are homologous to each other and to the given dyadics. Show also that the reciprocals of homologous dyadics are homologous. Justify the statement that if $\Phi \cdot \Psi^{-1}$ or $\Psi^{-1} \cdot \Phi$, which are equal, be called the quotient of Φ by Ψ, then the rules governing addition, subtraction, multiplication and *division* of *homologous* dyadics are identical with the rules governing these operations in ordinary algebra — it being understood that incomplete dyadics are analogous to zero, and the idemfactor, to unity. Hence the algebra and higher analysis of *homologous* dyadics is practically identical with that of scalar quantities.

9. Show that $(\mathrm{I} \times \mathbf{c}) \cdot \Phi = \mathbf{c} \times \Psi$ and $(\mathbf{c} \times \mathrm{I}) \cdot \Phi = \mathbf{c} \times \Phi$.

10. Show that whether or not **a**, **b**, **c** be coplanar

$$\mathbf{a}\,\mathbf{b} \times \mathbf{c} + \mathbf{b}\,\mathbf{c} \times \mathbf{a} + \mathbf{c}\,\mathbf{a} \times \mathbf{b} = [\mathbf{a}\,\mathbf{b}\,\mathbf{c}]\,\mathrm{I}$$

and

$$\mathbf{b} \times \mathbf{c}\,\mathbf{a} + \mathbf{c} \times \mathbf{a}\,\mathbf{b} + \mathbf{a} \times \mathbf{b}\,\mathbf{c} = [\mathbf{a}\,\mathbf{b}\,\mathbf{c}]\,\mathrm{I}.$$

11. If **a**, **b**, **c** are coplanar use the above relation to prove the law of sines for the triangle and to obtain the relation with scalar coefficients which exists between three coplanar vectors. This may be done by multiplying the equation by a unit normal to the plane of **a**, **b**, and **c**.

12. What is the condition which must subsist between the coefficients in the expansion of a dyadic into nonion form if the dyadic be self-conjugate? What, if the dyadic be anti-self-conjugate?

13. Prove the statements made in Art. 116 concerning the number of ways in which a dyadic may be reduced to its normal form.

14. The necessary and sufficient condition that an anti-self-conjugate dyadic Φ be zero is that the vector of the dyadic shall be zero.

15. Show that if Φ be any dyadic the product $\Phi \cdot \Phi_C$ is self-conjugate.

16. Show how to make use of the relation $\Phi_\times = 0$ to demonstrate that the antecedents and consequents of a self-conjugate dyadic are the same (Art. 116).

17. Show that $$\Phi_2 {}^\times_\times \Phi_2 = \Phi_3{}^2\, \Phi$$
and $$(\Phi + \Psi)_2 = \Phi_2 + \Phi {}^\times_\times \Psi + \Psi_2.$$

18. Show that if the double dot product $\Phi : \Phi$ of a dyadic by itself vanishes, the dyadic vanishes. Hence obtain the condition for a linear dyadic in the form $\Phi_2 : \Phi_2 = 0$.

19. Show that $(\Phi + \mathbf{e\,f})_3 = \Phi_3 + \mathbf{e} \cdot \Phi_2 \cdot \mathbf{f}$.

20. Show that $(\Phi + \Psi)_3 = \Phi_3 + \Phi_2 : \Psi + \Phi : \Psi_2 + \Psi_3$.

21. Show that the scalar of a product of dyadics is unchanged by cyclic permutation of the dyadics. That is

$$(\Phi \cdot \Psi \cdot \Omega)_S = (\Omega \cdot \Phi \cdot \Psi)_S = (\Psi \cdot \Omega \cdot \Phi)_S.$$

CHAPTER VI

ROTATIONS AND STRAINS

123.] In the foregoing chapter the *analytical* theory of dyadics has been dealt with and brought to a state of completeness which is nearly final for practical purposes. There are, however, a number of new questions which present themselves and some old questions which present themselves under a new form when the dyadic is applied to physics or geometry. Moreover it was for the sake of the applications of dyadics that the theory of them was developed. It is then the object of the present chapter to supply an extended application of dyadics to the theory of rotations and strains and to develop, as far as may appear necessary, the further analytical theory of dyadics.

That the dyadic Φ may be used to denote a transformation of space has already been mentioned. A knowledge of the precise nature of this transformation, however, was not needed at the time. Consider $\mathbf{r}$ as drawn from a fixed origin, and $\mathbf{r}'$ as drawn from the same origin. Let now

$$\mathbf{r}' = \Phi \cdot \mathbf{r}.$$

This equation therefore may be regarded as defining a transformation of the points P of space situated at the terminus of $\mathbf{r}$ into the point P', situated at the terminus of $\mathbf{r}'$. The origin remains fixed. Points in the finite regions of space remain in the finite regions of space. Any point upon a line

$$\mathbf{r} = \mathbf{b} + x\,\mathbf{a}$$

becomes a point $$\mathbf{r}' = \Phi \cdot \mathbf{b} + x\,\Phi \cdot \mathbf{a}.$$

Hence straight lines go over into straight lines and lines parallel to the same line $\mathbf{a}$ go over by the transformation into lines parallel to the same line $\Phi \cdot \mathbf{a}$. In like manner planes go over into planes and the quality of parallelism is invariant.

Such a transformation is known as a *homogeneous strain*. Homogeneous strain is of frequent occurrence in physics. For instance, the deformation of the infinitesimal sphere in a fluid (Art. 76) is a homogeneous strain. In geometry the homogeneous strain is generally known by different names. It is called an affine collineation with the origin fixed. Or it is known as a linear homogeneous transformation. The equations of such a transformation are

$$x' = a_{11}\, x + a_{12}\, y + a_{13}\, z$$

$$y' = a_{21}\, x + a_{21}\, y + a_{23}\, z$$

$$z' = a_{31}\, x + a_{32}\, y + a_{33}\, z.$$

124.] *Theorem:* If the dyadic Φ gives the transformation of the *points* of space which is due to a homogeneous strain, Φ_2, the second of Φ, gives the transformation of *plane areas* which is due to that strain and all *volumes* are magnified by that strain in the ratio of Φ_3, the third or determinant of Φ to unity.

Let $$\Phi = \mathbf{a\,l} + \mathbf{b\,m} + \mathbf{c\,n}$$

$$\mathbf{r}' = \Phi \cdot \mathbf{r} = \mathbf{a\,l} \cdot \mathbf{r} + \mathbf{b\,m} \cdot \mathbf{r} + \mathbf{c\,n} \cdot \mathbf{r}.$$

The vectors $\mathbf{l}'$, $\mathbf{m}'$, $\mathbf{n}'$ are changed by Φ into $\mathbf{a}$, $\mathbf{b}$, $\mathbf{c}$. Hence the planes determined by $\mathbf{m}'$ and $\mathbf{n}'$, $\mathbf{n}'$ and $\mathbf{l}'$, $\mathbf{l}'$ and $\mathbf{m}'$ are transformed into the planes determined by $\mathbf{b}$ and $\mathbf{c}$, $\mathbf{c}$ and $\mathbf{a}$, $\mathbf{a}$ and $\mathbf{b}$. The dyadic which accomplishes this result is

$$\Phi_2 = \mathbf{b} \times \mathbf{c}\;\; \mathbf{m} \times \mathbf{n} + \mathbf{c} \times \mathbf{a}\;\; \mathbf{n} \times \mathbf{l} + \mathbf{a} \times \mathbf{b}\;\; \mathbf{l} \times \mathbf{m}.$$

Hence if $\mathbf{s}$ denote any plane area in space, the transformation due to Φ replaces $\mathbf{s}$ by the area $\mathbf{s}'$ such that

$$\mathbf{s}' = \Phi_2 \cdot \mathbf{s}.$$

It is important to notice that the vector **s** denoting a plane area is not transformed into the same vector $\mathbf{s}'$ as it would be if it denoted a line. This is evident from the fact that in the latter case Φ acts on **s** whereas in the former case Φ_2 acts upon **s**.

To show that volumes are magnified in the ratio of Φ_3 to unity choose any three vectors **d**, **e**, **f** which determine the volume of a parallelopiped $[\mathbf{d}\ \mathbf{e}\ \mathbf{f}]$. Express Φ with the vectors which form the reciprocal system to **d**, **e**, **f** as consequents.

$$\Phi = \mathbf{a}\,\mathbf{d}' + \mathbf{b}\,\mathbf{c}' + \mathbf{c}\,\mathbf{f}'.$$

The dyadic Φ changes **d**, **e**, **f** into **a**, **b**, **c** (which are different from the **a**, **b**, **c** above unless **d**, **e**, **f** are equal to $\mathbf{l}'$, $\mathbf{m}'$, $\mathbf{n}'$). Hence the volume $[\mathbf{d}\ \mathbf{e}\ \mathbf{f}]$ is changed into the volume $[\mathbf{a}\ \mathbf{b}\ \mathbf{c}]$.

$$\Phi_3 = [\mathbf{a}\,\mathbf{b}\,\mathbf{c}]\,[\mathbf{d}'\,\mathbf{e}'\,\mathbf{f}']$$

$$[\mathbf{d}'\,\mathbf{e}'\,\mathbf{f}']^{-1} = [\mathbf{d}\,\mathbf{e}\,\mathbf{f}].$$

Hence $$[\mathbf{a}\,\mathbf{b}\,\mathbf{c}] = [\mathbf{d}\,\mathbf{e}\,\mathbf{f}]\,\Phi_3.$$

The ratio of the volume $[\mathbf{a}\ \mathbf{b}\ \mathbf{c}]$ to $[\mathbf{d}\ \mathbf{e}\ \mathbf{f}]$ is as Φ_3 is to unity. But the vectors **d**, **e**, **f** were *any* three vectors which determine a parallelopiped. Hence all volumes are changed by the action of Φ in the same ratio and this ratio is as Φ_3 is to 1.

Rotations about a Fixed Point. Versors

125.] *Theorem:* The necessary and sufficient condition that a dyadic represent a *rotation* about some axis is that it be reducible to the form

$$\Phi = \mathbf{i}'\,\mathbf{i} + \mathbf{j}'\,\mathbf{j} + \mathbf{k}'\,\mathbf{k} \qquad (1)$$

where $\mathbf{i}'$, $\mathbf{j}'$, $\mathbf{k}'$ and **i**, **j**, **k** are two right-handed rectangular systems of unit vectors.

Let $$\mathbf{r} = x\,\mathbf{i} + y\,\mathbf{j} + z\,\mathbf{k}$$

$$\Phi \cdot \mathbf{r} = x\,\mathbf{i}' + y\,\mathbf{j}' + z\,\mathbf{k}'.$$

Hence if Φ is reducible to the given form the vectors $\mathbf{i}, \mathbf{j}, \mathbf{k}$ are changed into the vectors $\mathbf{i}', \mathbf{j}', \mathbf{k}'$ and any vector $\mathbf{r}$ is changed from its position relative to $\mathbf{i}, \mathbf{j}, \mathbf{k}$ into the same position relative to $\mathbf{i}', \mathbf{j}', \mathbf{k}'$. Hence by the transformation no change of shape is effected. The strain reduces to a rotation which carries $\mathbf{i}, \mathbf{j}, \mathbf{k}$ into $\mathbf{i}', \mathbf{j}', \mathbf{k}'$. Conversely suppose the body suffers no change of shape — that is, suppose it subjected to a rotation. The vectors $\mathbf{i}, \mathbf{j}, \mathbf{k}$ must be carried into another right-handed rectangular system of unit vectors. Let these be $\mathbf{i}', \mathbf{j}', \mathbf{k}'$. The dyadic Φ may therefore be reduced to the form

$$\Phi = \mathbf{i}'\,\mathbf{i} + \mathbf{j}'\,\mathbf{j} + \mathbf{k}'\,\mathbf{k}.$$

Definition: A dyadic which is reducible to the form

$$\mathbf{i}'\,\mathbf{i} + \mathbf{j}'\,\mathbf{j} + \mathbf{k}'\,\mathbf{k}$$

and which consequently represents a rotation is called a *versor*.

Theorem: The conjugate and reciprocal of a versor are equal, and conversely if the conjugate and reciprocal of a dyadic are equal the dyadic reduces to a versor or a versor multiplied by the negative sign.

Let

$$\Phi = \mathbf{i}'\,\mathbf{i} + \mathbf{j}'\,\mathbf{j} + \mathbf{k}'\,\mathbf{k},$$

$$\Phi_C = \mathbf{i}\,\mathbf{i}' + \mathbf{j}\,\mathbf{j}' + \mathbf{k}\,\mathbf{k}',$$

$$\Phi \cdot \Phi_C = \mathbf{i}'\,\mathbf{i}' + \mathbf{j}'\,\mathbf{j}' + \mathbf{k}'\,\mathbf{k}' = \mathrm{I}$$

$$\Phi^{-1} = \Phi_C.$$

Hence the first part of the theorem is proved. To prove the second part let

$$\Phi = \mathbf{a}\,\mathbf{i} + \mathbf{b}\,\mathbf{j} + \mathbf{c}\,\mathbf{k},$$

$$\Phi_C = \mathbf{i}\,\mathbf{a} + \mathbf{j}\,\mathbf{b} + \mathbf{k}\,\mathbf{c},$$

$$\Phi \cdot \Phi_C = \mathbf{a}\,\mathbf{a} + \mathbf{b}\,\mathbf{b} + \mathbf{c}\,\mathbf{c}.$$

If

$$\Phi^{-1} = \Phi_C, \qquad \Phi \cdot \Phi_C = \mathrm{I}.$$

Hence

$$\mathbf{a}\,\mathbf{a} + \mathbf{b}\,\mathbf{b} + \mathbf{c}\,\mathbf{c} = \mathrm{I}.$$

Hence (Art. 108) the antecedents **a**, **b**, **c** and the consequents **a**, **b**, **c** must be reciprocal systems. Hence (page 87) they must be either a right-handed or a left-handed rectangular system of unit vectors. The left-handed system may be changed to a right-handed one by prefixing the negative sign to each vector. Then

$$\Phi = \mathbf{i}'\,\mathbf{i} + \mathbf{j}'\,\mathbf{j} + \mathbf{k}'\,\mathbf{k},$$

or
$$\Phi = -\,(\mathbf{i}'\,\mathbf{i} + \mathbf{j}'\,\mathbf{j} + \mathbf{k}'\,\mathbf{k}). \qquad (1)'$$

The third or determinant of a versor is evidently equal to unity; that of the versor with a negative sign, to minus one. Hence the criterion for a versor may be stated in the form

$$\Phi \cdot \Phi_C = \mathrm{I}, \qquad \Phi_3 = |\,\Phi\,| = 1. \qquad (2)$$

Or inasmuch as the determinant of Φ is plus or minus one if $\Phi \cdot \Phi_C = \mathrm{I}$, it is only necessary to state that if

$$\Phi \cdot \Phi_C = \mathrm{I}, \qquad \Phi_3 = |\,\Phi\,| > 0, \qquad (2)'$$

Φ is a versor.

There are two geometric interpretations of the transformation due to a dyadic Φ such that

$$\Phi \cdot \Phi_C = \mathrm{I} \qquad \Phi_3 = |\,\Phi\,| = -1 \qquad (3)$$

$$\Phi = -\,(\mathbf{i}'\,\mathbf{i} + \mathbf{j}'\,\mathbf{j} + \mathbf{k}'\,\mathbf{k}).$$

The transformation due to Φ is one of rotation combined with reflection in the origin. The dyadic $\mathbf{i}'\,\mathbf{i} + \mathbf{j}'\,\mathbf{j} + \mathbf{k}'\,\mathbf{k}$ causes a rotation about a definite axis — it is a versor. The negative sign then reverses the direction of every vector in space and replaces each figure by a figure symmetrical to it with respect to the origin. By reversing the directions of $\mathbf{i}'$ and $\mathbf{j}'$ the system $\mathbf{i}'$, $\mathbf{j}'$, $\mathbf{k}'$ still remains right-handed and rectangular, but the dyadic takes the form

$$\Phi = \mathbf{i}'\,\mathbf{i} + \mathbf{j}'\,\mathbf{j} - \mathbf{k}'\,\mathbf{k},$$

or
$$\Phi = (\mathbf{i}'\,\mathbf{i}' + \mathbf{j}'\,\mathbf{j}' - \mathbf{k}'\,\mathbf{k}') \cdot (\mathbf{i}'\,\mathbf{i} + \mathbf{j}'\,\mathbf{j} + \mathbf{k}'\,\mathbf{k}).$$

Hence the transformation due to Φ is a rotation due to $\mathbf{i}'\mathbf{i}+\mathbf{j}'\mathbf{j}+\mathbf{k}'\mathbf{k}$ followed by a reflection in the plane of $\mathbf{i}'$ and $\mathbf{j}'$. For the dyadic $\mathbf{i}'\mathbf{i}'+\mathbf{j}'\mathbf{j}'-\mathbf{k}'\mathbf{k}'$ causes such a transformation of space that each point goes over into a point symmetrically situated to it with respect to the plane of $\mathbf{i}'$ and $\mathbf{j}'$. Each figure is therefore replaced by a symmetrical figure.

Definition: A transformation that replaces each figure by a symmetrical figure is called a *perversion* and the dyadic which gives the transformation is called a *perversor*.

The criterion for a perversor is that the conjugate of a dyadic shall be equal to its reciprocal and that the determinant of the dyadic shall be equal to minus one.

$$\Phi \cdot \Phi_C = \mathrm{I}, \quad |\Phi| = -1. \tag{3}$$

Or inasmuch as if $\Phi \cdot \Phi_C = \mathrm{I}$, the determinant must be plus or minus one the criterion may take the form

$$\Phi \cdot \Phi_C = \mathrm{I}, \quad |\Phi| < 0, \tag{3$'$}$$

$$\Phi \text{ is a perversor.}$$

It is evident from geometrical considerations that the product of two versors is a versor; of two perversors, a versor; but of a versor and a perversor taken in either order, a perversor.

126.] If the axis of rotation be the **i**-axis and if the angle of rotation be the angle q measured positive in the positive trigonometric direction, then by the rotation the vectors $\mathbf{i}, \mathbf{j}, \mathbf{k}$ are changed into the vectors $\mathbf{i}', \mathbf{j}', \mathbf{k}'$ such that

$$\mathbf{i}' = \mathbf{i}$$

$$\mathbf{j}' = \mathbf{j} \cos q + \mathbf{k} \sin q,$$

$$\mathbf{k}' = -\mathbf{j} \sin q + \mathbf{k} \cos q.$$

The dyadic $\Phi = \mathbf{i}'\mathbf{i} + \mathbf{j}'\mathbf{j} + \mathbf{k}'\mathbf{k}$ which accomplishes this rotation is

$$\Phi = \mathbf{i\,i} + \cos q\,(\mathbf{j\,j} + \mathbf{k\,k}) + \sin q\,(\mathbf{k\,j} - \mathbf{j\,k}). \qquad (4)$$

$$\mathbf{j\,j} + \mathbf{k\,k} = \mathbf{I} - \mathbf{i\,i},$$

$$\mathbf{k\,j} - \mathbf{j\,k} = \mathbf{I} \times \mathbf{i}.$$

Hence $$\Phi = \mathbf{i\,i} + \cos q\,(\mathbf{I} - \mathbf{i\,i}) + \sin q\;\mathbf{I} \times \mathbf{i}. \qquad (5)$$

If more generally in place of the **i**-axis any axis denoted by the unit vector **a** be taken as the axis of rotation and if as before the angle of rotation about that axis be denoted by q, the dyadic Φ which accomplishes the rotation is

$$\Phi = \mathbf{a\,a} + \cos q\,(\mathbf{I} - \mathbf{a\,a}) + \sin q\;\mathbf{I} \times \mathbf{a}. \qquad (6)$$

To show that this dyadic actually does accomplish the rotation apply it to a vector **r**. The dyad **a a** is an idemfactor for all vectors parallel to **a**; but an annihilator for vectors perpendicular to **a**. The dyadic $\mathbf{I} - \mathbf{a\,a}$ is an idemfactor for all vectors in the plane perpendicular to **a**; but an annihilator for all vectors parallel to **a**. The dyadic $\mathbf{I} \times \mathbf{a}$ is a quadrantal versor (Art. 113) for vectors perpendicular to **a**; but an annihilator for vectors parallel to **a**. If then **r** be parallel to **a**

$$\Phi \cdot \mathbf{r} = \mathbf{a\,a} \cdot \mathbf{r} = \mathbf{r}.$$

Hence Φ leaves unchanged all vectors (or components of vectors) which are parallel to **a**. If **r** is perpendicular to **a**

$$\Phi \cdot \mathbf{r} = \cos q\,\mathbf{r} + \sin q\;\mathbf{a} \times \mathbf{r}.$$

Hence the vector **r** has been rotated in its plane through the angle q. If **r** were any vector in space its component parallel to **a** suffers no change; but its component perpendicular to **a** is rotated about **a** through an angle of q degrees. The whole vector is therefore rotated about **a** through that angle.

Let **a** be given in terms of **i**, **j**, **k** as

$$\mathbf{a} = a_1\,\mathbf{i} + a_2\,\mathbf{j} + a_3\,\mathbf{k},$$

$$\mathbf{a\,a} = a_1^2\,\mathbf{i\,i} + a_1 a_2\,\mathbf{i\,j} + a_1 a_3\,\mathbf{i\,k}$$

$$+ a_2 a_1 \mathbf{j i} + a_2^2 \mathbf{j j} + a_2 a_3 \mathbf{j k}$$

$$+ a_3 a_1 \mathbf{k i} + a_3 a_2 \mathbf{k j} + a_3^2 \mathbf{k k},$$

$$\mathbf{I} = \mathbf{i i} + \mathbf{j j} + \mathbf{k k},$$

$$\mathbf{I} \times \mathbf{a} = 0 \, \mathbf{i i} - a_3 \mathbf{i j} + a_2 \mathbf{i k},$$

$$+ a_3 \mathbf{j i} + 0 \, \mathbf{j j} - a_1 \mathbf{j k},$$

$$- a_2 \mathbf{k i} + a_1 \mathbf{k j} + 0 \, \mathbf{k k}.$$

Hence

$$\Phi = \{a_1^2 (1 - \cos q) + \cos q\} \, \mathbf{i i}$$

$$+ \{a_1 a_2 (1 - \cos q) - a_3 \sin q\} \, \mathbf{i j}$$

$$+ \{a_1 a_3 (1 - \cos q) + a_2 \sin q\} \, \mathbf{i k}$$

$$+ \{a_2 a_1 (1 - \cos q) + a_3 \sin q\} \, \mathbf{j i}$$

$$+ \{a_2^2 (1 - \cos q) + \cos q\} \, \mathbf{j j}$$

$$+ \{a_2 a_3 (1 - \cos q) - a_1 \sin q\} \, \mathbf{j k}$$

$$+ \{a_3 a_1 (1 - \cos q) - a_2 \sin q\} \, \mathbf{k i}$$

$$+ \{a_3 a_2 (1 - \cos q) + a_1 \sin q\} \, \mathbf{k j}$$

$$+ \{a_3^2 (1 + \cos q) + \cos q\} \, \mathbf{k k}. \qquad (7)$$

127.] If Φ be written as in equation (4) the vector of Φ and the scalar of Φ may be found.

$$\Phi_\times = \mathbf{i} \times \mathbf{i} + \cos q \, (\mathbf{j} \times \mathbf{j} + \mathbf{k} \times \mathbf{k}) + \sin q \, (\mathbf{k} \times \mathbf{j} - \mathbf{j} \times \mathbf{k})$$

$$\Phi_\times = - 2 \sin q \, \mathbf{i}$$

$$\Phi_S = \mathbf{i} \cdot \mathbf{i} + \cos q \, (\mathbf{j} \cdot \mathbf{j} + \mathbf{k k}) + \sin q \, (\mathbf{k} \circ \mathbf{j} - \mathbf{j} \cdot \mathbf{k}),$$

$$\Phi_S = 1 + 2 \cos q.$$

The axis of rotation $\mathbf{i}$ is seen to have the direction of $-\Phi_\times$, the negative of the vector of Φ. This is true in general. The direction of the axis of rotation of any versor is the negative of the vector of Φ. The proof of this statement depends on the invariant property of $\Phi_\times$. Any versor Φ may be reduced to the form (4) by taking the direction of $\mathbf{i}$

coincident with the direction of the axis of rotation. After this reduction has been made the direction of the axis is seen to be the negative of $\Phi_\times$. But $\Phi_\times$ is not altered by the reduction of Φ to any particular form — nor is the axis of rotation altered by such a reduction. Hence the direction of the axis of rotation is always coincident with $-\Phi_\times$, the direction of the negative of the vector of Φ.

The tangent of one-half the angle of version q is

$$\tan\frac{q}{2} = \frac{\sin q}{1 + \cos q} = \frac{\sqrt{\Phi_\times \cdot \Phi_\times}}{1 + \Phi_S}. \tag{8}$$

The tangent of one-half the angle of version is therefore determined when the values of $\Phi_\times$ and Φ_S are known. The vector $\Phi_\times$ and the scalar Φ_S, which are invariants of Φ, determine completely the versor Φ. Let $\mathbf{Q}$ be a vector drawn in the direction of the axis of rotation. Let the magnitude of $\mathbf{Q}$ be equal to the tangent of one-half the angle q of version.

$$\mathbf{Q} = \frac{-\Phi_\times}{1 + \Phi_S}, \qquad \mathbf{Q} \cdot \mathbf{Q} = \tan^2 \frac{1}{2} q.$$

The vector $\mathbf{Q}$ determines the versor Φ completely. $\mathbf{Q}$ will be called the *vector semi-tangent of version.*

By (6) a versor Φ was expressed in terms of a unit vector parallel to the axis of rotation.

$$\Phi = \mathbf{a}\,\mathbf{a} + \cos q\,(\mathrm{I} - \mathbf{a}\,\mathbf{a}) + \sin q\, \mathrm{I} \times \mathbf{a}.$$

Hence if $\mathbf{Q}$ be the vector semi-tangent of version

$$\Phi = \frac{\mathbf{Q}\,\mathbf{Q}}{\mathbf{Q} \cdot \mathbf{Q}} + \cos q \left(1 - \frac{\mathbf{Q}\,\mathbf{Q}}{\mathbf{Q} \cdot \mathbf{Q}}\right) + \sin q\, \mathrm{I} \times \frac{\mathbf{Q}}{\sqrt{\mathbf{Q} \cdot \mathbf{Q}}}. \tag{10}$$

There is a more compact expression for a versor Φ in terms of the vector semi-tangent of version. Let $\mathbf{c}$ be any vector in space. The version represented by $\mathbf{Q}$ carries

$$\mathbf{c} - \mathbf{Q} \times \mathbf{c} \text{ into } \mathbf{c} + \mathbf{Q} \times \mathbf{c}.$$

It will be sufficient to show this in case $\mathbf{c}$ is perpendicular to $\mathbf{Q}$. For if $\mathbf{c}$ (or any component of it) were parallel to $\mathbf{Q}$ the result of multiplying by $\mathbf{Q}\times$ would be zero and the statement would be that $\mathbf{c}$ is carried into $\mathbf{c}$. In the first place the magnitudes of the two vectors are equal. For

$$(\mathbf{c} - \mathbf{Q} \times \mathbf{c}) \cdot (\mathbf{c} - \mathbf{Q} \times \mathbf{c}) = \mathbf{c} \cdot \mathbf{c} + \mathbf{Q} \times \mathbf{c} \cdot \mathbf{Q} \times \mathbf{c} - 2\, \mathbf{c} \cdot \mathbf{Q} \times \mathbf{c}$$

$$(\mathbf{c} + \mathbf{Q} \times \mathbf{c}) \cdot (\mathbf{c} + \mathbf{Q} \times \mathbf{c}) = \mathbf{c} \cdot \mathbf{c} + \mathbf{Q} \times \mathbf{c} \cdot \mathbf{Q} \times \mathbf{c} + 2\, \mathbf{c} \cdot \mathbf{Q} \times \mathbf{c}$$

$$\mathbf{c} \cdot \mathbf{c} + \mathbf{Q} \times \mathbf{c} \cdot \mathbf{Q} \times \mathbf{c} = \mathbf{c} \cdot \mathbf{c} + \mathbf{Q} \cdot \mathbf{Q} \;\; \mathbf{c} \cdot \mathbf{c} - \mathbf{Q} \cdot \mathbf{c} \;\; \mathbf{Q} \cdot \mathbf{c}.$$

Since $\mathbf{Q}$ and $\mathbf{c}$ are by hypothesis perpendicular

$$\mathbf{c} \cdot \mathbf{c} + \mathbf{Q} \times \mathbf{c} \cdot \mathbf{Q} \times \mathbf{c} = c^2 \left(1 + \tan^2 \tfrac{1}{2} q\right).$$

The term $\mathbf{c} \cdot \mathbf{Q} \times \mathbf{c}$ vanishes. Hence the equality. In the second place the angle between the two vectors is equal to q.

$$\frac{(\mathbf{c} - \mathbf{Q} \times \mathbf{c}) \cdot (\mathbf{c} + \mathbf{Q} \times \mathbf{c})}{c^2 \left(1 + \tan^2 \frac{1}{2} q\right)} = \frac{\mathbf{c} \cdot \mathbf{c} - \mathbf{Q} \times \mathbf{c} \cdot \mathbf{Q} \times \mathbf{c}}{c^2 \left(1 + \tan^2 \frac{1}{2} q\right)}$$

$$\frac{c^2 \left(1 - \tan^2 \frac{1}{2} q\right)}{c^2 \left(1 + \tan^2 \frac{1}{2} q\right)} = \cos q$$

$$\frac{(\mathbf{c} - \mathbf{Q} \times \mathbf{c}) \times (\mathbf{c} + \mathbf{Q} \times \mathbf{c})}{c^2 \left(1 + \tan^2 \frac{1}{2} q\right)} = \frac{2\, \mathbf{c} \times (\mathbf{Q} \times \mathbf{c})}{c^2 \left(1 + \tan^2 \frac{1}{2} q\right)} =$$

$$\frac{2\, c^2 \tan \frac{1}{2} q}{c^2 \left(1 + \tan^2 \frac{1}{2} q\right)} = \sin q.$$

Hence the cosine and sine of the angle between $\mathbf{c} - \mathbf{Q} \times \mathbf{c}$ and $\mathbf{c} + \mathbf{Q} \times \mathbf{c}$ are equal respectively to the cosine and sine of the angle q: and consequently the angle between the vectors must equal the angle q. Now

$$\mathbf{c} - \mathbf{Q} \times \mathbf{c} = (\mathrm{I} - \mathrm{I} \times \mathbf{Q}) \cdot \mathbf{c}$$

and
$$(\mathbf{c} + \mathbf{Q} \times \mathbf{c}) = (\mathrm{I} + \mathrm{I} \times \mathbf{Q}) \cdot \mathbf{c}$$

$$(\mathrm{I} + \mathrm{I} \times \mathbf{Q}) \cdot (\mathrm{I} - \mathrm{I} \times \mathbf{Q})^{-1} \cdot (\mathrm{I} - \mathrm{I} \times \mathbf{Q}) = \mathrm{I} + \mathrm{I} \times \mathbf{Q}.$$

Multiply by $\mathbf{c}$

$$(\mathrm{I} + \mathrm{I} \times \mathbf{Q}) \cdot (\mathrm{I} - \mathrm{I} \times \mathbf{Q})^{-1} \cdot (\mathbf{c} - \mathbf{Q} \times \mathbf{c}) = \mathbf{c} + \mathbf{Q} \times \mathbf{c}.$$

Hence the dyadic

$$\Phi = (\mathrm{I} + \mathrm{I} \times \mathbf{Q}) \cdot (\mathrm{I} - \mathrm{I} \times \mathbf{Q})^{-1} \qquad (10)'$$

carries the vector $\mathbf{c} - \mathbf{Q} \times \mathbf{c}$ into the vector $\mathbf{c} + \mathbf{Q} \times \mathbf{c}$ no matter what the value of $\mathbf{c}$. Hence the dyadic Φ determines the version due to the vector semi-tangent of version $\mathbf{Q}$.

The dyadic $\mathrm{I} + \mathrm{I} \times \mathbf{Q}$ carries the vector $\mathbf{c} - \mathbf{Q} \times \mathbf{c}$ into $(\mathrm{I} + \mathbf{Q} \cdot \mathbf{Q})\, \mathbf{c}$.

$$(\mathrm{I} + \mathrm{I} \times \mathbf{Q}) \cdot (\mathbf{c} - \mathbf{Q} \times \mathbf{c}) = \mathbf{c} + \mathbf{Q} \times \mathbf{c} - \mathbf{Q} \times \mathbf{c} - \mathbf{Q} \times (\mathbf{Q} \times \mathbf{c})$$

$$(\mathrm{I} + \mathrm{I} \times \mathbf{Q}) \cdot (\mathbf{c} - \mathbf{Q} + \mathbf{c}) = \mathbf{c} + \mathbf{Q} \cdot \mathbf{Q}\, \mathbf{c} = (1 + \mathbf{Q} \cdot \mathbf{Q})\, \mathbf{c}.$$

Hence the dyadic

$$\frac{\mathrm{I} + \mathrm{I} \times \mathbf{Q}}{1 + \mathbf{Q} \cdot \mathbf{Q}} = (\mathrm{I} - \mathrm{I} \times \mathbf{Q})^{-1}$$

carries the vector $\mathbf{c} - \mathbf{Q} \times \mathbf{c}$ into the vector $\mathbf{c}$, if $\mathbf{c}$ be perpendicular to $\mathbf{Q}$ as has been supposed. Consequently the dyadic

$$\frac{(\mathrm{I} + \mathrm{I} \times \mathbf{Q})^2}{1 + \mathbf{Q} \cdot \mathbf{Q}}$$

produces a rotation of all vectors in the plane perpendicular to $\mathbf{Q}$. If, however, it be applied to a vector $x\, \mathbf{Q}$ parallel to $\mathbf{Q}$ the result is not equal to $x\, \mathbf{Q}$.

$$\frac{(\mathrm{I} + \mathrm{I} \times \mathbf{Q}) \cdot (\mathrm{I} + \mathrm{I} \times \mathbf{Q})}{1 + \mathbf{Q} \cdot \mathbf{Q}}\, x\, \mathbf{Q} = x\, \frac{(\mathrm{I} + \mathrm{I} \times \mathbf{Q})}{1 + \mathbf{Q} \cdot \mathbf{Q}}\, \mathbf{Q} = \frac{x\, \mathbf{Q}}{1 + \mathbf{Q} \cdot \mathbf{Q}}.$$

To obviate this difficulty the dyad $\mathbf{Q}\,\mathbf{Q}$, which is an annihilator for all vectors perpendicular to $\mathbf{Q}$, may be added to the numerator. The versor Φ may then be written

$$\Phi = \frac{\mathbf{Q}\,\mathbf{Q} + (\mathrm{I} + \mathrm{I} \times \mathbf{Q})^2}{1 + \mathbf{Q} \cdot \mathbf{Q}} \qquad (10)''$$

$$(\mathrm{I} + \mathrm{I} \times \mathbf{Q}) \cdot (\mathrm{I} + \mathrm{I} \times \mathbf{Q}) = \mathrm{I} + 2\ \mathrm{I} \times \mathbf{Q} + (\mathrm{I} \times \mathbf{Q}) \cdot (\mathrm{I} \times \mathbf{Q})$$

$$(\mathrm{I} \times \mathbf{Q}) \cdot (\mathrm{I} \times \mathbf{Q}) = (\mathrm{I} \times \mathbf{Q}) \times \mathbf{Q} = \mathrm{I} \cdot \mathbf{Q}\,\mathbf{Q} - \mathbf{Q} \cdot \mathbf{Q}\,\mathrm{I}.$$

Hence substituting:

$$\Phi = \frac{(1 - \mathbf{Q} \cdot \mathbf{Q})\ \mathrm{I} + 2\,\mathbf{Q}\,\mathbf{Q} + 2\ \mathrm{I} \times \mathbf{Q}}{1 + \mathbf{Q} \cdot \mathbf{Q}}. \qquad (10)'''$$

This may be expanded in nonion form. Let

$$\mathbf{Q} = a\,\mathbf{i} + b\,\mathbf{j} + c\,\mathbf{k}.$$

$$\Phi = \frac{\left\{\begin{array}{l} (1 + a^2 - b^2 - c^2)\,\mathbf{ii} + (2\,ab - 2c)\,\mathbf{ij} + (2\,ac + 2\,b)\,\mathbf{ik} \\ + (2ab + 2c)\,\mathbf{ji} + (1 - a^2 + b^2 - c^2)\,\mathbf{jj} + (2\,bc - 2\,a)\,\mathbf{jk} \\ + (2ac - 2\,b)\,\mathbf{ki} + (2bc + 2\,a)\,\mathbf{kj} + (1 - a^2 - b^2 + c^2)\,\mathbf{kk} \end{array}\right\} (11)}{1 + a^2 + b^2 + c^2}.$$

128.] If $\mathbf{a}$ is a unit vector a dyadic of the form

$$\Phi = 2\,\mathbf{a}\,\mathbf{a} - \mathrm{I} \qquad (12)$$

is a *biquadrantal* versor. That is, the dyadic Φ turns the points of space about the axis $\mathbf{a}$ through two right angles. This may be seen by setting q equal to π in the general expression for a versor

$$\Phi = \mathbf{a}\,\mathbf{a} + \cos q\ (\mathrm{I} - \mathbf{a}\,\mathbf{a}) + \sin q\ \mathrm{I} \times \mathbf{a},$$

or it may be seen directly from geometrical considerations. The dyadic Φ leaves a vector parallel to $\mathbf{a}$ unchanged but reverses every vector perpendicular to $\mathbf{a}$ in direction.

Theorem: The product of two biquadrantal versors is a versor the axis of which is perpendicular to the axes of the biquadrantal versors and the angle of which is twice the angle from the axis of the second to the axis of the first.

Let $\mathbf{a}$ and $\mathbf{b}$ be the axes of two biquadrantal versors. The product

$$\Omega = (2\,\mathbf{b}\,\mathbf{b} - \mathrm{I}) \cdot (2\,\mathbf{a}\,\mathbf{a} - \mathrm{I})$$

is certainly a versor; for the product of any two versors is a versor. Consider the common perpendicular to $\mathbf{a}$ and $\mathbf{b}$. The biquadrantal versor $2\,\mathbf{a}\,\mathbf{a} - \mathrm{I}$ reverses this perpendicular in direction. $(2\,\mathbf{b}\,\mathbf{b} - \mathrm{I})$ again reverses it in direction and consequently brings it back to its original position. Hence the product Ω leaves the common perpendicular to $\mathbf{a}$ and $\mathbf{b}$ unchanged. Ω is therefore a rotation about this line as axis.

$$\Omega \cdot \mathbf{a} = (2\,\mathbf{b}\,\mathbf{b} - \mathrm{I}) \cdot (2\,\mathbf{a}\,\mathbf{a} - \mathrm{I}) \cdot \mathbf{a} = (2\,\mathbf{b}\,\mathbf{b} - \mathrm{I}) \cdot \mathbf{a} = 2\,\mathbf{b}\,\mathbf{b} \cdot \mathbf{a} - \mathbf{a}.$$

The cosine of the angle from $\mathbf{a}$ to $\Omega \cdot \mathbf{a}$ is

$$\mathbf{a} \cdot \Omega \cdot \mathbf{a} = 2\,\mathbf{b} \cdot \mathbf{a}\;\mathbf{b} \cdot \mathbf{a} - \mathbf{a} \cdot \mathbf{a} = 2\,(\mathbf{b} \cdot \mathbf{a})^2 - 1 = \cos 2\,(\mathbf{b}, \mathbf{a}).$$

Hence the angle of the versor Ω is equal to twice the angle from $\mathbf{a}$ to $\mathbf{b}$.

Theorem: Conversely any given versor may be expressed as the product of two biquadrantal versors, of which the axes lie in the plane perpendicular to the axis of the given versor and include between them an angle equal to one half the angle of the given versor.

For let Ω be the given versor. Let $\mathbf{a}$ and $\mathbf{b}$ be unit vectors perpendicular to the axis $-\Omega_\times$ of this versor. Furthermore let the angle from $\mathbf{a}$ to $\mathbf{b}$ be equal to one half the angle of this versor. Then by the foregoing theorem

$$\Omega = (2\,\mathbf{b}\,\mathbf{b} - \mathrm{I}) \cdot (2\,\mathbf{a}\,\mathbf{a} - \mathrm{I}). \qquad (14)$$

The resolution of versors into the product of two biquadrantal versors affords an immediate and simple method for compounding two finite rotations about a fixed point. Let Φ and Ψ be two given versors. Let $\mathbf{b}$ be a unit vector perpendicular to the axes of Φ and Ψ. Let **a** be a unit vector perpendicular to the axis of Φ and such that the angle from **a** to **b** is equal to one half the angle of Φ. Let **c** be a unit vector perpendicular to the axis of Ψ and such that the angle from **b** to **c** is equal to one half the angle of Ψ. Then

$$\Phi = (2\,\mathbf{b}\mathbf{b} - \mathrm{I}) \cdot (2\,\mathbf{a}\mathbf{a} - \mathrm{I})$$

$$\Psi = (2\,\mathbf{c}\mathbf{c} - \mathrm{I}) \cdot (2\,\mathbf{b}\mathbf{b} - \mathrm{I})$$

$$\Psi \cdot \Phi = (2\,\mathbf{c}\mathbf{c} - \mathrm{I}) \cdot (2\,\mathbf{b}\mathbf{b} - \mathrm{I})^2 \cdot (2\,\mathbf{a}\mathbf{a} - \mathrm{I}).$$

But $(2\,\mathbf{bb} - \mathrm{I})^2$ is equal to the idemfactor, as may be seen from the fact that it represents a rotation through four right angles or from the expansion

$$(2\,\mathbf{b}\mathbf{b} - \mathrm{I}) \cdot (2\,\mathbf{b}\mathbf{b} - \mathrm{I}) = 4\,\mathbf{b} \cdot \mathbf{b}\ \ \mathbf{b}\mathbf{b} - 4\,\mathbf{b}\mathbf{b} + \mathrm{I} = \mathrm{I}.$$

Hence
$$\Psi \cdot \Phi = (2\,\mathbf{c}\mathbf{c} - \mathrm{I}) \cdot (2\,\mathbf{a}\mathbf{a} - \mathrm{I}).$$

The product of Ψ into Φ is a versor the axis of which is perpendicular to **a** and **c** and the angle of which is equal to one half the angle from **a** to **c**.

If Φ and Ψ are two versors of which the vector semi-tangents of version are respectively $\mathbf{Q}_1$ and $\mathbf{Q}_2$, the vector semi-tangent of version $\mathbf{Q}_3$ of the product $\Psi \cdot \Phi$ is

$$\mathbf{Q}_3 = \frac{\mathbf{Q}_1 + \mathbf{Q}_2 + \mathbf{Q}_2 \times \mathbf{Q}_1}{1 - \mathbf{Q}_1 \cdot \mathbf{Q}_2}. \qquad (15)$$

Let
$$\Phi = (2\,\mathbf{b}\mathbf{b} - \mathrm{I}) \cdot (2\,\mathbf{a}\mathbf{a} - \mathrm{I})$$

and
$$\Psi = (2\,\mathbf{c}\mathbf{c} - \mathrm{I}) \cdot (2\,\mathbf{b}\mathbf{b} - \mathrm{I}).$$

$$\Psi \cdot \Phi = (2\,\mathbf{c}\mathbf{c} - \mathrm{I}) \cdot (2\,\mathbf{a}\mathbf{a} - \mathrm{I}).$$

$$\mathbf{Q}_1 = \frac{-\Phi_\times}{1 + \Phi_S}, \qquad \mathbf{Q}_2 = \frac{-\Psi_\times}{1 + \Phi_S}, \qquad \mathbf{Q}_3 = \frac{-(\Psi \cdot \Phi)_\times}{1 + (\Psi \cdot \Phi)_S}$$

$$\Phi = 4\,\mathbf{a} \cdot \mathbf{b}\ \ \mathbf{b}\mathbf{a} - 2\,\mathbf{a}\mathbf{a} - 2\,\mathbf{b}\ \mathbf{b} + \mathrm{I},$$

$$\Phi_\times = 4\,\mathbf{a} \cdot \mathbf{b}\ \ \mathbf{b} \times \mathbf{a},$$

$$\Phi_S = 4\,(\mathbf{a}\cdot\mathbf{b})^2 - 1,$$

$$\Psi = 4\,\mathbf{c}\cdot\mathbf{b}\ \mathbf{c}\mathbf{b} - 2\,\mathbf{b}\mathbf{b} - 2\,\mathbf{c}\mathbf{c} + \mathrm{I},$$

$$\Psi_\times = 4\,\mathbf{c}\cdot\mathbf{b}\ \mathbf{c}\times\mathbf{b},$$

$$\Psi_S = 4\,(\mathbf{c}\cdot\mathbf{b})^2 - 1$$

$$\Psi\cdot\Phi = 4\,\mathbf{c}\cdot\mathbf{a}\ \mathbf{c}\mathbf{a} - 2\,\mathbf{c}\mathbf{c} - 2\,\mathbf{a}\mathbf{a} + \mathrm{I},$$

$$(\Psi\cdot\Phi)_\times = 4\,\mathbf{c}\cdot\mathbf{a}\ \mathbf{c}\times\mathbf{a},$$

$$(\Psi\cdot\Phi)_S = 4\,(\mathbf{c}\cdot\mathbf{a})^2 - 1.$$

Hence $$\mathrm{Q}_1 = \frac{\mathbf{a}\times\mathbf{b}}{\mathbf{a}\cdot\mathbf{b}},\quad \mathrm{Q}_2 = \frac{\mathbf{b}\times\mathbf{c}}{\mathbf{b}\cdot\mathbf{c}},\quad \mathrm{Q}_3 = \frac{\mathbf{a}\times\mathbf{c}}{\mathbf{a}\cdot\mathbf{c}}$$

$$\mathrm{Q}_2\times\mathrm{Q}_1 = \frac{(\mathbf{b}\times\mathbf{c})\times(\mathbf{a}\times\mathbf{b})}{\mathbf{a}\cdot\mathbf{b}\ \mathbf{b}\cdot\mathbf{c}} = -\frac{[\mathbf{a}\,\mathbf{b}\,\mathbf{c}]\,\mathbf{b}}{\mathbf{a}\cdot\mathbf{b}\,\mathbf{b}\cdot\mathbf{c}}.$$

But $$[\mathbf{a}\,\mathbf{b}\,\mathbf{c}]\,\mathbf{r} = \mathbf{b}\times\mathbf{c}\ \mathbf{a}\cdot\mathbf{r} + \mathbf{c}\times\mathbf{a}\ \mathbf{b}\cdot\mathbf{r} + \mathbf{a}\times\mathbf{b}\ \mathbf{c}\cdot\mathbf{r},$$

Hence $$[\mathbf{a}\,\mathbf{b}\,\mathbf{c}]\,\mathbf{b} = \mathbf{b}\times\mathbf{c}\ \mathbf{a}\cdot\mathbf{b} + \mathbf{c}\times\mathbf{a}\ \mathbf{b}\cdot\mathbf{b} + \mathbf{a}\times\mathbf{b}\ \mathbf{c}\cdot\mathbf{b}.$$

$$\mathrm{Q}_2\times\mathrm{Q}_1 = -\frac{\mathbf{b}\times\mathbf{c}}{\mathbf{b}\cdot\mathbf{c}} - \frac{\mathbf{a}\times\mathbf{b}}{\mathbf{a}\cdot\mathbf{b}} + \frac{\mathbf{a}\times\mathbf{c}}{\mathbf{a}\cdot\mathbf{b}\,\mathbf{b}\cdot\mathbf{c}}.$$

Hence $$\mathrm{Q}_2\times\mathrm{Q}_1 = -\mathrm{Q}_1 - \mathrm{Q}_2 + \frac{\mathbf{a}\cdot\mathbf{c}\ \mathrm{Q}_3}{\mathbf{a}\cdot\mathbf{b}\mathbf{b}\cdot\mathbf{c}}$$

$$\mathrm{Q}_3 = \frac{\mathrm{Q}_1\times\mathrm{Q}_2 + \mathrm{Q}_2\times\mathrm{Q}_1}{\dfrac{\mathbf{a}\cdot\mathbf{c}}{\mathbf{a}\cdot\mathbf{b}\,\mathbf{b}\cdot\mathbf{c}}}$$

$$\mathrm{Q}_1\cdot\mathrm{Q}_2 = \frac{(\mathbf{a}\times\mathbf{b})\cdot(\mathbf{b}\times\mathbf{c})}{\mathbf{a}\cdot\mathbf{b}\,\mathbf{b}\cdot\mathbf{c}} = \frac{\mathbf{a}\cdot\mathbf{b}\ \mathbf{b}\cdot\mathbf{c}}{\mathbf{a}\cdot\mathbf{b}\ \mathbf{b}\cdot\mathbf{c}} - \frac{\mathbf{a}\cdot\mathbf{c}\,\mathbf{b}\cdot\mathbf{b}}{\mathbf{a}\cdot\mathbf{b}\ \mathbf{b}\cdot\mathbf{c}}.$$

Hence $$\frac{\mathbf{a}\cdot\mathbf{c}}{\mathbf{a}\cdot\mathbf{b}\ \mathbf{b}\cdot\mathbf{c}} = 1 - \mathrm{Q}_2\cdot\mathrm{Q}_1.$$

Hence $$\mathrm{Q}_3 = \frac{\mathrm{Q}_1\times\mathrm{Q}_2 + \mathrm{Q}_2 + \mathrm{Q}_1}{1 - \mathrm{Q}_1\cdot\mathrm{Q}_2}.$$

This formula gives the composition of two finite rotations. If the rotations be infinitesimal $\mathbf{Q}_1$ and $\mathbf{Q}_2$ are both infinitesimal. Neglecting infinitesimals of the second order the formula reduces to

$$\mathbf{Q}_3 = \mathbf{Q}_1 + \mathbf{Q}_2.$$

The infinitesimal rotations combine according to the law of vector addition. This demonstrates the parallelogram law for angular velocities. The subject was treated from different standpoints in Arts. 51 and 60.

Cyclics, Right Tensors, Tonics, and Cyclotonics

129.] If the dyadic Φ be a versor it may be written in the form (4)

$$\Phi = \mathbf{i}\,\mathbf{i} + \cos q\,(\mathbf{j}\,\mathbf{j} + \mathbf{k}\,\mathbf{k}) + \sin q\,(\mathbf{k}\,\mathbf{j} - \mathbf{j}\,\mathbf{k}).$$

The axis of rotation is $\mathbf{i}$ and the angle of rotation about that axis is q. Let Ψ be another versor with the same axis and an angle of rotation equal to q'.

$$\Psi = \mathbf{i}\,\mathbf{i} + \cos q'\,(\mathbf{j}\,\mathbf{j} + \mathbf{k}\,\mathbf{k}) + \sin q'\,(\mathbf{k}\,\mathbf{j} - \mathbf{j}\,\mathbf{k}).$$

Multiplying:

$$\Phi \cdot \Psi = \Psi \cdot \Phi = \mathbf{i}\,\mathbf{i} + \cos(q + q')\,(\mathbf{j}\,\mathbf{j} + \mathbf{k}\,\mathbf{k}) + \sin(q + q')\,(\mathbf{k}\,\mathbf{j} - \mathbf{j}\,\mathbf{k}). \qquad (16)$$

This is the result which was to be expected — the product of two versors of which the axes are coincident is a versor with the same axis and with an angle equal to the sum of the angles of the two given versors.

If a versor be multiplied by itself, geometric and analytic considerations alike make it evident that

$$\Phi^2 = \mathbf{i}\,\mathbf{i} + \cos 2q\,(\mathbf{j}\,\mathbf{j} + \mathbf{k}\,\mathbf{k}) + \sin 2q\,(\mathbf{k}\,\mathbf{j} - \mathbf{j}\,\mathbf{k}),$$

and $\Phi^n = \mathbf{i}\,\mathbf{i} + \cos nq\,(\mathbf{j}\,\mathbf{j} + \mathbf{k}\,\mathbf{k}) + \sin nq\,(\mathbf{k}\,\mathbf{j} - \mathbf{j}\,\mathbf{k}).$

On the other hand let Φ_1 equal $\mathbf{j\,j} + \mathbf{k\,k}$; and Φ_2 equal $\mathbf{k\,j} - \mathbf{j\,k}$. Then

$$\Phi^n = (\mathbf{i\,i} + \cos q\, \Phi_1 + \sin q\, \Phi_2)^n.$$

The product of $\mathbf{i\,i}$ into either Φ_1 or Φ_2 is zero and into itself is $\mathbf{i\,i}$. Hence

$$\Phi^n = \mathbf{i\,i} + (\cos q\, \Phi_1 + \sin q\, \Phi_2)^n$$

$$\Phi^n = \mathbf{i\,i} + \cos^n q\, \Phi_1{}^n + n \cos^{n-1} q \sin q\, \Phi_1{}^{n-1} \cdot \Phi_2 + \cdots$$

The dyadic Φ_1 raised to any power reproduces itself. $\Phi_1{}^n = \Phi_1$. The dyadic Φ_2 raised to the second power gives the negative of Φ_1; raised to the third power, the negative of Φ_2; raised to the fourth power, Φ_1; raised to the fifth power, Φ_2 and so on (Art. 114). The dyadic Φ_1 multiplied by Φ_2 is equal to Φ_2. Hence

$$\Phi^n = \mathbf{i\,i} + \cos^n q\, \Phi_1 + n \cos^{n-1} q \sin q\, \Phi_2$$

$$- \frac{n(n-1)}{2!} \cos^{n-2} q \sin_2 \Phi_2 + \cdots$$

But
$$\Phi^n = \mathbf{i\,i} + \cos n\, q\, \Phi_1 + \sin n\, q\, \Phi_2.$$

Equating coefficients of Φ_1 and Φ_2 in these two expressions for Φ^n

$$\cos n\, q = \cos^n q - \frac{n\,(n-1)}{2!} \cos^{n-2} q \sin_2 q + \cdots$$

$$\sin n\, q = n \cos^{n-1} q \sin q - \frac{n\,(n-1)\,(n-2)}{3!} \cos^{n-3} q \sin^3 q + \cdots$$

Thus the ordinary expansions for $\cos nq$ and $\sin nq$ are obtained in a manner very similar to the manner in which they are generally obtained.

The expression for a versor may be generalized as follows. Let $\mathbf{a}, \mathbf{b}, \mathbf{c}$ be any three non-coplanar vectors; and $\mathbf{a}', \mathbf{b}', \mathbf{c}'$, the reciprocal system. Consider the dyadic

$$\Phi = \mathbf{a\,a}' + \cos q\, (\mathbf{b\,b}' + \mathbf{c\,c}') + \sin q\, (\mathbf{c\,b}' - \mathbf{b\,c}'). \quad (17)$$

This dyadic leaves vectors parallel to $\mathbf{a}$ unchanged. Vectors in the plane of $\mathbf{b}$ and $\mathbf{c}$ suffer a change similar to rotation. Let

$$\mathbf{r} = \cos p\, \mathbf{b} + \sin p\, \mathbf{c},$$

$$\mathbf{r}' = \Phi \cdot \mathbf{r} = \cos (p + q)\, \mathbf{b} + \sin (p + q)\, \mathbf{c}.$$

This transformation may be given a definite geometrical interpretation as follows. The vector $\mathbf{r}$, when p is regarded as a variable scalar parameter, describes an ellipse of which $\mathbf{b}$ and $\mathbf{c}$ are two conjugate semi-diameters (page 117). Let this ellipse be regarded as the parallel projection of the unit circle

$$\bar{\mathbf{r}} = \cos p\, \mathbf{i} + \sin q\, \mathbf{j}.$$

That is, the ellipse and the circle are cut from the same cylinder. The two semi-diameters $\mathbf{i}$ and $\mathbf{j}$ of the circle project into the conjugate semi-diameters $\mathbf{a}$ and $\mathbf{b}$ of the ellipse. The radius vector $\mathbf{r}$ in the ellipse projects into the radius vector $\bar{\mathbf{r}}$ in the unit circle. The radius vector $\mathbf{r}'$ in the ellipse which is equal to $\Phi \cdot \mathbf{r}$, projects into a radius vector $\bar{\mathbf{r}}'$ in the circle such that

$$\bar{\mathbf{r}}' = \cos (p + q)\, \mathbf{i} + \sin (p + q)\, \mathbf{j}.$$

Thus the vector $\mathbf{r}$ in the ellipse is so changed by the application of Φ as a prefactor that its projection $\bar{\mathbf{r}}$ in the unit circle is rotated through an angle q.

This statement may be given a neater form by making use of the fact that in parallel projection areas are changed in a definite constant ratio. The vector $\bar{\mathbf{r}}$ in the unit circle may be regarded as describing a sector of which the area is to the area of the whole circle as q is to 2π. The radius vector $\bar{\mathbf{r}}$ then describes a sector of the ellipse. The area of this sector is to the area of the whole ellipse as q is to 2π. Hence *the dyadic* Φ *applied as a prefactor to a radius vector* $\mathbf{r}$ *in an ellipse of which* $\mathbf{b}$ *and* $\mathbf{c}$ *are two conjugate semi-diameters advances that vector through a sector the area of which is to the area of* *the whole ellipse as q is to 2π.*[1] Such a displacement of the radius vector **r** may be called an *elliptic rotation* through a sector q from its similarity to an ordinary rotation of which it is the projection.

Definition: A dyadic Φ of the form

$$\Phi = \mathbf{a}\,\mathbf{a}' + \cos q\,(\mathbf{b}\,\mathbf{b}' + \mathbf{c}\,\mathbf{c}') + \sin q\,(\mathbf{c}\,\mathbf{b}' - \mathbf{b}\,\mathbf{c}') \quad (17)$$

is called a *cyclic* dyadic. The versor is a special case of a cyclic dyadic.

It is evident from geometric or analytic considerations that the powers of a cyclic dyadic are formed, as the powers of a versor were formed, by multiplying the scalar q by the power to which the dyadic is to be raised.

$$\Phi^n = \mathbf{a}\,\mathbf{a}' + \cos nq\,(\mathbf{b}\,\mathbf{b}' + \mathbf{c}\,\mathbf{c}') + \sin nq\,(\mathbf{c}\,\mathbf{b}' - \mathbf{b}\,\mathbf{c}').$$

If the scalar q is an integral sub-multiple of 2π, that is, if

$$\frac{2\pi}{q} = m,$$

it is possible to raise the dyadic Φ to such an integral power, namely, the power m, that it becomes the idemfactor

$$\Phi^m = \mathrm{I}$$

Φ may then be regarded as the mth root of the idemfactor. In like manner if q and 2π are commensurable it is possible to raise Φ to such a power that it becomes equal to the idemfactor and even if q and 2π are incommensurable a power of Φ may be found which differs by as little as one pleases from the idemfactor. Hence any cyclic dyadic may be regarded as a root of the idemfactor.

[1] It is evident that fixing the result of the application of Φ to all radii vectors in an ellipse practically fixes it for all vectors in the plane of **b** and **c**. For any vector in that plane may be regarded as a scalar multiple of a radius vector of the ellipse.

130.] *Definition:* The transformation represented by the dyadic

$$\Phi = a\,\mathbf{i}\,\mathbf{i} + b\,\mathbf{j}\,\mathbf{j} + c\,\mathbf{k}\,\mathbf{k} \tag{18}$$

where a, b, c are positive scalars is called a *pure strain.* The dyadic itself is called a *right tensor.*

A right tensor may be factored into three factors

$$\Phi = (a\,\mathbf{i}\,\mathbf{i} + \mathbf{j}\,\mathbf{j} + \mathbf{k}\,\mathbf{k}) \cdot (\mathbf{i}\,\mathbf{i} + b\,\mathbf{j}\,\mathbf{j} + \mathbf{k}\,\mathbf{k}) \cdot (\mathbf{i}\,\mathbf{i} + \mathbf{j}\,\mathbf{j} + c\,\mathbf{k}\,\mathbf{k}).$$

The order in which these factors occur is immaterial. The transformation

$$\mathbf{r}' = (\mathbf{i}\,\mathbf{i} + \mathbf{j}\,\mathbf{j} + c\,\mathbf{k}\,\mathbf{k}) \cdot \mathbf{r}$$

is such that the **i** and **j** components of a vector remain unaltered but the **k**-component is altered in the ratio of c to 1. The transformation may therefore be described as a stretch or elongation along the direction **k**. If the constant c is greater than unity the elongation is a true elongation: but if c is less than unity the elongation is really a compression, for the ratio of elongation is less than unity. Between these two cases comes the case in which the constant is unity. The lengths of the **k**-components are then not altered.

The transformation due to the dyadic Φ may be regarded as the successive or simultaneous elongation of the components of **r** parallel to **i**, **j**, and **k** respectively in the ratios a to 1, b to 1, c to 1. If one or more of the constants a, b, c is less than unity the elongation in that or those directions becomes a compression. If one or more of the constants is unity, components parallel to that direction are not altered. The directions **i**, **j**, **k** are called the *principal axes of the strain.* Their directions are not altered by the strain whereas, if the constants a, b, c be different, every other direction is altered. The scalars a, b, c are known as the *principal ratios of elongation.*

In Art. 115 it was seen that any complete dyadic was reducible to the normal form

$$\Phi = \pm\,(a\,\mathbf{i}'\,\mathbf{i} + b\,\mathbf{j}'\,\mathbf{j} + c\,\mathbf{k}'\,\mathbf{k})$$

where a, b, c are positive constants. This expression may be factored into the product of two dyadics.

$$\Phi = \pm (a\, \mathbf{i}'\mathbf{i}' + b\, \mathbf{j}'\mathbf{j}' + c\, \mathbf{k}'\mathbf{k}') \cdot (\mathbf{i}'\mathbf{i} + \mathbf{j}'\mathbf{j} + \mathbf{k}'\mathbf{k}), \quad (19)$$

or $$\Phi = \pm (\mathbf{i}'\mathbf{i} + \mathbf{j}'\mathbf{j} + \mathbf{k}'\mathbf{k}) \cdot (a\, \mathbf{i}\mathbf{i} + b\, \mathbf{j}\mathbf{j} + c\, \mathbf{k}\mathbf{k}).$$

The factor $$\mathbf{i}'\mathbf{i} + \mathbf{j}'\mathbf{j} + \mathbf{k}'\mathbf{k}$$

which is the same in either method of factoring is a versor. It turns the vectors $\mathbf{i}, \mathbf{j}, \mathbf{k}$ into the vectors $\mathbf{i}', \mathbf{j}', \mathbf{k}'$. This versor may be represented by its vector semi-tangent of version as

$$\mathbf{i}'\mathbf{i} + \mathbf{j}'\mathbf{j} + \mathbf{k}'\mathbf{k} = \frac{\mathbf{i} \times \mathbf{i}' + \mathbf{j} \times \mathbf{j}' + \mathbf{k} \times \mathbf{k}'}{1 + \mathbf{i} \cdot \mathbf{i}' + \mathbf{j} \cdot \mathbf{j}' + \mathbf{k} \cdot \mathbf{k}'}.$$

The other factor

$$a\, \mathbf{i}'\mathbf{i}' + b\, \mathbf{j}'\mathbf{j}' + c\, \mathbf{k}'\mathbf{k}',$$

or $$a\, \mathbf{i}\mathbf{i} + b\, \mathbf{j}\mathbf{j} + c\, \mathbf{k}\mathbf{k}$$

is a right tensor and represents a pure strain. In the first case the strain has the lines $\mathbf{i}', \mathbf{j}', \mathbf{k}'$ for principal axes: in the second, $\mathbf{i}, \mathbf{j}, \mathbf{k}$. In both cases the ratios of elongation are the same, — a to 1, b to 1, c to 1. If the negative sign occurs before the product the version and pure strain must have associated with them a reversal of directions of all vectors in space — that is, a perversion. Hence

Theorem: Any dyadic is reducible to the product of a versor and a right tensor taken in either order and a positive or negative sign. Hence the most general transformation representable by a dyadic consists of the product of a rotation or version about a definite axis through a definite angle accompanied by a pure strain either with or without perversion. The rotation and strain may be performed in either order. In the two cases the rotation and the ratios of elongation of the strain are the same; but the principal axes of the strain differ according as it is performed before or after the rotation, either system of axes being derivable from the other by the application of the versor as a prefactor or postfactor respectively.

If a dyadic Φ be given the product of Φ and its conjugate is a right tensor the ratios of elongation of which are the squares of the ratios of elongation of Φ and the axes of which are respectively the antecedents or consequents of Φ according as Φ_C follows or precedes Φ in the product.

$$\Phi = \pm (a\, \mathbf{i}'\mathbf{i} + b\, \mathbf{j}'\mathbf{j} + c\, \mathbf{k}'\mathbf{k}),$$

$$\Phi_C = \pm (a\, \mathbf{i}\mathbf{i}' + b\, \mathbf{j}\mathbf{j}' + c\, \mathbf{k}\mathbf{k}'),$$

$$\Phi \cdot \Phi_C = a^2\, \mathbf{i}'\mathbf{i}' + b^2\, \mathbf{j}'\mathbf{j}' + c^2\, \mathbf{k}'\mathbf{k}', \qquad (20)$$

$$\Phi_C \cdot \Phi = a^2\, \mathbf{i}\mathbf{i} + b^2\, \mathbf{j}\mathbf{j} + c^2\, \mathbf{k}\mathbf{k}.$$

The general problem of finding the principal ratios of elongation, the antecedents, and consequents of a dyadic in its normal form, therefore reduces to the simpler problem of finding the principal ratios of elongation and the principal axes of a pure strain.

131.] The natural and immediate generalization of the right tensor

$$a\, \mathbf{i}\mathbf{i} + b\, \mathbf{j}\mathbf{j} + c\, \mathbf{k}\mathbf{k},$$

is the dyadic
$$\Phi = a\, \mathbf{a}\mathbf{a}' + b\, \mathbf{b}\mathbf{b}' + c\, \mathbf{c}\mathbf{c}' \qquad (21)$$

where a, b, c are positive or negative scalars and where $\mathbf{a}$, $\mathbf{b}$, $\mathbf{c}$ and $\mathbf{a}'$, $\mathbf{b}'$, $\mathbf{c}'$ are two reciprocal systems of vectors. Necessarily $\mathbf{a}$, $\mathbf{b}$, $\mathbf{c}$ and $\mathbf{a}'$, $\mathbf{b}'$, $\mathbf{c}'$ are each three non-coplanar.

Definition: A dyadic that may be reduced to the form

$$\Phi = a\, \mathbf{a}\mathbf{a}' + b\, \mathbf{b}\mathbf{b}' + c\, \mathbf{c}\mathbf{c}' \qquad (21)$$

is called a *tonic*.

The effect of a tonic is to leave unchanged three non-coplanar directions $\mathbf{a}$, $\mathbf{b}$, $\mathbf{c}$ in space. If a vector be resolved into its components parallel to $\mathbf{a}$, $\mathbf{b}$, $\mathbf{c}$ respectively these components are stretched in the ratios a to 1, b to 1, c to 1. If one or more of the constants a, b, c are negative the components parallel to the corresponding vector **a**, **b**, **c** are reversed in direction as well as changed in magnitude. The tonic may be factored into three factors of which each stretches the components parallel to one of the vectors **a**, **b**, **c** but leaves unchanged the components parallel to the other two.

$$\Phi = (a\,\mathbf{a}\,\mathbf{a}' + \mathbf{b}\,\mathbf{b}' + \mathbf{c}\,\mathbf{c}') \cdot (\mathbf{a}\,\mathbf{a}' + b\,\mathbf{b}\,\mathbf{b}' + \mathbf{c}\,\mathbf{c}')\,(\mathbf{a}\,\mathbf{a}' + \mathbf{b}\,\mathbf{b}' + c\,\mathbf{c}\,\mathbf{c}').$$

The value of a tonic Φ is not altered if in place of **a**, **b**, **c** any three vectors respectively collinear with them be substituted, provided of course that the corresponding changes which are necessary be made in the reciprocal system **a**′, **b**′, **c**′. But with the exception of this change, a dyadic which is expressible in the form of a tonic is so expressible in only one way if the constants a, b, c are different. If two of the constants say b and c are equal, any two vectors coplanar with the corresponding vectors **b** and **c** may be substituted in place of **b** and **c**. If all the constants are equal the tonic reduces to a constant multiple of the idemfactor. Any three non-coplanar vectors may be taken for **a**, **b**, **c**.

The product of two tonics of which the axes **a**, **b**, **c** are the same is commutative and is a tonic with these axes and with scalar coefficients equal respectively to the products of the corresponding coefficients of the two dyadics.

$$\Phi = a_1\,\mathbf{a}\,\mathbf{a}' + b_1\,\mathbf{b}\,\mathbf{b}' + c_1\,\mathbf{c}\,\mathbf{c}'$$

$$\Psi = a_2\,\mathbf{a}\,\mathbf{a}' + b_2\,\mathbf{b}\,\mathbf{b}' + c_2\,\mathbf{c}\,\mathbf{c}'$$

$$\Phi \cdot \Psi = \Psi \cdot \Phi = a_1\,a_2\,\mathbf{a}\,\mathbf{a}' + b_1\,b_2\,\mathbf{b}\,\mathbf{b}' + c_1\,c_2\,\mathbf{c}\,\mathbf{c}'. \qquad (22)$$

The generalization of the cyclic dyadic

$$\mathbf{a}\,\mathbf{a}' + \cos q\,(\mathbf{b}\,\mathbf{b}' + \mathbf{c}\,\mathbf{c}') + \sin q\,(\mathbf{c}\,\mathbf{b}' - \mathbf{b}\,\mathbf{c}')$$

is

$$\Phi = a\,\mathbf{a}\,\mathbf{a}' + b\,(\mathbf{b}\,\mathbf{b}' + \mathbf{c}\,\mathbf{c}') + c\,(\mathbf{c}\,\mathbf{b}' - \mathbf{b}\,\mathbf{c}'), \qquad (23)$$

where $\mathbf{a}, \mathbf{b}, \mathbf{c}$ are three non-coplanar vectors of which $\mathbf{a}', \mathbf{b}', \mathbf{c}'$ is the reciprocal system and where the quantities a, b, c, are positive or negative scalars. This dyadic may be changed into a more convenient form by determining the positive scalar p and the positive or negative scalar q (which may always be chosen between the limits $\pm \pi$) so that

$$b = p \cos q$$

and
$$c = p \sin q. \tag{24}$$

That is,
$$p = + \sqrt{b^2 + c^2}$$

and
$$\tan \tfrac{1}{2} q = \frac{p - b}{p + b}. \tag{24$'$}$$

Then

$$\Phi = a\, \mathbf{a}\, \mathbf{a}' + p \cos q\, (\mathbf{b}\, \mathbf{b}' + \mathbf{c}\, \mathbf{c}') + p \sin q\, (\mathbf{c}\, \mathbf{b}' - \mathbf{b}\, \mathbf{c}'). \tag{25}$$

This may be factored into the product of three dyadics

$$\Phi = (a\, \mathbf{a}\, \mathbf{a}' + \mathbf{b}\, \mathbf{b}' + \mathbf{c}\, \mathbf{c}') \cdot (\mathbf{a}\, \mathbf{a}' + p\, \mathbf{b}\, \mathbf{b}' + p\, \mathbf{c}\, \mathbf{c}') \cdot$$
$$\{\mathbf{a}\, \mathbf{a}' + \cos q\, (\mathbf{b}\, \mathbf{b}' + \mathbf{c}\, \mathbf{c}') + \sin q\, (\mathbf{c}\, \mathbf{b}' - \mathbf{b}\, \mathbf{c}')\}.$$

The order of these factors is immaterial. The first is a tonic which leaves unchanged vectors parallel to $\mathbf{b}$ and $\mathbf{c}$ but stretches those parallel to $\mathbf{a}$ in the ratio of a to 1. If a is negative the stretching must be accompanied by reversal in direction. The second factor is also a tonic. It leaves unchanged vectors parallel to $\mathbf{a}$ but stretches all vectors in the plane of $\mathbf{b}$ and $\mathbf{c}$ in the ratio p to 1. The third is a cyclic factor. Vectors parallel to $\mathbf{a}$ remain unchanged; but radii vectors in the ellipse of which $\mathbf{b}$ and $\mathbf{c}$ are conjugate semi-diameters are rotated through a sector such that the area of the sector is to the area of the whole ellipse as q to 2π. Other vectors in the plane of $\mathbf{b}$ and $\mathbf{c}$ may be regarded as scalar multiples of the radii vectors of the ellipse.

Definition: A dyadic which is reducible to the form

$$\Phi = a\,\mathbf{a}\,\mathbf{a}' + p \cos q\,(\mathbf{b}\,\mathbf{b}' + \mathbf{c}\,\mathbf{c}') + p \sin q\,(\mathbf{c}\,\mathbf{b}' - \mathbf{b}\,\mathbf{c}'), \qquad (25)$$

owing to the fact that it combines the properties of the cyclic dyadic and the tonic is called a *cyclotonic.*

The product of two cyclotonics which have the same three vectors, **a**, **b**, **c** as antecedents and the reciprocal system **a**′, **b**′, **c**′ for consequents is a third cyclotonic and is commutative.

$$\Phi = a_1\,\mathbf{a}\,\mathbf{a}' + p_1 \cos q_1\,(\mathbf{b}\,\mathbf{b}' + \mathbf{c}\,\mathbf{c}') + p_1 \sin q_1\,(\mathbf{c}\,\mathbf{b}' - \mathbf{b}\,\mathbf{c}')$$

$$\Psi = a_2\,\mathbf{a}\,\mathbf{a}' + p_2 \cos q_2\,(\mathbf{b}\,\mathbf{b}' + \mathbf{c}\,\mathbf{c}') + p_2 \sin q_2\,(\mathbf{c}\,\mathbf{b}' - \mathbf{b}\,\mathbf{c}')$$

$$\Phi \cdot \Psi = \Psi \cdot \Phi = a_1\,a_2\,\mathbf{a}\,\mathbf{a}' + p_1\,p_2 \cos (q_1 + q_2)\,(\mathbf{b}\,\mathbf{b}' + \mathbf{c}\,\mathbf{c}') + p_1\,p_2 \sin (q_1 + q_2)\,(\mathbf{c}\,\mathbf{b}' - \mathbf{b}\,\mathbf{c}'). \qquad (26)$$

Reduction of Dyadics to Canonical Forms

132.] *Theorem:* In general any dyadic Φ may be reduced either to a tonic or to a cyclotonic. The dyadics for which the reduction is impossible may be regarded as limiting cases which may be represented to any desired degree of approximation by tonics or cyclotonics.

From this theorem the importance of the tonic and cyclotonic which have been treated as natural generalizations of the right tensor and the cyclic dyadic may be seen. The proof of the theorem, including a discussion of all the special cases that may arise, is long and somewhat tedious. The method of proving the theorem in general however is patent. If three directions **a**, **b**, **c** may be found which are left unchanged by the application of Φ then Φ must be a tonic. If only one such direction can be found, there exists a plane in which the vectors suffer a change such as that due to the cyclotonic and the dyadic indeed proves to be such.

The question is to find the directions which are unchanged by the application of the dyadic Φ.

If the direction $\mathbf{a}$ is unchanged, then

$$\Phi \cdot \mathbf{a} = a\, \mathbf{a} \qquad (27)$$

or

$$(\Phi - a\, \mathrm{I}) \cdot \mathbf{a} = 0.$$

The dyadic $\Phi - a\,\mathrm{I}$ is therefore planar since it reduces vectors in the direction $\mathbf{a}$ to zero. In special cases, which are set aside for the present, the dyadic may be linear or zero. In any case if the dyadic

$$\Phi - a\, \mathrm{I}$$

reduces vectors collinear with $\mathbf{a}$ to zero it possesses at least one degree of nullity and the third or determinant of Φ vanishes.

$$(\Phi - a\, \mathrm{I})_3 = 0. \qquad (28)$$

Now (page 331) $\quad (\Phi + \Psi)_3 = \Phi_3 + \Phi_2 : \Psi + \Phi : \Psi_2 + \Psi_3.$

Hence $\quad (\Phi - a\, \mathrm{I})_3 = \Phi_3 - a\, \Phi_2 : \mathrm{I} + a^2\, \Phi : \mathrm{I}_2 - a^3\, \mathrm{I}_3$

$$\mathrm{I}_2 = \mathrm{I} \text{ and } \mathrm{I}_3 = 1.$$

But

$$\Phi : \mathrm{I} = \Phi_S$$

$$\Phi_2 : \mathrm{I} = \Phi_{2S}.$$

Hence the equation becomes

$$a^3 - a^2\, \Phi_S + a\, \Phi_{2S} - \Phi_3 = 0. \qquad (29)$$

The value of a which satisfies the condition that

$$\Phi \cdot \mathbf{a} = a\, \mathbf{a}$$

is a solution of a cubic equation. Let x replace a. The cubic equation becomes

$$x^3 - x^2\, \Phi_S + x\, \Phi_{2S} - \Phi_3 = 0. \qquad (29)$$

Any value of x which satisfies this equation will be such that

$$(\Phi - x\,\mathrm{I})_3 = 0. \tag{28'}$$

That is to say, the dyadic $\Phi - x\,\mathrm{I}$ is planar. A vector perpendicular to its consequents is reduced to zero. Hence Φ leaves such a direction unchanged. The further discussion of the reduction of a dyadic to the form of a tonic or a cyclotonic depends merely upon whether the cubic equation in x has one or three real roots.

133.] *Theorem:* If the cubic equation

$$x^3 - x^2\,\Phi_S + x\,\Phi_{2S} - \Phi_3 = 0 \tag{29'}$$

has three real roots the dyadic Φ may in general be reduced to a tonic.

For let $$x = a, \quad x = b, \quad x = c$$

be the three roots of the equation. The dyadics

$$\Phi - a\,\mathrm{I}, \quad \Phi - b\,\mathrm{I}, \quad \Phi - c\,\mathrm{I}$$

are in general planar. Let **a**, **b**, **c** be respectively three vectors drawn perpendicular to the planes of the consequents of these dyadics.

$$\begin{aligned} (\Phi - a\,\mathrm{I}) \cdot \mathbf{a} &= 0, \\ (\Phi - b\,\mathrm{I}) \cdot \mathbf{b} &= 0, \\ (\Phi - c\,\mathrm{I}) \cdot \mathbf{c} &= 0. \end{aligned} \tag{30}$$

Then $$\begin{aligned} \Phi \cdot \mathbf{a} &= a\,\mathbf{a}, \\ \Phi \cdot \mathbf{b} &= b\,\mathbf{b}, \\ \Phi \cdot \mathbf{c} &= c\,\mathbf{c}. \end{aligned} \tag{30'}$$

If the roots a, b, c are distinct the vectors **a**, **b**, **c** are non-coplanar. For suppose

$$\mathbf{c} = m\,\mathbf{a} + n\,\mathbf{b}$$

$$(\Phi - c\,\mathrm{I}) \cdot (m\,\mathbf{a} + n\,\mathbf{b}) = 0,$$

$$m\,\Phi \cdot \mathbf{a} - m\,c\,\mathbf{a} + n\,\Phi \cdot \mathbf{b} - n\,c\,\mathbf{b} = 0.$$

But $\Phi \cdot \mathbf{a} = a\,\mathbf{a}, \quad \Phi \cdot \mathbf{b} = b\,\mathbf{b}.$

Hence $m\,(a - c)\,\mathbf{a} + n\,(b - c)\,\mathbf{b} = 0,$

and $m\,(\mathbf{a} - \mathbf{c}) = 0, \quad n\,(\mathbf{b} - \mathbf{c}) = 0.$

Hence $m = 0$ or $a = c, \quad n = 0$ or $b = c.$

Consequently if the vectors **a**, **b**, **c** are coplanar, the roots are not distinct; and therefore if the roots are distinct, the vectors **a**, **b**, **c** are necessarily non-coplanar. In case the roots are not distinct it is still always possible to choose three non-coplanar vectors **a**, **b**, **c** in such a manner that the equations (30) hold. This being so, there exists a system **a**′, **b**′, **c**′ reciprocal to **a**, **b**, **c** and the dyadic which carries **a**, **b**, **c** into $a\,\mathbf{a}$, $b\,\mathbf{b}$, $c\,\mathbf{c}$ is the tonic

$$\Phi = a\,\mathbf{a}\,\mathbf{a}' + b\,\mathbf{b}\,\mathbf{b}' + c\,\mathbf{c}\,\mathbf{c}.$$

Theorem: If the cubic equation

$$x^3 - x^2\,\Phi_S + x\,\Phi_{2S} - \Phi_3 = 0 \qquad (29)'$$

has one real root the dyadic Φ may in general be reduced to a cyclotonic.

The cubic equation has one real root. This must be positive or negative according as Φ_3 is positive or negative. Let the root be a. Determine a perpendicular to the plane of the consequents of $\Phi - a\,\mathrm{I}$.

$$(\Phi - a\,\mathrm{I}) \cdot \mathbf{a} = 0.$$

Determine **a**′ also so that

$$\mathbf{a}' \cdot (\Phi - a\,\mathrm{I}) = 0$$

and let the lengths of **a** and **a**′ be so adjusted that $\mathbf{a}' \cdot \mathbf{a} = 1$. This cannot be accomplished in the special case in which **a** and $\mathbf{a}'$ are mutually perpendicular. Let $\mathbf{b}$ be any vector in the plane perpendicular to $\mathbf{a}'$.

$$\mathbf{a}' \cdot (\Phi - a\,\mathrm{I}) \cdot \mathbf{b} = 0.$$

Hence $(\Phi - a\mathrm{I}) \cdot \mathbf{b}$ is perpendicular to $\mathbf{a}'$. Hence $\Phi \cdot \mathbf{b}$ is perpendicular to $\mathbf{a}'$. In a similar manner $\Phi^2 \cdot \mathbf{b}$, $\Phi^3 \cdot \mathbf{b}$, and $\Phi^{-1} \cdot \mathbf{b}$, $\Phi^{-2} \cdot \mathbf{b}$, etc., will all be perpendicular to $\mathbf{a}'$ and lie in one plane. The vectors $\Phi \cdot \mathbf{b}$ and $\mathbf{b}$ cannot be parallel or Φ would have the direction $\mathbf{b}$ as well as $\mathbf{a}$ unchanged and thus the cubic would have more than one real root.

The dyadic Φ changes $\mathbf{a}$, $\Phi \cdot \mathbf{b}$, $\mathbf{b}$ into $\Phi \cdot \mathbf{a}$, $\Phi^2 \cdot \mathbf{b}$, $\Phi \cdot \mathbf{b}$ respectively. The volume of the parallelopiped

$$[\Phi \cdot \mathbf{a} \quad \Phi^2 \cdot \mathbf{b} \quad \Phi \cdot \mathbf{b}] = \Phi_3 \, [\mathbf{a} \quad \Phi \cdot \mathbf{b} \quad \mathbf{b}]. \tag{31}$$

But
$$\Phi \cdot \mathbf{a} = a\,\mathbf{a}.$$

Hence
$$a\,\mathbf{a} \cdot (\Phi^2 \cdot \mathbf{b}) \times (\Phi \cdot \mathbf{b}) = \Phi_3 \, \mathbf{a} \cdot (\Phi \cdot \mathbf{b}) \times \mathbf{b}. \tag{31)'$$

The vectors $\Phi^2 \cdot \mathbf{b}$, $\Phi \cdot \mathbf{b}$, $\mathbf{b}$ all lie in the same plane. Their vector products are parallel to $\mathbf{a}'$ and to each other. Hence

$$a\,(\Phi^2 \cdot \mathbf{b}) \times (\Phi \cdot \mathbf{b}) = \Phi_3 \quad \Phi \cdot \mathbf{b} \times \mathbf{b}. \tag{31)''$$

Inasmuch as a and Φ_3 have the same sign, let

$$p^2 = a^{-1} \, \Phi_3. \tag{32}$$

Let also
$$\mathbf{b}_3 = p^{-1} \, \Phi \cdot \mathbf{b} \qquad \mathbf{b}_2 = p^{-2} \, \Phi^2 \cdot \mathbf{b}, \quad \text{etc.} \tag{33}$$

and
$$\mathbf{b}_{-1} = p^2 \, \Phi^{-1} \cdot \mathbf{b} \qquad \mathbf{b}_{-2} = p^2 \, \Phi^{-2} \cdot \mathbf{b}, \quad \text{etc.}$$

$$\mathbf{b}_2 \times \mathbf{b}_1 = \mathbf{b}_1 \times \mathbf{b}_2,$$

or
$$(\mathbf{b}_2 + \mathbf{b}) \times \mathbf{b}_1 = 0.$$

The vectors $\mathbf{b}_2 + \mathbf{b}$ and $\mathbf{b}_1$ are parallel. Let

$$\mathbf{b}_2 + \mathbf{b} = 2\,n\,\mathbf{b}_1. \tag{34}$$

Then
$$\left.\begin{array}{lll} \mathbf{b}_3 + \mathbf{b}_1 = 2\,n\,\mathbf{b}_2 & \mathbf{b}_1 + \mathbf{b}_2 = 2\,n\,\mathbf{b}_3 & \text{etc.,} \\ \mathbf{b}_1 + \mathbf{b}_{-1} = 2\,n\,\mathbf{b} & \mathbf{b}_{-1} + \mathbf{b}_{-2} = 2\,n\,\mathbf{b}_{-1} & \text{etc.} \end{array}\right. \tag{35}$$

Lay off from a common origin the vectors

$$\mathbf{b},\ \mathbf{b}_1,\ \mathbf{b}_2,\ \text{etc.,} \qquad \mathbf{b}_{-1},\ \mathbf{b}_{-2},\ \text{etc.}$$

Since Φ is not a tonic, that is, since there is no direction in the plane perpendicular to $\mathbf{a}'$ which is left unchanged by Φ these vectors $\mathbf{b}_m$ pass round and round the origin as m takes on all positive and negative values. The value of n must therefore lie between plus one and minus one. Let

$$n = \cos q. \qquad (36)$$

Then $$\mathbf{b}_{-1} + \mathbf{b}_1 = 2 \cos q\ \mathbf{b}.$$

Determine $\mathbf{c}$ from the equation

$$\mathbf{b}_1 = \cos q\ \mathbf{b} + \sin q\ \mathbf{c}.$$

Then $$\mathbf{b}_{-1} = \cos q\ \mathbf{b} - \sin q\ \mathbf{c}.$$

Let $\mathbf{a}'$, $\mathbf{b}'$, $\mathbf{c}'$ be the reciprocal system of $\mathbf{a}$, $\mathbf{b}$, $\mathbf{c}$. This is possible since $\mathbf{a}'$ was so determined that $\mathbf{a}' \cdot \mathbf{a} = 1$ and since $\mathbf{a}$, $\mathbf{b}$, $\mathbf{c}$ are non-coplanar. Let

$$\Psi = \cos q\ (\mathbf{b}\,\mathbf{b}' + \mathbf{c}\,\mathbf{c}') + \sin q\ (\mathbf{c}\,\mathbf{b}' - \mathbf{b}\,\mathbf{c}').$$

Then $$\Psi \cdot \mathbf{a} = 0, \qquad \Psi \cdot \mathbf{b} = \mathbf{b}_1, \qquad \Psi \cdot \mathbf{b}_{-1} = \mathbf{b}.$$

Hence $$(a\ \mathbf{a}\,\mathbf{a}' + p\ \Psi) \cdot \mathbf{a} = a\ \mathbf{a} = \Phi \cdot \mathbf{a},$$

$$(a\ \mathbf{a}\,\mathbf{a}' + p\ \Psi) \cdot \mathbf{b} = p\ \mathbf{b}_2 = \Phi \cdot \mathbf{b},$$

$$(a\ \mathbf{a}\,\mathbf{a}' + p\ \Psi) \cdot \mathbf{b}_{-1} = p\ \mathbf{b} = \Phi \cdot \mathbf{b}_{-1}.$$

The dyadic $a\ \mathbf{a}\,\mathbf{a}' + p\ \Psi$ changes the vectors $\mathbf{a}$, $\mathbf{b}$ and $\mathbf{b}_{-1}$ into the vectors $\Phi \cdot \mathbf{a}$, $\Phi \cdot \mathbf{b}$, and $\Phi \cdot \mathbf{b}_{-2}$ respectively. Hence

$$\Phi = (a\ \mathbf{a}\,\mathbf{a}' + p\ \Psi) = a\ \mathbf{a}\,\mathbf{a}' + p \cos q\ (\mathbf{b}\,\mathbf{b}' + \mathbf{c}\,\mathbf{c}')$$
$$+ p \sin q\ (\mathbf{c}\,\mathbf{b}' - \mathbf{b}\,\mathbf{c}').$$

The dyadic Φ in case the cubic equation has only one real root is reducible except in special cases to a cyclotonic. The theorem that a dyadic in general is reducible to a tonic or cyclotonic has therefore been demonstrated.

134.] There remain two cases[1] in which the reduction is impossible, as can be seen by looking over the proof. In the first place if the constant n used in the reduction to cyclotonic form be ± 1 the reduction falls through. In the second place if the plane of the antecedents of

$$\Phi - a\,\mathrm{I}$$

and the plane of the consequents are perpendicular the vectors $\mathbf{a}$ and $\mathbf{a}'$ used in the reduction to cyclotonic form are perpendicular and it is impossible to determine $\mathbf{a}'$ such that $\mathbf{a} \cdot \mathbf{a}'$ shall be unity. The reduction falls through.

If $$n = \pm 1, \qquad \mathbf{b}_{-1} + \mathbf{b}_1 = \pm 2\,\mathbf{b}.$$

Let $$\mathbf{b}_{-1} + \mathbf{b}_1 = 2\,\mathbf{b}.$$

Choose $$\mathbf{c} = \mathbf{b}_1 - \mathbf{b} = \mathbf{b} - \mathbf{b}_{-1}.$$

Consider the dyadic $\Psi = a\ \mathbf{a}\mathbf{a}' + p\ (\mathbf{b}\mathbf{b}' + \mathbf{c}\mathbf{c}') + p\ \mathbf{c}\,\mathbf{b}'$

$$\Psi \cdot \mathbf{a} = a\,\mathbf{a} = \Phi \cdot \mathbf{a},$$

$$\Psi \cdot \mathbf{b} = p\,\mathbf{b} + p\,\mathbf{b}_1 = \Phi \cdot \mathbf{b},$$

$$\Psi \cdot \mathbf{c} = p\,\mathbf{c} = p\,\mathbf{b}_1 - p\,\mathbf{b} = \Phi \cdot \mathbf{c}.$$

Hence $$\Phi = a\,\mathbf{a}\mathbf{a}' + p\,(\mathbf{b}\mathbf{b}' + \mathbf{c}\mathbf{c}') + p\,\mathbf{c}\,\mathbf{b}_1. \qquad (37)$$

The transformation due to this dyadic may be seen best by factoring it into three factors which are independent of the order or arrangement

$$\Phi = (a\,\mathbf{a}\mathbf{a}' + \mathbf{b}\mathbf{b}' + \mathbf{c}\mathbf{c}') \cdot \{\mathbf{a}\mathbf{a}' + p\,(\mathbf{b}\mathbf{b}' + \mathbf{c}\mathbf{c}')\}$$
$$\cdot\,(\mathbf{a}\mathbf{a}' + \mathbf{b}\mathbf{b}' + \mathbf{c}\mathbf{c}' + \mathbf{c}\,\mathbf{b}').$$

[1] In these cases it will be seen that the cubic equation has three real roots. In one case two of them are equal and in the other case three of them. Thus these dyadics may be regarded as limiting cases lying between the cyclotonic in which two of the roots are imaginary and the tonic in which all the roots are real and distinct. The limit may be regarded as taking place either by the pure imaginary part of the two imaginary roots of the cyclotonic becoming zero or by two of the roots of the tonic approaching each other.

The first factor represents an elongation in the direction **a** in a ratio a to 1. The plane of **b** and **c** is undisturbed. The second factor represents a stretching of the plane of **b** and **c** in the ratio p to 1. The last factor takes the form

$$\mathbf{I} + \mathbf{c}\,\mathbf{b}'.$$

$$(\mathbf{I} + \mathbf{c}\,\mathbf{b}') \cdot x\,\mathbf{a} = x\,\mathbf{a},$$

$$(\mathbf{I} + \mathbf{c}\,\mathbf{b}') \cdot x\,\mathbf{b} = x\,\mathbf{b} + x\,\mathbf{c},$$

$$(\mathbf{I} + \mathbf{c}\,\mathbf{b}') \cdot x\,\mathbf{c} = x\,\mathbf{c}.$$

A dyadic of the form $\mathbf{I} + \mathbf{c}\,\mathbf{b}'$ leaves vectors parallel to **a** and **c** unaltered. A vector $x\,\mathbf{b}$ parallel to **b** is increased by the vector **c** multiplied by the ratio of the vector $x\,\mathbf{b}$ to **b**. In other words the transformation of points in space is such that the plane of **a** and **c** remains fixed point for point but the points in planes parallel to that plane are shifted in the direction **c** by an amount proportional to the distance of the plane in which they lie from the plane of **a** and **c**.

Definition: A dyadic reducible to the form

$$\mathbf{I} + \mathbf{c}\,\mathbf{b}'$$

is called a *shearing* dyadic or *shearer* and the geometrical transformation which it causes is called a *shear*. The more general dyadic

$$\Phi = a\,\mathbf{a}\,\mathbf{a}' + p\,(\mathbf{b}\,\mathbf{b}' + \mathbf{c}\,\mathbf{c}') + \mathbf{c}\,\mathbf{b}' \qquad (37)$$

will also be called a shearing dyadic or *shearer*. The transformation to which it gives rise is a shear combined with elongations in the direction of **a** and is in the plane of **b** and **c**.

If $\mathbf{n} = -1$ instead of $\mathbf{n} = +1$, the result is much the same. The dyadic then becomes

$$\Phi = a\,\mathbf{a}\,\mathbf{a}' - p\,(\mathbf{b}\,\mathbf{b}' + \mathbf{c}\,\mathbf{c}') - \mathbf{c}\,\mathbf{b}' \qquad (37)'$$

$$\Phi = (a\,\mathbf{a}\,\mathbf{a}' + \mathbf{b}\,\mathbf{b}' + \mathbf{c}\,\mathbf{c}') \cdot \{\mathbf{a}\,\mathbf{a}' - p\,(\mathbf{b}\,\mathbf{b}' + \mathbf{c}\,\mathbf{c}')\} \cdot (\mathbf{I} + \mathbf{c}\,\mathbf{b}').$$

The factors are the same except the second which now represents a stretching of the plane of **b** and **c** combined with a reversal of all the vectors in that plane. The shearing dyadic Φ then represents an elongation in the direction **a**, an elongation combined with a reversal of direction in the plane of **b** and **c**, and a shear.

Suppose that the plane of the antecedents and the plane of the consequents of the dyadic $\Phi - a\,\mathrm{I}$ are perpendicular. Let these planes be taken respectively as the plane of **j** and **k** and the plane of **i** and **j k**. The dyadic then takes the form

$$\Phi - a\,\mathrm{I} = A\,\mathbf{j}\,\mathbf{i} + B\,\mathbf{j}\,\mathbf{j} + C\,\mathbf{k}\,\mathbf{k} + D\,\mathbf{k}\,\mathbf{j}.$$

The coefficient B must vanish. For otherwise the dyadic

$$\Phi - a\mathrm{I} - B\mathrm{I} = (-B\,\mathbf{i} + A\,\mathbf{j} + C\,\mathbf{k})\,\mathbf{i} + \mathbf{k}\,(D\,\mathbf{i} - B\,\mathbf{k})$$

is planar and the scalar $a + B$ is a root of the cubic equation. With this root the reduction to the form of a tonic may be carried on as before. Nothing new arises. But if B vanishes a new case occurs. Let

$$\Psi = \Phi - a\,\mathrm{I} = A\,\mathbf{j}\,\mathbf{i} + C\,\mathbf{k}\,\mathbf{k} + D\,\mathbf{k}\,\mathbf{j}.$$

This may be reduced as follows to the form

$$\mathbf{a}\,\mathbf{b}' + \mathbf{b}\,\mathbf{c}'$$

where $\quad \mathbf{a} \cdot \mathbf{b}' = \mathbf{a} \cdot \mathbf{c}' = \mathbf{b} \cdot \mathbf{c}' = 0 \quad$ and $\mathbf{b} \cdot \mathbf{b}' = 1.$

Square Ψ $\qquad \Psi^2 = A\,D\ \mathbf{k}\,\mathbf{i} = \mathbf{a}\,\mathbf{c}'.$

Hence **a** must be chosen parallel to **k**; and **c**′, parallel to **i**. The dyadic Ψ may then be transformed into

$$\Psi = A\,D\,\mathbf{k}\left(\frac{C\,\mathbf{i} + D\,\mathbf{j}}{A\,D}\right) + A\,\mathbf{j}\ \mathbf{i}.$$

Then $\qquad = A\,D\,\mathbf{k}, \qquad \mathbf{b}' = \dfrac{C\,\mathbf{i} + D\,\mathbf{i}}{A\,D}$

$$\mathbf{b} = A\,\mathbf{j} \qquad \mathbf{c}' = \mathbf{i}.$$

With this choice of $\mathbf{a}$, $\mathbf{b}$, $\mathbf{b}'$, $\mathbf{c}'$ the dyadic Ψ reduces to the desired form $\mathbf{a}\,\mathbf{b}' + \mathbf{b}\,\mathbf{c}'$ and hence the dyadic Φ is reduced to

$$\Phi = a\,\mathrm{I} + \mathbf{a}\,\mathbf{b}' + \mathbf{b}\mathbf{c}' \tag{38}$$

or
$$\Phi = a\,\mathbf{a}\mathbf{a}' + a\,\mathbf{b}\,\mathbf{b}' + a\,\mathbf{c}\,\mathbf{c}' + \mathbf{a}\,\mathbf{b}' + \mathbf{b}\,\mathbf{c}'.$$

This may be factored into the product of two dyadics the order of which is immaterial.

$$\Phi = a\,\mathrm{I} \cdot (\mathrm{I} + \mathbf{a}\,\mathbf{b}' + \mathbf{b}\,\mathbf{c}').$$

The first factor $a\mathrm{I}$ represents a stretching of space in all directions in the ratio a to 1. The second factor

$$\Omega = \mathrm{I} + \mathbf{a}\,\mathbf{b}' + \mathbf{b}\,\mathbf{c}'$$

represents what may be called *a complex shear*. For

$$\mathbf{r}' = \mathrm{I} \cdot \mathbf{r} + \mathbf{a}\,\mathbf{b}' \cdot \mathbf{r} + \mathbf{b}\,\mathbf{c}' \cdot \mathbf{r} = \mathbf{r} + \mathbf{a}\,\mathbf{b}' \cdot \mathbf{r} + \mathbf{b}\,\mathbf{c}' \cdot \mathbf{r}.$$

If $\mathbf{r}$ is parallel to $\mathbf{a}$ it is left unaltered by the dyadic Ω. If $\mathbf{r}$ is parallel to $\mathbf{b}$ it is changed by the addition of a term which is in direction equal to $\mathbf{a}$ and in magnitude proportional to the magnitude of the vector $\mathbf{r}$. In like manner if $\mathbf{r}$ is parallel to $\mathbf{c}$ it is changed by the addition of a term which in direction is equal to $\mathbf{b}$ and which in magnitude is proportional to the magnitude of the vector $\mathbf{r}$.

$$\Omega \cdot x\,\mathbf{b} = (\mathrm{I} + \mathbf{a}\,\mathbf{b}' + \mathbf{b}\,\mathbf{c}') \cdot x\,\mathbf{b} = x\,\mathbf{b} + x\,\mathbf{a}$$

$$\Omega \cdot x\mathbf{c} = (\mathrm{I} + \mathbf{a}\,\mathbf{b}' + \mathbf{b}\,\mathbf{c}') \cdot x\,\mathbf{c} = x\mathbf{c} + x\,\mathbf{b}.$$

Definition: A dyadic which may be reduced to the form

$$\Phi = a\,\mathrm{I} + \mathbf{a}\,\mathbf{b}' + \mathbf{b}\,\mathbf{c}' \tag{38}$$

is called a *complex shearer*.

The complex shearer as well as the simple shearer mentioned before are limiting cases of the cyclotonic and tonic dyadics.

135.] A more systematic treatment of the various kinds of dyadics which may arise may be given by means of the Hamilton-Cayley equation

$$\Phi^3 - \Phi_S\, \Phi^2 + \Phi_{2S}\, \Phi - \Phi_3\, \mathrm{I} = 0 \qquad (39)$$

and the cubic equation in x

$$x^3 - \Phi_S\, x^2 + \Phi_{2S}\, x - \Phi_3 = 0. \qquad (29)'$$

If a, b, c are the roots of this cubic the Hamilton-Cayley equation may be written as

$$(\Phi - a\,\mathrm{I}) \cdot (\Phi - b\,\mathrm{I}) \cdot (\Phi - c\,\mathrm{I}) = 0. \qquad (40)$$

If, however, the cubic has only one root the Hamilton-Cayley equation takes the form

$$(\Phi - a\,\mathrm{I}) \cdot (\Phi^2 - 2\,p \cos q\ \Phi + p^2\,\mathrm{I}) = 0. \qquad (41)$$

In general the Hamilton-Cayley equation which is an equation of the third degree in Φ is the equation of lowest degree which is satisfied by Φ. In general therefore one of the above equations and the corresponding reductions to the tonic or cyclotonic form hold. In special cases, however, the dyadic Φ may satisfy an equation of lower degree. That equation of lowest degree which may be satisfied by a dyadic is called its *characteristic* equation. The following possibilities occur.

I. $(\Phi - a\,\mathrm{I}) \cdot (\Phi - b\,\mathrm{I}) \cdot (\Phi - c\,\mathrm{I}) = 0.$

II. $(\Phi - a\,\mathrm{I}) \cdot (\Phi^2 - 2\,p \cos q\ \Phi + p^2\,\mathrm{I}) = 0.$

III. $(\Phi - a\,\mathrm{I}) \cdot (\Phi - b\,\mathrm{I})^2 = 0.$

IV. $(\Phi - a\,\mathrm{I}) \cdot (\Phi - b\,\mathrm{I}) = 0.$

V. $(\Phi - a\,\mathrm{I})^3 = 0.$

VI. $(\Phi - a\,\mathrm{I})^2 = 0.$

VII. $(\Phi - a\,\mathrm{I}) = 0.$

In the first case the dyadic is a tonic and may be reduced to the form

$$\Phi = a\,\mathbf{a}\,\mathbf{a}' + b\,\mathbf{b}\,\mathbf{b}' + c\,\mathbf{c}\,\mathbf{c}'.$$

In the second case the dyadic is a cyclotonic and may be reduced to the form

$$\Phi = a\,\mathbf{a}\,\mathbf{a}' + p\cos q\,(\mathbf{b}\,\mathbf{b}' + \mathbf{c}\mathbf{c}') + p\sin q\,(\mathbf{c}\,\mathbf{b}' - \mathbf{b}\,\mathbf{c}').$$

In the third case the dyadic is a simple shearer and may be reduced to the form

$$\Phi = a\,\mathbf{a}\,\mathbf{a}' + b\,(\mathbf{b}\,\mathbf{b}' + \mathbf{c}\,\mathbf{c}') + \mathbf{c}\,\mathbf{b}'.$$

In the fourth case the dyadic is again a tonic. Two of the ratios of elongation are the same. The following reduction may be accomplished in an infinite number of ways.

$$\Phi = a\,\mathbf{a}\,\mathbf{a}' + b\,(\mathbf{b}\,\mathbf{b}' + \mathbf{c}\mathbf{c}').$$

In the fifth case the dyadic is a complex shearer and may be so expressed that

$$\Phi = a\,\mathrm{I} + \mathbf{a}\,\mathbf{b}' + \mathbf{b}\,\mathbf{c}'.$$

In the sixth case the dyadic is again a simple shearer which may be reduced to the form

$$\Phi = a\,\mathrm{I} + \mathbf{c}\,\mathbf{b}' = a\,(\mathbf{a}\,\mathbf{a}' + \mathbf{b}\,\mathbf{b}' + \mathbf{c}\,\mathbf{c}') + \mathbf{c}\,\mathbf{b}'.$$

In the seventh case the dyadic is again a tonic which may be reduced in a doubly infinite number of ways to the form

$$\Phi = a\,\mathrm{I} = a\,(\mathbf{a}\,\mathbf{a}' + \mathbf{b}\,\mathbf{b}' + \mathbf{c}\,\mathbf{c}').$$

These seven are the only essentially different forms which a dyadic may take. There are then only seven really different kinds of dyadics — three tonics in which the ratios of elongation are all different, two alike, or all equal, and the cyclotonic together with three limiting cases, the two simple and the one complex shearer.

Summary of Chapter VI

The transformation due to a dyadic is a linear homogeneous strain. The dyadic itself gives the transformation of the points in space. The second of the dyadic gives the transformation of plane areas. The third of the dyadic gives the ratio in which volumes are changed.

$$\mathbf{r}' = \Phi \cdot \mathbf{r}, \qquad \mathbf{s}' = \Phi_2 \cdot \mathbf{s}, \qquad v' = \Phi_3\, v.$$

The necessary and sufficient condition that a dyadic represent a rotation about a definite axis is that it be reducible to the form

$$\Phi = \mathbf{i}'\mathbf{i} + \mathbf{j}'\mathbf{j} + \mathbf{k}'\mathbf{k} \tag{1}$$

or that
$$\Phi \cdot \Phi_C = \mathrm{I} \quad \Phi_3 = +1 \tag{2}$$

or that
$$\Phi \cdot \Phi_C = \mathrm{I} \quad \Phi_3 > 0$$

The necessary and sufficient condition that a dyadic represent a rotation combined with a transformation of reflection by which each figure is replaced by one symmetrical to it is that

$$\Phi = -(\mathbf{i}'\mathbf{i} + \mathbf{j}'\mathbf{j} + \mathbf{k}'\mathbf{k}) \tag*{(1)$'$}$$

or that
$$\Phi \cdot \Phi_C = \mathrm{I}, \quad \Phi_3 = -1$$

or that
$$\Phi \cdot \Phi_C = \mathrm{I}, \quad \Phi_3 < 0. \tag{3}$$

A dyadic of the form (1) is called a versor; one of the form (1)$'$, a perversor.

If the axis of rotation of a versor be chosen as the **i**-axis the versor reduces to

$$\Phi = \mathbf{i}\mathbf{i} + \cos q\,(\mathbf{j}\mathbf{j} + \mathbf{k}\mathbf{k}) + \sin q\,(\mathbf{k}\mathbf{j} - \mathbf{j}\mathbf{k}) \tag{4}$$

or
$$\Phi = \mathbf{i}\mathbf{i} + \cos q\,(\mathrm{I} - \mathbf{i}\mathbf{i}) + \sin q\,\mathrm{I} \times \mathbf{i}. \tag{5}$$

If any unit vector **a** is directed along the axis of rotation

$$\Phi = \mathbf{a}\mathbf{a} + \cos q\,(\mathrm{I} - \mathbf{a}\mathbf{a}) + \sin q\,\mathrm{I} \times \mathbf{a} \tag{6}$$

The axis of the versor coincides in direction with $-\Phi_\times$.

If a vector be drawn along the axis and if the magnitude of the vector be taken equal to the tangent of one-half the angle of rotation, the vector determines the rotation completely. This vector is called the vector semi-tangent of version.

$$\mathbf{Q} = \frac{-\Phi_\times}{1 + \Phi_S} \qquad \mathbf{Q} \cdot \mathbf{Q} = \tan^2 \tfrac{1}{2} q \tag{9}$$

In terms of $\mathbf{Q}$ the versor Φ may be expressed in a number of ways.

$$\Phi = \frac{\mathbf{Q}\,\mathbf{Q}}{\mathbf{Q} \cdot \mathbf{Q}} + \cos q \left(\mathrm{I} - \frac{\mathbf{Q}\,\mathbf{Q}}{\mathbf{Q} \cdot \mathbf{Q}} \right) + \sin q \;\; \mathrm{I} \times \frac{\mathbf{Q}}{\sqrt{\mathbf{Q} \cdot \mathbf{Q}}} \tag{10}$$

or

$$\Phi = (\mathrm{I} + \mathrm{I} \times \mathbf{Q}) \cdot (\mathrm{I} - \mathrm{I} \times \mathbf{Q})^{-1} \tag{10$'$}$$

or

$$\Phi = \frac{\mathbf{Q}\,\mathbf{Q} \pm (\mathrm{I} + \mathrm{I} \times \mathbf{Q})^2}{1 + \mathbf{Q}\,\mathbf{Q}} \tag{10$''$}$$

or

$$\Phi = \frac{(1 - \mathbf{Q} \cdot \mathbf{Q})\,\mathrm{I} + 2\,\mathbf{Q}\,\mathbf{Q} + 2\,\mathrm{I} \times \mathbf{Q}}{1 + \mathbf{Q} \cdot \mathbf{Q}} \tag{10$'''$}$$

If $\mathbf{a}$ is a unit vector a dyadic of the form

$$\Phi = 2\,\mathbf{a}\,\mathbf{a} - \mathrm{I} \tag{11}$$

is a biquadrantal versor. Any versor may be resolved into the product of two biquadrantal versors and by means of such resolutions any two versors may be combined into another. The law of composition for the vector semi-tangents of version is

$$\mathbf{Q}_3 = \frac{\mathbf{Q}_1 + \mathbf{Q}_2 \; + \mathbf{Q}_2 \times \mathbf{Q}_1}{1 - \mathbf{Q}_1 \cdot \mathbf{Q}_2}.$$

A dyadic reducible to the form

$$\Phi = \mathbf{a}\,\mathbf{a}' + \cos q \; (\mathbf{b}\,\mathbf{b}' + \mathbf{c}\,\mathbf{c}') + \sin q \; (\mathbf{c}\,\mathbf{b}' - \mathbf{b}\,\mathbf{c}') \tag{17}$$

is called a cyclic dyadic. It produces a generalization of simple rotation — an elliptic rotation, so to speak. The product of two cyclic dyadics which have the same antecedents **a, b, c** and consequents **a′ b′ c′** is obtained by adding their angles q. A cyclic dyadic may be regarded as a root of the idemfactor. A dyadic reducible to the form

$$\Phi = a\,\mathbf{i}\,\mathbf{i} + b\,\mathbf{j}\,\mathbf{j} + c\,\mathbf{k}\,\mathbf{k} \tag{18}$$

where a, b, c are positive scalars is called a right tensor. It represents a stretching along the principal axis **i, j, k** in the ratio a to 1, b to 1, c to 1 which are called the principal ratios of elongation. This transformation is a pure strain.

Any dyadic may be expressed as the product of a versor, a right tensor, and a positive or negative sign.

$$\Phi = \pm\,(a\,\mathbf{i}'\mathbf{i}' + b\,\mathbf{j}'\mathbf{j}' + c\,\mathbf{k}'\mathbf{k}')\,(\mathbf{i}'\mathbf{i} + \mathbf{j}'\mathbf{j} + \mathbf{k}'\mathbf{k})$$

or $$\Phi = \pm\,(\mathbf{i}'\mathbf{i} + \mathbf{j}'\mathbf{j} + \mathbf{k}'\mathbf{k})\cdot(a\,\mathbf{i}\,\mathbf{i} + b\,\mathbf{j}\,\mathbf{j} + c\,\mathbf{k}\,\mathbf{k}). \tag{19}$$

Consequently any linear homogeneous strain may be regarded as a combination of a rotation and a pure strain accompanied or unaccompanied by a perversion.

The immediate generalizations of the right tensor and the cyclic dyadic is to the tonic

$$\Phi = a\,\mathbf{a}\,\mathbf{a}' + b\,\mathbf{b}\,\mathbf{b}' + \mathbf{c}\,c\,\mathbf{c}' \tag{21}$$

and cyclotonic

$$\Phi = a\,\mathbf{a}\,\mathbf{a}' + b\,(\mathbf{b}\,\mathbf{b}' + \mathbf{c}\,\mathbf{c}') + c\,(\mathbf{c}\,\mathbf{b}' - \mathbf{b}\,\mathbf{c}) \tag{23}$$

or $$\Phi = a\,\mathbf{a}\,\mathbf{a}' + p\cos q\,(\mathbf{b}\,\mathbf{b}' + \mathbf{c}\,\mathbf{c}') + p\sin q\,(\mathbf{c}\,\mathbf{b}' - \mathbf{b}\,\mathbf{c}') \tag{25}$$

where $$p = +\sqrt{b^2 + c^2} \text{ and } \tan\tfrac{1}{2}q = \frac{p-b}{p+b}. \tag{24$'$}$$

Any dyadic in general may be reduced either to the form (21), and is therefore a tonic, or to the form (25), and is therefore a cyclotonic. The condition that a dyadic be a tonic is that the cubic equation

$$x^3 - \Phi_S\,x^2 + \Phi_{2S}\,x - \Phi_3 = 0 \tag{29$'$}$$

shall have three real roots. Special cases in which the reduction may be accomplished in more ways than one arise when the equation has equal roots. The condition that a dyadic be a cyclotonic is that this cubic equation shall have only one real root. There occur two limiting cases in which the dyadic cannot be reduced to cyclotonic form. In these cases it may be written as

$$\Phi = a\,\mathbf{a}\,\mathbf{a}' + p\,(\mathbf{b}\,\mathbf{b}' + \mathbf{c}\,\mathbf{c}') + \mathbf{c}\,\mathbf{b}' \tag{37}$$

and is a simple shearer, or it takes the form

$$\Phi = a\,\mathrm{I} + \mathbf{a}\,\mathbf{b}' + \mathbf{b}\,\mathbf{c}' \tag{38}$$

and is a complex shearer. Dyadics may be classified according to their characteristic equations

$$(\Phi - a\,\mathrm{I}) \cdot (\Phi - b\,\mathrm{I}) \cdot (\Phi - c\,\mathrm{I}) = 0 \qquad \text{tonic}$$

$$(\Phi - a\,\mathrm{I}) \cdot (\Phi^2 - 2\,p\cos q\,\Phi + p^2\,\mathrm{I}) = 0 \qquad \text{cyclotonic}$$

$$(\Phi - a\,\mathrm{I}) \cdot (\Phi - b\,\mathrm{I})^2 = 0 \qquad \text{simple shearer}$$

$$(\Phi - a\,\mathrm{I}) \cdot (\Phi - b\,\mathrm{I}) = 0 \qquad \text{special tonic}$$

$$(\Phi - a\,\mathrm{I})^3 = 0 \qquad \text{complex shearer}$$

$$(\Phi - a\,\mathrm{I})^2 = 0 \qquad \text{special simple shearer}$$

$$(\Phi - a\,\mathrm{I}) = 0 \qquad \text{special tonic.}$$

CHAPTER VII

MISCELLANEOUS APPLICATIONS

Quadric Surfaces

136.] If Φ be any constant dyadic the equation

$$\mathbf{r} \cdot \Phi \cdot \mathbf{r} = \text{const.} \tag{1}$$

is quadratic in $\mathbf{r}$. The constant, in case it be not zero, may be divided into the dyadic Φ and hence the equation takes the form

$$\mathbf{r} \cdot \Phi \cdot \mathbf{r} = 1, \tag{1'}$$

or

$$\mathbf{r} \cdot \Phi \cdot \mathbf{r} = 0. \tag{2}$$

The dyadic Φ may be assumed to be self-conjugate. For if Ψ is an anti-self-conjugate dyadic, the product $\mathbf{r} \cdot \Psi \cdot \mathbf{r}$ is identically zero for all values of $\mathbf{r}$. The proof of this statement is left as an exercise. By Art. 116 any self-conjugate dyadic is reducible to the form

$$\Phi = \pm \frac{\mathbf{i}\,\mathbf{i}}{a^2} \pm \frac{\mathbf{j}\mathbf{j}}{b^2} \pm \frac{\mathbf{k}\,\mathbf{k}}{c^2}. \tag{3}$$

If

$$\mathbf{r} = x\,\mathbf{i} + y\,\mathbf{j} + z\,\mathbf{k},$$

$$\mathbf{r} \cdot \Phi \cdot \mathbf{r} = \pm \frac{x^2}{a^2} \pm \frac{y^2}{b^2} \pm \frac{z^2}{c^2}. \tag{4}$$

Hence the equation $\mathbf{r} \cdot \Phi \cdot \mathbf{r} = 1$

represents a quadric surface real or imaginary.

The different cases which arise are four in number. If the signs are all positive, the quadric is a real ellipsoid. If one sign is negative it is an hyperboloid of one sheet; if two are negative, a hyperboloid of two sheets. If the three signs are all negative the quadric is imaginary. In like manner the equation

$$\mathbf{r} \cdot \Phi \cdot \mathbf{r} = 0$$

is seen to represent a cone which may be either real or imaginary according as the signs are different or all alike. Thus the equation

$$\mathbf{r} \cdot \Phi \cdot \mathbf{r} = \text{const.}$$

represents a central quadric surface. The surface reduces to a cone in case the constant is zero. Conversely any central quadric surface may be represented by a suitably chosen self-conjugate dyadic Φ in the form

$$\mathbf{r} \cdot \Phi \cdot \mathbf{r} = \text{const.}$$

This is evident from the equations of the central quadric surfaces when reduced to the normal form. They are

$$\pm \frac{x^2}{a^2} \pm \frac{y^2}{b^2} \pm \frac{z^2}{c^2} = \text{const.}$$

The corresponding dyadic Φ is $\quad \Phi = \pm \frac{\mathbf{i}\,\mathbf{i}}{a^2} \pm \frac{\mathbf{j}\,\mathbf{j}}{b^2} \pm \frac{\mathbf{k}\,\mathbf{k}}{c^2}.$

The most general scalar expression which is quadratic in the vector $\mathbf{r}$ and which consequently when set equal to a constant represents a quadric surface, contains terms like

$$\mathbf{r} \cdot \mathbf{r}, \quad (\mathbf{r} \cdot \mathbf{a})\,(\mathbf{b} \cdot \mathbf{r}), \quad \mathbf{r} \cdot \mathbf{c}, \quad \mathbf{d} \cdot \mathbf{e},$$

where $\mathbf{a}$, $\mathbf{b}$, $\mathbf{c}$, $\mathbf{d}$, $\mathbf{e}$ are constant vectors. The first two terms are of the second order in $\mathbf{r}$; the third, of the first order; and the last, independent of $\mathbf{r}$. Moreover, it is evident that these four sorts of terms are the only ones which can occur in a scalar expression which is quadratic in $\mathbf{r}$.

But
$$\mathbf{r} \cdot \mathbf{r} = \mathbf{r} \cdot \mathbf{I} \cdot \mathbf{r},$$
and
$$(\mathbf{r} \cdot \mathbf{a})\,(\mathbf{b} \cdot \mathbf{r}) = \mathbf{r} \cdot \mathbf{a}\,\mathbf{b} \cdot \mathbf{r}.$$

Hence the most general quadratic expression may be reduced to

$$\mathbf{r} \cdot \Phi \cdot \mathbf{r} + \mathbf{r} \cdot \mathbf{A} + C = 0,$$

where Φ is a constant dyadic, $\mathbf{A}$ a constant vector, and C a constant scalar. The dyadic may be regarded as self-conjugate if desired.

To be rid of the linear term $\mathbf{r} \cdot \mathbf{A}$, make a change of origin by replacing $\mathbf{r}$ by $\mathbf{r}' - \mathbf{t}$.

$$(\mathbf{r}' - \mathbf{t}) \cdot \Phi \cdot (\mathbf{r}' - \mathbf{t}) + (\mathbf{r}' - \mathbf{t}) \cdot \mathbf{A} + C = 0$$

$$\mathbf{r}' \cdot \Phi \cdot \mathbf{r}' - \mathbf{t} \cdot \Phi \cdot \mathbf{r}' - \mathbf{r}' \cdot \Phi \cdot \mathbf{t} + \mathbf{t} \cdot \Phi \cdot \mathbf{t}$$
$$+ \mathbf{r}' \cdot \mathbf{A} - \mathbf{t} \cdot \mathbf{A} + C = 0.$$

Since Φ is self-conjugate the second and third terms are equal. Hence

$$\mathbf{r}' \cdot \Phi \cdot \mathbf{r}' + 2\, \mathbf{r}' \cdot (\tfrac{1}{2} \mathbf{A} - \Phi \cdot \mathbf{t}) + C' = 0.$$

If now Φ is complete the vector $\mathbf{t}$ may be chosen so that

$$\frac{1}{2} \mathbf{A} = \Phi \cdot \mathbf{t} \text{ or } \mathbf{t} = \frac{1}{2} \Phi^{-1} \cdot \mathbf{A}.$$

Hence the quadric is reducible to the central form

$$\mathbf{r}' \cdot \Phi \cdot \mathbf{r}' = \text{const.}$$

In case Φ is incomplete it is *uni*planar or *uni*linear because Φ is self-conjugate. If $\mathbf{A}$ lies in the plane of Φ or in the line of Φ as the case may be the equation

$$\frac{1}{2} \mathbf{A} = \Phi \cdot \mathbf{t}$$

is soluble for $\mathbf{t}$ and the reduction to central form is still possible. But unless $\mathbf{A}$ is so situated the reduction is impossible. The quadric surface is not a central surface.

The discussion and classification of the various non-central quadrics is an interesting exercise. It will not be taken up here. The present object is to develop so much of the theory of quadric surfaces as will be useful in applications to mathematical physics with especial reference to non-isotropic media. Hereafter therefore the central quadrics and in particular the ellipsoid will be discussed.

137.] The tangent plane may be found by differentiation.

$$\mathbf{r} \cdot \Phi \cdot \mathbf{r} = 1.$$

$$d\,\mathbf{r} \cdot \Phi \cdot \mathbf{r} + \mathbf{r} \cdot \Phi \cdot d\,\mathbf{r} = 0.$$

Since Φ is self-conjugate these two terms are equal and

$$d\,\mathbf{r} \cdot \Phi \cdot \mathbf{r} = 0. \tag{5}$$

The increment $d\,\mathbf{r}$ is perpendicular to $\Phi \cdot \mathbf{r}$. Hence $\Phi \cdot \mathbf{r}$ is normal to the surface at the extremity of the vector $\mathbf{r}$. Let this normal be denoted by $\mathbf{N}$ and let the unit normal be $\mathbf{n}$.

$$\mathbf{N} = \Phi \cdot \mathbf{r} \tag{6}$$

$$\mathbf{n} = \frac{\Phi \cdot \mathbf{r}}{\sqrt{(\Phi \cdot \mathbf{r}) \cdot (\Phi \cdot \mathbf{r})}} = \frac{\Phi \cdot \mathbf{r}}{\sqrt{\mathbf{r} \cdot \Phi^2 \cdot \mathbf{r}}}.$$

Let $\mathbf{p}$ be the vector drawn from the origin perpendicular to the tangent plane. $\mathbf{p}$ is parallel to $\mathbf{n}$. The perpendicular distance from the origin to the tangent plane is the square root of $\mathbf{p} \cdot \mathbf{p}$. It is also equal to the square root of $\mathbf{r} \cdot \mathbf{p}$.

$$\mathbf{r} \cdot \mathbf{p} = r \cos(\mathbf{r}, \mathbf{p})\, p = p^2.$$

Hence
$$\mathbf{r} \cdot \mathbf{p} = \mathbf{p} \cdot \mathbf{p}.$$

Or
$$\frac{\mathbf{p} \cdot \mathbf{p}}{\mathbf{p} \cdot \mathbf{p}} = \mathbf{r} \cdot \frac{\mathbf{p}}{\mathbf{p} \cdot \mathbf{p}} = 1.$$

But
$$\mathbf{r} \cdot \Phi \cdot \mathbf{r} = \mathbf{r} \cdot \mathbf{N} = 1.$$

Hence inasmuch as $\mathbf{p}$ and $\mathbf{N}$ are parallel, they are equal.

$$\Phi \cdot \mathbf{r} = \mathbf{N} = \frac{\mathbf{p}}{\mathbf{p} \cdot \mathbf{p}}. \tag{7}$$

On page 108 it was seen that the vector which has the direction of the normal to a plane and which is in magnitude equal to the reciprocal of the distance from the origin to the plane may be taken as the vector coördinate of that plane. Hence the above equation shows that $\Phi \cdot \mathbf{r}$ is not merely normal to the tangent plane, but is also the coördinate of the plane. That is, the length of $\Phi \cdot \mathbf{r}$ is the reciprocal of the distance from the origin to the plane tangent to the ellipsoid at the extremity of the vector $\mathbf{r}$.

The equation of the ellipsoid in plane coördinates may be found by eliminating $\mathbf{r}$ from the two equations.

$$\begin{cases} \mathbf{r} \cdot \Phi \cdot \mathbf{r} = 1, \\ \Phi \cdot \mathbf{r} = \mathbf{N}. \end{cases}$$

$$\mathbf{r} = \Phi^{-1} \cdot \mathbf{N} = \mathbf{N} \cdot \Phi^{-1}.$$

Hence $\mathbf{r} \cdot \Phi \cdot \mathbf{r} = \mathbf{N} \cdot \Phi^{-1} \cdot \Phi \cdot \Phi^{-1} \cdot \mathbf{N} = \mathbf{N} \cdot \Phi^{-1} \cdot \mathbf{N}.$

Hence the desired equation is

$$\mathbf{N} \cdot \Phi^{-1} \cdot \mathbf{N} = 1. \tag{8}$$

If $$\Phi = \frac{\mathbf{i\,i}}{a^2} + \frac{\mathbf{j\,j}}{b^2} + \frac{\mathbf{k\,k}}{c^2},$$

$$\Phi^{-1} = a^2\, \mathbf{i\,i} + b^2\, \mathbf{j\,j} + c^2\, \mathbf{k\,k}.$$

Let $$\mathbf{r} = x\, \mathbf{i} + y\, \mathbf{j} + z\, \mathbf{k},$$

and $$\mathbf{N} = u\, \mathbf{i} + v\, \mathbf{j} + w\, \mathbf{k},$$

where u, v, w are the reciprocals of the intercepts of the plane $\mathbf{N}$ upon the axes $\mathbf{i}, \mathbf{j}, \mathbf{k}$. Then the ellipsoid may be written in either of the two forms familiar in Cartesian geometry.

$$\mathbf{r} \cdot \Phi \cdot \mathbf{r} = \frac{x^2}{a^2} + \frac{y^2}{b^2} + \frac{z^2}{c^2} = 1 \tag{9}$$

or $$\mathbf{N} \cdot \Phi^{-1} \cdot \mathbf{N} = a^2\, u^2 + b^2\, v^2 + c^2\, w^2 = 1. \tag{10}$$

138.] The locus of the middle points of a system of parallel chords in an ellipsoid is a plane. This plane is called the diametral plane conjugate with the system of chords. It is parallel to the plane drawn tangent to the ellipsoid at the extremity of that one of the chords which passes through the center.

Let $\mathbf{r}$ be any radius vector in the ellipsoid. Let $\mathbf{s}$ be the vector drawn to the middle point of a chord parallel to $\mathbf{a}$.

Let
$$\mathbf{r} = \mathbf{s} + x\,\mathbf{a}.$$

If $\mathbf{r}$ is a radius vector of the ellipsoid

$$\mathbf{r} \cdot \Phi \cdot \mathbf{r} = (\mathbf{s} + x\,\mathbf{a}) \cdot \Phi \cdot (\mathbf{s} + x\,\mathbf{a}) = 1.$$

Hence
$$\mathbf{s} \cdot \Phi \cdot \mathbf{s} + 2\,x\,\mathbf{s} \cdot \Phi \cdot \mathbf{a} + x^2\,\mathbf{a} \cdot \Phi \cdot \mathbf{a} = 1.$$

Inasmuch as the vector $\mathbf{s}$ bisects the chord parallel to $\mathbf{a}$ the two solutions of x given by this equation are equal in magnitude and opposite in sign. Hence the coefficient of the linear term x vanishes.

$$\mathbf{s} \cdot \Phi \cdot \mathbf{a} = 0.$$

Consequently the vector $\mathbf{s}$ is perpendicular to $\Phi \cdot \mathbf{a}$. The locus of the terminus of $\mathbf{s}$ is therefore a plane passed through the center of the ellipsoid, perpendicular to $\Phi \cdot \mathbf{a}$, and parallel to the tangent plane at the extremity of $\mathbf{a}$.

If $\mathbf{b}$ is any radius vector in the diametral plane conjugate with $\mathbf{a}$,

$$\mathbf{b} \cdot \Phi \cdot \mathbf{a} = 0.$$

The symmetry of this equation shows that $\mathbf{a}$ is a radius vector in the plane conjugate with $\mathbf{b}$. Let $\mathbf{c}$ be a third radius vector in the ellipsoid and let it be chosen as the line of intersection of the diametral planes conjugate respectively with $\mathbf{a}$ and $\mathbf{b}$. Then

$$\begin{aligned} \mathbf{a} \cdot \Phi \cdot \mathbf{b} &= 0, \\ \mathbf{b} \cdot \Phi \cdot \mathbf{c} &= 0, \\ \mathbf{c} \cdot \Phi \cdot \mathbf{a} &= 0. \end{aligned} \tag{11}$$

The vectors $\mathbf{a}$, $\mathbf{b}$, $\mathbf{c}$ are changed into $\Phi \cdot \mathbf{a}$, $\Phi \cdot \mathbf{b}$, $\Phi \cdot \mathbf{c}$ by the dyadic Φ. Let

$$\mathbf{a}' = \Phi \cdot \mathbf{a}, \quad \mathbf{b}' = \Phi \cdot \mathbf{b}, \quad \mathbf{c}' = \Phi \cdot \mathbf{c}.$$

The vectors $\mathbf{a}'$, $\mathbf{b}'$, $\mathbf{c}'$ form the system reciprocal to $\mathbf{a}$, $\mathbf{b}$, $\mathbf{c}$.

For $$\mathbf{a} \cdot \mathbf{a}' = \mathbf{a} \cdot \Phi \cdot \mathbf{a} = 1, \quad \mathbf{b} \cdot \mathbf{b}' = \mathbf{b} \cdot \Phi \cdot \mathbf{b} = 1,$$

$$\mathbf{c} \cdot \mathbf{c}' = \mathbf{c} \cdot \Phi \cdot \mathbf{c} = 1,$$

and $$\mathbf{a} \cdot \mathbf{b}' = \mathbf{a} \cdot \Phi \cdot \mathbf{b} = 0, \quad \mathbf{b} \cdot \mathbf{c}' = \mathbf{b} \cdot \Phi \cdot \mathbf{c} = 0,$$

$$\mathbf{c} \cdot \mathbf{a}' = \mathbf{c} \cdot \Phi \cdot \mathbf{a} = 0.$$

The dyadic Φ may be therefore expressed in the forms

$$\Phi = \mathbf{a}'\mathbf{a}' + \mathbf{b}'\mathbf{b}' + \mathbf{c}'\mathbf{c}', \tag{12}$$

and $$\Phi^{-1} = \mathbf{a}\,\mathbf{a} + \mathbf{b}\,\mathbf{b} + \mathbf{c}\,\mathbf{c}.$$

If for convenience the three directions $\mathbf{a}$, $\mathbf{b}$, $\mathbf{c}$, be called a system of three conjugate radii vectors, and if in a similar manner the three tangent planes at their extremities be called a system of three conjugate tangent planes, a number of geometric theorems may be obtained from interpreting the invariants of Φ. A system of three conjugate radii vectors may be obtained in a doubly infinite number of ways.

The volume of a parallelopiped of which three concurrent edges constitute a system of three conjugate radii vectors is constant and equal in magnitude to the rectangular parallelopiped constructed upon the three semi-axes of the ellipsoid.

For let $\mathbf{a}$, $\mathbf{b}$, $\mathbf{c}$ be any system of three conjugate axes.

$$\Phi^{-1} = \mathbf{a}\,\mathbf{a} + \mathbf{b}\,\mathbf{b} + \mathbf{c}\,\mathbf{c}.$$

The determinant or third of Φ^{-1} is an invariant and independent of the form in which Φ is expressed.

$$\Phi_3^{-1} = [\mathbf{a}\,\mathbf{b}\,\mathbf{c}]^2.$$

But if $\Phi^{-1} = a^2\, \mathbf{i\,i} + b^2\, \mathbf{j\,j} + c^2\, \mathbf{k\,k}$,

$$\Phi_3^{-1} = a^2\, b^2\, c^2.$$

Hence $$[\mathbf{a\,b\,c}] = a\, b\, c.$$

This demonstrates the theorem. In like manner by interpreting Φ_3, Φ_S^{-1}, and Φ_S it is possible to show that:

The sum of the squares of the radii vectors drawn to an ellipsoid in a system of three conjugate directions is constant and equal to the sum of the squares of the semi-axes.

The volume of the parallelopiped, whose three concurrent edges are in the directions of the perpendiculars upon a system of three conjugate tangent planes and in magnitude equal to the reciprocals of the distances of those planes from the center of the ellipsoid, is constant and equal to the reciprocal of the parallelopiped constructed upon the semi-axes of the ellipsoid.

The sum of the squares of the reciprocals of the three perpendiculars dropped from the origin upon a system of three conjugate tangent planes is constant and equal to the sum of the squares of the reciprocals of the semi-axes.

If **i**, **j**, **k** be three mutually perpendicular unit vectors

$$\Phi_S = \mathbf{i} \cdot \Phi \cdot \mathbf{i} + \mathbf{j} \cdot \Phi \cdot \mathbf{j} + \mathbf{k} \cdot \Phi \cdot \mathbf{k},$$

$$\Phi_S^{-1} = \mathbf{i} \cdot \Phi^{-1} \cdot \mathbf{i} + \mathbf{j} \cdot \Phi^{-1} \cdot \mathbf{j} + \mathbf{k} \cdot \Phi^{-1} \cdot \mathbf{k}.$$

Let **a**, **b**, **c** be three radii vectors in the ellipsoid drawn respectively parallel to **i**, **j**, **k**.

$$\mathbf{a} \cdot \Phi \cdot \mathbf{a} = \mathbf{b} \cdot \Phi \cdot \mathbf{b} = \mathbf{c} \cdot \Phi \cdot \mathbf{c} = 1.$$

Hence $$\Phi_S = \frac{\mathbf{i} \cdot \Phi \cdot \mathbf{i}}{\mathbf{a} \cdot \Phi \cdot \mathbf{a}} + \frac{\mathbf{j} \cdot \Phi \cdot \mathbf{j}}{\mathbf{b} \cdot \Phi \cdot \mathbf{b}} + \frac{\mathbf{k} \cdot \Phi \cdot \mathbf{k}}{\mathbf{c} \cdot \Phi \cdot \mathbf{c}}.$$

But the three terms in this expression are the squares of the reciprocals of the radii vectors drawn respectively in the **i**, **j**, **k** directions. Hence:

The sum of the squares of the reciprocals of three mutually perpendicular radii vectors in an ellipsoid is constant. And in a similar manner: the sum of the squares of the perpendiculars dropped from the origin upon three mutually perpendicular tangent planes is constant.

139.] The equation of the polar plane of the point determined by the vector $\mathbf{a}$ is [1]

$$\mathbf{s} \cdot \Phi \cdot \mathbf{a} = 1. \tag{13}$$

For let $\mathbf{s}$ be the vector of a point in the polar plane. The vector of any point upon the line which joins the terminus of $\mathbf{s}$ and the terminus of $\mathbf{a}$ is

$$\frac{y\,\mathbf{s} + x\,\mathbf{a}}{x + y}.$$

If this point lies upon the surface

$$\frac{y\,\mathbf{s} + x\,\mathbf{a}}{x + y} \cdot \Phi \cdot \frac{y\,\mathbf{s} + x\,\mathbf{a}}{x + y} = 1$$

$$\frac{y^2}{(x+y)^2}\,\mathbf{s} \cdot \Phi \cdot \mathbf{s} + \frac{2\,x\,y}{(x+y)^2}\,\mathbf{s} \cdot \Phi \cdot \mathbf{a} + \frac{x^2}{(x+y)^2}\,\mathbf{a} \cdot \Phi \cdot \mathbf{a} = 1.$$

If the terminus of $\mathbf{s}$ lies in the polar plane of $\mathbf{a}$ the two values of the ratio $x:y$ determined by this equation must be equal in magnitude and opposite in sign. Hence the term in $x\,y$ vanishes.

Hence $$\mathbf{s} \cdot \Phi \cdot \mathbf{a} = 1$$

is the desired equation of the polar plane of the terminus of $\mathbf{a}$.

Let $\mathbf{a}$ be replaced by $z\,\mathbf{a}$. The polar plane becomes

$$\mathbf{s} \cdot \Phi \cdot z\,\mathbf{a} = 1,$$

or $$\mathbf{s} \cdot \Phi \cdot \mathbf{a} = \frac{1}{z}.$$

[1] It is evidently immaterial whether the central quadric determined by Φ be real or imaginary, ellipsoid or hyperboloid.

When z increases the polar plane of the terminus of $z\mathbf{a}$ approaches the origin. In the limit when z becomes infinite the polar plane becomes

$$\mathbf{s} \cdot \Phi \cdot \mathbf{a} = 0.$$

Hence the polar plane of the point at infinity in the direction $\mathbf{a}$ is the same as the diametral plane conjugate with $\mathbf{a}$. This statement is frequently taken as the definition of the diametral plane conjugate with $\mathbf{a}$. In case the vector $\mathbf{a}$ is a radius vector of the surface the polar plane becomes identical with the tangent plane at the terminus of $\mathbf{a}$. The equation

$$\mathbf{s} \cdot \Phi \cdot \mathbf{a} = 1 \quad \text{or } \mathbf{s} \cdot \mathbf{N} = 1$$

therefore represents the tangent plane.

The polar plane may be obtained from another standpoint which is important. If a quadric Q and a plane P are given,

$$Q \equiv \mathbf{r} \cdot \Phi \cdot \mathbf{r} - 1 = 0$$

and

$$P \equiv \mathbf{r} \cdot \mathbf{c} - C = 0,$$

the equation

$$(\mathbf{r} \cdot \Phi \cdot \mathbf{r} - 1) + k\,(\mathbf{r} \cdot \mathbf{c} - C)^2 = 0$$

represents a quadric surface which passes through the curve of intersection of Q and P and is tangent to Q along that curve. In like manner if two quadrics Q and Q' are given,

$$Q \equiv \mathbf{r} \cdot \Phi \cdot \mathbf{r} - 1 = 0$$

$$Q' \equiv \mathbf{r} \cdot \Phi' \cdot \mathbf{r} - 1 = 0,$$

the equation

$$(\mathbf{r} \cdot \Phi \cdot \mathbf{r} - 1) + k\,(\mathbf{r} \cdot \Phi' \cdot \mathbf{r} - 1) = 0$$

represents a quadric surface which passes through the curves of intersection of Q and Q' and which cuts Q and Q' at no other points. In case this equation is factorable into two equations which are linear in $\mathbf{r}$, and which consequently represent two planes, the curves of intersection of Q and Q' become plane and lie in those two planes.

If A is any point outside of the quadric and if all the tangent planes which pass through A are drawn, these planes envelop a cone. This cone touches the quadric along a plane curve — the plane of the curve being the polar plane of the point A. For let **a** be the vector drawn to the point A. The equation of any tangent plane to the quadric is

$$\mathbf{s} \cdot \Phi \cdot \mathbf{r} = 1.$$

If this plane contains A, its equation is satisfied by **a**. Hence the conditions which must be satisfied by **r** if its tangent plane passes through A are

$$\mathbf{a} \cdot \Phi \cdot \mathbf{r} = 1,$$

$$\mathbf{r} \cdot \Phi \cdot \mathbf{r} = 1.$$

The points **r** therefore lie in a plane $\mathbf{r} \cdot (\Phi \cdot \mathbf{a}) = 1$ which on comparison with (13) is seen to be the polar plane of A. The quadric which passes through the curve of intersection of this polar plane with the given quadric and which touches the quadric along that curve is

$$(\mathbf{r} \cdot \Phi \cdot \mathbf{r} - 1) + k\,(\mathbf{a} \cdot \Phi \cdot \mathbf{r} - 1)^2 = 0.$$

If this passes through the point A,

$$(\mathbf{a} \cdot \Phi \cdot \mathbf{a} - 1) + k\,(\mathbf{a} \cdot \Phi \cdot \mathbf{a} - 1)^2 = 0.$$

Hence $\quad (\mathbf{r} \cdot \Phi \cdot \mathbf{r} - 1)\,(\mathbf{a} \cdot \Phi \cdot \mathbf{a} - 1) - (\mathbf{a} \cdot \Phi \cdot \mathbf{r} - 1)^2 = 0.$

By transforming the origin to the point A this is easily seen to be a cone whose vertex is at that point.

140.] Let Φ be any self-conjugate dyadic. It is expressible in the form

$$\Phi = A\,\mathbf{i}\,\mathbf{i} + B\,\mathbf{j}\,\mathbf{j} + C\,\mathbf{k}\,\mathbf{k}$$

where A, B, C are positive or negative scalars. Furthermore let

$$A < B < C$$

$$\Phi - B\,\mathrm{I} = (C - B)\,\mathbf{k}\,\mathbf{k} - (B - A)\,\mathbf{i}\,\mathbf{i}.$$

Let $\sqrt{C-B}\,\mathbf{k}=\mathbf{c}$ and $\sqrt{B-A}\,\mathbf{i}=\mathbf{a}$.

Then $\Phi - B\mathrm{I} = \mathbf{c}\,\mathbf{c} - \mathbf{a}\,\mathbf{a} = \frac{1}{2}\left\{(\mathbf{c}+\mathbf{a})(\mathbf{c}-\mathbf{a}) + (\mathbf{c}-\mathbf{a})(\mathbf{c}+\mathbf{a})\right\}.$

Let $$\mathbf{c}+\mathbf{a}=\mathbf{p} \quad \text{and} \quad \mathbf{c}-\mathbf{a}=\mathbf{q}.$$

Then $$\Phi = B\mathrm{I} + \tfrac{1}{2}(\mathbf{p}\,\mathbf{q} + \mathbf{q}\,\mathbf{p}). \tag{14}$$

The dyadic Φ has been expressed as the sum of a constant multiple of the idemfactor and one half the sum

$$\mathbf{p}\,\mathbf{q} + \mathbf{q}\,\mathbf{p}.$$

The reduction has assumed tacitly that the constants A, B, C are different from each other and from zero.

This expression for Φ is closely related to the circular sections of the quadric surface

$$\mathbf{r}\cdot\Phi\cdot\mathbf{r} = 1.$$

Substituting the value of Φ, $\mathbf{r}\cdot\Phi\cdot\mathbf{r} = 1$ becomes

$$B\,\mathbf{r}\cdot\mathbf{r} + \mathbf{r}\cdot\mathbf{p}\;\mathbf{q}\cdot\mathbf{r} = 1.$$

Let $$\mathbf{r}\cdot\mathbf{p} = n$$

be any plane perpendicular to $\mathbf{p}$. By substitution

$$B\;\mathbf{r}\cdot\mathbf{r} + n\;\mathbf{q}\cdot\mathbf{r} - 1 = 0.$$

This is a sphere because the terms of the second order all have the same coefficient B. If the equation of this sphere be subtracted from that of the given quadric, the resulting equation is that of a quadric which passes through the intersection of the sphere and the given quadric. The difference is

$$\mathbf{q}\cdot\mathbf{r}\;(\mathbf{r}\cdot\mathbf{p} - n) = 0.$$

Hence the sphere and the quadric intersect in two plane curves lying in the planes

$$\mathbf{q}\cdot\mathbf{r} = 0 \quad \text{and} \quad \mathbf{r}\cdot\mathbf{p} = n.$$

Inasmuch as these curves lie upon a sphere they are circles. Hence planes perpendicular to $\mathbf{p}$ cut the quadric in circles. In like manner it may be shown that planes perpendicular to $\mathbf{q}$ cut the quadric in circles. The proof may be conducted as follows:

$$B\, \mathbf{r} \cdot \mathbf{r} + \mathbf{r} \cdot \mathbf{p}\, \mathbf{q} \cdot \mathbf{r} = 1.$$

If $\mathbf{r}$ is a radius vector in the plane passed through the center of the quadric perpendicular to $\mathbf{p}$ or $\mathbf{q}$, the term $\mathbf{r} \cdot \mathbf{p}\, \mathbf{q} \cdot \mathbf{r}$ vanishes. Hence the vector $\mathbf{r}$ in this plane satisfies the equation

$$B\, \mathbf{r} \cdot \mathbf{r} = 1$$

and is of constant length. The section is therefore a circular section. The radius of the section is equal in length to the mean semi-axis of the quadric.

For convenience let the quadric be an ellipsoid. The constants A, B, C are then positive. The reciprocal dyadic Φ^{-1} may be reduced in a similar manner.

$$\Phi^{-1} = \frac{\mathbf{i}\,\mathbf{i}}{A} + \frac{\mathbf{j}\,\mathbf{j}}{B} + \frac{\mathbf{k}\,\mathbf{k}}{C}$$

$$\Phi^{-1} - \frac{1}{B}\mathrm{I} = \left(\frac{1}{B} - \frac{1}{C}\right)\mathbf{k}\,\mathbf{k} - \left(\frac{1}{A} - \frac{1}{B}\right)\mathbf{i}\,\mathbf{i}.$$

Let $$\mathbf{f} = \sqrt{\frac{1}{B} - \frac{1}{C}}\,\mathbf{k} \quad \text{and} \quad \mathbf{d} = \sqrt{\frac{1}{A} - \frac{1}{B}}\,\mathbf{i}.$$

Then $$\Phi^{-1} - \frac{1}{B}\mathrm{I} = \mathbf{f}\,\mathbf{f} - \mathbf{d}\,\mathbf{d} = \tfrac{1}{2}\Big\{(\mathbf{f} + \mathbf{d})(\mathbf{f} - \mathbf{d}) + (\mathbf{f} - \mathbf{d})(\mathbf{f} + \mathbf{d})\Big\}.$$

Let $$\mathbf{f} + \mathbf{d} = \mathbf{u} \quad \text{and} \quad \mathbf{f} - \mathbf{d} = \mathbf{v}.$$

Then $$\Phi^{-1} = \frac{1}{B}\mathrm{I} + \tfrac{1}{2}(\mathbf{u}\,\mathbf{v} + \mathbf{v}\,\mathbf{u}). \qquad (15)$$

The vectors $\mathbf{u}$ and $\mathbf{v}$ are connected intimately with the circular cylinders which envelop the ellipsoid

$$\mathbf{r} \cdot \Phi \cdot \mathbf{r} = 1 \quad \text{or} \quad \mathbf{N} \cdot \Phi^{-1} \cdot \mathbf{N} = 1.$$

For
$$\frac{1}{B} \mathbf{N} \cdot \mathbf{N} + \mathbf{N} \cdot \mathbf{u}\,\mathbf{v} \cdot \mathbf{N} = 1.$$

If now $\mathbf{N}$ be perpendicular to $\mathbf{u}$ or $\mathbf{v}$ the second term, namely, $\mathbf{N} \cdot \mathbf{u}\,\mathbf{v} \cdot \mathbf{N}$, vanishes and hence the equation becomes

$$\mathbf{N} \cdot \mathbf{N} = B.$$

That is, the vector $\mathbf{N}$ is of constant length. But the equation

$$\mathbf{N} \cdot \mathbf{u} = 0$$

is the equation of a cylinder of which the elements and tangent planes are parallel to $\mathbf{u}$. If then $\mathbf{N} \cdot \mathbf{N}$ is constant the cylinder is a circular cylinder enveloping the ellipsoid. The radius of the cylinder is equal in length to the mean semi-axis of the ellipsoid.

There are consequently two planes passing through the origin and cutting out circles from the ellipsoid. The normals to these planes are $\mathbf{p}$ and $\mathbf{q}$. The circles pass through the extremities of the mean axis of the ellipsoid. There are also two circular cylinders enveloping the ellipsoid. The direction of the axes of these cylinders are $\mathbf{u}$ and $\mathbf{v}$. Two elements of these cylinders pass through the extremities of the mean axis of the ellipsoid.

These results can be seen geometrically as follows. Pass a plane through the mean axis and rotate it about that axis from the major to the minor axis. The section is an ellipse. One axis of this ellipse is the mean axis of the ellipsoid. This remains constant during the rotation. The other axis of the ellipse varies in length from the major to the minor axis of the ellipsoid and hence at some stage must pass through a length equal to the mean axis. At this stage of the rotation the section is a circle. In like manner consider the projection or shadow of the ellipsoid cast upon a plane parallel to the mean axis by a point at an infinite distance from that plane and in a direction perpendicular to it. As the ellipsoid is rotated about its mean axis, from the position in which the major axis is perpendicular to the plane of projection to the position in which the minor axis is perpendicular to that plane, the shadow and the projecting cylinder have the mean axis of the ellipsoid as one axis. The other axis changes from the minor axis of the ellipsoid to the major and hence at some stage of the rotation it passes through a value equal to the mean axis. At this stage the shadow and projecting cylinder are circular.

The necessary and sufficient condition that $\mathbf{r}$ be the major or minor semi-axis of the section of the ellipsoid $\mathbf{r} \cdot \Phi \cdot \mathbf{r} = 1$ by a plane passing through the center and perpendicular to $\mathbf{a}$ is that $\mathbf{a}$, $\mathbf{r}$, and $\Phi \cdot \mathbf{r}$ be coplanar.

Let $$\mathbf{r} \cdot \Phi \cdot \mathbf{r} = 1$$

and $$\mathbf{r} \cdot \mathbf{a} = 0.$$

Differentiate: $$d\,\mathbf{r} \cdot \Phi \cdot \mathbf{r} = 0,$$

$$d\,\mathbf{r} \circ \mathbf{a} = 0.$$

Furthermore $$d\,\mathbf{r} \cdot \mathbf{r} = 0,$$

if $\mathbf{r}$ is to be a major or minor axis of the section; for $\mathbf{r}$ is a maximum or a mininum and hence is perpendicular to $d\,\mathbf{r}$. These three equations show that $\mathbf{a}$, $\mathbf{r}$, and $\Phi \cdot \mathbf{r}$ are all orthogonal to the same vector $d\,\mathbf{r}$. Hence they are coplanar.

$$[\mathbf{a} \;\; \mathbf{r} \;\; \Phi \cdot \mathbf{r}] = 0. \tag{16}$$

Conversely if $$[\mathbf{a} \;\; \mathbf{r} \;\; \Phi \cdot \mathbf{r}] = 0,$$

$d\,\mathbf{r}$ may be chosen perpendicular to their common plane. Then

$$d\,\mathbf{r} \cdot \mathbf{r} = 0.$$

Hence $\mathbf{r}$ is a maximum or a mininum and is one of the principal semi-axes of the section perpendicular to $\mathbf{a}$.

141.] It is frequently an advantage to write the equation of an ellipsoid in the form

$$\mathbf{r} \cdot \Psi^2 \cdot \mathbf{r} = 1, \tag{17}$$

instead of

$$\mathbf{r} \cdot \Phi \cdot \mathbf{r} = 1.$$

This may be done; because if

$$\Phi = \frac{\mathbf{i}\,\mathbf{i}}{a^2} + \frac{\mathbf{j}\,\mathbf{j}}{b^2} + \frac{\mathbf{k}\,\mathbf{k}}{c^2},$$

$$\Psi = \frac{\mathbf{i}\,\mathbf{i}}{a} + \frac{\mathbf{j}\,\mathbf{j}}{b} + \frac{\mathbf{k}\,\mathbf{k}}{c} \tag{18}$$

is a dyadic such that Ψ^2 is equal to Φ. Ψ may be regarded as a square root of Φ and written as $\Phi^{\frac{1}{2}}$. But it must be remembered that there are other square roots of Φ — for example,

$$\frac{\mathbf{i}\,\mathbf{i}}{a} + \frac{\mathbf{j}\,\mathbf{j}}{b} - \frac{\mathbf{k}\,\mathbf{k}}{c}$$

and

$$\frac{\mathbf{i}\,\mathbf{i}}{a} - \frac{\mathbf{j}\,\mathbf{j}}{b} - \frac{\mathbf{k}\,\mathbf{k}}{c}.$$

For this reason it is necessary to bear in mind that the square root which is meant by $\Phi^{\frac{1}{2}}$ is that particular one which has been denoted by Ψ.

The equation of the ellipsoid may be written in the form

$$\mathbf{r} \cdot \Psi \cdot \Psi \cdot \mathbf{r} = 1,$$

or

$$(\Psi \cdot \mathbf{r}) \cdot (\Psi \cdot \mathbf{r}) = 1.$$

Let $\mathbf{r}'$ be the radius vector of a unit sphere. The equation of the sphere is

$$\mathbf{r}' \cdot \mathbf{r}' = 1.$$

If $\mathbf{r}' = \Psi \cdot \mathbf{r}$ it becomes evident that an ellipsoid may be transformed into a unit sphere by applying the operator Ψ to each radius vector $\mathbf{r}$, and vice versa, the unit sphere may be transformed into an ellipsoid by applying the inverse operator Ψ^{-1} to each radius vector $\mathbf{r}'$. Furthermore if $\mathbf{a}, \mathbf{b}, \mathbf{c}$ are a system of three conjugate radii vectors in an ellipsoid

$$\mathbf{a} \cdot \Psi^2 \cdot \mathbf{a} = \mathbf{b} \cdot \Psi^2 \cdot \mathbf{b} = \mathbf{c} \cdot \Psi^2 \cdot \mathbf{c} = 1,$$

$$\mathbf{a} \cdot \Psi^2 \cdot \mathbf{b} = \mathbf{b} \cdot \Psi^2 \cdot \mathbf{c} = \mathbf{c} \cdot \Psi^2 \cdot \mathbf{a} = 0.$$

If for the moment $\mathbf{a}', \mathbf{b}', \mathbf{c}'$ denote respectively $\Psi \cdot \mathbf{a}$, $\Psi \cdot \mathbf{b}$, $\Psi \cdot \mathbf{c}$,

$$\mathbf{a}' \cdot \mathbf{a}' = \mathbf{b}' \cdot \mathbf{b}' = \mathbf{c}' \cdot \mathbf{c}' = 1,$$

$$\mathbf{a}' \cdot \mathbf{b}' = \mathbf{b}' \cdot \mathbf{c}' = \mathbf{c}' \cdot \mathbf{a}' = 0.$$

Hence the three radii vectors $\mathbf{a}', \mathbf{b}', \mathbf{c}'$ of the unit sphere into which three conjugate radii vectors in the ellipsoid are transformed by the operator Ψ^{-1} are mutually orthogonal. They form a right-handed or left-handed system of three mutually perpendicular unit vectors.

Theorem: Any ellipsoid may be transformed into any other ellipsoid by means of a homogeneous strain.

Let the equations of the ellipsoids be

$$\mathbf{r} \cdot \Phi \cdot \mathbf{r} = 1,$$

and

$$\bar{\mathbf{r}} \cdot \Psi \cdot \bar{\mathbf{r}} = 1.$$

By means of the strain $\Phi^{\frac{1}{2}}$ the radii vectors $\mathbf{r}$ of the first ellipsoid are changed into the radii vectors $\mathbf{r}'$ of a unit sphere

$$\mathbf{r}' = \Phi^{\frac{1}{2}} \cdot \mathbf{r}, \qquad \mathbf{r}' \cdot \mathbf{r}' = 1.$$

By means of the strain $\Psi^{-\frac{1}{2}}$ the radii vectors $\mathbf{r}'$ of this unit sphere are transformed in like manner into the radii vectors $\bar{\mathbf{r}}$ of the second ellipsoid. Hence by the product $\mathbf{r}$ is changed into $\bar{\mathbf{r}}$.

$$\bar{\mathbf{r}} = \Psi^{-\frac{1}{2}} \cdot \Phi^{\frac{1}{2}} \cdot \mathbf{r}. \qquad (19)$$

The transformation may be accomplished in more ways than one. The radii vectors $\mathbf{r}'$ of the unit sphere may be transformed among themselves by means of a rotation with or without a perversion. Any three mutually orthogonal unit vectors in the sphere may be changed into any three others. Hence the semi-axes of the first ellipsoid may be carried over by a suitable strain into the semi-axes of the second. The strain is then completely determined and the transformation can be performed in only one way.

142.] The equation of a family of confocal quadric surfaces is

$$\frac{x^2}{a^2 - n} + \frac{y^2}{b^2 - n} + \frac{z^2}{c^2 - n} = 1. \qquad (20)$$

If $\mathbf{r} \cdot \Phi \cdot \mathbf{r} = 1$ and $\mathbf{r} \cdot \Psi \cdot \mathbf{r} = 1$ are two surfaces of the family,

$$\Phi = \frac{\mathbf{i}\,\mathbf{i}}{a^2 - n_1} + \frac{\mathbf{j}\,\mathbf{j}}{b^2 - n_1} + \frac{\mathbf{k}\,\mathbf{k}}{c^2 - n_1}$$

$$\Psi = \frac{\mathbf{i}\,\mathbf{i}}{a^2 - n_2} + \frac{\mathbf{j}\,\mathbf{j}}{b^2 - n_2} + \frac{\mathbf{k}\,\mathbf{k}}{c^2 - n_2}$$

$$\Phi^{-1} = (a^2 - n_1)\,\mathbf{i}\,\mathbf{i} + (b^2 - n_1)\,\mathbf{j}\,\mathbf{j} + (c^2 - n_1)\,\mathbf{k}\,\mathbf{k},$$

$$\Psi^{-1} = (a^2 - n_2)\,\mathbf{i}\,\mathbf{i} + (b^2 - n_2)\,\mathbf{j}\,\mathbf{j} + (c^2 - n_2)\,\mathbf{k}\,\mathbf{k}.$$

Hence $\quad \Phi^{-1} - \Psi^{-1} = (n_2 - n_1)\,(\mathbf{i}\,\mathbf{i} + \mathbf{j}\,\mathbf{j} + \mathbf{k}\,\mathbf{k})$

$$= (n_2 - n_1)\,\mathrm{I}.$$

The necessary and sufficient condition that the two quadrics

$$\mathbf{r} \cdot \Phi \cdot \mathbf{r} = 1$$

and

$$\mathbf{r} \cdot \Psi \cdot \mathbf{r} = 1$$

be confocal, is that the reciprocals of Φ and Ψ differ by a multiple of the idemfactor

$$\Phi^{-1} - \Psi^{-1} = x\,\mathrm{I}. \qquad (21)$$

If two confocal quadrics intersect, they do so at right angles.

Let the quadrics be $\quad \mathbf{r} \cdot \Phi \cdot \mathbf{r} = 1,$

and $\quad \mathbf{r} \cdot \Psi \cdot \mathbf{r} = 1.$

Let $\quad \mathbf{s} = \Phi \cdot \mathbf{r}$ and $\mathbf{s}' = \Psi \cdot \mathbf{r},$

$$\mathbf{r} = \Phi^{-1} \cdot \mathbf{s} \text{ and } \mathbf{r} = \Psi^{-1} \cdot \mathbf{s}'.$$

Then the quadrics may be written in terms of $\mathbf{s}$ and $\mathbf{s}'$ as

$$\mathbf{s} \cdot \Phi^{-1} \cdot \mathbf{s} = 1,$$

and $\quad \mathbf{s}' \cdot \Psi^{-1} \cdot \mathbf{s}' = 1,$

where by the confocal property,

$$\Phi^{-1} - \Psi^{-1} = x \, \mathrm{I}.$$

If the quadrics intersect at $\mathbf{r}$ the condition for perpendicularity is that the normals $\Phi \cdot \mathbf{r}$ and $\Psi \cdot \mathbf{r}$ be perpendicular. That is,

$$\mathbf{s} \cdot \mathbf{s}' = 0.$$

But $\quad \mathbf{r} = \Psi^{-1} \cdot \mathbf{s}' = \Phi^{-1} \cdot \mathbf{s} = (\Psi^{-1} + x \, \mathrm{I}) \cdot \mathbf{s}$

$$= \Psi^{-1} \cdot \mathbf{s} + x \, \mathbf{s},$$

$$x \, \mathbf{s} \cdot \mathbf{s}' = \mathbf{s}' \cdot \Psi^{-1} \cdot \mathbf{s}' - \mathbf{s} \cdot \Psi^{-1} \cdot \mathbf{s}' = 1 - \mathbf{s} \cdot \Psi^{-1} \cdot \mathbf{s}'.$$

In like manner

$$\mathbf{r} = \Phi^{-1} \cdot \mathbf{s} = \Psi^{-1} \cdot \mathbf{s}' = (\Phi^{-1} - x \, \mathrm{I}) \cdot \mathbf{s}' = \Phi^{-1} \cdot \mathbf{s}' - x \, \mathbf{s}'.$$

$$x \, \mathbf{s} \cdot \mathbf{s}' = \mathbf{s} \cdot \Phi^{-1} \cdot \mathbf{s}' - \mathbf{s} \cdot \Phi^{-1} \cdot \mathbf{s} = \mathbf{s} \cdot \Phi^{-1} \cdot \mathbf{s}' - 1.$$

Add: $\quad 2 \, x \, \mathbf{s} \cdot \mathbf{s}' = \mathbf{s} \cdot (\Phi^{-1} - \Psi^{-1}) \cdot \mathbf{s}' = x \, \mathbf{s} \cdot \mathbf{s}'.$

Hence $\quad \mathbf{s} \cdot \mathbf{s}' = 0,$

and the theorem is proved.

If the parameter n be allowed to vary from $-\infty$ to $+\infty$ the resulting confocal quadrics will consist of three families of which one is ellipsoids; another, hyperboloids of one sheet; and the third, hyperboloids of two sheets. By the foregoing theorem each surface of any one family cuts every surface of the other two orthogonally. The surfaces form a triply orthogonal system. The lines of intersection of two families (say the family of one-sheeted and the family of two-sheeted hyperboloids) cut orthogonally the other family — the family of ellipsoids. The points in which two ellipsoids are cut by these lines are called corresponding points upon the two ellipsoids. It may be shown that the ratios of the components of the radius vector of a point to the axes of the ellipsoid through that point are the same for any two corresponding points.

For let any ellipsoid be given by the dyadic

$$\Phi = \frac{\mathbf{i}\,\mathbf{i}}{a^2} + \frac{\mathbf{j}\,\mathbf{j}}{b^2} + \frac{\mathbf{k}\,\mathbf{k}}{c^2}.$$

The neighboring ellipsoid in the family is represented by the dyadic

$$\Psi = \frac{\mathbf{i}\,\mathbf{i}}{a^2 - d\,n} + \frac{\mathbf{j}\,\mathbf{j}}{b^2 - d\,n} + \frac{\mathbf{k}\,\mathbf{k}}{c^2 - d\,n},$$

$$\Psi^{-1} = \Phi^{-1} + \mathrm{I}\,d\,n.$$

Inasmuch as Φ and Ψ are homologous (see Ex. 8, p. 330) dyadics they may be treated as ordinary scalars in algebra. Therefore if terms of order higher than the first in $d\,n$ be omitted,

$$\Psi = \Phi + \mathrm{I}\,d\,n.$$

The two neighboring ellipsoids are then

$$\mathbf{r} \cdot \Phi \cdot \mathbf{r} = 1,$$

and $$\bar{\mathbf{r}} \cdot (\Phi + \mathrm{I}\,d\,n) \cdot \bar{\mathbf{r}} = 1.$$

By (19) $$\bar{\mathbf{r}} = (\Phi + \mathrm{I}\,d\,n)^{-\frac{1}{2}} \cdot \Phi^{\frac{1}{2}} \cdot \mathbf{r},$$

$$\bar{\mathbf{r}} = (\mathrm{I} + \Phi\,d\,n)^{-\frac{1}{2}} \cdot \mathbf{r},$$

$$\bar{\mathbf{r}} - (\mathrm{I} - \tfrac{1}{2}\,\Phi\,d\,n) \cdot \mathbf{r} = \mathbf{r} - \frac{d\,n}{2}\,\Phi \cdot \mathbf{r}.$$

The vectors $\bar{\mathbf{r}}$ and $\mathbf{r}$ differ by a multiple of $\Phi \cdot \mathbf{r}$ which is perpendicular to the ellipsoid Φ. Hence the termini of $\bar{\mathbf{r}}$ and $\mathbf{r}$ are corresponding points, for they lie upon one of the lines which cut the family of ellipsoids orthogonally. The components of $\mathbf{r}$ and $\bar{\mathbf{r}}$ in the direction $\mathbf{i}$ are $\mathbf{r} \cdot \mathbf{i} = x$ and

$$\bar{\mathbf{r}} \cdot \mathbf{i} = \bar{x} = \mathbf{r} \cdot \mathbf{i} - \frac{d\,n}{2} \mathbf{i} \cdot \Phi \cdot \mathbf{r} = x - \frac{d\,n}{2} \frac{x}{a^2}.$$

The ratio of these components is $$\frac{\bar{x}}{x} = 1 - \frac{d\,n}{2\,a^2}.$$

The axes of the ellipsoids in the direction $\mathbf{i}$ are $\sqrt{a^2 - d\,n}$ and a. Their ratio is

$$\frac{\sqrt{a^2 - dn}}{a} = \frac{a - \frac{1}{2}\frac{dn}{a}}{a} = 1 - \frac{dn}{2\,a^2} = \frac{\bar{x}}{x}.$$

In like manner $$\frac{\sqrt{b^2 - d\,n}}{b} = \frac{\bar{y}}{y} \text{ and } \frac{\sqrt{c^2 - d\,n}}{c} = \frac{\bar{z}}{z}.$$

Hence the ratios of the components of the vectors $\bar{\mathbf{r}}$ and $\mathbf{r}$ drawn to corresponding points upon two neighboring ellipsoids only differ at most by terms of the second order in $d\,n$ from the ratios of the axes of those ellipsoids. It follows immediately that the ratios of the components of the vectors drawn to corresponding points upon any two ellipsoids, separated by a finite variation in the parameter n, only differ at most by terms of the *first* order in $d\,n$ from the ratios of the axes of the ellipsoids and hence must be identical with them. This completes the demonstration.

The Propagation of Light in Crystals[1]

143.] The electromagnetic equations of the ether or of any infinite isotropic medium which is transparent to electromagnetic waves may be written in the form

[1] The following discussion must be regarded as mathematical not physical. To treat the subject from the standpoint of physics would be out of place here.

$$\text{Pot}\,\frac{d^2\,\mathbf{D}}{d\,t^2} + E\,\mathbf{D} + \nabla V = 0, \qquad \nabla \cdot \mathbf{D} = 0 \qquad (1)$$

where $\mathbf{D}$ is the electric displacement satisfying the hydrodynamic equation $\nabla \cdot \mathbf{D} = 0$, E a constant of the dielectric measured in electromagnetic units, and ∇V the electrostatic force due to the function V. In case the medium is not isotropic the constant E becomes a linear vector function Φ. This function is self-conjugate as is evident from physical considerations. For convenience it will be taken as $4\,\pi\,\Phi$. The equations then become

$$\text{Pot}\,\frac{d^2\,\mathbf{D}}{d\,t^2} + 4\,\pi\,\Phi \cdot \mathbf{D} + \nabla V = 0, \qquad \nabla \cdot \mathbf{D} = 0. \quad (2)$$

Operate by $\nabla \times \nabla \times$.

$$\nabla \times \nabla \times \text{Pot}\,\frac{d^2\,\mathbf{D}}{d\,t^2} + 4\,\pi\,\nabla \times \nabla \times \Phi \cdot \mathbf{D} = 0. \qquad (3)$$

The last term disappears owing to the fact that the curl of the derivative ∇V vanishes (page 167). The equation may also be written as

$$\text{Pot}\,\nabla \times \nabla \times \frac{d^2\,\mathbf{D}}{d\,t^2} + 4\,\pi\,\nabla \times \nabla \times \Phi \cdot \mathbf{D} = 0. \qquad (3)'$$

But $$\nabla \times \nabla \times = \nabla\,\nabla \cdot - \nabla \cdot \nabla.$$

Remembering that $\nabla \cdot \mathbf{D}$ and consequently $\nabla \cdot \dfrac{d\,\mathbf{D}}{d\,t}$ and $\nabla \cdot \dfrac{d^2\,\mathbf{D}}{d\,t^2}$ vanish and that Pot $\nabla \cdot \nabla$ is equal to $-\,4\,\pi$ the equation reduces at once to

$$\frac{d^2\,\mathbf{D}}{d\,t^2} = \nabla \cdot \nabla\,\Phi \cdot \mathbf{D} - \nabla\,\nabla \cdot \Phi \cdot \mathbf{D}, \qquad \nabla \cdot \mathbf{D} = 0. \qquad (4)$$

Suppose that the vibration $\mathbf{D}$ is harmonic. Let $\mathbf{r}$ be the vector drawn from a fixed origin to any point of space.

Then $$\mathbf{D} = \mathbf{A} \cos (\mathbf{m} \cdot \mathbf{r} - n\, t)$$

where **A** and **m** are constant vectors and n a constant scalar represents a train of waves. The vibrations take place in the direction **A**. That is, the wave is plane polarized. The wave advances in the direction **m**. The velocity v of that advance is the quotient of n by m, the magnitude of the vector **m**. If this wave is an electromagnetic wave in the medium considered it must satisfy the two equations of that medium. Substitute the value of **D** in those equations.

The value of $\nabla \cdot \mathbf{D}$, $\nabla \cdot \nabla \;\Phi \cdot \mathbf{D}$, and $\nabla \nabla \cdot \Phi \cdot \mathbf{D}$ may be obtained most easily by assuming the direction **i** to be coincident with **m**. $\mathbf{m} \cdot \mathbf{r}$ then reduces to $m\, \mathbf{i} \cdot \mathbf{r}$ which is equal to $m\, x$. The variables y and z no longer occur in **D**. Hence

$$\mathbf{D} = \mathbf{A} \cos (m\, x - n\, t)$$

$$\nabla \cdot \mathbf{D} = \mathbf{i} \cdot \frac{\partial \mathbf{D}}{\partial x} = - \mathbf{i} \cdot \mathbf{A} \quad m \sin (m\, x - n\, t)$$

$$\nabla \cdot \nabla \;\Phi \cdot \mathbf{D} = - m^2 \;\Phi \cdot \mathbf{A} \quad \cos (m\, x - n\, t)$$

$$\nabla \nabla \cdot \Phi \cdot \mathbf{D} = - m^2 \mathbf{i} \;\; \mathbf{i} \cdot \Phi \cdot \mathbf{A} \quad \cos (m\, x - n\, t).$$

Hence $$\nabla \cdot \mathbf{D} = - \mathbf{m} \cdot \mathbf{A} \quad \sin (\mathbf{m} \cdot \mathbf{r} - n\, t)$$

$$\nabla \cdot \nabla \;\Phi \cdot \mathbf{D} = - \mathbf{m} \cdot \mathbf{m} \quad \Phi \cdot \mathbf{D}$$

$$\nabla \nabla \cdot \Phi \cdot \mathbf{D} = - \mathbf{m}\, \mathbf{m} \cdot \Phi \cdot \mathbf{D}.$$

Moreover $$\frac{d^2 \mathbf{D}}{d\, t^2} = - n^2\, \mathbf{D}.$$

Hence if the harmonic vibration **D** is to satisfy the equations (4) of the medium

$$n^2\, \mathbf{D} = \mathbf{m} \cdot \mathbf{m} \quad \Phi \cdot \mathbf{D} - \mathbf{m}\, \mathbf{m} \cdot \Phi \cdot \mathbf{D} \tag{5}$$

and $$\mathbf{m} \cdot \mathbf{A} = 0. \tag{6}$$

The latter equation states at once that *the vibrations must be transverse* to the direction **m** of propagation of the waves. The former equation may be put in the form

$$\mathbf{D} = \frac{\mathbf{m} \cdot \mathbf{m}}{n^2}\, \Phi \cdot \mathbf{D} - \frac{\mathbf{m}\,\mathbf{m}}{n^2} \cdot \Phi \cdot \mathbf{D}. \qquad (5)'$$

Introduce $$\mathbf{s} = \frac{\mathbf{m}}{n}.$$

The vector **s** is in the direction of advance **m**. The magnitude of **s** is the quotient of m by n. This is the reciprocal of the velocity of the wave. The vector **s** may therefore be called the wave-slowness.

$$\mathbf{D} = \mathbf{s} \cdot \mathbf{s}\ \ \Phi \cdot \mathbf{D} - \mathbf{s}\,\mathbf{s} \cdot \Phi \cdot \mathbf{D}.$$

This may also be written as

$$\mathbf{D} = -(\mathbf{s} \times \mathbf{s} \times \Phi \cdot \mathbf{D}) = \mathbf{s} \times (\Phi \cdot \mathbf{D}) \times \mathbf{s}.$$

Dividing by the scalar factor $\cos(m\,x - n\,t)$,

$$\mathbf{A} = \mathbf{s} \times (\Phi \cdot \mathbf{A}) \times \mathbf{s} = \mathbf{s} \cdot \mathbf{s}\ \ \Phi \cdot \mathbf{A} - \mathbf{s}\,\mathbf{s} \cdot \Phi \cdot \mathbf{A}. \qquad (7)$$

It is evident that the wave slowness **s** depends not at all upon the phrase of the vibration but only upon its direction. The motion of a wave not plane polarized may be discussed by decomposing the wave into waves which are plane polarized.

144.] Let **a** be a vector drawn in the direction **A** of the displacement and let the magnitude of **a** be so determined that

$$\mathbf{a} \cdot \Phi \cdot \mathbf{a} = 1. \qquad (8)$$

The equation (7) then becomes reduced to the form

$$\mathbf{a} = \mathbf{s} \times (\Phi \cdot \mathbf{a}) \times \mathbf{s} = \mathbf{s} \cdot \mathbf{s}\ \ \Phi \cdot \mathbf{a} = \mathbf{s}\,\mathbf{s} \cdot \Phi \cdot \mathbf{a} \qquad (9)$$

$$\mathbf{a} \cdot \Phi \cdot \mathbf{a} = 1. \qquad (8)$$

These are the equations by which the discussion of the velocity or rather the slowness of propagation of a wave in different directions in a non-isotropic medium may be carried on.

$$\mathbf{a} \cdot \mathbf{a} = \mathbf{s} \cdot \mathbf{s} \quad \mathbf{a} \cdot \Phi \cdot \mathbf{a} = \mathbf{s} \cdot \mathbf{s}. \qquad (10)$$

Hence the wave slowness $\mathbf{s}$ due to a displacement in the direction $\mathbf{a}$ is equal in magnitude (but not in direction) to the radius vector drawn in the ellipsoid $\mathbf{a}\cdot\Phi\cdot\mathbf{a} = 1$ in that direction.

$$\mathbf{a}\times\mathbf{a} = 0 = \mathbf{s}\cdot\mathbf{s}\;\;\mathbf{a}\times\Phi\cdot\mathbf{a} - \mathbf{a}\times\mathbf{s}\;\;\mathbf{s}\cdot\Phi\cdot\mathbf{a}$$

$$0 = \mathbf{s}\cdot\mathbf{s}\,(\mathbf{a}\times\Phi\cdot\mathbf{a})\cdot\Phi\cdot\mathbf{a} = \mathbf{a}\times\mathbf{s}\cdot\Phi\cdot\mathbf{a}\;\;\mathbf{s}\cdot\Phi\cdot\mathbf{a}.$$

But the first term contains $\Phi\cdot\mathbf{a}$ twice and vanishes. Hence

$$\mathbf{a}\times\mathbf{s}\cdot\Phi\cdot\mathbf{a} = [\mathbf{a}\;\;\mathbf{s}\;\;\Phi\cdot\mathbf{a}] = 0. \qquad (11)$$

The wave-slowness $\mathbf{s}$ therefore lies in a plane with the direction $\mathbf{a}$ of displacement and the normal $\Phi\cdot\mathbf{a}$ drawn to the ellipsoid $\mathbf{a}\cdot\Phi\cdot\mathbf{a} = 1$ at the terminus of $\mathbf{a}$. Since $\mathbf{s}$ is perpendicular to $\mathbf{a}$ and equal in magnitude to $\mathbf{a}$ it is evidently completely determined except as regards sign when the direction $\mathbf{a}$ is known. Given the direction of displacement the line of advance of the wave compatible with the displacement is completely determined, the velocity of the advance is likewise known. The wave however may advance in either direction along that line. By reference to page 386, equation (11) is seen to be the condition that $\mathbf{a}$ shall be one of the principal axes of the ellipsoid formed by passing a plane through the ellipsoid perpendicular to $\mathbf{s}$. Hence for any given direction of advance there are two possible lines of displacement. These are the principal axes of the ellipse cut from the ellipsoid $\mathbf{a}\cdot\Phi\cdot\mathbf{a} = 1$ by a plane passed through the center perpendicular to the line of advance. To these statements concerning the determinateness of $\mathbf{s}$ when $\mathbf{a}$ is given and of $\mathbf{a}$ when $\mathbf{s}$ is given just such exceptions occur as are obvious geometrically. If $\mathbf{a}$ and $\Phi\cdot\mathbf{a}$ are parallel $\mathbf{s}$ may have any direction perpendicular to $\mathbf{a}$. This happens when $\mathbf{a}$ is directed along one of the principal axes of the ellipsoid. If $\mathbf{s}$ is perpendicular to one of the circular sections of the ellipsoid $\mathbf{a}$ may have any direction in the plane of the section.

When the direction of displacement is allowed to vary the slowness $\mathbf{s}$ varies. To obtain the locus of the terminus of $\mathbf{s}$, $\mathbf{a}$ must be eliminated from the equation

$$\mathbf{a} = \mathbf{s} \cdot \mathbf{s} \ \Phi \cdot \mathbf{a} - \mathbf{s}\mathbf{s} \cdot \Phi \cdot \mathbf{a}$$

or

$$(\mathrm{I} - \mathbf{s} \cdot \mathbf{s} \ \Phi + \mathbf{s}\mathbf{s} \cdot \Phi) \cdot \mathbf{a} = 0. \qquad (12)$$

The dyadic in the parenthesis is planar because it annihilates vectors parallel to $\mathbf{a}$. The third or determinant is zero. This gives immediately

$$(\mathrm{I} - \mathbf{s} \cdot \mathbf{s} \ \Phi + \mathbf{s}\mathbf{s} \cdot \Phi)_3 = 0,$$

or

$$(\Phi^{-1} - \mathbf{s} \cdot \mathbf{s} \ \mathrm{I} + \mathbf{s}\mathbf{s})_3 = 0. \qquad (13)$$

This is a scalar equation in the vector $\mathbf{s}$. It is the locus of the extremity of $\mathbf{s}$ when $\mathbf{a}$ is given all possible directions. A number of transformations may be made. By Ex. 19, p. 331,

$$(\Phi + \mathbf{e}\,\mathbf{f})_3 = \Phi_3 + \mathbf{e} \cdot \Phi_2 \cdot \mathbf{f} = \Phi_3 + \mathbf{e} \cdot \Phi_C^{-1} \cdot \mathbf{f} \ \Phi_3.$$

Hence

$$(\Phi^{-1} - \mathbf{s} \cdot \mathbf{s}\mathrm{I})_3 + \mathbf{s} \cdot (\Phi^{-1} - \mathbf{s} \cdot \mathbf{s}\mathrm{I})_C^{-1} \cdot \mathbf{s} \, (\Phi^{-1} - \mathbf{s} \cdot \mathbf{s}\mathrm{I})_3 = 0.$$

Dividing out the common factor and remembering that Φ is self-conjugate.

$$1 + \mathbf{s} \cdot (\Phi^{-1} - \mathbf{s} \cdot \mathbf{s}\,\mathrm{I})^{-1} \cdot \mathbf{s} = 0.$$

or

$$1 + \mathbf{s} \cdot \frac{\Phi}{\mathrm{I} - \mathbf{s} \cdot \mathbf{s} \ \Phi} \cdot \mathbf{s} = 0,$$

$$\frac{\mathbf{s} \cdot \mathrm{I} \cdot \mathbf{s}}{\mathbf{s} \cdot \mathbf{s}} + \mathbf{s} \cdot \frac{\Phi}{\mathrm{I} - \mathbf{s} \cdot \mathbf{s} \ \Phi} \cdot \mathbf{s} = 0$$

$$\mathbf{s} \cdot \left(\mathrm{I} + \frac{\mathbf{s} \cdot \mathbf{s} \ \Phi}{\mathrm{I} - \mathbf{s} \cdot \mathbf{s} \ \Phi}\right) \cdot \mathbf{s} = 0.$$

Hence

$$\mathbf{s} \cdot \frac{\mathrm{I}}{\mathrm{I} - \mathbf{s} \cdot \mathbf{s} \ \Phi} \cdot \mathbf{s} = 0. \qquad (14)$$

Let

$$\Phi = \frac{\mathbf{i}\,\mathbf{i}}{a^2} + \frac{\mathbf{j}\,\mathbf{j}}{b^2} + \frac{\mathbf{k}\,\mathbf{k}}{c^2}$$

$$\frac{\mathrm{I}}{\mathrm{I}-\mathbf{s}\cdot\mathbf{s}\,\Phi} = \left(\frac{1}{1-\frac{s^2}{a^2}}\right)\mathbf{i}\,\mathbf{i} + \left(\frac{1}{1-\frac{s^2}{b^2}}\right)\mathbf{j}\,\mathbf{j} + \left(\frac{1}{1-\frac{s^2}{c^2}}\right)\mathbf{k}\,\mathbf{k}.$$

Let $\quad \mathbf{s} = x\,\mathbf{i} + y\,\mathbf{j} + z\,\mathbf{k}$ and $s^2 = x^2 + y^2 + z^2$.

Then the equation of the surface in Cartesian coördinates is

$$\frac{x^2}{1-\frac{s^2}{a^2}} + \frac{y^2}{1-\frac{s^2}{b^2}} + \frac{z^2}{1-\frac{s^2}{c^2}} = 0. \qquad (14)'$$

The equation in Cartesian coördinates may be obtained directly from

$$(\Phi^{-1} - \mathbf{s}\cdot\mathbf{s}\ \mathrm{I} + \mathbf{s}\,\mathbf{s})_3 = 0.$$

The determinant of this dyadic is

$$\begin{vmatrix} a^2 - s^2 + x^2 & x\,y & x\,z \\ x\,y & b^2 - s^2 + y^2 & y\,z \\ x\,z & y\,z & c^2 - s^2 + z^2 \end{vmatrix} = 0. \qquad (13)'$$

By means of the relation $s^2 = x^2 + y^2 + z^2$ this assumes the forms

$$\frac{x^2}{s^2 - a^2} + \frac{y^2}{s^2 - b^2} + \frac{z^2}{s^2 - c^2} = 1$$

or

$$\frac{a^2\,x^2}{s^2 - a^2} + \frac{b^2\,y^2}{s^2 - b^2} + \frac{c^2\,z^2}{s^2 - c^2} = 0,$$

or

$$\frac{x^2}{1-\frac{s^2}{a^2}} + \frac{y^2}{1-\frac{s^2}{b^2}} + \frac{z^2}{1-\frac{s^2}{c^2}} = 0.$$

This equation appears to be of the sixth degree. It is however of only the fourth. The terms of the sixth order cancel out.

The vector $\mathbf{s}$ represents the wave-slowness. Suppose that a plane wave polarized in the direction $\mathbf{a}$ passes the origin at a certain instant of time with this slowness. At the end of a unit of time it will have travelled in the direction $\mathbf{s}$, a distance equal to the reciprocal of the magnitude of $\mathbf{s}$. The plane will be in this position represented by the vector $\mathbf{s}$ (page 108).

If
$$\mathbf{s} = u\,\mathbf{i} + v\,\mathbf{j} + w\,\mathbf{k}$$

the plane at the expiration of the unit time cuts off intercepts upon the axes equal to the reciprocals of u, v, w. These quantities are therefore the plane coördinates of the plane. They are connected with the coördinates of the points in the plane by the relation

$$u\,x + v\,y + w\,z = 1.$$

If different plane waves polarized in all possible different directions $\mathbf{a}$ be supposed to pass through the origin at the same instant they will envelop a surface at the end of a unit of time. This surface is known as the wave-surface. The perpendicular upon a tangent plane of the wave-surface is the reciprocal of the slowness and gives the velocity with which the wave travels in that direction. The equation of the wave-surface in plane coördinates u, v, w is identical with the equation for the locus of the terminus of the slowness vector $\mathbf{s}$. The equation is

$$\frac{u^2}{1 - \frac{s^2}{a^2}} + \frac{v^2}{1 - \frac{s^2}{b^2}} + \frac{w^2}{1 - \frac{s^2}{c^2}} = 0 \qquad (15)$$

where $s^2 = u^2 + v^2 + w^2$. This may be written in any of the forms given previously. The surface is known as *Fresnel's Wave-Surface.* The equations in vector form are given on page 397 if the variable vector $\mathbf{s}$ be regarded as determining a plane instead of a point.

145.] In an isotropic medium the direction of a *ray* of light is perpendicular to the wave-front. It is the same as the direction of the wave's advance. The velocity of the ray is equal to the velocity of the wave. In a non-isotropic medium this is no longer true. The ray does not travel perpendicular to the wave-front — that is, in the direction of the wave's advance. And the velocity with which the ray travels is greater than the velocity of the wave. In fact, whereas the wave-front travels off always tangent to the wave-surface, the ray travels along the radius vector drawn to the point of tangency of the wave-plane. The wave-planes envelop the wave-surface; the termini of the rays are situated upon it. Thus in the wave-surface the radius vector represents in magnitude and direction the velocity of a ray and the perpendicular upon the tangent plane represents in magnitude and direction the velocity of the wave. If instead of the wave-surface the surface which is the locus of the extremity of the wave slowness be considered it is seen that the radius vector represents the slowness of the wave; and the perpendicular upon the tangent plane, the slowness of the ray.

Let $\mathbf{v}'$ be the velocity of the ray. Then $\mathbf{s}\cdot\mathbf{v}'=1$ because the extremity of $\mathbf{v}'$ lies in the plane denoted by $\mathbf{s}$. Moreover the condition that $\mathbf{v}'$ be the point of tangency gives $d\,\mathbf{v}'$ perpendicular to $\mathbf{s}$. In like manner if $\mathbf{s}'$ be the slowness of the ray and $\mathbf{v}$ the velocity of the wave, $\mathbf{s}'\cdot\mathbf{v}=1$ and the condition of tangency gives $d\,\mathbf{s}'$ perpendicular to $\mathbf{v}$. Hence

$$\mathbf{s}\cdot\mathbf{v}'=1 \text{ and } \mathbf{s}'\cdot\mathbf{v}=1, \tag{16}$$

and $\mathbf{s}\cdot d\mathbf{v}'=0,\ \mathbf{v}\cdot d\,\mathbf{s}'=0,\ \mathbf{v}'\cdot d\,\mathbf{s}=0,\ \mathbf{s}'\cdot d\,\mathbf{v}=0,$

$\mathbf{v}'$ may be expressed in terms of $\mathbf{a}$, $\mathbf{s}$, and Φ as follows.

$$\mathbf{a}=\mathbf{s}\cdot\mathbf{s}\,\Phi\cdot\mathbf{a}-\mathbf{s}\mathbf{s}\cdot\Phi\cdot\mathbf{a},$$

$$d\,\mathbf{a}=2\,\mathbf{s}\cdot d\,\mathbf{s}\,\Phi\cdot\mathbf{a}-\mathbf{s}\cdot\Phi\cdot\mathbf{a}\,d\,\mathbf{s}+\mathbf{s}\cdot\mathbf{s}\,\Phi\cdot d\,\mathbf{a}$$
$$-\,\mathbf{s}\,d\,\mathbf{s}\cdot\Phi\cdot\mathbf{a}=\mathbf{s}\mathbf{s}\cdot\Phi\cdot d\,\mathbf{a}.$$

Multiply by $\mathbf{a}$ and take account of the relations $\mathbf{a}\cdot\mathbf{s}=0$ and $\mathbf{a}\cdot\Phi\cdot d\,\mathbf{a}=0$ and $\mathbf{a}\cdot\mathbf{a}=\mathbf{s}\cdot\mathbf{s}$. Then

$$\mathbf{s} \cdot d\mathbf{s} - \mathbf{a} \cdot d\mathbf{s}\ \mathbf{s} \cdot \Phi \cdot \mathbf{a} = 0,$$

or

$$d\mathbf{s} \cdot (\mathbf{s} - \mathbf{a}\,\mathbf{s} \cdot \Phi \cdot \mathbf{a}) = 0.$$

But since $\mathbf{v}' \cdot d\mathbf{s} = 0$, $\mathbf{v}'$ and $\mathbf{s} - \mathbf{a}\,\mathbf{s} \cdot \Phi \cdot \mathbf{a}$ have the same direction.

$$\mathbf{v}' = x\,(\mathbf{s} - \mathbf{a}\,\mathbf{s} \cdot \Phi \cdot \mathbf{a}),$$

$$\mathbf{s} \cdot \mathbf{v}' = x\,(\mathbf{s} \cdot \mathbf{s} - \mathbf{s} \cdot \mathbf{a}\,\mathbf{s} \cdot \Phi \cdot \mathbf{a}) = x\,\mathbf{s} \cdot \mathbf{s}.$$

Hence

$$\mathbf{v}' = \frac{\mathbf{s} - \mathbf{a}\,\mathbf{s} \cdot \Phi \cdot \mathbf{a}}{\mathbf{s} \cdot \mathbf{s}}, \qquad (17)$$

$$\mathbf{v}' \cdot \Phi \cdot \mathbf{a} = \frac{\mathbf{s} \cdot \Phi \cdot \mathbf{a} - \mathbf{a} \cdot \Phi \cdot \mathbf{a}\,\mathbf{s} \cdot \Phi \cdot \mathbf{a}}{\mathbf{s} \cdot \mathbf{s}} = 0.$$

Hence the ray velocity $\mathbf{v}'$ is perpendicular to $\Phi \cdot \mathbf{a}$, that is, the ray velocity lies in the tangent plane to the ellipsoid at the extremity of the radius vector $\mathbf{a}$ drawn in the direction of the displacement. Equation (17) shows that $\mathbf{v}'$ is coplanar with $\mathbf{a}$ and $\mathbf{s}$. The vectors $\mathbf{a}$, $\mathbf{s}$, $\Phi \cdot \mathbf{a}$, and $\mathbf{v}'$ therefore lie in one plane. In that plane $\mathbf{s}$ is perpendicular to $\mathbf{a}$; and $\mathbf{v}'$, to $\Phi \cdot \mathbf{a}$. The angle from $\mathbf{s}$ to $\mathbf{v}'$ is equal to the angle from $\mathbf{a}$ to $\Phi \cdot \mathbf{a}$.

Making use of the relations already found (8) (9) (11) (16) (17), it is easy to show that the two systems of vectors

$$\mathbf{a},\quad \mathbf{v}',\quad \mathbf{a} \times \mathbf{v}' \quad \text{and} \quad \Phi \cdot \mathbf{a},\quad \mathbf{s},\quad (\Phi \cdot \mathbf{a}) \times \mathbf{s}$$

are reciprocal systems. If $\Phi \cdot \mathbf{a}$ be replaced by $\mathbf{a}'$ the equations take on the symmetrical form

$$\begin{array}{lll}
\mathbf{s} \cdot \mathbf{a} = 0 & \mathbf{s} \cdot \mathbf{s} = \mathbf{a} \cdot \mathbf{a} & \mathbf{a} \cdot \mathbf{a}' = 1, \\
\mathbf{v}' \cdot \mathbf{a}' = 0 & \mathbf{v}' \cdot \mathbf{v}' = \mathbf{a}' \cdot \mathbf{a}' & \mathbf{s} \cdot \mathbf{v}' = 1, \\
\mathbf{a} = \mathbf{s} \times \mathbf{a}' \times \mathbf{s} & & \mathbf{a}' = \mathbf{v}' \times \mathbf{a} \times \mathbf{v}' \\
\mathbf{s} = \mathbf{a} \times \mathbf{v}' \times \mathbf{a} & & \mathbf{v}' = \mathbf{a}' \times \mathbf{s} \times \mathbf{a}' \\
\mathbf{a} \cdot \Phi \cdot \mathbf{a} = 1 & & \mathbf{a}' \cdot \Phi^{-1} \cdot \mathbf{a}' = 1.
\end{array} \qquad (18)$$

Thus a dual relation exists between the direction of displacement, the ray-velocity, and the ellipsoid Φ on the one hand;

and the normal to the ellipsoid, the wave-slowness, and the ellipsoid Φ^{-1} on the other.

146.] It was seen that if **s** was normal to one of the circular sections of Φ the displacement **a** could take place in any direction in the plane of that section. For all directions in this plane the wave-slowness had the same direction and the same magnitude. Hence the wave-surface has a singular plane perpendicular to **s**. This plane is tangent to the surface along a curve instead of at a single point. Hence if a wave travels in the direction **s** the ray travels along the elements of the cone drawn from the center of the wave-surface to this curve in which the singular plane touches the surface. The two directions **s** which are normal to the circular sections of Φ are called the *primary optic axes*. These are the axes of equal wave velocities but unequal ray velocities.

In like manner **v′** being coplanar with **a** and $\Phi \cdot \mathbf{a}$

$$[\Phi \cdot \mathbf{a} \ \mathbf{v}' \ \mathbf{a}] = [\mathbf{a}' \ \mathbf{v}' \ \Phi^{-1} \cdot \mathbf{a}'] = 0.$$

The last equation states that if a plane be passed through the center of the ellipsoid Φ^{-1} perpendicular to **v′**, then **a′** which is equal to $\Phi \cdot \mathbf{a}$ will be directed along one of the principal axes of the section. Hence if a ray is to take a definite direction **a′** may have one of two directions. It is more convenient however to regard **v′** as a vector determining a plane. The first equation

$$[\Phi \cdot \mathbf{a} \ \mathbf{v}' \ \mathbf{a}] = 0$$

states that **a** is the radius vector drawn in the ellipsoid Φ to the point of tangency of one of the principal elements of the cylinder circumscribed about Φ parallel to **v′**: if by a principal element is meant an element passing through the extremities of the major or minor axes of orthogonal plane sections of that cylinder. Hence given the direction **v′** of the ray, the two possible directions of displacement are those radii vectors of the ellipsoid which lie in the principal planes of the cylinder circumscribed about the ellipsoid parallel to $\mathbf{v}'$.

If the cylinder is one of the two circular cylinders which may be circumscribed about Φ the direction of displacement may be any direction in the plane passed through the center of the ellipsoid and containing the common curve of tangency of the cylinder with the ellipsoid. The ray-velocity for all these directions of displacement has the same direction and the same magnitude. It is therefore a line drawn to one of the singular points of the wave-surface. At this singular point there are an infinite number of tangent planes enveloping a cone. The wave-velocity may be equal in magnitude and direction to the perpendicular drawn from the origin to any of these planes. The directions of the axes of the two circular cylinders circumscriptible about the ellipsoid Φ are the directions of equal ray-velocity but unequal wave-velocity. They are the radii drawn to the singular points of the wave-surface and are called the *secondary optic axes*. If a ray travels along one of the secondary optic axes the wave planes travel along the elements of a cone.

Variable Dyadics. The Differential and Integral Calculus

147.] Hitherto the dyadics considered have been constant. The vectors which entered into their make up and the scalar coefficients which occurred in the expansion in nonion form have been constants. For the elements of the theory and for elementary applications these constant dyadics suffice. The introduction of variable dyadics, however, leads to a simplification and unification of the differential and integral calculus of vectors, and furthermore variable dyadics become a necessity in the more advanced applications — for instance, in the theory of the curvature of surfaces and in the dynamics of a rigid body one point of which is fixed.

Let **W** be a vector function of position in space. Let **r** be the vector drawn from a fixed origin to any point in space.

$$\mathbf{r} = x\,\mathbf{i} + y\,\mathbf{j} + z\,\mathbf{k},$$

$$d\,\mathbf{r} = d\,x\,\mathbf{i} + d\,y\,\mathbf{j} + d\,z\,\mathbf{k},$$

$$d\,\mathbf{W} = d\,x\,\frac{\partial\,\mathbf{W}}{\partial\,x} + d\,y\,\frac{\partial\,\mathbf{W}}{\partial\,y} + d\,z\,\frac{\partial\,\mathbf{W}}{\partial\,z}.$$

Hence $$d\,\mathbf{W} = d\,\mathbf{r} \cdot \left\{ \mathbf{i}\,\frac{\partial\,\mathbf{W}}{\partial\,x} + \mathbf{j}\,\frac{\partial\,\mathbf{W}}{\partial\,y} + \mathbf{k}\,\frac{\partial\,\mathbf{W}}{\partial\,z} \right\}.$$

The expression enclosed in the braces is a dyadic. It thus appears that the differential of **W** is a linear function of $d\,\mathbf{r}$, the differential change of position. The antecedents are **i**, **j**, **k**, and the consequents the first partial derivatives of **W** with respect of x, y, z. The expression is found in a manner precisely analogous to *del* and will in fact be denoted by $\nabla\,\mathbf{W}$.

$$\nabla\,\mathbf{W} = \mathbf{i}\,\frac{\partial\,\mathbf{W}}{\partial\,x} + \mathbf{j}\,\frac{\partial\,\mathbf{W}}{\partial\,y} + \mathbf{k}\,\frac{\partial\,\mathbf{W}}{\partial\,z}. \qquad (1)$$

Then $$d\,\mathbf{W} = d\,\mathbf{r} \cdot \nabla\,\mathbf{W}. \qquad (2)$$

This equation is like the one for the differential of a scalar function V.

$$d\,V = d\,\mathbf{r} \cdot \nabla\,V.$$

It may be regarded as defining $\nabla\,\mathbf{W}$. If expanded into nonion form $\nabla\,\mathbf{W}$ becomes

$$\begin{aligned} \nabla\,\mathbf{W} = {} & \mathbf{i}\,\mathbf{i}\,\frac{\partial X}{\partial\,x} + \mathbf{i}\,\mathbf{j}\,\frac{\partial Y}{\partial\,x} + \mathbf{i}\,\mathbf{k}\,\frac{\partial Z}{\partial\,x} \\ & + \mathbf{j}\,\mathbf{i}\,\frac{\partial X}{\partial\,y} + \mathbf{j}\,\mathbf{j}\,\frac{\partial Y}{\partial\,y} + \mathbf{i}\,\mathbf{k}\,\frac{\partial Z}{\partial\,y} \qquad (3) \\ & + \mathbf{k}\,\mathbf{i}\,\frac{\partial X}{\partial\,z} + \mathbf{k}\,\mathbf{j}\,\frac{\partial Y}{\partial\,z} + \mathbf{k}\,\mathbf{k}\,\frac{\partial Z}{\partial\,z}, \end{aligned}$$

if $$\mathbf{W} = X\,\mathbf{i} + Y\,\mathbf{j} + Z\,\mathbf{k}.$$

The operators $\nabla \cdot$ and $\nabla \times$ which were applied to a vector function now become superfluous from a purely analytic standpoint. For they are nothing more nor less than the scalar and the vector of the dyadic $\nabla \mathbf{W}$.

$$\operatorname{div} \mathbf{W} = \nabla \cdot \mathbf{W} = (\nabla \mathbf{W})_S, \tag{4}$$

$$\operatorname{curl} \mathbf{W} = \nabla \times \mathbf{W} = (\nabla \mathbf{W})_\times. \tag{5}$$

The analytic advantages of the introduction of the variable dyadic $\nabla \mathbf{W}$ are therefore these. In the first place the operator ∇ may be applied to a vector function just as to a scalar function. In the second place the two operators $\nabla \cdot$ and $\nabla \times$ are reduced to positions as functions of the dyadic $\nabla \mathbf{W}$. On the other hand from the standpoint of physics nothing is to be gained and indeed much may be lost if the important interpretations of $\nabla \cdot \mathbf{W}$ and $\nabla \times \mathbf{W}$ as the divergence and curl of $\mathbf{W}$ be forgotten and their places taken by the analytic idea of the scalar and vector of $\nabla \mathbf{W}$.

If the vector function $\mathbf{W}$ be the derivative of a scalar function V,

$$d\,\mathbf{W} = d \nabla V = d\,\mathbf{r} \cdot \nabla \nabla V,$$

where

$$\begin{aligned} \nabla \nabla V = {} & \mathbf{i}\,\mathbf{i} \frac{\partial^2 V}{\partial x^2} + \mathbf{i}\,\mathbf{j} \frac{\partial^2 V}{\partial x\, \partial y} + \mathbf{i}\,\mathbf{k} \frac{\partial^2 V}{\partial x\, \partial z}, \\ & + \mathbf{j}\,\mathbf{i} \frac{\partial^2 V}{\partial y\, \partial x} + \mathbf{j}\,\mathbf{j} \frac{\partial^2 V}{\partial y^2} + \mathbf{j}\,\mathbf{k} \frac{\partial^2 V}{\partial y\, \partial z}, \\ & + \mathbf{k}\,\mathbf{j} \frac{\partial^2 V}{\partial z\, \partial x} + \mathbf{k}\,\mathbf{j} \frac{\partial^2 V}{\partial z\, \partial y} + \mathbf{k}\,\mathbf{k} \frac{\partial^2 V}{\partial z^2}. \end{aligned} \tag{6}$$

The result of applying ∇ twice to a scalar function is seen to be a dyadic. This dyadic is self-conjugate. Its vector $\nabla \times \nabla V$ is zero; its scalar $\nabla \cdot \nabla V$ is evidently

$$\nabla \cdot \nabla V = (\nabla \nabla V)_S = \frac{\partial^2 V}{\partial x^2} + \frac{\partial^2 V}{\partial y^2} + \frac{\partial^2 V}{\partial z^2}.$$

If an attempt were made to apply the operator ∇ symbolically to a scalar function V *three* times, the result would be a sum of twenty-seven terms like

$$\mathbf{i}\,\mathbf{i}\,\mathbf{i}\,\frac{\partial^3 V}{\partial x^3}, \quad \mathbf{i}\,\mathbf{j}\,\mathbf{k}\,\frac{\partial^3 V}{\partial x\,\partial y\,\partial z}, \text{ etc.}$$

This is a *triadic*. Three vectors are placed in juxtaposition without any sign of multiplication. Such expressions will not be discussed here. In a similar manner if the operator ∇ be applied *twice* to a vector function, or *once* to a dyadic function of position in space, the result will be a triadic and hence outside the limits set to the discussion here. The operators $\nabla\times$ and $\nabla\cdot$ may however be applied to a dyadic Φ to yield respectively a dyadic and a vector.

$$\nabla\times\Phi = \mathbf{i}\times\frac{\partial\,\Phi}{\partial x} + \mathbf{j}\times\frac{\partial\,\Phi}{\partial y} + \mathbf{k}\times\frac{\partial\,\Phi}{\partial z}, \tag{7}$$

$$\nabla\cdot\Phi = \mathbf{i}\cdot\frac{\partial\,\Phi}{\partial x} + \mathbf{j}\cdot\frac{\partial\,\Phi}{\partial y} + \mathbf{k}\cdot\frac{\partial\,\Phi}{\partial z}. \tag{8}$$

If $$\Phi = \mathbf{u}\,\mathbf{i} + \mathbf{v}\,\mathbf{j} + \mathbf{w}\,\mathbf{k},$$

where $\mathbf{u}$, $\mathbf{v}$, $\mathbf{w}$ are vector functions of position in space,

$$\nabla\times\Phi = \nabla\times\mathbf{u}\;\mathbf{i} + \nabla\times\mathbf{v}\;\mathbf{j} + \nabla\times\mathbf{w}\;\mathbf{k}, \tag{7$'$}$$

and $$\nabla\cdot\Phi = \nabla\cdot\mathbf{u}\;\mathbf{i} + \nabla\cdot\mathbf{v}\;\mathbf{j} + \nabla\cdot\mathbf{w}\;\mathbf{k}. \tag{8$'$}$$

Or if $$\Phi = \mathbf{i}\;\mathbf{u} + \mathbf{j}\;\mathbf{v} + \mathbf{k}\;\mathbf{w},$$

$$\nabla\times\Phi = \mathbf{i}\left(\frac{\partial\,\mathbf{w}}{\partial y} - \frac{\partial\,\mathbf{v}}{\partial z}\right) + \mathbf{j}\left(\frac{\partial\,\mathbf{u}}{\partial z} - \frac{\partial\,\mathbf{w}}{\partial x}\right) + \mathbf{k}\left(\frac{\partial\,\mathbf{v}}{\partial x} - \frac{\partial\,\mathbf{u}}{\partial y}\right) \tag{7$''$}$$

and $$\nabla\cdot\Phi = \frac{\partial\,\mathbf{u}}{\partial x} + \frac{\partial\,\mathbf{v}}{\partial y} + \frac{\partial\,\mathbf{w}}{\partial z}. \tag{8$''$}$$

In a similar manner the scalar operators $(\mathbf{a}\cdot\nabla)$ and $(\nabla\cdot\nabla)$ may be applied to Φ. The result is in each case a dyadic,

$$(\mathbf{a} \cdot \nabla)\, \Phi = a_1 \frac{\partial \Phi}{\partial x} + a_2 \frac{\partial \Phi}{\partial y} + a_3 \frac{\partial \Phi}{\partial z}, \qquad (9)$$

$$(\nabla \cdot \nabla)\, \Phi = \frac{\partial^2 \Phi}{\partial x^2} + \frac{\partial^2 \Phi}{\partial y^2} + \frac{\partial^2 \Phi}{\partial z^2}. \qquad (10)$$

The operators $\mathbf{a} \cdot \nabla$ and $\nabla \cdot \nabla$ as applied to vector functions are no longer necessarily to be regarded as single operators. The individual steps may be carried out by means of the dyadic $\nabla \mathbf{W}$.

$$(\mathbf{a} \cdot \nabla)\, \mathbf{W} = \mathbf{a} \cdot (\nabla \mathbf{W}) = \mathbf{a} \cdot \nabla \mathbf{W},$$

$$(\nabla \cdot \nabla)\, \mathbf{W} = \nabla \cdot (\nabla \mathbf{W}) = \nabla \cdot \nabla \mathbf{W}.$$

But when applied to a dyadic the operators cannot be interpreted as made up of two successive steps without making use of the triadic $\nabla \Phi$. The parentheses however may be removed without danger of confusion just as they were removed in case of a vector function before the introduction of the dyadic.

Formulæ similar to those upon page 176 may be given for differentiating products in the case that the differentiation lead to dyadics.

$$\nabla (u \mathbf{v}) = \nabla u \mathbf{v} + u \nabla \mathbf{v},$$

$$\nabla (\mathbf{v} \times \mathbf{w}) = \nabla \mathbf{v} \times \mathbf{w} - \nabla \mathbf{w} \times \mathbf{v},$$

$$\nabla \times (\mathbf{v} \times \mathbf{w}) = \mathbf{w} \cdot \nabla \mathbf{v} - \nabla \cdot \mathbf{v} \mathbf{w} - \mathbf{v} \circ \nabla \mathbf{w} + \nabla \cdot \mathbf{w} \mathbf{v},$$

$$\nabla (\mathbf{v} \cdot \mathbf{w}) = \nabla \mathbf{v} \cdot \mathbf{w} + \nabla \mathbf{w} \cdot \mathbf{v},$$

$$\nabla \cdot (\mathbf{v} \mathbf{w}) = \nabla \cdot \mathbf{v} \mathbf{w} + \mathbf{v} \cdot \nabla \mathbf{w}.$$

$$\nabla \times (\mathbf{v} \mathbf{w}) = \nabla \times \mathbf{v} \mathbf{w} - \mathbf{v} \times \nabla \mathbf{w},$$

$$\nabla \cdot (u \Phi) = \nabla u \cdot \Phi + u \nabla \cdot \Phi,$$

$$\nabla \times \nabla \times \Phi = \nabla \nabla \cdot \Phi - \nabla \cdot \nabla \Phi, \text{ etc.}$$

The principle in these and all similar cases is that enunciated before, namely: The operator ∇ may be treated symbolically as a vector. The differentiations which it implies must be carried out in turn upon each factor of a product to which it is applied. Thus

$$\nabla \times (\mathbf{v}\,\mathbf{w}) = [\nabla \times (\mathbf{v}\,\mathbf{w})]_{\mathbf{v}} + [\nabla \times (\mathbf{v}\,\mathbf{w})]_{\mathbf{w}},$$

$$[\nabla \times (\mathbf{v}\,\mathbf{w})]_{\mathbf{w}} = \nabla \times \mathbf{v}\,\mathbf{w},$$

$$[\nabla \times (\mathbf{v}\,\mathbf{w})]_{\mathbf{v}} = - [\mathbf{v} \times \nabla\,\mathbf{w}]_{\mathbf{v}} = - \mathbf{v} \times \nabla\,\mathbf{w}.$$

Hence $$\nabla \times (\mathbf{v}\,\mathbf{w}) = \nabla \times \mathbf{v}\,\mathbf{w} - \mathbf{v} \times \nabla\,\mathbf{w}.$$

Again $$\nabla\,(\mathbf{v} \times \mathbf{w}) = [\nabla\,(\mathbf{v} \times \mathbf{w})]_{\mathbf{v}} + [\nabla\,(\mathbf{v} \times \mathbf{w})]_{\mathbf{w}},$$

$$[\nabla\,(\mathbf{v} \times \mathbf{w})]_{\mathbf{w}} = \nabla\,\mathbf{v} \times \mathbf{w},$$

$$[\nabla\,(\mathbf{v} \times \mathbf{w})]_{\mathbf{v}} = [- \nabla\,(\mathbf{w} \times \mathbf{v})]_{\mathbf{v}} = - \nabla\,\mathbf{w} \times \mathbf{v}.$$

Hence $$\nabla\,(\mathbf{v} \times \mathbf{w}) = \nabla\,\mathbf{v} \times \mathbf{w} - \nabla\,\mathbf{w} \times \mathbf{v}.$$

148.] It was seen (Art. 79) that if C denote an arc of a curve of which the initial point is $\mathbf{r}_0$ and the final point is $\mathbf{r}$ the line integral of the derivative of a scalar function taken along the curve is equal to the difference between the values of that function at $\mathbf{r}$ and $\mathbf{r}_0$.

$$\int_C d\,\mathbf{r} \cdot \nabla V = V(\mathbf{r}) - V(\mathbf{r}_0).$$

In like manner $$\int_C d\,\mathbf{r} \cdot \nabla \mathbf{W} = \mathbf{W}\,(\mathbf{r}) - \mathbf{W}\,(\mathbf{r}_0),$$

and $$\int_{\mathrm{O}} d\,\mathbf{r} \cdot \nabla \mathbf{W} = 0.$$

It may be well to note that the integrals

$$\int d\,\mathbf{r} \cdot \nabla \mathbf{W} \quad \text{and} \quad \int \nabla \mathbf{W} \cdot d\,\mathbf{r}$$

are by no means the same thing. $\nabla \mathbf{W}$ is a dyadic. The vector $d\,\mathbf{r}$ cannot be placed arbitrarily upon either side of it.

Owing to the fundamental equation (2) the differential $d\mathbf{r}$ necessarily precedes $\nabla \mathbf{W}$. The differentials must be written before the integrands in most cases. For the sake of uniformity they always will be so placed.

Passing to surface integrals, the following formulæ, some of which have been given before and some of which are new, may be mentioned.

$$\iint d\mathbf{a} \times \nabla V = \int d\mathbf{r}\ V$$

$$\iint d\mathbf{a} \times \nabla \mathbf{W} = \int d\mathbf{r}\ \mathbf{W}$$

$$\iint d\mathbf{a} \cdot \nabla \times \mathbf{W} = \int d\mathbf{r} \cdot \mathbf{W}$$

$$\iint d\mathbf{a} \cdot \nabla \times \Phi = \int d\mathbf{r} \cdot \Phi.$$

The line integrals are taken over the complete bounding curve of the surface over which the surface integrals are taken. In like manner the following relations exist between volume and surface integrals.

$$\iiint dv\ \nabla V = \iint d\mathbf{a}\ V$$

$$\iiint dv\ \nabla \mathbf{W} = \iint d\mathbf{a}\ \mathbf{W}$$

$$\iiint dv\ \nabla \cdot \mathbf{W} = \iint d\mathbf{a} \cdot \mathbf{W}$$

$$\iiint dv\ \nabla \times \mathbf{W} = \iint d\mathbf{a} \times \mathbf{W}$$

$$\iiint dv\ \nabla \cdot \Phi = \iint d\mathbf{a} \cdot \Phi$$

$$\iiint dv\ \nabla \times \Phi = \iint d\mathbf{a} \times \Phi.$$

The surface integrals are taken over the complete bounding surface of the region throughout which the volume integrals are taken.

Numerous formulæ of integration by parts like those upon page 250 might be added. The reader will find no difficulty in obtaining them for himself. The integrating operators may also be extended to other cases. To the potentials of scalar and vector functions the potential, *Pot* Φ, of a dyadic may be added. The Newtonian of a vector function and the Laplacian and Maxwellian of dyadics may be defined.

$$\text{Pot}\ \Phi = \iiint \frac{\Phi\,(x_2, y_2, z_2)}{r_{12}}\, d\,v_2.$$

$$\text{New}\ \mathbf{W} = \iiint \frac{\mathbf{r}_{12}\ \mathbf{W}\,(x_2, y_2, z_2)}{\mathbf{r}^3_{12}}\, d\,v_2,$$

$$\text{Lap}\ \Phi = \iiint \frac{\mathbf{r}_{12} \times \Phi\,(x_2, y_2, z_2)}{\mathbf{r}^3_{12}}\, d\,v_2,$$

$$\text{Max}\ \Phi = \iiint \frac{\mathbf{r}_{12} \cdot \Phi\,(x_2, y_2, z_2)}{\mathbf{r}^3_{12}}\, d\,v_2.$$

The analytic theory of these integrals may be developed as before. The most natural way in which the demonstrations may be given is by considering the vector function $\mathbf{W}$ as the sum of its components,

$$\mathbf{W} = X\,\mathbf{i} + Y\,\mathbf{j} + Z\,\mathbf{k}$$

and the dyadic Φ as expressed with the constant consequents $\mathbf{i}$, $\mathbf{j}$, $\mathbf{k}$ and variable antecedents $\mathbf{u}$, $\mathbf{v}$, $\mathbf{w}$, or *vice versa*,

$$\Phi = \mathbf{u}\,\mathbf{i} + \mathbf{v}\,\mathbf{j} + \mathbf{w}\,\mathbf{k}.$$

These matters will be left at this point. The object of entering upon them at all was to indicate the natural extensions which occur when variable dyadics are considered. These extensions differ so slightly from the simple cases which have gone before that it is far better to leave the details to be worked out or assumed from analogy whenever they may be needed rather than to attempt to develop them in advance. It is sufficient merely to mention what the extensions are and how they may be treated.

The Curvature of Surfaces[1]

149.] There are two different methods of treating the curvature of surfaces. In one the surface is expressed in parametic form by three equations

$$x = f_1(u, v), \quad y = f_2(u, v), \quad z = f_3(u, v),$$

or

$$\mathbf{r} = \mathbf{f}(u, v).$$

This is analogous to the method followed (Art. 57) in dealing with curvature and torsion of curves and it is the method employed by Fehr in the book to which reference was made. In the second method the surface is expressed by a single equation connecting the variables x, y, z — thus

$$F(x, y, z) = 0.$$

The latter method of treatments affords a simple application of the differential calculus of variable dyadics. Moreover, the dyadics lead naturally to the most important results connected with the elementary theory of surfaces.

Let $\mathbf{r}$ be a radius vector drawn from an arbitrary fixed origin to a variable point of the surface. The increment $d\,\mathbf{r}$ lies in the surface or in the tangent plane drawn to the surface at the terminus of $\mathbf{r}$.

$$d\,F = d\,\mathbf{r} \cdot \nabla F = 0.$$

Hence the derivative ∇F is collinear with the normal to the surface. Moreover, inasmuch as F and the negative of F when

[1] Much of what follows is practically free from the use of dyadics. This is especially true of the treatment of geodetics, Arts. 155-157.

equated to zero give the same geometric surface, ∇F may be considered as the normal upon either side of the surface. In case the surface belongs to the family defined by

$$F(x, y, z) = \text{const.}$$

the normal ∇F lies upon that side upon which the constant increases. Let ∇F be represented by $\mathbf{N}$ the magnitude of which may be denoted by N, and let $\mathbf{n}$ be a unit normal drawn in the direction of $\mathbf{N}$. Then

$$\mathbf{N} = \nabla F,$$

$$\mathbf{N} \cdot \mathbf{N} = N^2 = \nabla F \cdot \nabla F, \tag{1}$$

$$\mathbf{n} = \frac{1}{N} \nabla F.$$

If $\mathbf{s}$ is the vector drawn to any point in the tangent plane at the terminus of $\mathbf{r}$, $\mathbf{s}-\mathbf{r}$ and $\mathbf{n}$ are perpendicular. Consequently the equation of the tangent plane is

$$(\mathbf{s} - \mathbf{r}) \cdot \nabla F = 0.$$

and in like manner the equation of the normal line is

$$(\mathbf{s} - \mathbf{r}) \times \nabla F = 0,$$

or

$$\mathbf{s} = \mathbf{r} + k \nabla F$$

where k is a variable parameter. These equations may be translated into Cartesian form and give the familiar results.

150.] The variation $d\mathbf{n}$ of the unit normal to a surface plays an important part in the theory of curvature. $d\mathbf{n}$ is perpendicular to $\mathbf{n}$ because $\mathbf{n}$ is a unit vector.

$$\mathbf{n} = \frac{1}{N} \nabla F$$

$$d\mathbf{n} = -\frac{dN}{N^2} \nabla F + \frac{1}{N} d \nabla F,$$

$$d\mathbf{n} = \frac{1}{N} d\mathbf{r} \cdot \nabla \nabla F - \frac{1}{N^2} d\mathbf{r} \cdot \nabla N \nabla F.$$

The dyadic $\mathrm{I} - \mathbf{n}\mathbf{n}$ is an idemfactor for all vectors perpendicular to $\mathbf{n}$ and an annihilator for vectors parallel to $\mathbf{n}$. Hence

$$d\mathbf{n} \cdot (\mathrm{I} - \mathbf{n}\mathbf{n}) = d\mathbf{n},$$

and

$$\nabla F \cdot (\mathrm{I} - \mathbf{n}\mathbf{n}) = 0,$$

$$d\mathbf{n} = d\left(\frac{1}{N} \nabla F\right) = \nabla F \, d\frac{1}{N} + \frac{1}{N} d \nabla F.$$

Hence

$$d\mathbf{n} = \frac{1}{N} d\mathbf{r} \cdot \nabla \nabla F \cdot (\mathrm{I} - \mathbf{n}\mathbf{n}).$$

But

$$d\mathbf{r} = d\mathbf{r} \cdot (\mathrm{I} - \mathbf{n}\mathbf{n}).$$

Hence

$$d\mathbf{n} = d\mathbf{r} \cdot \frac{(\mathrm{I} - \mathbf{n}\mathbf{n}) \cdot \nabla \nabla F \cdot (\mathrm{I} - \mathbf{n}\mathbf{n})}{N}. \tag{2}$$

Let

$$\Phi = \frac{(\mathrm{I} - \mathbf{n}\mathbf{n}) \cdot \nabla \nabla F \cdot (\mathrm{I} - \mathbf{n}\mathbf{n})}{N}. \tag{3}$$

Then

$$d\mathbf{n} = d\mathbf{r} \cdot \Phi. \tag{4}$$

In the vicinity of any point upon a surface the variation $d\mathbf{n}$ of the unit normal is a linear function of the variation of the radius vector $\mathbf{r}$.

The dyadic Φ is self-conjugate. For

$$N \Phi_C = (\mathrm{I} - \mathbf{n}\mathbf{n})_C \cdot (\nabla \nabla F)_C \cdot (\mathrm{I} - \mathbf{n}\mathbf{n})_C.$$

Evidently $(\mathrm{I} - \mathbf{n}\mathbf{n})_C = (\mathrm{I} - \mathbf{n}\mathbf{n})$ and by (6) Art. 147 $\nabla \nabla F$ is self-conjugate. Hence Φ_C is equal to Φ. When applied to a vector parallel to $\mathbf{n}$, the dyadic Φ produces zero. It is therefore planar and in fact uniplanar because self-conjugate. The antecedents and the consequents lie in the tangent plane to the surface. It is possible (Art. 116) to reduce Φ to the form

$$\Phi = a\ \mathbf{i}'\mathbf{i}' + b\ \mathbf{j}'\mathbf{j}' \qquad (5)$$

where $\mathbf{i}'$ and $\mathbf{j}'$ are two perpendicular unit vectors lying in the tangent plane and a and b are positive or negative scalars.

$$d\,\mathbf{n} = d\,\mathbf{r} \cdot (a\ \mathbf{i}'\mathbf{i}' + b\ \mathbf{j}'\mathbf{j}').$$

The vectors $\mathbf{i}'$, $\mathbf{j}'$ and the scalars a, b vary from point to point of the surface. The dyadic Φ is variable.

151.] The conic $\mathbf{r} \cdot \Phi \cdot \mathbf{r} = 1$ is called the indicatrix of the surface at the point in question. If this conic is an ellipse, that is, if a and b have the same sign, the surface is convex at the point; but if the conic is an hyperbola, that is, if a and b have opposite signs the surface is concavo-convex. The curve $\mathbf{r} \cdot \Phi \cdot \mathbf{r} = 1$ may be regarded as approximately equal to the intersection of the surface with a plane drawn parallel to the tangent plane and near to it. If $\mathbf{r} \cdot \Phi \cdot \mathbf{r}$ be set equal to zero the result is a pair of straight lines. These are the asymptotes of the conic. If they are real the conic is an hyperbola; if imaginary, an ellipse. Two directions on the surface which are parallel to conjugate diameters of the conic are called conjugate directions. The directions on the surface which coincide with the directions of the principal axes $\mathbf{i}'$, $\mathbf{j}'$ of the indicatrix are known as the principal directions. They are a special case of conjugate directions. The directions upon the surface which coincide with the directions of the asymptotes of the indicatrix are known as asymptotic directions. In case the surface is convex, the indicatrix is an ellipse and the asymptotic directions are imaginary.

In special cases the dyadic Φ may be such that the coefficients a and b are equal. Φ may then be reduced to the form

$$\Phi = a\ (\mathbf{i}'\mathbf{i}' + \mathbf{j}'\mathbf{j}') \qquad (5)'$$

in an infinite number of ways. The directions $\mathbf{i}'$ and $\mathbf{j}'$ may be any two perpendicular directions. The indicatrix becomes a circle. Any pair of perpendicular diameters of this circle give principal directions upon the surface. Such a point is called an *umbilic*. The surface in the neighborhood of an umbilic is convex. The asymptotic directions are imaginary. In another special case the dyadic Φ becomes linear and reducible to the form

$$\Phi = a\ \mathbf{i}'\mathbf{i}'. \qquad (5)''$$

The indicatrix consists of a pair of parallel lines perpendicular to $\mathbf{i}'$. Such a point is called a parabolic point of the surface. The further discussion of these and other special cases will be omitted.

The quadric surfaces afford examples of the various kinds of points. The ellipsoid and the hyperboloid of two sheets are convex. The indicatrix of points upon them is an ellipse. The hyperboloid of one sheet is concavo-convex. The indicatrix of points upon it is an hyperbola. The indicatrix of any point upon a sphere is a circle. The points are all umbilics. The indicatrix of any point upon a cone or cylinder is a pair of parallel lines. The points are parabolic. A surface in general may have upon it points of all types—elliptic, hyperbolic, parabolic, and umbilical.

152.] A line of principal curvature upon a surface is a curve which has at each point the direction of one of the principal axes of the indicatrix. The direction of the curve at a point is always one of the principal directions on the surface at that point. Through any given point upon a surface two perpendicular lines of principal curvature pass. Thus the lines of curvature divide the surface into a system of infinitesimal rectangles. An asymptotic line upon a surface is a curve which has at each point the direction of the asymptotes of the indicatrix. The direction of the curve at a point is always one of the asymptotic directions upon the surface. Through any given point of a surface two asymptotic lines pass. These lines are imaginary if the surface is convex. Even when real they do not in general intersect at right angles. The angle between the two asymptotic lines at any point is bisected by the lines of curvature which pass through that point.

The necessary and sufficient condition that a curve upon a surface be a line of principal curvature is that as one advances along that curve, the increment of $d\mathbf{n}$, the unit normal to the surface is parallel to the line of advance. For

$$d\mathbf{n} = \Phi \cdot d\mathbf{r} = (a\ \mathbf{i}'\mathbf{i}' + b\ \mathbf{j}'\mathbf{j}') \cdot d\mathbf{r}$$

$$d\mathbf{r} = x\ \mathbf{i}' + y\mathbf{j}'.$$

Then evidently $d\mathbf{n}$ and $d\mathbf{r}$ are parallel when and only when $d\mathbf{r}$ is parallel to $\mathbf{i}'$ or $\mathbf{j}'$. The statement is therefore proved. It is frequently taken as the definition of lines of curvature. The differential equation of a line of curvature is

$$d\mathbf{n} \times d\mathbf{r} = 0. \tag{6}$$

Another method of statement is that the normal to the surface, the increment $d\mathbf{n}$ of the normal, and the element $d\mathbf{r}$ of the surface lie in one plane when and only when the element $d\mathbf{r}$ is an element of a line of principal curvature. The differential equation then becomes

$$[\mathbf{n}\quad d\mathbf{n}\quad d\mathbf{r}] = 0. \tag{7}$$

The necessary and sufficient condition that a curve upon a surface be an asymptotic line, is that as one advances along that curve the increment of the unit normal to the surface is perpendicular to the line of advance. For

$$d\mathbf{n} = d\mathbf{r} \cdot \Phi$$

$$d\mathbf{n} \cdot d\mathbf{r} = d\mathbf{r} \cdot \Phi \cdot d\mathbf{r}.$$

If then $d\mathbf{n} \cdot d\mathbf{r}$ is zero $d\mathbf{r} \cdot \Phi \cdot d\mathbf{r}$ is zero. Hence $d\mathbf{r}$ is an asymptotic direction. The statement is therefore proved. It is frequently taken as the definition of asymptotic lines. The differential equation of an asymptotic line is

$$d\,\mathbf{n} \cdot d\,\mathbf{r} = 0. \tag{8}$$

153.] Let P be a given point upon a surface and $\mathbf{n}$ the normal to the surface at P. Pass a plane p through $\mathbf{n}$. This plane p is normal to the surface and cuts out a plane section. Consider the curvature of this plane section at the point P. Let $\mathbf{n}'$ be normal to the plane section in the plane of the section. $\mathbf{n}'$ coincides with $\mathbf{n}$ at the point P. But unless the plane p cuts the surface everywhere orthogonally, the normal $\mathbf{n}'$ to the plane section and the normal $\mathbf{n}$ to the surface will not coincide. $d\,\mathbf{n}$ and $d\,\mathbf{n}'$ will also be different. The curvature of the plane section lying in p is (Art. 57).

$$\mathbf{C} = \frac{d\,\mathbf{t}}{d\,s} = \frac{d^2\,\mathbf{r}}{d\,s^2}.$$

As far as numerical value is concerned the increment of the unit tangent $\mathbf{t}$ and the increment of the unit normal $\mathbf{n}'$ are equal. Moreover, the quotient of $d\,\mathbf{r}$ by $d\,s$ is a unit vector in the direction of $d\,\mathbf{n}'$. Consequently the scalar value of $\mathbf{C}$ is

$$C = \frac{d\,\mathbf{n}'}{d\,s} \cdot \frac{d\,\mathbf{r}}{d\,s} = \frac{d\,\mathbf{n}' \cdot d\,\mathbf{r}}{d\,s^2}.$$

By hypothesis $\quad \mathbf{n} = \mathbf{n}'$ at P $\quad$ and $\quad \mathbf{n} \cdot d\,\mathbf{r} = \mathbf{n}' \cdot d\,\mathbf{r} = 0,$

$$d\,(\mathbf{n} \cdot d\,\mathbf{r}) = d\,\mathbf{n} \cdot d\,\mathbf{r} + \mathbf{n} \cdot d^2\,\mathbf{r} = 0,$$

$$d\,(\mathbf{n}' \cdot d\,\mathbf{r}) = d\,\mathbf{n}' \cdot d\,\mathbf{r} + \mathbf{n}' \cdot d^2\,\mathbf{r} = 0.$$

Hence $\quad d\,\mathbf{n} \cdot d\,\mathbf{r} + \mathbf{n} \cdot d^2\,\mathbf{r} = d\,\mathbf{n}' \cdot d\,\mathbf{r} + \mathbf{n}' \cdot d^2\,\mathbf{r}.$

Since $\mathbf{n}$ and $\mathbf{n}'$ are equal at P,

$$d\,\mathbf{n} \cdot d\,\mathbf{r} = d\,\mathbf{n}' \cdot d\,\mathbf{r}.$$

Hence $$C = \frac{d\,\mathbf{n} \cdot d\,\mathbf{r}}{d\,s^2} = \frac{d\,\mathbf{r} \cdot \Phi \cdot d\,\mathbf{r}}{d\,s^2} = \frac{d\,\mathbf{r} \cdot \Phi \cdot d\,\mathbf{r}}{d\,\mathbf{r} \cdot d\,\mathbf{r}} \tag{9}$$

$$C = a \frac{(\mathbf{i}' \cdot d\,\mathbf{r})^2}{d\,\mathbf{r} \cdot d\,\mathbf{r}} + b \frac{(\mathbf{j}' \cdot d\,\mathbf{r})^2}{d\,\mathbf{r} \cdot d\,\mathbf{r}}.$$

Hence $$C = a \cos^2 (\mathbf{i}', d\,\mathbf{r}) + b \cos^2 (\mathbf{j}', d\,\mathbf{r}),$$

or $$C = a \cos^2 (\mathbf{i}', d\,\mathbf{r}) + b \sin^2 (\mathbf{i}', d\,\mathbf{r}). \qquad (10)$$

The interpretation of this formula for the curvature of a normal section is as follows: When the plane p turns about the normal to the surface from $\mathbf{i}'$ to $\mathbf{j}'$, the curvature C of the plane section varies from the value a when the plane passes through the principal direction $\mathbf{i}'$, to the value b when it passes through the other principal direction $\mathbf{j}'$. The values of the curvature have algebraically a maximum and minimum in the directions of the principal lines of curvature. If a and b have unlike signs, that is, if the surface is concavo-convex at P, there exist two directions for which the curvature of a normal section vanishes. These are the asymptotic directions.

154.] The sum of the curvatures in two normal sections at right angles to one another is constant and independent of the actual position of those sections. For the curvature in one section is

$$C_1 = a \cos^2 (\mathbf{i}', d\,\mathbf{r}) + b \sin^2 (\mathbf{i}', d\,\mathbf{r}),$$

and in the section at right angles to this

$$C_2 = a \sin^2 (\mathbf{i}', d\,\mathbf{r}) + b \cos^2 (\mathbf{i}', d\,\mathbf{r}).$$

Hence $$C_1 + C_2 = a + b = \Phi_S \qquad (11)$$

which proves the statement.

It is easy to show that the invariant Φ_{2S} is equal to the product of the curvatures a and b of the lines of principal curvature.

$$\Phi_{2S} = a\,b$$

Hence the equation $$x^2 - \Phi_S\, x + \Phi_{2S} = 0 \qquad (12)$$

is the quadratic equation which determines the principal curvatures a and b at any point of the surface. By means of this equation the scalar quantities a and b may be found in terms of $F(x, y, z)$.

$$\Phi = \frac{(\mathrm{I} - \mathbf{n}\mathbf{n}) \cdot \nabla\nabla F \cdot (\mathrm{I} - \mathbf{n}\mathbf{n})}{N}$$

$$\Phi = \frac{\nabla\nabla F - 2\,\mathbf{n}\mathbf{n} \cdot \nabla\nabla F + \mathbf{n}\mathbf{n} \cdot \nabla\nabla F \cdot \mathbf{n}\mathbf{n}}{N}$$

$$(\mathbf{n}\mathbf{n} \cdot \nabla\nabla F \cdot \mathbf{n}\mathbf{n})_S = (\mathbf{n}\mathbf{n} \cdot \mathbf{n}\mathbf{n} \cdot \nabla\nabla F)_S = (\mathbf{n}\mathbf{n} \cdot \nabla\nabla F)_S$$

Hence $$\Phi_S = \frac{(\nabla\nabla F)_S}{N} - \frac{(\mathbf{n}\mathbf{n} \cdot \nabla\nabla F)_S}{N}$$

$$(\nabla\nabla F)_S = \nabla \cdot \nabla F,$$

$$(\mathbf{n}\mathbf{n} \cdot \nabla\nabla F)_S = \mathbf{n}\mathbf{n} : \nabla\nabla F = \mathbf{n} \cdot \nabla\nabla F \cdot \mathbf{n}.$$

Hence $$\Phi_S = \frac{\nabla \cdot \nabla F}{N} - \frac{\nabla F\,\nabla F : \nabla\nabla F}{N^3} \qquad (13)$$

or $$\Phi_S = \frac{\nabla \cdot \nabla F}{N} - \frac{\nabla F \cdot \nabla\nabla F \cdot \nabla F}{N^3}. \qquad (13)'$$

These expressions may be written out in Cartesian coördinates, but they are extremely long. The Cartesian expressions for Φ_{2S} are even longer. The vector expression may be obtained as follows:

$$\Phi_S = \frac{(\mathrm{I} - \mathbf{n}\mathbf{n})_2 \cdot (\nabla\nabla F)_2 \cdot (\mathrm{I} - \mathbf{n}\mathbf{n})_2}{N^2}$$

$$(\mathrm{I} - \mathbf{n}\mathbf{n})_2 = \mathbf{n}\mathbf{n}.$$

Hence

$$\Phi_{2S} = \frac{\nabla F \cdot (\nabla\nabla F)_2 \cdot \nabla F}{N^4} = \frac{\nabla F\,\nabla F : (\nabla\nabla F)_2}{N^4}. \qquad (14)$$

155.] Given any curve upon a surface. Let $\mathbf{t}$ be a unit tangent to the curve, $\mathbf{n}$ a unit normal to the surface and $\mathbf{m}$ a vector defined as $\mathbf{n} \times \mathbf{t}$. The three vectors $\mathbf{n}$, $\mathbf{t}$, $\mathbf{m}$ constitute an $\mathbf{i}$, $\mathbf{j}$, $\mathbf{k}$ system. The vector $\mathbf{t}$ is parallel to the element $d\,\mathbf{r}$. Hence the condition for a line of curvature becomes

$$\mathbf{t} \times d\,\mathbf{n} = 0. \tag{15}$$

Hence
$$\mathbf{m} \cdot d\,\mathbf{n} = 0$$

$$d\,(\mathbf{m} \cdot \mathbf{n}) = 0 = \mathbf{m} \cdot d\mathbf{n} + \mathbf{n} \cdot d\,\mathbf{m}.$$

Hence
$$\mathbf{n} \cdot d\,\mathbf{m} = 0.$$

Moreover
$$\mathbf{m} \cdot d\,\mathbf{m} = 0.$$

Hence
$$\mathbf{t} \times d\,\mathbf{m} = 0, \tag{16}$$

or
$$d\,\mathbf{m} \times d\,\mathbf{n} = 0. \tag{16)'}$$

The increments of $\mathbf{m}$ and of $\mathbf{n}$ and of $\mathbf{r}$ are all parallel in case of a line of principal curvature.

A geodetic line upon a surface is a curve whose osculating plane at each point is perpendicular to the surface. That the geodetic line is the shortest line which can be drawn between two points upon a surface may be seen from the following considerations of mechanics. Let the surface be smooth and let a smooth elastic string which is constrained to lie in the surface be stretched between any two points of it. The string acting under its own tensions will take a position of equilibrium along the shortest curve which can be drawn upon the surface between the two given points. Inasmuch as the string is at rest upon the surface the normal reactions of the surface must lie in the osculating plane of the curve. Hence that plane is normal to the surface at every point of the curve and the curve itself is a geodetic line.

The vectors $\mathbf{t}$ and $d\,\mathbf{t}$ lie in the osculating plane and determine that plane. In case the curve is a geodetic, the normal to the osculating plane lies in the surface and consequently is perpendicular to the normal $\mathbf{n}$. Hence

$$\mathbf{n} \cdot \mathbf{t} \times d\, \mathbf{t} = 0,$$

$$\mathbf{n} \times \mathbf{t} \cdot d\, \mathbf{t} = 0 \qquad (17)$$

or

$$\mathbf{m} \cdot d\, \mathbf{t} = 0.$$

The differential equation of a geodetic line is therefore

$$[\mathbf{n} \;\; d\,\mathbf{r} \;\; d^2\,\mathbf{r}] = 0. \qquad (18)$$

Unlike the differential equations of the lines of curvature and the asymptotic line, this equation is of the second order. The surface is therefore covered over with a doubly infinite system of geodetics. Through any two points of the surface one geodetic may be drawn.

As one advances along any curve upon a surface there is necessarily some turning up and down, that is, around the axis **m**, due to the fact that the surface is curved. There may or may not be any turning to the right or left. If one advances along a curve such that there is no turning to the right or left, but only the unavoidable turning up and down, it is to be expected that the advance is along the shortest possible route — that is, along a geodetic. Such is in fact the case. The total amount of deviation from a straight line is $d\,\mathbf{t}$. Since **n**, **t**, **m** form an **i**, **j**, **k** system

$$\mathbf{I} = \mathbf{t}\mathbf{t} + \mathbf{n}\mathbf{n} + \mathbf{m}\mathbf{m}.$$

Hence

$$d\,\mathbf{t} = \mathbf{t}\,\mathbf{t} \cdot d\,\mathbf{t} + \mathbf{n}\,\mathbf{n} \cdot d\,\mathbf{t} + \mathbf{m}\,\mathbf{m} \cdot d\,\mathbf{t}.$$

Since **t** is a unit vector the first term vanishes. The second term represents the amount of turning up and down; the third term, the amount to the right or left. Hence $\mathbf{m} \cdot d\,\mathbf{t}$ is the proper measure of this part of the deviation from a straightest line. In case the curve is a geodetic this term vanishes as was expected.

156.] A curve or surface may be mapped upon a unit sphere by the method of parallel normals. A fixed origin is assumed, from which the unit normal **n** at the point P of a given surface is laid off. The terminus P' of this normal lies upon the surface of a sphere. If the normals to a surface at all points P of a curve are thus constructed from the same origin, the points P' will trace a curve upon the surface of a unit sphere. This curve is called the spherical image of the given curve. In like manner a whole region T of the surface may be mapped upon a region T' the sphere. The region T' upon the sphere has been called the *hodogram* of the region T upon the surface. If $d\mathbf{r}$ be an element of arc upon the surface the corresponding element upon the unit sphere is

$$d\mathbf{n} = \Phi \cdot d\mathbf{r}.$$

If $d\mathbf{a}$ be an element of area upon the surface, the corresponding element upon the sphere is $d\mathbf{a}'$ where (Art. 124).

$$d\mathbf{a}' = \Phi_2 \cdot d\mathbf{a}.$$

$$\Phi = a\ \mathbf{i}'\mathbf{i}' + b\ \mathbf{j}'\mathbf{j}'$$

$$\Phi_2 = a\,b\ \ \mathbf{i}' \times \mathbf{j}'\ \mathbf{i}' \times \mathbf{j}' = a\,b\ \ \mathbf{n}\mathbf{n}.$$

Hence
$$d\mathbf{a}' = a\,b\ \ \mathbf{n}\mathbf{n} \cdot d\mathbf{a}. \tag{19}$$

The ratio of an element of surface at a point P to the area of its hodogram is equal to the product of the principal radii of curvature at P or to the reciprocal of the product of the principal curvatures at P.

It was seen that the measure of turning to the right or left is $\mathbf{m} \cdot d\mathbf{t}$. If then C is any curve drawn upon a surface the total amount of turning in advancing along the curve is the integral.

$$\int_C \mathbf{m} \cdot d\mathbf{t}. \tag{20}$$

For any closed curve this integral may be evaluated in a manner analogous to that employed (page 190) in the proof of Stokes's theorem. Consider two curves C and C' near together. The variation which the integral undergoes when the curve of integration is changed from C to C' is

$$\delta \int \mathbf{m} \cdot d\mathbf{t}.$$

$$\delta \int \mathbf{m} \cdot d\mathbf{t} = \int \delta (\mathbf{m} \cdot d\mathbf{t}) = \int \delta \mathbf{m} \cdot d\mathbf{t} + \int \mathbf{m} \cdot \delta\, d\mathbf{t}$$

$$d(\mathbf{m} \cdot \delta \mathbf{t}) = d\mathbf{m} \cdot \delta \mathbf{t} + \mathbf{m} \cdot d\, \delta \mathbf{t}$$

$$\delta \int \mathbf{m} \cdot d\mathbf{t} = \int \delta \mathbf{m} \cdot d\mathbf{t} - \int d\mathbf{m} \cdot \delta \mathbf{t} + \int d(\mathbf{m} \cdot \delta \mathbf{t}).$$

The integral of the perfect differential $d(\mathbf{m} \cdot \delta \mathbf{t})$ vanishes when taken around a closed curve. Hence

$$\delta \int \mathbf{m} \cdot d\mathbf{t} = \int \delta \mathbf{m} \cdot d\mathbf{t} - \int d\mathbf{m} \cdot \delta \mathbf{t}.$$

The idemfactor is $\quad \mathrm{I} = \mathbf{t}\mathbf{t} + \mathbf{n}\mathbf{n} + \mathbf{m}\mathbf{m},$

$$\delta \mathbf{m} \cdot d\mathbf{t} = \delta \mathbf{m} \cdot \mathrm{I} \cdot d\mathbf{t} = \delta \mathbf{m} \cdot \mathbf{n}\mathbf{n} \cdot d\mathbf{t},$$

for $\mathbf{t} \cdot d\mathbf{t}$ and $\delta \mathbf{m} \cdot \mathbf{m}$ vanish. A similar transformation may be effected upon the term $d\mathbf{m} \cdot \delta \mathbf{t}$. Then

$$\delta \int \mathbf{m} \cdot d\mathbf{t} = \int (\delta \mathbf{m} \cdot \mathbf{n}\ \mathbf{n} \cdot d\mathbf{t} - d\mathbf{m} \cdot \mathbf{n}\ \mathbf{n} \cdot \delta \mathbf{t}).$$

By differentiating the relations $\mathbf{m} \cdot \mathbf{n} = 0$ and $\mathbf{n} \cdot \mathbf{t} = 0$ it is seen that

$$\delta \mathbf{m} \cdot \mathbf{n} = -\mathbf{m} \cdot \delta \mathbf{n} \qquad \mathbf{n} \cdot \delta \mathbf{t} = -\delta \mathbf{n} \cdot \mathbf{t}$$

$$d\mathbf{m} \cdot \mathbf{n} = -\mathbf{m} \cdot d\mathbf{n} \qquad \mathbf{n} \cdot d\mathbf{t} = -d\mathbf{n} \cdot \mathbf{t}.$$

Hence $\quad \delta \int \mathbf{m} \cdot d\mathbf{t} = \int (\mathbf{m} \cdot \delta \mathbf{n}\ \ \mathbf{t} \cdot d\mathbf{n} - \mathbf{m} \cdot d\mathbf{n}\ \ \mathbf{t} \cdot \delta \mathbf{n})$

$$\delta \int \mathbf{m} \cdot d\mathbf{t} = \int (\mathbf{m} \times \mathbf{t} \cdot \delta \mathbf{n} \times d\mathbf{n}) = -\int \mathbf{n} \cdot \delta \mathbf{n} \times d\mathbf{n}.$$

The differential $\delta \mathbf{n} \times d \mathbf{n}$ represents the element of area in the hodogram upon the unit sphere. The integral

$$\int \mathbf{n} \cdot \delta \mathbf{n} \times d \mathbf{n} = \int \mathbf{n} \cdot d \mathbf{a}'$$

represents the total area of the hodogram of the strip of surface which lies between the curves C and C'. Let the curve C start at a point upon the surface and spread out to any desired size. The total amount of turning which is required in making an infinitesimal circuit about the point is 2π. The total variation in the integral is

$$\int \delta \int \mathbf{m} \cdot d\mathbf{t} = \int \mathbf{m} \cdot d\mathbf{t} - 2\pi.$$

$$\int\int \mathbf{n} \cdot d\mathbf{a}' = H, \tag{21}$$

But if H denote the total area of the hodogram.

Hence $$\int \mathbf{m} \cdot d\mathbf{t} = 2\pi - H,$$

or $$H = 2\pi - \int \mathbf{m} \cdot d\mathbf{t}, \tag{22}$$

or $$H + \int \mathbf{m} \cdot d\mathbf{t} = 2\pi.$$

The area of the hodogram of the region enclosed by any closed curve plus the total amount of turning along that curve is equal to 2π. If the surface in question is convex the area upon the sphere will appear positive when the curve upon the surface is so described that the enclosed area appears positive. If, however, the surface is concavo-convex the area upon the sphere will appear negative. This matter of the sign of the hodogram must be taken into account in the statement made above.

157.] If the closed curve is a polygon whose sides are geodetic lines the amount of turning along each side is zero. The total turning is therefore equal to the sum of the exterior angles of the polygon. The statement becomes: the sum of the exterior angles of a geodetic polygon and of the area of the hodogram of that polygon (taking account of sign) is equal to 2π. Suppose that the polygon reduces to a triangle. If the surface is convex the area of the hodogram is positive and the sum of the exterior angles of the triangle is less than 2π. The sum of the interior angles is therefore greater than π. The sphere or ellipsoid is an example of such a surface. If the surface is concavo-convex the area of the hodogram is negative. The sum of the interior angles of a triangle is in this case less than π. Such a surface is the hyperboloid of one sheet or the pseudosphere. There is an intermediate case in which the hodogram of any geodetic triangle is traced twice in opposite directions and hence the total area is zero. The sum of the interior angles of a triangle upon such a surface is equal to π. Examples of this surface are afforded by the cylinder, cone, and plane.

A surface is said to be *developed* when it is so deformed that lines upon the surface retain their length. Geodetics remain geodetics. One surface is said to be developable or applicable upon another when it can be so deformed as to coincide with the other without altering the lengths of lines. Geodetics upon one surface are changed into geodetics upon the other. The sum of the angles of any geodetic triangle remain unchanged by the process of developing. From this it follows that the total amount of turning along any curve or the area of the hodogram of any portion of a surface are also invariant of the process of developing.

Harmonic Vibrations and Bivectors

158.] The differential equation of rectilinear harmonic motion is

$$\frac{d^2 x}{d t^2} = - n^2 x.$$

The integral of this equation may be reduced by a suitable choice of the constants to the form

$$x = A \sin n t.$$

This represents a vibration back and forth along the X-axis about the point $x = 0$. Let the displacement be denoted by $\mathbf{D}$ in place of x. The equation may be written

$$\mathbf{D} = \mathbf{i} \ A \sin n t.$$

Consider $$\mathbf{D} = \mathbf{i} \ A \sin n t \cos m x.$$

This is a displacement not merely near the point $x = 0$ but along the entire axis of x. At points $x = \frac{2 k \pi}{m}$, where k is a positive or negative integer, the displacement is at all times equal to zero. The equation represents a stationary wave with nodes at these points. At points midway between these the wave has points of maximum vibration. If the equation be regarded as in three variables x, y, z it represents a plane wave the plane of which is perpendicular to the axis of the variable x.

The displacement given by the equation

$$\mathbf{D}_1 = \mathbf{i} \ A_1 \cos (m x - n t) \tag{1}$$

is likewise a plane wave perpendicular to the axis of x but not stationary. The vibration is harmonic and advances along the direction $\mathbf{i}$ with a velocity equal to the quotient of

n by m. If v be the velocity; p the period; and l the wave length,

$$v = \frac{n}{m}, \qquad p = \frac{2\pi}{n}, \qquad l = \frac{2\pi}{m}, \qquad v = \frac{l}{p}. \qquad (2)$$

The displacement

$$\mathbf{D}_2 = \mathbf{j}\ A_2 \cos(m\,x - n\,t)$$

differs from $\mathbf{D}_1$ in the particular that the displacement takes place in the direction $\mathbf{j}$, not in the direction $\mathbf{i}$. The wave as before proceeds in the direction of x with the same velocity. This vibration is transverse instead of longitudinal. By a simple extension it is seen that

$$\mathbf{D} = \mathbf{A} \cos(m\,x - n\,t)$$

is a displacement in the direction $\mathbf{A}$. The wave advances along the direction of x. Hence the vibration is oblique to the wave-front. A still more general form may be obtained by substituting $\mathbf{m} \cdot \mathbf{r}$ for $m\,x$. Then

$$\mathbf{D} = \mathbf{A} \cos(\mathbf{m} \cdot \mathbf{r} - n\,t). \qquad (3)$$

This is a displacement in the direction $\mathbf{A}$. The maximum amount of that displacement is the magnitude of $\mathbf{A}$. The wave advances in the direction $\mathbf{m}$ oblique to the displacement; the velocity, period, and wave-length are as before.

So much for rectilinear harmonic motion. Elliptic harmonic motion may be defined by the equation (p. 117).

$$\frac{d^2\,\mathbf{r}}{d\,t^2} = -\,n^2\,\mathbf{r}.$$

The general integral is obtained as

$$\mathbf{r} = \mathbf{A} \cos n\,t + \mathbf{B} \sin n\,t.$$

The discussion of waves may be carried through as previously. The general wave of elliptic harmonic motion advancing in the direction $\mathbf{m}$ is seen to be

$$\mathbf{D} = \mathbf{A} \cos (\mathbf{m} \cdot \mathbf{r} - n\,t) - \mathbf{B} \sin (\mathbf{m} \cdot \mathbf{r} - n\,t). \qquad (4)$$

$$\frac{d\,\mathbf{D}}{d\,t} = n \left\{ \mathbf{A} \sin (\mathbf{m} \cdot \mathbf{r} - n\,t) + \mathbf{B} \cos (\mathbf{m} \cdot \mathbf{r} - n\,t) \right\} \qquad (5)$$

is the velocity of the displaced point at any moment in the ellipse in which it vibrates. This is of course entirely different from the velocity of the wave.

An interesting result is obtained by adding up the displacement and the velocity multiplied by the imaginary unit $\sqrt{-1}$ and divided by n.

$$\mathbf{D} + \frac{\sqrt{-1}}{n} \frac{d\,\mathbf{D}}{d\,t} = \mathbf{A} \cos (\mathbf{m} \cdot \mathbf{r} - n\,t) - \mathbf{B} \sin (\mathbf{m} \cdot \mathbf{r} - n\,t) \qquad (6)$$

$$+ \sqrt{-1} \left\{ \mathbf{A} \sin (\mathbf{m} \cdot \mathbf{r} - n\,t) + \mathbf{B} \cos (\mathbf{m} \cdot \mathbf{r} - n\,t) \right\}.$$

$$\mathbf{D} + \frac{\sqrt{-1}}{n} \frac{d\,\mathbf{D}}{d\,t} = (\mathbf{A} + \sqrt{-1}\,\mathbf{B})\, e^{\sqrt{-1}(\mathbf{m} \cdot \mathbf{r} - n\,t)} \qquad (6)'$$

The expression here obtained, as far as its form is concerned, is an imaginary vector. It is the sum of two real vectors of which one has been multiplied by the imaginary scalar $\sqrt{-1}$. Such a vector is called a *bivector* or *imaginary* vector. The ordinary imaginary scalars may be called *biscalars*. The use of bivectors is found very convenient in the discussion of elliptic harmonic motion. Indeed any undamped elliptic harmonic plane wave may be represented as above by the product of a bivector and an exponential factor. The real part of the product gives the displacement of any point and the pure imaginary part gives the velocity of displacement divided by n.

159.] The analytic theory of bivectors differs from that of real vectors very much as the analytic theory of biscalars differs from that of real scalars. It is unnecessary to have any distinguishing character for bivectors just as it is needless to have a distinguishing notation for biscalars. The bivector may be regarded as a natural and inevitable extension of the real vector. It is the formal sum of two real vectors of which one has been multiplied by the imaginary unit $\sqrt{-1}$. The usual symbol i will be maintained for $\sqrt{-1}$. There is not much likelihood of confusion with the vector **i** for the reason that the two could hardly be used in the same place and for the further reason that the Italic i and the Clarendon **i** differ considerably in appearance. Whenever it becomes especially convenient to have a separate alphabet for bivectors the small Greek or German letters may be called upon.

A bivector may be expressed in terms of **i, j, k** with complex coefficients.

If
$$\mathbf{r} = \mathbf{r}_1 + i\,\mathbf{r}_2$$
and
$$\mathbf{r}_1 = x_1\,\mathbf{i} + y_1\,\mathbf{j} + z_1\,\mathbf{k},$$
$$\mathbf{r}_2 = x_2\,\mathbf{i} + y_2\,\mathbf{j} + z_2\,\mathbf{k},$$
$$\mathbf{r} = (x_1 + i\,x_2)\,\mathbf{i} + (y_1 + i\,y_2)\,\mathbf{j} + (z_1 + i\,z_2)\,\mathbf{k},$$
or
$$\mathbf{r} = x\,\mathbf{i} + y\,\mathbf{j} + z\,\mathbf{k}.$$

Two bivectors are equal when their real and their imaginary parts are equal. Two bivectors are parallel when one is the product of the other by a scalar (real or imaginary). If a bivector is parallel to a real vector it is said to have a real direction. In other cases it has a complex or imaginary direction. The value of the sum, difference, direct, skew, and indeterminate products of two bivectors is obvious without special definition. These statements may be put into analytic form as follows.

Let $\quad \mathbf{r} = \mathbf{r}_1 + i\,\mathbf{r}_2 \quad$ and $\quad \mathbf{s} = \mathbf{s}_2 + i\,\mathbf{s}_2.$

Then if $\quad \mathbf{r} = \mathbf{s}, \quad \mathbf{r}_1 = \mathbf{s}_1$ and $\mathbf{r}_2 = \mathbf{s}_2$

if $\quad \mathbf{r} \parallel \mathbf{s} \quad \mathbf{r} = x\,\mathbf{s} = (x_1 + i\,x_2)\,\mathbf{s},$

$$\mathbf{r} + \mathbf{s} = (\mathbf{r}_1 + \mathbf{s}_1) + i\,(\mathbf{r}_2 + \mathbf{s}_2),$$

$$\mathbf{r} \cdot \mathbf{s} = (\mathbf{r}_1 \cdot \mathbf{s}_1 - \mathbf{r}_2 \cdot \mathbf{s}_2) + i\,(\mathbf{r}_1 \cdot \mathbf{s}_2 + \mathbf{r}_2 \cdot \mathbf{s}_1),$$

$$\mathbf{r} \times \mathbf{s} = (\mathbf{r}_1 \times \mathbf{s}_1 - \mathbf{r}_2 \times \mathbf{s}_2) + i\,(\mathbf{r}_1 \times \mathbf{s}_2 + \mathbf{r}_2 \times \mathbf{s}_1)$$

$$\mathbf{r}\,\mathbf{s} = (\mathbf{r}_1\,\mathbf{s}_1 + \mathbf{r}_2\,\mathbf{s}_2) + i\,(\mathbf{r}_1\,\mathbf{s}_2 + \mathbf{r}_2\,\mathbf{s}_1).$$

Two bivectors or biscalars are said to be conjugate when their real parts are equal and their pure imaginary parts differ only in sign. The conjugate of a real scalar or vector is equal to the scalar or vector itself. The conjugate of any sort of product of bivectors and biscalars is equal to the product of the conjugates taken in the same order. A similar statement may be made concerning sums and differences.

$$(\mathbf{r}_1 + i\,\mathbf{r}_2) \cdot (\mathbf{r}_1 - i\,\mathbf{r}_2) = \mathbf{r}_1 \cdot \mathbf{r}_1 + \mathbf{r}_2 \cdot \mathbf{r}_2,$$

$$(\mathbf{r}_1 + i\,\mathbf{r}_2) \times (\mathbf{r}_1 - i\,\mathbf{r}_2) = 2\,i\,\mathbf{r}_2 \times \mathbf{r}_1,$$

$$(\mathbf{r}_1 + i\,\mathbf{r}_2)\,(\mathbf{r}_1 - i\,\mathbf{r}_2) = (\mathbf{r}_1\,\mathbf{r}_1 + \mathbf{r}_2\,\mathbf{r}_2) + i\,(\mathbf{r}_2\,\mathbf{r}_1 - \mathbf{r}_1\,\mathbf{r}_2).$$

If the bivector $\mathbf{r} = \mathbf{r}_1 + i\,\mathbf{r}_2$ be multiplied by a root of unity or cyclic factor as it is frequently called, that is, by an imaginary scalar of the form

$$\cos q + i \sin q = a + i b, \tag{7}$$

where $$a^2 + b^2 = 1,$$

the conjugate is multiplied by $a - i\,b$, and hence the four products

$$\mathbf{r}_1 \cdot \mathbf{r}_1 + \mathbf{r}_2 \cdot \mathbf{r}_2, \quad \mathbf{r}_2 \times \mathbf{r}_1, \quad \mathbf{r}_1\,\mathbf{r}_2 + \mathbf{r}_2\,\mathbf{r}_2, \quad \mathbf{r}_2\,\mathbf{r}_1 - \mathbf{r}_1\,\mathbf{r}_2$$

are unaltered by multiplying the bivector $\mathbf{r}$ by such a factor. Thus if

$$\mathbf{r}' = \mathbf{r}_1' + i\,\mathbf{r}_2' = (a + i b)\,(\mathbf{r}_1 + i\,\mathbf{r}_2),$$

$$\mathbf{r}_1' \cdot \mathbf{r}_1' + \mathbf{r}_2' \cdot \mathbf{r}_2' = \mathbf{r}_1 \cdot \mathbf{r}_1 + \mathbf{r}_2 \cdot \mathbf{r}_2, \text{ etc.}$$

160.] A closer examination of the effect of multiplying a bivector by a cyclic factor yields interesting and important geometric results. Let

$$\mathbf{r}_1' + i\,\mathbf{r}_2' = (\cos q + i \sin q)\,(\mathbf{r}_1 + i\,\mathbf{r}_2). \qquad (8)$$

Then

$$\mathbf{r}_1' = \mathbf{r}_1 \cos q - \mathbf{r}_2 \sin q,$$

$$\mathbf{r}_2' = \mathbf{r}_2 \cos q + \mathbf{r}_1 \sin q.$$

By reference to Art. 129 it will be seen that the change produced in the real and imaginary vector parts of a bivector by multiplication with a cyclic factor, is precisely the same as would be produced upon those vectors by a cyclic dyadic

$$\Phi = \mathbf{a}\,\mathbf{a}' + \cos q\,(\mathbf{b}\,\mathbf{b}' + \mathbf{c}\,\mathbf{c}') - \sin q\,(\mathbf{c}\,\mathbf{b}' - \mathbf{b}\,\mathbf{c}')$$

used as a prefactor. $\mathbf{b}$ and $\mathbf{c}$ are supposed to be two vectors collinear respectively with $\mathbf{r}_1$ and $\mathbf{r}_2$. $\mathbf{a}$ is any vector not in their plane. Consider the ellipse of which $\mathbf{r}_1$ and $\mathbf{r}_2$ are a pair of conjugate semi-diameters. It then appears that $\mathbf{r}_1'$ and $\mathbf{r}_2'$ are also a pair of conjugate semi-diameters of that ellipse. They are rotated in the ellipse from $\mathbf{r}_2$ toward $\mathbf{r}_1$, by a sector of which the area is to the area of the whole ellipse as q is to 2π. Such a change of position has been called an elliptic rotation through the sector q.

The ellipse of which $\mathbf{r}_1$ and $\mathbf{r}_2$ are a pair of conjugate semi-diameters is called the *directional ellipse of the bivector* $\mathbf{r}$. When the bivector has a real direction the directional ellipse reduces to a right line in that direction. When the bivector has a complex direction the ellipse is a true ellipse. The angular direction from the real part $\mathbf{r}_1$ to the complex part $\mathbf{r}_2$ is considered as the positive direction in the directional ellipse, and must always be known. If the real and imaginary parts of a bivector turn in the positive direction in the ellipse they are said to be advanced; if in the negative direction they are said to be retarded. Hence *multiplication of a bivector by a cyclic factor retards it in its directional ellipse by a sector equal to the angle of the cyclic factor.*

It is always possible to multiply a bivector by such a cyclic factor that the real and imaginary parts become coincident with the axes of the ellipse and are perpendicular.

$$\mathbf{r} = (\cos q + i \sin q)\,(\mathbf{a} + i\,\mathbf{b}) \text{ where } \mathbf{a} \cdot \mathbf{b} = 0.$$

To accomplish the reduction proceed as follows: Form

$$\mathbf{r} \cdot \mathbf{r} = (\cos 2\,q + i \sin 2\,q)\,(\mathbf{a} + i\,\mathbf{b}) \cdot (\mathbf{a} + i\,\mathbf{b}).$$

If $\mathbf{a} \cdot \mathbf{b} = 0$,

$$\mathbf{r} \cdot \mathbf{r} = (\cos 2\,q + i \sin 2\,q)\,(\mathbf{a} \cdot \mathbf{a} - \mathbf{b} \cdot \mathbf{b}).$$

Let $$\mathbf{r} \cdot \mathbf{r} = a + i\,b,$$

and $$\tan 2\,q = \frac{b}{a}.$$

With this value of q the axes of the directional ellipse are given by the equation

$$\mathbf{a} + i\,\mathbf{b} = (\cos q - i \sin q)\,\mathbf{r}.$$

In case the real and imaginary parts $\mathbf{a}$ and $\mathbf{b}$ of a bivector are equal in magnitude and perpendicular in direction both a and b in the expression for $\mathbf{r} \cdot \mathbf{r}$ vanish. Hence the angle q is indeterminate. The directional ellipse is a circle. A bivector whose directional ellipse is a circle is called a *circular bivector.* The necessary and sufficient condition that a non-vanishing bivector $\mathbf{r}$ be circular is

$$\mathbf{r} \cdot \mathbf{r} = 0, \qquad \mathbf{r} \text{ circular.}$$

If $$\mathbf{r} = x\,\mathbf{i} + y\,\mathbf{j} + z\,\mathbf{k},$$

$$\mathbf{r} \cdot \mathbf{r} = x^2 + y^2 + z^2 = 0.$$

The condition $\mathbf{r} \cdot \mathbf{r} = 0$, which for real vectors implies $\mathbf{r} = 0$, is not sufficient to ensure the vanishing of a bivector. The bivector is circular, not necessarily zero. The condition that a bivector vanish is that the direct product of it by its conjugate vanishes.

$$(\mathbf{r}_1 + i\,\mathbf{r}_2) \cdot (\mathbf{r}_1 - i\,\mathbf{r}_2) = \mathbf{r}_1 \cdot \mathbf{r}_1 + \mathbf{r}_2 \cdot \mathbf{r}_2 = 0,$$

then $$\mathbf{r}_1 = \mathbf{r}_2 = 0 \text{ and } \mathbf{r} = 0.$$

In case the bivector has a real direction it becomes equal to its conjugate and their product becomes equal to $\mathbf{r} \cdot \mathbf{r}$.

161.] The condition that two bivectors be parallel is that one is the product of the other by a scalar factor. Any biscalar factor may be expressed as the product of a cyclic factor and a positive scalar, the modulus of the biscalar. If two bivectors differ by only a cyclic factor their directional ellipses are the same. Hence two parallel vectors have their directional ellipse similar and similarly placed — the ratio of similitude being the modulus of the biscalar. It is evident that any two circular bivectors whose planes coincide are parallel. A circular vector and a non-circular vector cannot be parallel.

The condition that two bivectors be perpendicular

is $$\mathbf{r} \cdot \mathbf{s} = 0,$$

or $$\mathbf{r}_1 \cdot \mathbf{s}_1 - \mathbf{r}_2 \cdot \mathbf{s}_2 = \mathbf{r}_1 \cdot \mathbf{s}_2 + \mathbf{r}_2 \cdot \mathbf{s}_1 = 0.$$

Consider first the case in which the planes of the bivectors coincide. Let

$$\mathbf{r} = a\,(\mathbf{r}_1 + i\,\mathbf{r}_2), \quad \mathbf{s} = b\,(\mathbf{s}_1 + i\,\mathbf{s}_2).$$

The scalars a and b are biscalars. $\mathbf{r}_1$ may be chosen perpendicular to $\mathbf{r}_2$, and $\mathbf{s}_1$ may be taken in the direction of $\mathbf{r}_2$. The condition $\mathbf{r} \cdot \mathbf{s} = 0$ then gives

$$\mathbf{r}_2 \cdot \mathbf{s}_2 = 0 \text{ and } \mathbf{r}_1 \cdot \mathbf{s}_2 + \mathbf{r}_2 \cdot \mathbf{s}_1 = 0.$$

The first equation shows that $\mathbf{r}_2$ and $\mathbf{s}_2$ are perpendicular and hence $\mathbf{s}_1$ and $\mathbf{s}_2$ are perpendicular. Moreover, the second shows that the angular directions from $\mathbf{r}_1$ to $\mathbf{r}_2$ and from $\mathbf{s}_1$ to $\mathbf{s}_2$ are the same, and that the axes of the directional ellipses of $\mathbf{r}$ and $\mathbf{s}$ are proportional.

Hence the conditions for perpendicularity of two bivectors whose planes coincide are that their directional ellipses are similar, the angular direction in both is the same, and the major axes of the ellipses are perpendicular.[1] If both vectors have real directions the conditions degenerate into the perpendicularity of those directions. The conditions therefore hold for real as well as for imaginary vectors.

Let $\mathbf{r}$ and $\mathbf{s}$ be two perpendicular bivectors the planes of which do not coincide. Resolve $\mathbf{r}_1$ and $\mathbf{r}_2$ each into two components respectively parallel and perpendicular to the plane of $\mathbf{s}$. The components perpendicular to that plane contribute nothing to the value of $\mathbf{r} \cdot \mathbf{s}$. Hence the components of $\mathbf{r}_1$ and $\mathbf{r}_2$ parallel to the plane of $\mathbf{s}$ form a bivector $\mathbf{r}'$ which is perpendicular to $\mathbf{s}$. To this bivector and $\mathbf{s}$ the conditions stated above apply. The directional ellipse of the bivector $\mathbf{r}'$ is evidently the projection of the directional ellipse of $\mathbf{r}$ upon the plane of $\mathbf{s}$.

Hence, if two bivectors are perpendicular the directional ellipse of either bivector and the directional ellipse of the other projected upon the plane of that one are similar, have the same angular direction, and have their major axes perpendicular.

162.] Consider a bivector of the type

$$\mathbf{D} = \mathbf{A}\, e,^{i(\mathbf{m} \cdot \mathbf{r} - nt)}, \qquad (9)$$

where $\mathbf{A}$ and $\mathbf{m}$ are bivectors and n is a biscalar. $\mathbf{r}$ is the position vector of a point in space. It is therefore to be con-

[1] It should be noted that the condition of perpendicularity of major axes is not the same as the condition of perpendicularity of real parts and imaginary parts.

sidered as real. t is the scalar variable time and is also to be considered as real. Let

$$\mathbf{A} = \mathbf{A}_1 + i\,\mathbf{A}_2,$$

$$\mathbf{m} = \mathbf{m}_1 + i\,\mathbf{m}_2,$$

$$n = n_1 + i\,n_2,$$

$$\mathbf{D} = (\mathbf{A}_1 + i\,\mathbf{A}_2)\, e^{i\,(\mathbf{m}_1 \cdot \mathbf{r} + i\,\mathbf{m}_2 \cdot \mathbf{r} - n_1 t - i\,n_2 t)}$$

$$\mathbf{D} = (\mathbf{A}_1 + i\,\mathbf{A}_2)\, e^{-\mathbf{m}_2 \cdot \mathbf{r}}\, e^{n_2 t}\, e^{i\,(\mathbf{m}_1 \cdot \mathbf{r} - n_1 t)}. \qquad (10)$$

As has been seen before, the factor $(\mathbf{A}_1 + i\,\mathbf{A}_2)\, e^{i\,(\mathbf{m}_1 \cdot \mathbf{r} - n_1 t)}$ represents a train of plane waves of elliptic harmonic vibrations. The vibrations take place in the plane of $\mathbf{A}_1$ and $\mathbf{A}_2$, in an ellipse of which $\mathbf{A}_1$ and $\mathbf{A}_2$ are conjugate semi-diameters. The displacement of the vibrating point from the center of the ellipse is given by the real part of the factor. The velocity of the point after it has been divided by n_1 is given by the pure imaginary part. The wave advances in the direction $\mathbf{m}_1$. The other factors in the expression are dampers. The factor $e^{-\mathbf{m}_2 \cdot \mathbf{r}}$ is a damper in the direction $\mathbf{m}_2$. As the wave proceeds in the direction $\mathbf{m}_2$ it dies away. The factor $e^{n_2 t}$ is a damper in time. If n_2 is negative the wave dies away as time goes on. If n_2 is positive the wave increases in energy as time increases. The presence (for unlimited time) of any such factor in an expression which represents an actual vibration is clearly inadmissible. It contradicts the law of conservation of energy. In any physical vibration of a conservative system n_2 is necessarily negative or zero.

The general expression (9) therefore represents a train of plane waves of elliptic harmonic vibrations damped in a definite direction and in time. Two such waves may be compounded by adding the bivectors which represent them. If the exponent $\mathbf{m} \cdot \mathbf{r} - n\,t$ is the same for both the resulting train of waves advances in the same direction and has the same period and wave-length as the individual waves. The vibrations, however, take place in a different ellipse. If the waves are

$$\mathbf{A}\, e^{i(\mathbf{m}\cdot\mathbf{r}-nt)} \text{ and } \mathbf{B}\, e^{i(\mathbf{m}\cdot\mathbf{r}-nt)}$$

the resultant is $$(\mathbf{A}+\mathbf{B})\, e^{i(\mathbf{m}\cdot\mathbf{r}-nt)}.$$

By combining two trains of waves which advance in opposite directions but which are in other respects equal a system of stationary waves is obtained.

$$\mathbf{A}\, e^{-\mathbf{m}_2\cdot\mathbf{r}}\, e^{i(\mathbf{m}_1\cdot\mathbf{r}-nt)} + \mathbf{A}\, e^{-\mathbf{m}_2\cdot\mathbf{r}}\, e^{i(-\mathbf{m}_1\cdot\mathbf{r}-nt)} =$$

$$\mathbf{A}\, e^{-\mathbf{m}_2\cdot\mathbf{r}}\, e^{-int}\left(e^{i\mathbf{m}_1\cdot\mathbf{r}} + e^{-i\mathbf{m}_1\cdot\mathbf{r}}\right) = 2\,\mathbf{A}\cos(\mathbf{m}_1\cdot\mathbf{r})\, e^{-\mathbf{m}_2\cdot\mathbf{r}}\, e^{-int}$$

The theory of bivectors and their applications will not be carried further. The object in entering at all upon this very short and condensed discussion of bivectors was first to show the reader how the simple idea of a *direction* has to give way to the more complicated but no less useful idea of a *directional ellipse* when the generalization from real to imaginary vectors is made, and second to set forth the manner in which a single bivector **D** may be employed to represent a train of plane waves of elliptic harmonic vibrations. This application of bivectors may be used to give the Theory of Light a wonderfully simple and elegant treatment.[1]

[1] Such use of bivectors is made by Professor Gibbs in his course of lectures on "*The Electromagnetic Theory of Light*," delivered biannually at Yale University. Bivectors were not used in the second part of this chapter, because in the opinion of the present author they possess no essential advantage over real vectors until the more advanced parts of the theory, rotation of the plane of polarization by magnets and crystals, total and metallic reflection, etc., are reached.

CATALOG OF DOVER BOOKS

BOOKS EXPLAINING SCIENCE AND MATHEMATICS

THE COMMON SENSE OF THE EXACT SCIENCES, W. K. Clifford. Introduction by James Newman, edited by Karl Pearson. For 70 years this has been a guide to classical scientific and mathematical thought. Explains with unusual clarity basic concepts, such as extension of meaning of symbols, characteristics of surface boundaries, properties of plane figures, vectors, Cartesian method of determining position, etc. Long preface by Bertrand Russell. Bibliography of Clifford. Corrected, 130 diagrams redrawn. 249pp. 5⅜ x 8.
T61 Paperbound **$1.60**

SCIENCE THEORY AND MAN, Erwin Schrödinger. This is a complete and unabridged reissue of SCIENCE AND THE HUMAN TEMPERAMENT plus an additional essay: "What is an Elementary Particle?" Nobel Laureate Schrödinger discusses such topics as nature of scientific method, the nature of science, chance and determinism, science and society, conceptual models for physical entities, elementary particles and wave mechanics. Presentation is popular and may be followed by most people with little or no scientific training. "Fine practical preparation for a time when laws of nature, human institutions . . . are undergoing a critical examination without parallel," Waldemar Kaempffert, N. Y. TIMES. 192pp. 5⅜ x 8.
T428 Paperbound **$1.35**

PIONEERS OF SCIENCE, O. Lodge. Eminent scientist-expositor's authoritative, yet elementary survey of great scientific theories. Concentrating on individuals—Copernicus, Brahe, Kepler, Galileo, Descartes, Newton, Laplace, Herschel, Lord Kelvin, and other scientists—the author presents their discoveries in historical order adding biographical material on each man and full, specific explanations of their achievements. The clear and complete treatment of the post-Newtonian astronomers is a feature seldom found in other books on the subject. Index. 120 illustrations. xv + 404pp. 5⅜ x 8. T716 Paperbound **$1.50**

THE EVOLUTION OF SCIENTIFIC THOUGHT FROM NEWTON TO EINSTEIN, A. d'Abro. Einstein's special and general theories of relativity, with their historical implications, are analyzed in non-technical terms. Excellent accounts of the contributions of Newton, Riemann, Weyl, Planck, Eddington, Maxwell, Lorentz and others are treated in terms of space and time, equations of electromagnetics, finiteness of the universe, methodology of science. 21 diagrams. 482pp. 5⅜ x 8. T2 Paperound **$2.00**

THE RISE OF THE NEW PHYSICS, A. d'Abro. A half-million word exposition, formerly titled THE DECLINE OF MECHANISM, for readers not versed in higher mathematics. The only thorough explanation, in everyday language, of the central core of modern mathematical physical theory, treating both classical and modern theoretical physics, and presenting in terms almost anyone can understand the equivalent of 5 years of study of mathematical physics. Scientifically impeccable coverage of mathematical-physical thought from the Newtonian system up through the electronic theories of Dirac and Heisenberg and Fermi's statistics. Combines both history and exposition; provides a broad yet unified and detailed view, with constant comparison of classical and modern views on phenomena and theories. "A must for anyone doing serious study in the physical sciences," JOURNAL OF THE FRANKLIN INSTITUTE. "Extraordinary faculty . . . to explain ideas and theories of theoretical physics in the language of daily life," ISIS. First part of set covers philosophy of science, drawing upon the practice of Newton, Maxwell, Poincaré, Einstein, others, discussing modes of thought, experiment, interpretations of causality, etc. In the second part, 100 pages explain grammar and vocabulary of mathematics, with discussions of functions, groups, series, Fourier series, etc. The remainder is devoted to concrete, detailed coverage of both classical and quantum physics, explaining such topics as analytic mechanics, Hamilton's principle, wave theory of light, electromagnetic waves, groups of transformations, thermodynamics, phase rule, Brownian movement, kinetics, special relativity, Planck's original quantum theory, Bohr's atom, Zeeman effect, Broglie's wave mechanics, Heisenberg's uncertainty, Eigen-values, matrices, scores of other important topics. Discoveries and theories are covered for such men as Alembert, Born, Cantor, Debye, Euler, Foucault, Galois, Gauss, Hadamard, Kelvin, Kepler, Laplace, Maxwell, Pauli, Rayleigh, Volterra, Weyl, Young, more than 180 others. Indexed. 97 illustrations. ix + 982pp. 5⅜ x 8. T3 Volume 1, Paperbound **$2.00**
T4 Volume 2, Paperbound **$2.00**

CONCERNING THE NATURE OF THINGS, Sir William Bragg. Christmas lectures delivered at the Royal Society by Nobel laureate. Why a spinning ball travels in a curved track; how uranium is transmuted to lead, etc. Partial contents: atoms, gases, liquids, crystals, metals, etc. No scientific background needed; wonderful for intelligent child. 32pp. of photos, 57 figures. xii + 232pp. 5⅜ x 8. T31 Paperbound **$1.35**

THE UNIVERSE OF LIGHT, Sir William Bragg. No scientific training needed to read Nobel Prize winner's expansion of his Royal Institute Christmas Lectures. Insight into nature of light, methods and philosophy of science. Explains lenses, reflection, color, resonance, polarization, x-rays, the spectrum, Newton's work with prisms, Huygens' with polarization, Crookes' with cathode ray, etc. Leads into clear statement of 2 major historical theories of light, corpuscle and wave. Dozens of experiments you can do. 199 illus., including 2 full-page color plates. 293pp. 5⅜ x 8. S538 Paperbound **$1.85**

PHYSICS, THE PIONEER SCIENCE, L. W. Taylor. First thorough text to place all important physical phenomena in cultural-historical framework; remains best work of its kind. Exposition of physical laws, theories developed chronologically, with great historical, illustrative experiments diagrammed, described, worked out mathematically. Excellent physics text for self-study as well as class work. Vol. 1: Heat, Sound: motion, acceleration, gravitation, conservation of energy, heat engines, rotation, heat, mechanical energy, etc. 211 illus. 407pp. 5⅜ x 8. Vol. 2: Light, Electricity: images, lenses, prisms, magnetism, Ohm's law, dynamos, telegraph, quantum theory, decline of mechanical view of nature, etc. Bibliography. 13 table appendix. Index. 551 illus. 2 color plates. 508pp. 5⅜ x 8.
Vol. 1 S565 Paperbound **$2.00**
Vol. 2 S566 Paperbound **$2.00**
The set **$4.00**

FROM EUCLID TO EDDINGTON: A STUDY OF THE CONCEPTIONS OF THE EXTERNAL WORLD, Sir Edmund Whittaker. A foremost British scientist traces the development of theories of natural philosophy from the western rediscovery of Euclid to Eddington, Einstein, Dirac, etc. The inadequacy of classical physics is contrasted with present day attempts to understand the physical world through relativity, non-Euclidean geometry, space curvature, wave mechanics, etc. 5 major divisions of examination: Space; Time and Movement; the Concepts of Classical Physics; the Concepts of Quantum Mechanics; the Eddington Universe. 212pp. 5⅜ x 8. T491 Paperbound **$1.35**

THE STORY OF ATOMIC THEORY AND ATOMIC ENERGY, J. G. Feinberg. Wider range of facts on physical theory, cultural implications, than any other similar source. Completely non-technical. Begins with first atomic theory, 600 B.C., goes through A-bomb, developments to 1959. Avogadro, Rutherford, Bohr, Einstein, radioactive decay, binding energy, radiation danger, future benefits of nuclear power, dozens of other topics, told in lively, related, informal manner. Particular stress on European atomic research. "Deserves special mention . . . authoritative," Saturday Review. Formerly "The Atom Story." New chapter to 1959. Index. 34 illustrations. 251pp. 5⅜ x 8. T625 Paperbound **$1.45**

THE STRANGE STORY OF THE QUANTUM, AN ACCOUNT FOR THE GENERAL READER OF THE GROWTH OF IDEAS UNDERLYING OUR PRESENT ATOMIC KNOWLEDGE, B. Hoffmann. Presents lucidly and expertly, with barest amount of mathematics, the problems and theories which led to modern quantum physics. Dr. Hoffmann begins with the closing years of the 19th century, when certain trifling discrepancies were noticed, and with illuminating analogies and examples takes you through the brilliant concepts of Planck, Einstein, Pauli, de Broglie, Bohr, Schroedinger, Heisenberg, Dirac, Sommerfeld, Feynman, etc. This edition includes a new, long postscript carrying the story through 1958. "Of the books attempting an account of the history and contents of our modern atomic physics which have come to my attention, this is the best," H. Margenau, Yale University, in "American Journal of Physics." 32 tables and line illustrations. Index. 275pp. 5⅜ x 8. T518 Paperbound **$1.45**

SPACE AND TIME, Emile Borel. An entirely non-technical introduction to relativity, by world-renowned mathematician, Sorbonne Professor. (Notes on basic mathematics are included separately.) This book has never been surpassed for insight, and extraordinary clarity of thought, as it presents scores of examples, analogies, arguments, illustrations, which explain such topics as: difficulties due to motion; gravitation a force of inertia; geodesic lines; wave-length and difference of phase; x-rays and crystal structure; the special theory of relativity; and much more. Indexes. 4 appendixes. 15 figures. xvi + 243pp. 5⅜ x 8.
T592 Paperbound **$1.45**

THE RESTLESS UNIVERSE, Max Born. New enlarged version of this remarkably readable account by a Nobel laureate. Moving from sub-atomic particles to universe, the author explains in very simple terms the latest theories of wave mechanics. Partial contents: air and its relatives, electrons & ions, waves & particles, electronic structure of the atom, nuclear physics. Nearly 1000 illustrations, including 7 animated sequences. 325pp. 6 x 9.
T412 Paperbound **$2.00**

SOAP SUBBLES, THEIR COLOURS AND THE FORCES WHICH MOULD THEM, C. V. Boys. Only complete edition, half again as much material as any other. Includes Boys' hints on performing his experiments, sources of supply. Dozens of lucid experiments show complexities of liquid films, surface tension, etc. Best treatment ever written. Introduction. 83 illustrations. Color plate. 202pp. 5⅜ x 8. T542 Paperbound **95¢**

SPINNING TOPS AND GYROSCOPIC MOTION, John Perry. Well-known classic of science still unsurpassed for lucid, accurate, delightful exposition. How quasi-rigidity is induced in flexible and fluid bodies by rapid motions; why gyrostat falls, top rises; nature and effect on climatic conditions of earth's precessional movement; effect of internal fluidity on rotating bodies, etc. Appendixes describe practical uses to which gyroscopes have been put in ships, compasses, monorail transportation. 62 figures. 128pp. 5⅜ x 8. T416 Paperbound **$1.00**

MATTER & LIGHT, THE NEW PHYSICS, L. de Broglie. Non-technical papers by a Nobel laureate explain electromagnetic theory, relativity, matter, light and radiation, wave mechan
quantum physics, philosophy of science. Einstein, Planck, Bohr, others explained s
that no mathematical training is needed for all but 2 of the 21 chapters. Unabri
300pp. 5⅜ x 8. T35 P

A SURVEY OF PHYSICAL THEORY, Max Planck. One of the greatest scientists of all time, creator of the quantum revolution in physics, writes in non-technical terms of his own discoveries and those of other outstanding creators of modern physics. Planck wrote this book when science had just crossed the threshold of the new physics, and he communicates the excitement felt then as he discusses electromagnetic theories, statistical methods, evolution of the concept of light, a step-by-step description of how he developed his own momentous theory, and many more of the basic ideas behind modern physics. Formerly "A Survey of Physics." Bibliography. Index. 128pp. 5⅜ x 8. S650 Paperbound **$1.15**

THE NATURE OF LIGHT AND COLOUR IN THE OPEN AIR, M. Minnaert. Why is falling snow sometimes black? What causes mirages, the fata morgana, multiple suns and moons in the sky? How are shadows formed? Prof. Minnaert of the University of Utrecht answers these and similar questions in optics, light, colour, for non-specialists. Particularly valuable to nature, science students, painters, photographers. Translated by H. M. Kremer-Priest, K. Jay. 202 illustrations, including 42 photos. xvi + 362pp. 5⅜ x 8. T196 Paperbound **$1.95**

THE STORY OF X-RAYS FROM RONTGEN TO ISOTOPES, A. R. Bleich. Non-technical history of x-rays, their scientific explanation, their applications in medicine, industry, research, and art, and their effect on the individual and his descendants. Includes amusing early reactions to Röntgen's discovery, cancer therapy, detections of art and stamp forgeries, potential risks to patient and operator, etc. Illustrations show x-rays of flower structure, the gall bladder, gears with hidden defects, etc. Original Dover publication. Glossary. Bibliography. Index. 55 photos and figures. xiv + 186pp. 5⅜ x 8. T662 Paperbound **$1.35**

TEACH YOURSELF ELECTRICITY, C. W. Wilman. Electrical resistance, inductance, capacitance, magnets, chemical effects of current, alternating currents, generators and motors, transformers, rectifiers, much more. 230 questions, answers, worked examples. List of units. 115 illus. 194pp. 6⅞ x 4¼. Clothbound **$2.00**

TEACH YOURSELF HEAT ENGINES, E. De Ville. Measurement of heat, development of steam and internal combustion engines, efficiency of an engine, compression-ignition engines, production of steam, the ideal engine, much more. 318 exercises, answers, worked examples. Tables. 76 illus. 220pp. 6⅞ x 4¼. Clothbound **$2.00**

TEACH YOURSELF MECHANICS, P. Abbott. The lever, centre of gravity, parallelogram of force, friction, acceleration, Newton's laws of motion, machines, specific gravity, gas, liquid pressure, much more. 280 problems, solutions. Tables. 163 illus. 271pp. 6⅞ x 4¼. Clothbound **$2.00**

GREAT IDEAS OF MODERN MATHEMATICS: THEIR NATURE AND USE, Jagjit Singh. Reader with only high school math will understand main mathematical ideas of modern physics, astronomy, genetics, psychology, evolution, etc., better than many who use them as tools, but comprehend little of their basic structure. Author uses his wide knowledge of non-mathematical fields in brilliant exposition of differential equations, matrices, group theory, logic, statistics, problems of mathematical foundations, imaginary numbers, vectors, etc. Original publication. 2 appendixes. 2 indexes. 65 illustr. 322pp. 5⅜ x 8. S587 Paperbound **$1.55**

MATHEMATICS IN ACTION, O. G. Sutton. Everyone with a command of high school algebra will find this book one of the finest possible introductions to the application of mathematics to physical theory. Ballistics, numerical analysis, waves and wavelike phenomena, Fourier series, group concepts, fluid flow and aerodynamics, statistical measures, and meteorology are discussed with unusual clarity. Some calculus and differential equations theory is developed by the author for the reader's help in the more difficult sections. 88 figures. Index. viii + 236pp. 5⅜ x 8. T440 Clothbound **$3.50**

FREE! All you do is ask for it!

A DOVER SCIENCE SAMPLER, edited by George Barkin. 64-page book, sturdily bound, containing excerpts from over 20 Dover books explaining science. Edwin Hubble, George Sarton, Ernst Mach, A. d'Abro, Galileo, Newton, others, discussing island universes, scientific truth, biological phenomena, stability in bridges, etc. Copies limited, no more than 1 to a customer. FREE

THE FOURTH DIMENSION SIMPLY EXPLAINED, edited by H. P. Manning. 22 essays, originally Scientific American contest entries, that use a minimum of mathematics to explain aspects of 4-dimensional geometry: analogues to 3-dimensional space, 4-dimensional absurdities and curiosities (such as removing the contents of an egg without puncturing its shell), possible measurements and forms, etc. Introduction by the editor. Only book of its sort on a truly elementary level, excellent introduction to advanced works. 82 figures. 251pp. 5⅜ x 8. T711 Paperbound **$1.35**

FAMOUS BRIDGES OF THE WORLD, D. B. Steinman. An up-to-the-minute revised edition of a book that explains the fascinating drama of how the world's great bridges came to be built. The author, designer of the famed Mackinac bridge, discusses bridges from all periods and all parts of the world, explaining their various types of construction, and describing the problems their builders faced. Although primarily for youngsters, this cannot fail to interest readers of all ages. 48 illustrations in the text. 23 photographs. 99pp. 6⅛ x 9¼. T161 Paperbound **$1.00**

BRIDGES AND THEIR BUILDERS, David Steinman and Sara Ruth Watson. Engineers, historians, everyone who has ever been fascinated by great spans will find this book an endless source of information and interest. Dr. Steinman, recipient of the Louis Levy medal, was one of the great bridge architects and engineers of all time, and his analysis of the great bridges of history is both authoritative and easily followed. Greek and Roman bridges, medieval bridges, Oriental bridges, modern works such as the Brooklyn Bridge and the Golden Gate Bridge, and many others are described in terms of history, constructional principles, artistry, and function. All in all this book is the most comprehensive and accurate semipopular history of bridges in print in English. New, greatly revised, enlarged edition. 23 photographs, 26 line drawings. Index. xvii + 401pp. 5⅜ x 8. T431 Paperbound **$2.00**

FADS AND FALLACIES IN THE NAME OF SCIENCE, Martin Gardner. Examines various cults, quack systems, frauds, delusions which at various times have masqueraded as science. Accounts of hollow-earth fanatics like Symmes; Velikovsky and wandering planets; Hoerbiger; Bellamy and the theory of multiple moons; Charles Fort; dowsing, pseudoscientific methods for finding water, ores, oil. Sections on naturopathy, iridiagnosis, zone therapy, food fads, etc. Analytical accounts of Wilhelm Reich and orgone sex energy; L. Ron Hubbard and Dianetics; A. Korzybski and General Semantics; many others. Brought up to date to include Bridey Murphy, others. Not just a collection of anecdotes, but a fair, reasoned appraisal of eccentric theory. Formerly titled IN THE NAME OF SCIENCE. Preface. Index. x + 384pp. 5⅜ x 8. T394 Paperbound **$1.50**

See also: **A PHILOSOPHICAL ESSAY ON PROBABILITIES, P. de Laplace; ON MATHEMATICS AND MATHEMATICIANS, R. E. Moritz; AN ELEMENTARY SURVEY OF CELESTIAL MECHANICS, Y. Ryabov; THE SKY AND ITS MYSTERIES, E. A. Beet; THE REALM OF THE NEBULAE, E. Hubble; OUT OF THE SKY, H. H. Nininger; SATELLITES AND SCIENTIFIC RESEARCH, D. King-Hele; HEREDITY AND YOUR LIFE, A. M. Winchester; INSECTS AND INSECT LIFE, S. W. Frost; PRINCIPLES OF STRATIGRAPHY, A. W. Grabau; TEACH YOURSELF SERIES.**

HISTORY OF SCIENCE AND MATHEMATICS

DIALOGUES CONCERNING TWO NEW SCIENCES, Galileo Galilei. This classic of experimental science, mechanics, engineering, is as enjoyable as it is important. A great historical document giving insights into one of the world's most original thinkers, it is based on 30 years' experimentation. It offers a lively exposition of dynamics, elasticity, sound, ballistics, strength of materials, the scientific method. "Superior to everything else of mine," Galileo. Trans. by H. Crew, A. Salvio. 126 diagrams. Index. xxi + 288pp. 5⅜ x 8. S99 Paperbound **$1.65**

A DIDEROT PICTORIAL ENCYCLOPEDIA OF TRADES AND INDUSTRY, Manufacturing and the Technical Arts in Plates Selected from "L'Encyclopédie ou Dictionnaire Raisonné des Sciences, des Arts, et des Métiers" of Denis Diderot. Edited with text by C. Gillispie. This first modern selection of plates from the high point of 18th century French engraving is a storehouse of valuable technological information to the historian of arts and science. Over 2000 illustrations on 485 full page plates, most of them original size, show the trades and industries of a fascinating era in such great detail that the processes and shops might very well be reconstructed from them. The plates teem with life, with men, women, and children performing all of the thousands of operations necessary to the trades before and during the early stages of the industrial revolution. Plates are in sequence, and show general operations, closeups of difficult operations, and details of complex machinery. Such important and interesting trades and industries are illustrated as sowing, harvesting, beekeeping, cheesemaking, operating windmills, milling flour, charcoal burning, tobacco processing, indigo, fishing, arts of war, salt extraction, mining, smelting, casting iron, steel, extracting mercury, zinc, sulphur, copper, etc., slating, tinning, silverplating, gilding, making gunpowder, cannons, bells, shoeing horses, tanning, papermaking, printing, dyeing, and more than 40 other categories. Professor Gillispie, of Princeton, supplies a full commentary on all the plates, identifying operations, tools, processes, etc. This material, presented in a lively and lucid fashion, is of great interest to the reader interested in history of science and technology. Heavy library cloth. 920pp. 9 x 12. T421 Two volume set **$18.50**

DE MAGNETE, William Gilbert. This classic work on magnetism founded a new science. Gilbert was the first to use the word "electricity", to recognize mass as distinct from weight, to discover the effect of heat on magnetic bodies; invent an electroscope, differentiate between static electricity and magnetism, conceive of the earth as a magnet. Written by the first great experimental scientist, this lively work is valuable not only as an historical landmark, but as the delightfully easy to follow record of a perpetually searching, ingenious mind. Translated by P. F. Mottelay. 25 page biographical memoir. 90 figures. lix + 368pp. 5⅜ x 8. S470 Paperbound **$2.00**

A CONCISE HISTORY OF MATHEMATICS, D. Struik. Lucid study of development of mathematical ideas, techniques from Ancient Near East, Greece, Islamic science, Middle Ages, Renaissance, modern times. Important mathematicians are described in detail. Treatment is not anecdotal, but analytical development of ideas. "Rich in content, thoughtful in interpretation," U.S. QUARTERLY BOOKLIST. Non-technical; no mathematical training needed. Index. 60 illustrations, including Egyptian papyri, Greek mss., portraits of 31 eminent mathematicians. Bibliography. 2nd edition. xix + 299pp. 5⅜ x 8. T255 Paperbound **$1.75**

See also: **NON-EUCLIDEAN GEOMETRY, R. Bonola; THEORY OF DETERMINANTS IN HISTORICAL ORDER OF DEVELOPMENT, T. Muir; HISTORY OF THE THEORY OF ELASTICITY AND STRENGTH OF MATERIALS, I. Todhunter and K. Pearson; A SHORT HISTORY OF ASTRONOMY, A. Berry; CLASSICS OF SCIENCE.**

PHILOSOPHY OF SCIENCE AND MATHEMATICS

FOUNDATIONS OF SCIENCE: THE PHILOSOPHY OF THEORY AND EXPERIMENT, N. R. Campbell. A critique of the most fundamental concepts of science in general and physics in particular. Examines why certain propositions are accepted without question, demarcates science from philosophy, clarifies the understanding of the tools of science. Part One analyzes the presuppositions of scientific thought: existence of the material world, nature of scientific laws, multiplication of probabilities, etc.: Part Two covers the nature of experiment and the application of mathematics: conditions for measurement, relations between numerical laws and theories, laws of error, etc. An appendix covers problems arising from relativity, force, motion, space, and time. A classic in its field. Index. xiii + 565pp. 5⅝ x 8⅜. S372 Paperbound **$2.95**

WHAT IS SCIENCE?, Norman Campbell. This excellent introduction explains scientific method, role of mathematics, types of scientific laws. Contents: 2 aspects of science, science & nature, laws of science, discovery of laws, explanation of laws, measurement & numerical laws, applications of science. 192pp. 5⅜ x 8. S43 Paperbound **$1.25**

THE VALUE OF SCIENCE, Henri Poincaré. Many of the most mature ideas of the "last scientific universalist" covered with charm and vigor for both the beginning student and the advanced worker. Discusses the nature of scientific truth, whether order is innate in the universe or imposed upon it by man, logical thought versus intuition (relating to math, through the works of Weierstrass, Lie, Klein, Riemann), time and space (relativity, psychological time, simultaneity), Hertz's concept of force, interrelationship of mathematical physics to pure math, values within disciplines of Maxwell, Carnot, Mayer, Newton, Lorentz, etc. Index. iii + 147pp. 5⅜ x 8. S469 Paperbound **$1.35**

SCIENCE AND METHOD, Henri Poincaré. Procedure of scientific discovery, methodology, experiment, idea-germination—the intellectual processes by which discoveries come into being. Most significant and most interesting aspects of development, application of ideas. Chapters cover selection of facts, chance, mathematical reasoning, mathematics, and logic; Whitehead, Russell, Cantor; the new mechanics, etc. 288pp. 5⅜ x 8. S222 Paperbound **$1.35**

SCIENCE AND HYPOTHESIS, Henri Poincaré. Creative psychology in science. How such concepts as number, magnitude, space, force, classical mechanics were developed, and how the modern scientist uses them in his thought. Hypothesis in physics, theories of modern physics. Introduction by Sir James Larmor. "Few mathematicians have had the breadth of vision of Poincaré, and none is his superior in the gift of clear exposition," E. T. Bell. Index. 272pp. 5⅜ x 8. S221 Paperbound **$1.35**

PHILOSOPHY AND THE PHYSICISTS, L. S. Stebbing. The philosophical aspects of modern science examined in terms of a lively critical attack on the ideas of Jeans and Eddington. Discusses the task of science, causality, determinism, probability, consciousness, the relation of the world of physics to that of everyday experience. Probes the philosophical significance of the Planck-Bohr concept of discontinuous energy levels, the inferences to be drawn from Heisenberg's Uncertainty Principle, the implications of "becoming" involved in the 2nd law of thermodynamics, and other problems posed by the discarding of Laplacean determinism. 285pp. 5⅜ x 8. T480 Paperbound **$1.65**

EXPERIMENT AND THEORY IN PHYSICS, Max Born. A Nobel laureate examines the nature and value of the counterclaims of experiment and theory in physics. Synthetic versus analytical scientific advances are analyzed in the work of Einstein, Bohr, Heisenberg, Planck, Eddington, Milne, and others by a fellow participant. 44pp. 5⅜ x 8. S308 Paperbound **60¢**

THE NATURE OF PHYSICAL THEORY, P. W. Bridgman. Here is how modern physics looks to a highly unorthodox physicist—a Nobel laureate. Pointing out many absurdities of science, and demonstrating the inadequacies of various physical theories, Dr. Bridgman weighs and analyzes the contributions of Einstein, Bohr, Newton, Heisenberg, and many others. This is a non-technical consideration of the correlation of science and reality. Index. xi + 138pp. 5⅜ x 8. S33 Paperbound **$1.25**

THE PHILOSOPHY OF SPACE AND TIME, H. Reichenbach. An important landmark in the development of the empiricist conception of geometry, covering the problem of the foundations of geometry, the theory of time, the consequences of Einstein's relativity, including: relations between theory and observations; coordinate and metrical properties of space; the psychological problem of visual intuition of non-Euclidean structures; and many other important topics in modern science and philosophy. The majority of ideas require only a knowledge of intermediate math. Introduction by R. Carnap. 49 figures. Index. xviii + 296pp. 5⅜ x 8.
S443 Paperbound **$2.00**

MATTER & MOTION, James Clerk Maxwell, This excellent exposition begins with simple particles and proceeds gradually to physical systems beyond complete analysis: motion, force, properties of centre of mass of material system, work, energy, gravitation, etc. Written with all Maxwell's original insights and clarity. Notes by E. Larmor. 17 diagrams. 178pp. 5⅜ x 8.
S188 Paperbound **$1.35**

THE ANALYSIS OF MATTER, Bertrand Russell. How do our senses concord with the new physics? This volume covers such topics as logical analysis of physics, prerelativity physics, causality, scientific inference, physics and perception, special and general relativity, Weyl's theory, tensors, invariants and their physical interpretation, periodicity and qualitative series. "The most thorough treatment of the subject that has yet been published," THE NATION. Introduction by L. E. Denonn. 422pp. 5⅜ x 8. T231 Paperbound **$1.95**

SUBSTANCE AND FUNCTION, & EINSTEIN'S THEORY OF RELATIVITY, Ernst Cassirer. Two books bound as one. Cassirer establishes a philosophy of the exact sciences that takes into consideration newer developments in mathematics, and also shows historical connections. Partial contents: Aristotelian logic, Mill's analysis, Helmholtz & Kronecker, Russell & cardinal numbers, Euclidean vs. non-Euclidean geometry, Einstein's relativity. Bibliography. Index. xxi + 465pp. 5⅜ x 8. T50 Paperbound **$2.00**

PRINCIPLES OF MECHANICS, Heinrich Hertz. This last work by the great 19th century physicist is not only a classic, but of great interest in the logic of science. Creating a new system of mechanics based upon space, time, and mass, it returns to axiomatic analysis, to understanding of the formal or structural aspects of science, taking into account logic, observation, and a priori elements. Of great historical importance to Poincaré, Carnap, Einstein, Milne. A 20-page introduction by R. S. Cohen, Wesleyan University, analyzes the implications of Hertz's thought and the logic of science. Bibliography. 13-page introduction by Helmholtz. xlii + 274pp. 5⅜ x 8. S316 Clothbound **$3.50**
S317 Paperbound **$1.85**

THE PHILOSOPHICAL WRITINGS OF PEIRCE, edited by Justus Buchler. (Formerly published as THE PHILOSOPHY OF PEIRCE.) This is a carefully balanced exposition of Peirce's complete system, written by Peirce himself. It covers such matters as scientific method, pure chance vs. law, symbolic logic, theory of signs, pragmatism, experiment, and other topics. Introduction by Justus Buchler, Columbia University. xvi + 368pp. 5⅜ x 8.
T217 Paperbound **$1.95**

ESSAYS IN EXPERIMENTAL LOGIC, John Dewey. This stimulating series of essays touches upon the relationship between inquiry and experience, dependence of knowledge upon thought, character of logic; judgments of practice, data and meanings, stimuli of thought, etc. Index. viii + 444pp. 5⅜ x 8. T73 Paperbound **$1.95**

LANGUAGE, TRUTH AND LOGIC, A. Ayer. A clear introduction to the Vienna and Cambridge schools of Logical Positivism. It sets up specific tests by which you can evaluate validity of ideas, etc. Contents: Function of philosophy, elimination of metaphysics, nature of analysis, a priori, truth and probability, etc. 10th printing. "I should like to have written it myself," Bertrand Russell. Index. 160pp. 5⅜ x 8. T10 Paperbound **$1.25**

THE PSYCHOLOGY OF INVENTION IN THE MATHEMATICAL FIELD, J. Hadamard. Where do ideas come from? What role does the unconscious play? Are ideas best developed by mathematical reasoning, word reasoning, visualization? What are the methods used by Einstein, Poincaré, Galton, Riemann? How can these techniques be applied by others? Hadamard, one of the world's leading mathematicians, discusses these and other questions. xiii + 145pp. 5⅜ x 8.
T107 Paperbound **$1.25**

FOUNDATIONS OF GEOMETRY, Bertrand Russell. Analyzing basic problems in the overlap area between mathematics and philosophy, Nobel laureate Russell examines the nature of geometrical knowledge, the nature of geometry, and the application of geometry to space. It covers the history of non-Euclidean geometry, philosophic interpretations of geometry—especially Kant—projective and metrical geometry. This is most interesting as the solution offered in 1897 by a great mind to a problem still current. New introduction by Prof. Morris Kline of N. Y. University. xii + 201pp. 5⅜ x 8. S232 Clothbound **$3.25**
S233 Paperbound **$1.60**

BIBLIOGRAPHIES

GUIDE TO THE LITERATURE OF MATHEMATICS AND PHYSICS, N. G. Parke III. Over 5000 entries included under approximately 120 major subject headings, of selected most important books, monographs, periodicals, articles in English, plus important works in German, French, Italian, Spanish, Russian (many recently available works). Covers every branch of physics, math, related engineering. Includes author, title, edition, publisher, place, date, number of volumes, number of pages. A 40-page introduction on the basic problems of research and study provides useful information on the organization and use of libraries, the psychology of learning, etc. This reference work will save you hours of time. 2nd revised edition. Indices of authors, subjects. 464pp. 5⅜ x 8. S447 Paperbound **$2.49**

THE STUDY OF THE HISTORY OF MATHEMATICS & THE STUDY OF THE HISTORY OF SCIENCE, George Sarton. Scientific method & philosophy in 2 scholarly fields. Defines duty of historian of math., provides especially useful bibliography with best available biographies of modern mathematicians, editions of their collected works, correspondence. Observes combination of history & science, will aid scholar in understanding science today. Bibliography includes best known treatises on historical methods. 200-item critically evaluated bibliography. Index. 10 illustrations. 2 volumes bound as one. 113pp. + 75pp. 5⅜ x 8. T240 Paperbound **$1.25**

MATHEMATICAL PUZZLES

AMUSEMENTS IN MATHEMATICS, Henry Ernest Dudeney. The foremost British originator of mathematical puzzles is always intriguing, witty, and paradoxical in this classic, one of the largest collections of mathematical amusements. More than 430 puzzles, problems, and paradoxes. Mazes and games, problems on number manipulation, unicursal and other route problems, puzzles on measuring, weighing, packing, age, kinship, chessboards, joiners', crossing river, plane figure dissection, and many others. Solutions. More than 450 illustrations. vii + 258pp. 5⅜ x 8. T473 Paperbound **$1.25**

THE CANTERBURY PUZZLES, Henry Ernest Dudeney. Chaucer's pilgrims set one another problems in story form. Also Adventures of the Puzzle Club, the Strange Escape of the King's Jester, the Monks of Riddlewell, the Squire's Christmas Puzzle Party, and others. All puzzles are original, based on dissecting plane figures, arithmetic, algebra, elementary calculus, and other branches of mathematics, and purely logical ingenuity. "The limit of ingenuity and intricacy . . ." The Observer. Over 110 puzzles. Full solutions. 150 illustrations. viii + 225pp. 5⅜ x 8. T474 Paperbound **$1.25**

SYMBOLIC LOGIC and THE GAME OF LOGIC, Lewis Carroll. "Symbolic Logic" is not concerned with modern symbolic logic, but is instead a collection of over 380 problems posed with charm and imagination, using the syllogism, and a fascinating diagrammatic method of drawing conclusions. In "The Game of Logic," Carroll's whimsical imagination devises a logical game played with 2 diagrams and counters (included) to manipulate hundreds of tricky syllogisms. The final section, "Hit or Miss" is a lagniappe of 101 additional puzzles in the delightful Carroll manner. Until this reprint edition, both of these books were rarities costing up to $15 each. Symbolic Logic: Index, xxxi + 199pp. The Game of Logic: 96pp. Two vols. bound as one. 5⅜ x 8. T492 Paperbound **$1.50**

PILLOW PROBLEMS and A TANGLED TALE, Lewis Carroll. One of the rarest of all Carroll's works, "Pillow Problems" contains 72 original math puzzles, all typically ingenious. Particularly fascinating are Carroll's answers which remain exactly as he thought them out, reflecting his actual mental processes. The problems in "A Tangled Tale" are in story form, originally appearing as a monthly magazine serial. Carroll not only gives the solutions, but uses answers sent in by readers to discuss wrong approaches and misleading paths, and grades them for insight. Both of these books were rarities until this edition, "Pillow Problems" costing up to $25, and "A Tangled Tale" $15. Pillow Problems: Preface and introduction by Lewis Carroll. xx + 109pp. A Tangled Tale: 6 illustrations. 152pp. Two vols. bound as one. 5⅜ x 8. T493 Paperbound **$1.50**

DIVERSIONS AND DIGRESSIONS OF LEWIS CARROLL. A major new treasure for Carroll fans! Rare privately published puzzles, mathematical amusements and recreations, games. Includes the fragmentary Part III of "Curiosa Mathematica." Also contains humorous and satirical pieces: "The New Belfry," "The Vision of the Three T's," and much more. New 32-page supplement of rare photographs taken by Carroll. Formerly titled "The Lewis Carroll Picture Book." Edited by S. Collingwood. x + 375pp. 5⅜ x 8. T732 Paperbound **$1.50**

ANALYSIS & DESIGN OF EXPERIMENTS, H. B. Mann. Offers a method for grasping the analysis of variance and variance design within a short time. Partial contents: Chi-square distribution and analysis of variance distribution, matrices, quadratic forms, likelihood ration tests and tests of linear hypotheses, power of analysis, Galois fields, non-orthogonal data, interblock estimates, etc. 15pp. of useful tables. x + 195pp. 5 x 7⅜. S180 Paperbound **$1.45**

Numerical analysis, tables

PRACTICAL ANALYSIS, GRAPHICAL AND NUMERICAL METHODS, F. A. Willers. Translated by R. T. Beyer. Immensely practical handbook for engineers, showing how to interpolate, use various methods of numerical differentiation and integration, determine the roots of a single algebraic equation, system of linear equations, use empirical formulas, integrate differential equations, etc. Hundreds of shortcuts for arriving at numerical solutions. Special section on American calculating machines, by T. W. Simpson. 132 illustrations. 422pp. 5⅜ x 8. S273 Paperbound **$2.00**

NUMERICAL SOLUTIONS OF DIFFERENTIAL EQUATIONS, H. Levy & E. A. Baggott. Comprehensive collection of methods for solving ordinary differential equations of first and higher order. All must pass 2 requirements: easy to grasp and practical, more rapid than school methods. Partial contents: graphical integration of differential equations, graphical methods for detailed solution. Numerical solution. Simultaneous equations and equations of 2nd and higher orders. "Should be in the hands of all in research in applied mathematics, teaching," NATURE. 21 figures. viii + 238pp. 5⅜ x 8. S168 Paperbound **$1.75**

NUMERICAL INTEGRATION OF DIFFERENTIAL EQUATIONS, Bennett, Milne & Bateman. Unabridged republication of original monograph prepared for National Research Council. New methods of integration of differential equations developed by 3 leading mathematicians: THE INTERPOLATIONAL POLYNOMIAL and SUCCESSIVE APPROXIMATIONS by A. A. Bennett; STEP-BY-STEP METHODS OF INTEGRATION by W. W. Milne; METHODS FOR PARTIAL DIFFERENTIAL EQUATIONS by H. Bateman. Methods for partial differential equations, transition from difference equations to differential equations, solution of differential equations to non-integral values of a parameter will interest mathematicians and physicists. 288 footnotes, mostly bibliographic; 235-item classified bibliography. 108pp. 5⅜ x 8. S305 Paperbound **$1.35**

INTRODUCTION TO RELAXATION METHODS, F. S. Shaw. Fluid mechanics, design of electrical networks, forces in structural frameworks, stress distribution, buckling, etc. Solve linear simultaneous equations, linear ordinary differential equations, partial differential equations, Eigen-value problems by relaxation methods. Detailed examples throughout. Special tables for dealing with awkwardly-shaped boundaries. Indexes. 253 diagrams. 72 tables. 400pp. 5⅜ x 8. S244 Paperbound **$2.45**

TABLES OF INDEFINITE INTEGRALS, G. Petit Bois. Comprehensive and accurate, this orderly grouping of over 2500 of the most useful indefinite integrals will save you hours of laborious mathematical groundwork. After a list of 49 common transformations of integral expressions, with a wide variety of examples, the book takes up algebraic functions, irrational monomials, products and quotients of binomials, transcendental functions, natural logs, etc. You will rarely or never encounter an integral of an algebraic or transcendental function not included here; any more comprehensive set of tables costs at least $12 or $15. Index. 2544 integrals. xii + 154pp. 6⅛ x 9¼. S225 Paperbound **$1.65**

A TABLE OF THE INCOMPLETE ELLIPTIC INTEGRAL OF THE THIRD KIND, R. G. Selfridge, J. E. Maxfield. The first complete 6 place tables of values of the incomplete integral of the third kind, prepared under the auspices of the Research Department of the U.S. Naval Ordnance Test Station. Calculated on an IBM type 704 calculator and thoroughly verified by echo-checking and a check integral at the completion of each value of **a.** Of inestimable value in problems where the surface area of geometrical bodies can only be expressed in terms of the incomplete integral of the third and lower kinds; problems in aero-, fluid-, and thermodynamics involving processes where nonsymmetrical repetitive volumes must be determined; various types of seismological problems; problems of magnetic potentials due to circular current; etc. Foreword. Acknowledgment. Introduction. Use of table. xiv + 205pp. 5⅝ x 8⅜. S501 Clothbound **$7.50**

MATHEMATICAL TABLES, H. B. Dwight. Unique for its coverage in one volume of almost every function of importance in applied mathematics, engineering, and the physical sciences. Three extremely fine tables of the three trig functions and their inverse functions to thousandths of radians; natural and common logarithms; squares, cubes; hyperbolic functions and the inverse hyperbolic functions; $(a^2 + b^2)$ exp. ½a; complete elliptic integrals of the 1st and 2nd kind; sine and cosine integrals; exponential integrals Ei(x) and Ei(—x); binomial coefficients; factorials to 250; surface zonal harmonics and first derivatives; Bernoulli and Euler numbers and their logs to base of 10; Gamma function; normal probability integral; over 60 pages of Bessel functions; the Riemann Zeta function. Each table with formulae generally used, sources of more extensive tables, interpolation data, etc. Over half have columns of differences, to facilitate interpolation. Introduction. Index. viii + 231pp. 5⅜ x 8. S445 Paperbound **$1.75**

TABLES OF FUNCTIONS WITH FORMULAE AND CURVES, E. Jahnke & F. Emde. The world's most comprehensive 1-volume English-text collection of tables, formulae, curves of transcendent functions. 4th corrected edition, new 76-page section giving tables, formulae for elementary functions—not in other English editions. Partial contents: sine, cosine, logarithmic integral; factorial function; error integral; theta functions; elliptic integrals, functions; Legendre, Bessel, Riemann, Mathieu, hypergeometric functions, etc. Supplementary books. Bibliography. Indexed. "Out of the way functions for which we know no other source," SCIENTIFIC COMPUTING SERVICE, Ltd. 212 figures. 400pp. 5⅜ x 8. S133 Paperbound **$2.00**

JACOBIAN ELLIPTIC FUNCTION TABLES, L. M. Milne-Thomson. An easy to follow, practical book which gives not only useful numerical tables, but also a complete elementary sketch of the application of elliptic functions. It covers Jacobian elliptic functions and a description of their principal properties; complete elliptic integrals; Fourier series and power series expansions; periods, zeros, poles, residues, formulas for special values of the argument; transformations, approximations, elliptic integrals, conformal mapping, factorization of cubic and quartic polynomials; application to the pendulum problem; etc. Tables and graphs form the body of the book: Graph, 5 figure table of the elliptic function sn (u m); cn (u m); dn (u m). 8 figure table of complete elliptic integrals K, K′, E, E′, and the nome q. 7 figure table of the Jacobian zeta-function Z(u). 3 figures. xi + 123pp. 5⅜ x 8.
S194 Paperbound **$1.35**

PHYSICS

General physics

FOUNDATIONS OF PHYSICS, R. B. Lindsay & H. Margenau. Excellent bridge between semi-popular works & technical treatises. A discussion of methods of physical description, construction of theory; valuable for physicist with elementary calculus who is interested in ideas that give meaning to data, tools of modern physics. Contents include symbolism, mathematical equations; space & time foundations of mechanics; probability; physics & continua; electron theory; special & general relativity; quantum mechanics; causality. "Thorough and yet not overdetailed. Unreservedly recommended," NATURE (London). Unabridged, corrected edition. List of recommended readings. 35 illustrations. xi + 537pp. 5⅜ x 8.
S377 Paperbound **$2.45**

FUNDAMENTAL FORMULAS OF PHYSICS, ed. by D. H. Menzel. Highly useful, fully inexpensive reference and study text, ranging from simple to highly sophisticated operations. Mathematics integrated into text—each chapter stands as short textbook of field represented. Vol. 1: Statistics, Physical Constants, Special Theory of Relativity, Hydrodynamics, Aerodynamics, Boundary Value Problems in Math. Physics; Viscosity, Electromagnetic Theory, etc. Vol. 2: Sound, Acoustics, Geometrical Optics, Electron Optics, High-Energy Phenomena, Magnetism, Biophysics, much more. Index. Total of 800pp. 5⅜ x 8. Vol. 1 S595 Paperbound **$2.00**
Vol. 2 S596 Paperbound **$2.00**

MATHEMATICAL PHYSICS, D. H. Menzel. Thorough one-volume treatment of the mathematical techniques vital for classic mechanics, electromagnetic theory, quantum theory, and relativity. Written by the Harvard Professor of Astrophysics for junior, senior, and graduate courses, it gives clear explanations of all those aspects of function theory, vectors, matrices, dyadics, tensors, partial differential equations, etc., necessary for the understanding of the various physical theories. Electron theory, relativity, and other topics seldom presented appear here in considerable detail. Scores of definitions, conversion factors, dimensional constants, etc. "More detailed than normal for an advanced text . . . excellent set of sections on Dyadics, Matrices, and Tensors," JOURNAL OF THE FRANKLIN INSTITUTE. Index. 193 problems, with answers. x + 412pp. 5⅜ x 8. S56 Paperbound **$2.00**

THE SCIENTIFIC PAPERS OF J. WILLARD GIBBS. All the published papers of America's outstanding theoretical scientist (except for "Statistical Mechanics" and "Vector Analysis"). Vol I (thermodynamics) contains one of the most brilliant of all 19th-century scientific papers—the 300-page "On the Equilibrium of Heterogeneous Substances," which founded the science of physical chemistry, and clearly stated a number of highly important natural laws for the first time; 8 other papers complete the first volume. Vol II includes 2 papers on dynamics, 8 on vector analysis and multiple algebra, 5 on the electromagnetic theory of light, and 6 miscellaneous papers. Biographical sketch by H. A. Bumstead. Total of xxxvi + 718pp. 5⅝ x 8⅜.
S721 Vol I Paperbound **$2.00**
S722 Vol II Paperbound **$2.00**
The set **$4.00**

Relativity, quantum theory, nuclear physics

THE PRINCIPLE OF RELATIVITY, A. Einstein, H. Lorentz, M. Minkowski, H. Weyl. These are the 11 basic papers that founded the general and special theories of relativity, all translated into English. Two papers by Lorentz on the Michelson experiment, electromagnetic phenomena. Minkowski's SPACE & TIME, and Weyl's GRAVITATION & ELECTRICITY. 7 epoch-making papers by Einstein: ELECTROMAGNETICS OF MOVING BODIES, INFLUENCE OF GRAVITATION IN PROPAGATION OF LIGHT, COSMOLOGICAL CONSIDERATIONS, GENERAL THEORY, and 3 others. 7 diagrams. Special notes by A. Sommerfeld. 224pp. 5⅜ x 8.
S81 Paperbound **$1.75**

SPACE TIME MATTER, Hermann Weyl. "The standard treatise on the general theory of relativity," (Nature), written by a world-renowned scientist, provides a deep clear discussion of the logical coherence of the general theory, with introduction to all the mathematical tools needed: Maxwell, analytical geometry, non-Euclidean geometry, tensor calculus, etc. Basis is classical space-time, before absorption of relativity. Partial contents: Euclidean space, mathematical form, metrical continuum, relativity of time and space, general theory. 15 diagrams. Bibliography. New preface for this edition. xviii + 330pp. 5⅜ x 8.
S267 Paperbound **$1.85**

PRINCIPLES OF QUANTUM MECHANICS, W. V. Houston. Enables student with working knowledge of elementary mathematical physics to develop facility in use of quantum mechanics, understand published work in field. Formulates quantum mechanics in terms of Schroedinger's wave mechanics. Studies evidence for quantum theory, for inadequacy of classical mechanics, 2 postulates of quantum mechanics; numerous important, fruitful applications of quantum mechanics in spectroscopy, collision problems, electrons in solids; other topics. "One of the most rewarding features . . . is the interlacing of problems with text," Amer. J. of Physics. Corrected edition. 21 illus. Index. 296pp. 5⅜ x 8. S524 Paperbound **$1.85**

PHYSICAL PRINCIPLES OF THE QUANTUM THEORY, Werner Heisenberg. A Nobel laureate discusses quantum theory; Heisenberg's own work, Compton, Schroedinger, Wilson, Einstein, many others. Written for physicists, chemists who are not specialists in quantum theory, only elementary formulae are considered in the text; there is a mathematical appendix for specialists. Profound without sacrifice of clarity. Translated by C. Eckart, F. Hoyt. 18 figures. 192pp. 5⅜ x 8. S113 Paperbound **$1.25**

SELECTED PAPERS ON QUANTUM ELECTRODYNAMICS, edited by **J. Schwinger.** Facsimiles of papers which established quantum electrodynamics, from initial successes through today's position as part of the larger theory of elementary particles. First book publication in any language of these collected papers of Bethe, Bloch, Dirac, Dyson, Fermi, Feynman, Heisenberg, Kusch, Lamb, Oppenheimer, Pauli, Schwinger, Tomonoga, Weisskopf, Wigner, etc. 34 papers in all, 29 in English, 1 in French, 3 in German, 1 in Italian. Preface and historical commentary by the editor. xvii + 423pp. 6⅛ x 9¼. S444 Paperbound **$2.45**

THE FUNDAMENTAL PRINCIPLES OF QUANTUM MECHANICS, WITH ELEMENTARY APPLICATIONS, E. C. Kemble. An inductive presentation, for the graduate student or specialist in some other branch of physics. Assumes some acquaintance with advanced math; apparatus necessary beyond differential equations and advanced calculus is developed as needed. Although a general exposition of principles, hundreds of individual problems are fully treated, with applications of theory being interwoven with development of the mathematical structure. The author is the Professor of Physics at Harvard Univ. "This excellent book would be of great value to every student . . . a rigorous and detailed mathematical discussion of all of the principal quantum-mechanical methods . . . has succeeded in keeping his presentations clear and understandable," Dr. Linus Pauling, J. of the American Chemical Society. Appendices: calculus of variations, math. notes, etc. Indexes. 611pp. 5⅜ x 8.
S472 Paperbound **$2.95**

ATOMIC SPECTRA AND ATOMIC STRUCTURE, G. Herzberg. Excellent general survey for chemists, physicists specializing in other fields. Partial contents: simplest line spectra and elements of atomic theory, building-up principle and periodic system of elements, hyperfine structure of spectral lines, some experiments and applications. Bibilography. 80 figures. Index. xii + 257pp. 5⅜ x 8. S115 Paperbound **$1.95**

THE THEORY AND THE PROPERTIES OF METALS AND ALLOYS, N. F. Mott, H. Jones. Quantum methods used to develop mathematical models which show interrelationship of basic chemical phenomena with crystal structure, magnetic susceptibility, electrical, optical properties. Examines thermal properties of crystal lattice, electron motion in applied field, cohesion, electrical resistance, noble metals, para-, dia-, and ferromagnetism, etc. "Exposition . . . clear . . . mathematical treatment . . . simple," Nature. 138 figures. Bibliography. Index. xiii + 320pp. 5⅜ x 8. S456 Paperbound **$1.85**

FOUNDATIONS OF NUCLEAR PHYSICS, edited by **R. T. Beyer.** 13 of the most important papers on nuclear physics reproduced in facsimile in the original languages of their authors: the papers most often cited in footnotes, bibliographies. Anderson, Curie, Joliot, Chadwick, Fermi, Lawrence, Cockcroft, Hahn, Yukawa. UNPARALLELED BIBLIOGRAPHY. 122 double-columned pages, over 4,000 articles, books classified. 57 figures. 288pp. 6⅛ x 9¼.
S19 Paperbound **$1.75**

MESON PHYSICS, R. E. Marshak. Traces the basic theory, and explicity presents results of experiments with particular emphasis on theoretical significance. Phenomena involving mesons as virtual transitions are avoided, eliminating some of the least satisfactory predictions of meson theory. Includes production and study of π mesons at nonrelativistic nucleon energies, contrasts between π and μ mesons, phenomena associated with nuclear interaction of π mesons, etc. Presents early evidence for new classes of particles and indicates theoretical difficulties created by discovery of heavy mesons and hyperons. Name and subject indices. Unabridged reprint. viii + 378pp. 5⅜ x 8. S500 Paperbound **$1.95**

See also: **STRANGE STORY OF THE QUANTUM, B. Hoffmann; FROM EUCLID TO EDDINGTON, E. Whittaker; MATTER AND LIGHT, THE NEW PHYSICS, L. de Broglie; THE EVOLUTION OF SCIENTIFIC THOUGHT FROM NEWTON TO EINSTEIN, A. d'Abro; THE RISE OF THE NEW PHYSICS, A. d'Abro; THE THEORY OF GROUPS AND QUANTUM MECHANICS, H. Weyl; SUBSTANCE AND FUNCTION, & EINSTEIN'S THEORY OF RELATIVITY, E. Cassirer; FUNDAMENTAL FORMULAS OF PHYSICS, D. H. Menzel.**

Hydrodynamics

HYDRODYNAMICS, H. Dryden, F. Murnaghan, Harry Bateman. Published by the National Research Council in 1932 this enormous volume offers a complete coverage of classical hydrodynamics. Encyclopedic in quality. Partial contents: physics of fluids, motion, turbulent flow, compressible fluids, motion in 1, 2, 3 dimensions; viscous fluids rotating, laminar motion, resistance of motion through viscous fluid, eddy viscosity, hydraulic flow in channels of various shapes, discharge of gases, flow past obstacles, etc. Bibliography of over 2,900 items. Indexes. 23 figures. 634pp. 5⅜ x 8. S303 Paperbound **$2.75**

A TREATISE ON HYDRODYNAMICS, A. B. Basset. Favorite text on hydrodynamics for 2 generations of physicists, hydrodynamical engineers, oceanographers, ship designers, etc. Clear enough for the beginning student, and thorough source for graduate students and engineers on the work of d'Alembert, Euler, Laplace, Lagrange, Poisson, Green, Clebsch, Stokes, Cauchy, Helmholtz, J. J. Thomson, Love, Hicks, Greenhill, Besant, Lamb, etc. Great amount of documentation on entire theory of classical hydrodynamics. Vol I: theory of motion of frictionless liquids, vortex, and cyclic irrotational motion, etc. 132 exercises. Bibliography. 3 Appendixes. xii + 264pp. Vol II: motion in viscous liquids, harmonic analysis, theory of tides, etc. 112 exercises. Bibliography. 4 Appendixes. xv + 328pp. Two volume set. 5⅜ x 8.
S724 Vol I Paperbound **$1.75**
S725 Vol II Paperbound **$1.75**
The set **$3.50**

HYDRODYNAMICS, Horace Lamb. Internationally famous complete coverage of standard reference work on dynamics of liquids & gases. Fundamental theorems, equations, methods, solutions, background, for classical hydrodynamics. Chapters include Equations of Motion, Integration of Equations in Special Gases, Irrotational Motion, Motion of Liquid in 2 Dimensions, Motion of Solids through Liquid-Dynamical Theory, Vortex Motion, Tidal Waves, Surface Waves, Waves of Expansion, Viscosity, Rotating Masses of liquids. Excellently planned, arranged; clear, lucid presentation. 6th enlarged, revised edition. Index. Over 900 footnotes, mostly bibliographical. 119 figures. xv + 738pp. 6⅛ x 9¼. S256 Paperbound **$2.95**

See also: **FUNDAMENTAL FORMULAS OF PHYSICS, D. H. Menzel; THEORY OF FLIGHT, R. von Mises; FUNDAMENTALS OF HYDRO- AND AEROMECHANICS, L. Prandtl and O. G. Tietjens; APPLIED HYDRO- AND AEROMECHANICS, L. Prandtl and O. G. Tietjens; HYDRAULICS AND ITS APPLICATIONS, A. H. Gibson; FLUID MECHANICS FOR HYDRAULIC ENGINEERS, H. Rouse.**

Acoustics, optics, electromagnetics

ON THE SENSATIONS OF TONE, Hermann Helmholtz. This is an unmatched coordination of such fields as acoustical physics, physiology, experiment, history of music. It covers the entire gamut of musical tone. Partial contents: relation of musical science to acoustics, physical vs. physiological acoustics, composition of vibration, resonance, analysis of tones by sympathetic resonance, beats, chords, tonality, consonant chords, discords, progression of parts, etc. 33 appendixes discuss various aspects of sound, physics, acoustics, music, etc. Translated by A. J. Ellis. New introduction by Prof. Henry Margenau of Yale. 68 figures. 43 musical passages analyzed. Over 100 tables. Index. xix + 576pp. 6⅛ x 9¼.
S114 Paperbound **$2.95**

THE THEORY OF SOUND, Lord Rayleigh. Most vibrating systems likely to be encountered in practice can be tackled successfully by the methods set forth by the great Nobel laureate, Lord Rayleigh. Complete coverage of experimental, mathematical aspects of sound theory. Partial contents: Harmonic motions, vibrating systems in general, lateral vibrations of bars, curved plates or shells, applications of Laplace's functions to acoustical problems, fluid friction, plane vortex-sheet, vibrations of solid bodies, etc. This is the first inexpensive edition of this great reference and study work. Bibliography. Historical introduction by R. B. Lindsay. Total of 1040pp. 97 figures. 5⅜ x 8.
S292, S293, Two volume set, paperbound, **$4.00**

THE DYNAMICAL THEORY OF SOUND, H. Lamb. Comprehensive mathematical treatment of the physical aspects of sound, covering the theory of vibrations, the general theory of sound, and the equations of motion of strings, bars, membranes, pipes, and resonators. Includes chapters on plane, spherical, and simple harmonic waves, and the Helmholtz Theory of Audition. Complete and self-contained development for student and specialist; all fundamental differential equations solved completely. Specific mathematical details for such important phenomena as harmonics, normal modes, forced vibrations of strings, theory of reed pipes, etc. Index. Bibliography. 86 diagrams. viii + 307pp. 5⅜ x 8. S655 Paperbound **$1.50**

WAVE PROPAGATION IN PERIODIC STRUCTURES, L. Brillouin. A general method and application to different problems: pure physics, such as scattering of X-rays of crystals, thermal vibration in crystal lattices, electronic motion in metals; and also problems of electrical engineering. Partial contents: elastic waves in 1-dimensional lattices of point masses. Propagation of waves along 1-dimensional lattices. Energy flow. 2 dimensional, 3 dimensional lattices. Mathieu's equation. Matrices and propagation of waves along an electric line. Continuous electric lines. 131 illustrations. Bibliography. Index. xii + 253pp. 5⅜ x 8.
S34 Paperbound **$1.85**

THEORY OF VIBRATIONS, N. W. McLachlan. Based on an exceptionally successful graduate course given at Brown University, this discusses linear systems having 1 degree of freedom, forced vibrations of simple linear systems, vibration of flexible strings, transverse vibrations of bars and tubes, transverse vibration of circular plate, sound waves of finite amplitude, etc. Index. 99 diagrams. 160pp. 5⅜ x 8. S190 Paperbound **$1.35**

LOUD SPEAKERS: THEORY, PERFORMANCE, TESTING AND DESIGN, N. W. McLachlan. Most comprehensive coverage of theory, practice of loud speaker design, testing; classic reference, study manual in field. First 12 chapters deal with theory, for readers mainly concerned with math. aspects; last 7 chapters will interest reader concerned with testing, design. Partial contents: principles of sound propagation, fluid pressure on vibrators, theory of moving-coil principle, transients, driving mechanisms, response curves, design of horn type moving coil speakers, electrostatic speakers, much more. Appendix. Bibliography. Index. 165 illustrations, charts. 411pp. 5⅜ x 8. S588 Paperbound **$2.25**

MICROWAVE TRANSMISSION, J. S. Slater. First text dealing exclusively with microwaves, brings together points of view of field, circuit theory, for graduate student in physics, electrical engineering, microwave technician. Offers valuable point of view not in most later studies. Uses Maxwell's equations to study electromagnetic field, important in this area. Partial contents: infinite line with distributed parameters, impedance of terminated line, plane waves, reflections, wave guides, coaxial line, composite transmission lines, impedance matching, etc. Introduction. Index. 76 illus. 319pp. 5⅜ x 8.
S564 Paperbound **$1.50**

THE ANALYSIS OF SENSATIONS, Ernst Mach. Great study of physiology, psychology of perception, shows Mach's ability to see material freshly, his "incorruptible skepticism and independence." (Einstein). Relation of problems of psychological perception to classical physics, supposed dualism of physical and mental, principle of continuity, evolution of senses, will as organic manifestation, scores of experiments, observations in optics, acoustics, music, graphics, etc. New introduction by T. S. Szasz, M. D. 58 illus. 300-item bibliography. Index. 404pp. 5⅜ x 8. S525 Paperbound **$1.75**

APPLIED OPTICS AND OPTICAL DESIGN, A. E. Conrady. With publication of vol. 2, standard work for designers in optics is now complete for first time. Only work of its kind in English; only detailed work for practical designer and self-taught. Requires, for bulk of work, no math above trig. Step-by-step exposition, from fundamental concepts of geometrical, physical optics, to systematic study, design, of almost all types of optical systems. Vol. 1: all ordinary ray-tracing methods; primary aberrations; necessary higher aberration for design of telescopes, low-power microscopes, photographic equipment. Vol. 2: (Completed from author's notes by R. Kingslake, Dir. Optical Design, Eastman Kodak.) Special attention to high-power microscope, anastigmatic photographic objectives. "An indispensable work," J., Optical Soc. of Amer. "As a practical guide this book has no rival," Transactions, Optical Soc. Index. Bibliography. 193 diagrams. 852pp. 6⅛ x 9¼. Vol. 1 T611 Paperbound **$2.95**
Vol. 2 T612 Paperbound **$2.95**

THE THEORY OF OPTICS, Paul Drude. One of finest fundamental texts in physical optics, classic offers thorough coverage, complete mathematical treatment of basic ideas. Includes fullest treatment of application of thermodynamics to optics; sine law in formation of images, transparent crystals, magnetically active substances, velocity of light, apertures, effects depending upon them, polarization, optical instruments, etc. Introduction by A. A. Michelson. Index. 110 illus. 567pp. 5⅜ x 8. S532 Paperbound **$2.45**

OPTICKS, Sir Isaac Newton. In its discussions of light, reflection, color, refraction, theories of wave and corpuscular theories of light, this work is packed with scores of insights and discoveries. In its precise and practical discussion of construction of optical apparatus, contemporary understandings of phenomena it is truly fascinating to modern physicists, astronomers, mathematicians. Foreword by Albert Einstein. Preface by I. B. Cohen of Harvard University. 7 pages of portraits, facsimile pages, letters, etc. cxvi + 414pp. 5⅜ x 8. S205 Paperbound **$2.00**

OPTICS AND OPTICAL INSTRUMENTS: AN INTRODUCTION WITH SPECIAL REFERENCE TO PRACTICAL APPLICATIONS, B. K. Johnson. An invaluable guide to basic practical applications of optical principles, which shows how to set up inexpensive working models of each of the four main types of optical instruments—telescopes, microscopes, photographic lenses, optical projecting systems. Explains in detail the most important experiments for determining their accuracy, resolving power, angular field of view, amounts of aberration, all other necessary facts about the instruments. Formerly "Practical Optics." Index. 234 diagrams. Appendix. 224pp. 5⅜ x 8. S642 Paperbound **$1.65**

PRINCIPLES OF PHYSICAL OPTICS, Ernst Mach. This classical examination of the propagation of light, color, polarization, etc. offers an historical and philosophical treatment that has never been surpassed for breadth and easy readability. Contents: Rectilinear propagation of light. Reflection, refraction. Early knowledge of vision. Dioptrics. Composition of light. Theory of color and dispersion. Periodicity. Theory of interference. Polarization. Mathematical representation of properties of light. Propagation of waves, etc. 279 illustrations, 10 portraits. Appendix. Indexes. 324pp. 5⅜ x 8. S178 Paperbound **$1.75**

FUNDAMENTALS OF ELECTRICITY AND MAGNETISM, L. B. Loeb. For students of physics, chemistry, or engineering who want an introduction to electricity and magnetism on a higher level and in more detail than general elementary physics texts provide. Only elementary differential and integral calculus is assumed. Physical laws developed logically, from magnetism to electric currents, Ohm's law, electrolysis, and on to static electricity, induction, etc. Covers an unusual amount of material; one third of book on modern material: solution of wave equation, photoelectric and thermionic effects, etc. Complete statement of the various electrical systems of units and interrelations. 2 Indexes. 75 pages of problems with answers stated. Over 300 figures and diagrams. xix +669pp. 5⅜ x 8. S745 Paperbound **$2.75**

THE ELECTROMAGNETIC FIELD, Max Mason & Warren Weaver. Used constantly by graduate engineers. Vector methods exclusively: detailed treatment of electrostatics, expansion methods, with tables converting any quantity into absolute electromagnetic, absolute electrostatic, practical units. Discrete charges, ponderable bodies, Maxwell field equations, etc. Introduction. Indexes. 416pp. 5⅜ x 8. S185 Paperbound **$2.00**

ELECTRICAL THEORY ON THE GIORGI SYSTEM, P. Cornelius. A new clarification of the fundamental concepts of electricity and magnetism, advocating the convenient m.k.s. system of units that is steadily gaining followers in the sciences. Illustrating the use and effectiveness of his terminology with numerous applications to concrete technical problems, the author here expounds the famous Giorgi system of electrical physics. His lucid presentation and well-reasoned, cogent argument for the universal adoption of this system form one of the finest pieces of scientific exposition in recent years. 28 figures. Index. Conversion tables for translating earlier data into modern units. Translated from 3rd Dutch edition by L. J. Jolley. x + 187pp. 5½ x 8¾. S909 Clothbound **$6.00**

THEORY OF ELECTRONS AND ITS APPLICATION TO THE PHENOMENA OF LIGHT AND RADIANT HEAT, H. Lorentz. Lectures delivered at Columbia University by Nobel laureate Lorentz. Unabridged, they form a historical coverage of the theory of free electrons, motion, absorption of heat, Zeeman effect, propagation of light in molecular bodies, inverse Zeeman effect, optical phenomena in moving bodies, etc. 109 pages of notes explain the more advanced sections. Index. 9 figures. 352pp. 5⅜ x 8. S173 Paperbound **$1.85**

TREATISE ON ELECTRICITY AND MAGNETISM, James Clerk Maxwell. For more than 80 years a seemingly inexhaustible source of leads for physicists, mathematicians, engineers. Total of 1082pp. on such topics as Measurement of Quantities, Electrostatics, Elementary Mathematical Theory of Electricity, Electrical Work and Energy in a System of Conductors, General Theorems, Theory of Electrical Images, Electrolysis, Conduction, Polarization, Dielectrics, Resistance, etc. "The greatest mathematical physicist since Newton," Sir James Jeans. 3rd edition. 107 figures, 21 plates. 1082pp. 5⅜ x 8. S636-7, 2 volume set, paperbound **$4.00**

See also: **FUNDAMENTAL FORMULAS OF PHYSICS, D. H. Menzel; MATHEMATICAL ANALYSIS OF ELECTRICAL & OPTICAL WAVE MOTION, H. Bateman.**

Mechanics, dynamics, thermodynamics, elasticity

MECHANICS VIA THE CALCULUS, P. W. Norris, W. S. Legge. Covers almost everything, from linear motion to vector analysis: equations determining motion, linear methods, compounding of simple harmonic motions, Newton's laws of motion, Hooke's law, the simple pendulum, motion of a particle in 1 plane, centers of gravity, virtual work, friction, kinetic energy of rotating bodies, equilibrium of strings, hydrostatics, sheering stresses, elasticity, etc. 550 problems. 3rd revised edition. xii + 367pp. 6 x 9. S207 Clothbound **$3.95**

MECHANICS, J. P. Den Hartog. Already a classic among introductory texts, the M.I.T. professor's lively and discursive presentation is equally valuable as a beginner's text, an engineering student's refresher, or a practicing engineer's reference. Emphasis in this highly readable text is on illuminating fundamental principles and showing how they are embodied in a great number of real engineering and design problems: trusses, loaded cables, beams, jacks, hoists, etc. Provides advanced material on relative motion and gyroscopes not usual in introductory texts. "Very thoroughly recommended to all those anxious to improve their real understanding of the principles of mechanics." MECHANICAL WORLD. Index. List of equations. 334 problems, all with answers. Over 550 diagrams and drawings. ix + 462pp. 5⅜ x 8.
S754 Paperbound **$2.00**

THEORETICAL MECHANICS: AN INTRODUCTION TO MATHEMATICAL PHYSICS, J. S. Ames, F. D. Murnaghan. A mathematically rigorous development of theoretical mechanics for the advanced student, with constant practical applications. Used in hundreds of advanced courses. An unusually thorough coverage of gyroscopic and baryscopic material, detailed analyses of the Corilis acceleration, applications of Lagrange's equations, motion of the double pendulum, Hamilton-Jacobi partial differential equations, group velocity and dispersion, etc. Special relativity is also included. 159 problems. 44 figures. ix + 462pp. 5⅜ x 8.
S461 Paperbound **$2.00**

THEORETICAL MECHANICS: STATICS AND THE DYNAMICS OF A PARTICLE, W. D. MacMillan. Used for over 3 decades as a self-contained and extremely comprehensive advanced undergraduate text in mathematical physics, physics, astronomy, and deeper foundations of engineering. Early sections require only a knowledge of geometry; later, a working knowledge of calculus. Hundreds of basic problems, including projectiles to the moon, escape velocity, harmonic motion, ballistics, falling bodies, transmission of power, stress and strain, elasticity, astronomical problems. 340 practice problems plus many fully worked out examples make it possible to test and extend principles developed in the text. 200 figures. xvii + 430pp. 5⅜ x 8.
S467 Paperbound **$2.00**

THEORETICAL MECHANICS: THE THEORY OF THE POTENTIAL, W. D. MacMillan. A comprehensive, well balanced presentation of potential theory, serving both as an introduction and a reference work with regard to specific problems, for physicists and mathematicians. No prior knowledge of integral relations is assumed, and all mathematical material is developed as it becomes necessary. Includes: Attraction of Finite Bodies; Newtonian Potential Function; Vector Fields, Green and Gauss Theorems; Attractions of Surfaces and Lines; Surface Distribution of Matter; Two-Layer Surfaces; Spherical Harmonics; Ellipsoidal Harmonics; etc. "The great number of particular cases . . . should make the book valuable to geophysicists and others actively engaged in practical applications of the potential theory," Review of Scientific Instruments. Index. Bibliography. xiii + 469pp. 5⅜ x 8.
S486 Paperbound **$2.25**

THEORETICAL MECHANICS: DYNAMICS OF RIGID BODIES, W. D. MacMillan. Theory of dynamics of a rigid body is developed, using both the geometrical and analytical methods of instruction. Begins with exposition of algebra of vectors, it goes through momentum principles, motion in space, use of differential equations and infinite series to solve more sophisticated dynamics problems. Partial contents: moments of inertia, systems of free particles, motion parallel to a fixed plane, rolling motion, method of periodic solutions, much more. 82 figs. 199 problems. Bibliography. Indexes. xii + 476pp. 5⅜ x 8.
S641 Paperbound **$2.00**

MATHEMATICAL FOUNDATIONS OF STATISTICAL MECHANICS, A. I. Khinchin. Offering a precise and rigorous formulation of problems, this book supplies a thorough and up-to-date exposition. It provides analytical tools needed to replace cumbersome concepts, and furnishes for the first time a logical step-by-step introduction to the subject. Partial contents: geometry & kinematics of the phase space, ergodic problem, reduction to theory of probability, application of central limit problem, ideal monatomic gas, foundation of thermo-dynamics, dispersion and distribution of sum functions. Key to notations. Index. viii + 179pp. 5⅜ x 8.
S147 Paperbound **$1.35**

ELEMENTARY PRINCIPLES IN STATISTICAL MECHANICS, J. W. Gibbs. Last work of the great Yale mathematical physicist, still one of the most fundamental treatments available for advanced students and workers in the field. Covers the basic principle of conservation of probability of phase, theory of errors in the calculated phases of a system, the contributions of Clausius, Maxwell, Boltzmann, and Gibbs himself, and much more. Includes valuable comparison of statistical mechanics with thermodynamics: Carnot's cycle, mechanical definitions of entropy, etc. xvi + 208pp. 5⅜ x 8.
S707 Paperbound **$1.45**

THE DYNAMICS OF PARTICLES AND OF RIGID, ELASTIC, AND FLUID BODIES; BEING LECTURES ON MATHEMATICAL PHYSICS, A. G. Webster. The reissuing of this classic fills the need for a comprehensive work on dynamics. A wide range of topics is covered in unusually great depth, applying ordinary and partial differential equations. Part I considers laws of motion and methods applicable to systems of all sorts; oscillation, resonance, cyclic systems, etc. Part 2 is a detailed study of the dynamics of rigid bodies. Part 3 introduces the theory of potential; stress and strain, Newtonian potential functions, gyrostatics, wave and vortex motion, etc. Further contents: Kinematics of a point; Lagrange's equations; Hamilton's principle; Systems of vectors; Statics and dynamics of deformable bodies; much more, not easily found together in one volume. Unabridged reprinting of 2nd edition. 20 pages of notes on differential equations and the higher analysis. 203 illustrations. Selected bibliography. Index. xi + 588pp. 5⅜ x 8.
S522 Paperbound **$2.35**

A TREATISE ON DYNAMICS OF A PARTICLE, E. J. Routh. Elementary text on dynamics for beginning mathematics or physics student. Unusually detailed treatment from elementary definitions to motion in 3 dimensions, emphasizing concrete aspects. Much unique material important in recent applications. Covers impulsive forces, rectilinear and constrained motion in 2 dimensions, harmonic and parabolic motion, degrees of freedom, closed orbits, the conical pendulum, the principle of least action, Jacobi's method, and much more. Index. 559 problems, many fully worked out, incorporated into text. xiii + 418pp. 5⅜ x 8.
S696 Paperbound **$2.25**

DYNAMICS OF A SYSTEM OF RIGID BODIES (Elementary Section), E. J. Routh. Revised 7th edition of this standard reference. This volume covers the dynamical principles of the subject, and its more elementary applications: finding moments of inertia by integration, foci of inertia, d'Alembert's principle, impulsive forces, motion in 2 and 3 dimensions, Lagrange's equations, relative indicatrix, Euler's theorem, large tautochronous motions, etc. Index. 55 figures. Scores of problems. xv + 443pp. 5⅜ x 8. S664 Paperbound **$2.35**

DYNAMICS OF A SYSTEM OF RIGID BODIES (Advanced Section), E. J. Routh. Revised 6th edition of a classic reference aid. Much of its material remains unique. Partial contents: moving axes, relative motion, oscillations about equilibrium, motion. Motion of a body under no forces, any forces. Nature of motion given by linear equations and conditions of stability. Free, forced vibrations, constants of integration, calculus of finite differences, variations, precession and nutation, motion of the moon, motion of string, chain, membranes. 64 figures. 498pp. 5⅜ x 8. S229 Paperbound **$2.35**

DYNAMICAL THEORY OF GASES, James Jeans. Divided into mathematical and physical chapters for the convenience of those not expert in mathematics, this volume discusses the mathematical theory of gas in a steady state, thermodynamics, Boltzmann and Maxwell, kinetic theory, quantum theory, exponentials, etc. 4th enlarged edition, with new material on quantum theory, quantum dynamics, etc. Indexes. 28 figures. 444pp. 6⅛ x 9¼.
S136 Paperbound **$2.45**

FOUNDATIONS OF POTENTIAL THEORY, O. D. Kellogg. Based on courses given at Harvard this is suitable for both advanced and beginning mathematicians. Proofs are rigorous, and much material not generally avialable elsewhere is included. Partial contents: forces of gravity, fields of force, divergence theorem, properties of Newtonian potentials at points of free space, potentials as solutions of Laplace's equations, harmonic functions, electrostatics, electric images, logarithmic potential, etc. One of Grundlehren Series. ix + 384pp. 5⅜ x 8.
S144 Paperbound **$1.98**

THERMODYNAMICS, Enrico Fermi. Unabridged reproduction of 1937 edition. Elementary in treatment; remarkable for clarity, organization. Requires no knowledge of advanced math beyond calculus, only familiarity with fundamentals of thermometry, calorimetry. Partial Contents: Thermodynamic systems; First & Second laws of thermodynamics; Entropy; Thermodynamic potentials: phase rule, reversible electric cell; Gaseous reactions: van't Hoff reaction box, principle of LeChatelier; Thermodynamics of dilute solutions: osmotic & vapor pressures, boiling & freezing points; Entropy constant. Index. 25 problems. 24 illustrations. x + 160pp. 5⅜ x 8 S361 Paperbound **$1.75**

THE THERMODYNAMICS OF ELECTRICAL PHENOMENA IN METALS and A CONDENSED COLLECTION OF THERMODYNAMIC FORMULAS, P. W. Bridgman. Major work by the Nobel Prizewinner: stimulating conceptual introduction to aspects of the electron theory of metals, giving an intuitive understanding of fundamental relationships concealed by the formal systems of Onsager and others. Elementary mathematical formulations show clearly the fundamental thermodynamical relationships of the electric field, and a complete phenomenological theory of metals is created. This is the work in which Bridgman announced his famous "thermomotive force" and his distinction between "driving" and "working" electromotive force. We have added in this Dover edition the author's long unavailable tables of thermodynamic formulas, extremely valuable for the speed of reference they allow. Two works bound as one. Index. 33 figures. Bibliography. xviii + 256pp. 5⅜ x 8. S723 Paperbound **$1.65**

REFLECTIONS ON THE MOTIVE POWER OF FIRE, by Sadi Carnot, and other papers on the 2nd law of thermodynamics by E. Clapeyron and R. Clausius. Carnot's "Reflections" laid the groundwork of modern thermodynamics. Its non-technical, mostly verbal statements examine the relations between heat and the work done by heat in engines, establishing conditions for the economical working of these engines. The papers by Clapeyron and Clausius here reprinted added further refinements to Carnot's work, and led to its final acceptance by physicists. Selections from posthumous manuscripts of Carnot are also included. All papers in English. New introduction by E. Mendoza. 12 illustrations. xxii + 152pp. 5⅜ x 8.
S661 Paperbound **$1.50**

TREATISE ON THERMODYNAMICS, Max Planck. Based on Planck's original papers this offers a uniform point of view for the entire field and has been used as an introduction for students who have studied elementary chemistry, physics, and calculus. Rejecting the earlier approaches of Helmholtz and Maxwell, the author makes no assumptions regarding the nature of heat, but begins with a few empirical facts, and from these deduces new physical and chemical laws. 3rd English edition of this standard text by a Nobel laureate. xvi + 297pp. 5⅜ x 8. S219 Paperbound **$1.75**

THE THEORY OF HEAT RADIATION, Max Planck. A pioneering work in thermodynamics, providing basis for most later work. Nobel Laureate Planck writes on Deductions from Electrodynamics and Thermodynamics, Entropy and Probability, Irreversible Radiation Processes, etc. Starts with simple experimental laws of optics, advances to problems of spectral distribution of energy and irreversibility. Bibliography. 7 illustrations, xiv + 224pp. 5⅜ x 8.
S546 Paperbound **$1.50**

A HISTORY OF THE THEORY OF ELASTICITY AND THE STRENGTH OF MATERIALS, I. Todhunter and K. Pearson. For over 60 years a basic reference, unsurpassed in scope or authority. Both a history of the mathematical theory of elasticity from Galileo, Hooke, and Mariotte to Saint Venant, Kirchhoff, Clebsch, and Lord Kelvin and a detailed presentation of every important mathematical contribution during this period. Presents proofs of thousands of theorems and laws, summarizes every relevant treatise, many unavailable elsewhere. Practically a book apiece is devoted to modern founders: Saint Venant, Lame, Boussinesq, Rankine, Lord Kelvin, F. Neumann, Kirchhoff, Clebsch. Hundreds of pages of technical and physical treatises on specific applications of elasticity to particular materials. Indispensable for the mathematician, physicist, or engineer working with elasticity. Unabridged, corrected reprint of original 3-volume 1886-1893 edition. Three volume set. Two indexes. Appendix to Vol. I. Total of 2344pp. 5⅜ x 8⅜.
S914–916 The set, Clothbound **$12.50**

THE MATHEMATICAL THEORY OF ELASTICITY, A. E. H. Love. A wealth of practical illustration combined with thorough discussion of fundamentals—theory, application, special problems and solutions. Partial Contents: Analysis of Strain & Stress, Elasticity of Solid Bodies, Elasticity of Crystals, Vibration of Spheres, Cylinders, Propagation of Waves in Elastic Solid Media, Torsion, Theory of Continuous Beams, Plates. Rigorous treatment of Volterra's theory of dislocations, 2-dimensional elastic systems, other topics of modern interest. "For years the standard treatise on elasticity," AMERICAN MATHEMATICAL MONTHLY. 4th revised edition. Index. 76 figures. xviii + 643pp. 6⅛ x 9¼.
S174 Paperbound **$2.95**

RAYLEIGH'S PRINCIPLE AND ITS APPLICATIONS TO ENGINEERING, G. Temple & W. Bickley. Rayleigh's principle developed to provide upper and lower estimates of true value of fundamental period of a vibrating system, or condition of stability of elastic systems. Illustrative examples; rigorous proofs in special chapters. Partial contents: Energy method of discussing vibrations, stability. Perturbation theory, whirling of uniform shafts. Criteria of elastic stability. Application of energy method. Vibrating systems. Proof, accuracy, successive approximations, application of Rayleigh's principle. Synthetic theorems. Numerical, graphical methods. Equilibrium configurations, Ritz's method. Bibliography. Index. 22 figures. ix + 156pp. 5⅜ x 8.
S307 Paperbound **$1.50**

INVESTIGATIONS ON THE THEORY OF THE BROWNIAN MOVEMENT, Albert Einstein. Reprints from rare European journals. 5 basic papers, including the Elementary Theory of the Brownian Movement, written at the request of Lorentz to provide a simple explanation. Translated by A. D. Cowper. Annotated, edited by R. Fürth. 33pp. of notes elucidate, give history of previous investigations. Author, subject indexes. 62 footnotes. 124pp. 5⅜ x 8.
S304 Paperbound **$1.25**

See also: **FUNDAMENTAL FORMULAS OF PHYSICS, D. H. Menzel.**

ENGINEERING

THEORY OF FLIGHT, Richard von Mises. Remains almost unsurpassed as balanced, well-written account of fundamental fluid dynamics, and situations in which air compressibility effects are unimportant. Stressing equally theory and practice, avoiding formidable mathematical structure, it conveys a full understanding of physical phenomena and mathematical concepts. Contains perhaps the best introduction to general theory of stability. "Outstanding," Scientific, Medical, and Technical Books. New introduction by K. H. Hohenemser. Bibliographical, historical notes. Index. 408 illustrations. xvi + 620pp. 5⅜ x 8⅜. S541 Paperbound **$2.85**

THEORY OF WING SECTIONS, I. H. Abbott, A. E. von Doenhoff. Concise compilation of subsonic aerodynamic characteristics of modern NASA wing sections, with description of their geometry, associated theory. Primarily reference work for engineers, students, it gives methods, data for using wing-section data to predict characteristics. Particularly valuable: chapters on thin wings, airfoils; complete summary of NACA's experimental observations, system of construction families of airfoils. 350pp. of tables on Basic Thickness Forms, Mean Lines, Airfoil Ordinates, Aerodynamic Characteristics of Wing Sections. Index. Bibliography. 191 illustrations. Appendix. 705pp. 5⅜ x 8.
S558 Paperbound **$2.95**

SUPERSONIC AERODYNAMICS, E. R. C. Miles. Valuable theoretical introduction to the supersonic domain, with emphasis on mathematical tools and principles, for practicing aerodynamicists and advanced students in aeronautical engineering. Covers fundamental theory, divergence theorem and principles of circulation, compressible flow and Helmholtz laws, the Prandtl-Busemann graphic method for 2-dimensional flow, oblique shock waves, the Taylor-Maccoll method for cones in supersonic flow, the Chaplygin method for 2-dimensional flow, etc. Problems range from practical engineering problems to development of theoretical results. "Rendered outstanding by the unprecedented scope of its contents . . . has undoubtedly filled a vital gap," AERONAUTICAL ENGINEERING REVIEW. Index. 173 problems, answers. 106 diagrams. 7 tables. xii + 255pp. 5⅜ x 8.
S214 Paperbound **$1.45**

WEIGHT-STRENGTH ANALYSIS OF AIRCRAFT STRUCTURES, F. R. Shanley. Scientifically sound methods of analyzing and predicting the structural weight of aircraft and missiles. Deals directly with forces and the distances over which they must be transmitted, making it possible to develop methods by which the minimum structural weight can be determined for any material and conditions of loading. Weight equations for wing and fuselage structures. Includes author's original papers on inelastic buckling and creep buckling. "Particularly successful in presenting his analytical methods for investigating various optimum design principles," AERONAUTICAL ENGINEERING REVIEW. Enlarged bibliography. Index. 199 figures. xiv + 404pp. 5⅝ x 8⅜. S660 Paperbound **$2.45**

INTRODUCTION TO THE STATISTICAL DYNAMICS OF AUTOMATIC CONTROL SYSTEMS, V. V. Solodovnikov. First English publication of text-reference covering important branch of automatic control systems—random signals; in its original edition, this was the first comprehensive treatment. Examines frequency characteristics, transfer functions, stationary random processes, determination of minimum mean-squared error, of transfer function for a finite period of observation, much more. Translation edited by J. B. Thomas, L. A. Zadeh. Index. Bibliography. Appendix. xxii + 308pp. 5⅜ x 8. S420 Paperbound **$2.25**

TENSORS FOR CIRCUITS, Gabriel Kron. A boldly original method of analysing engineering problems, at center of sharp discussion since first introduced, now definitely proved useful in such areas as electrical and structural networks on automatic computers. Encompasses a great variety of specific problems by means of a relatively few symbolic equations. "Power and flexibility . . . becoming more widely recognized," Nature. Formerly "A Short Course in Tensor Analysis." New introduction by B. Hoffmann. Index. Over 800 diagrams. xix + 250pp. 5⅜ x 8. S534 Paperbound **$1.85**

DESIGN AND USE OF INSTRUMENTS AND ACCURATE MECHANISM, T. N. Whitehead. For the instrument designer, engineer; how to combine necessary mathematical abstractions with independent observation of actual facts. Partial contents: instruments & their parts, theory of errors, systematic errors, probability, short period errors, erratic errors, design precision, kinematic, semikinematic design, stiffness, planning of an instrument, human factor, etc. Index. 85 photos, diagrams. xii + 288pp. 5⅜ x 8. S270 Paperbound **$1.95**

APPLIED ELASTICITY, J. Prescott. Provides the engineer with the theory of elasticity usually lacking in books on strength of materials, yet concentrates on those portions useful for immediate application. Develops every important type of elasticity problem from theoretical principles. Covers analysis of stress, relations between stress and strain, the empirical basis of elasticity, thin rods under tension or thrust, Saint Venant's theory, transverse oscillations of thin rods, stability of thin plates, cylinders with thin walls, vibrations of rotating disks, elastic bodies in contact, etc. "Excellent and important contribution to the subject, not merely in the old matter which he has presented in new and refreshing form, but also in the many original investigations here published for the first time," NATURE. Index. 3 Appendixes. vi + 672pp. 5⅜ x 8. S726 Paperbound **$2.95**

STRENGTH OF MATERIALS, J. P. Den Hartog. Distinguished text prepared for M.I.T. course, ideal as introduction, refresher, reference, or self-study text. Full clear treatment of elementary material (tension, torsion, bending, compound stresses, deflection of beams, etc.), plus much advanced material on engineering methods of great practical value: full treatment of the Mohr circle, lucid elementary discussions of the theory of the center of shear and the "Myosotis" method of calculating beam deflections, reinforced concrete, plastic deformations, photoelasticity, etc. In all sections, both general principles and concrete applications are given. Index. 186 figures (160 others in problem section). 350 problems, all with answers. List of formulas. viii + 323pp. 5⅜ x 8. S755 Paperbound **$1.95**

PHOTOELASTICITY: PRINCIPLES AND METHODS, H. T. Jessop, F. C. Harris. For the engineer, for specific problems of stress analysis. Latest time-saving methods of checking calculations in 2-dimensional design problems, new techniques for stresses in 3 dimensions, and lucid description of optical systems used in practical photoelasticity. Useful suggestions and hints based on on-the-job experience included. Partial contents: strained and stress-strain relations, circular disc under thrust along diameter, rectangular block with square hole under vertical thrust, simply supported rectangular beam under central concentrated load, etc. Theory held to minimum, no advanced mathematical training needed. Index. 164 illustrations. viii + 184pp. 6⅛ x 9¼. S137 Clothbound **$3.75**

MECHANICS OF THE GYROSCOPE, THE DYNAMICS OF ROTATION, R. F. Deimel, Professor of Mechanical Engineering at Stevens Institute of Technology. Elementary general treatment of dynamics of rotation, with special application of gyroscopic phenomena. No knowledge of vectors needed. Velocity of a moving curve, acceleration to a point, general equations of motion, gyroscopic horizon, free gyro, motion of discs, the damped gyro, 103 similar topics. Exercises. 75 figures. 208pp 5⅜ x 8. S66 Paperbound **$1.65**
S144 Paperbound **$1.98**

A TREATISE ON GYROSTATICS AND ROTATIONAL MOTION: THEORY AND APPLICATIONS, Andrew Gray. Most detailed, thorough book in English, generally considered definitive study. Many problems of all sorts in full detail, or step-by-step summary. Classical problems of Bour, Lottner, etc.; later ones of great physical interest. Vibrating systems of gyrostats, earth as a top, calculation of path of axis of a top by elliptic integrals, motion of unsymmetrical top, much more. Index. 160 illus. 550pp. 5⅜ x 8. S589 Paperbound **$2.75**

FUNDAMENTALS OF HYDRO- AND AEROMECHANICS, L. Prandtl and O. G. Tietjens. The well-known standard work based upon Prandtl's lectures at Goettingen. Wherever possible hydrodynamics theory is referred to practical considerations in hydraulics, with the view of unifying theory and experience. Presentation is extremely clear and though primarily physical, mathematical proofs are rigorous and use vector analysis to a considerable extent. An Engineering Society Monograph, 1934. 186 figures. Index. xvi + 270pp. 5⅜ x 8.
S374 Paperbound **$1.85**

APPLIED HYDRO- AND AEROMECHANICS, L. Prandtl and O. G. Tietjens. Presents, for the most part, methods which will be valuable to engineers. Covers flow in pipes, boundary layers, airfoil theory, entry conditions, turbulent flow in pipes, and the boundary layer, determining drag from measurements of pressure and velocity, etc. "Will be welcomed by all students of aerodynamics," NATURE. Unabridged, unaltered. An Engineering Society Monograph, 1934. Index. 226 figures, 28 photographic plates illustrating flow patterns. xvi + 311pp. 5⅜ x 8.
S375 Paperbound **$1.85**

HYDRAULICS AND ITS APPLICATIONS, A. H. Gibson. Excellent comprehensive textbook for the student and thorough practical manual for the professional worker, a work of great stature in its area. Half the book is devoted to theory and half to applications and practical problems met in the field. Covers modes of motion of a fluid, critical velocity, viscous flow, eddy formation, Bernoulli's theorem, flow in converging passages, vortex motion, form of effluent streams, notches and weirs, skin friction, losses at valves and elbows, siphons, erosion of channels, jet propulsion, waves of oscillation, and over 100 similar topics. Final chapters (nearly 400 pages) cover more than 100 kinds of hydraulic machinery: Pelton wheel, speed regulators, the hydraulic ram, surge tanks, the scoop wheel, the Venturi meter, etc. A special chapter treats methods of testing theoretical hypotheses: scale models of rivers, tidal estuaries, siphon spillways, etc. 5th revised and enlarged (1952) edition. Index. Appendix. 427 photographs and diagrams. 95 examples, answers. xv + 813pp. 6 x 9.
S791 Clothbound **$8.00**

FLUID MECHANICS FOR HYDRAULIC ENGINEERS, H. Rouse. Standard work that gives a coherent picture of fluid mechanics from the point of view of the hydraulic engineer. Based on courses given to civil and mechanical engineering students at Columbia and the California Institute of Technology, this work covers every basic principle, method, equation, or theory of interest to the hydraulic engineer. Much of the material, diagrams, charts, etc., in this self-contained text are not duplicated elsewhere. Covers irrotational motion, conformal mapping, problems in laminar motion, fluid turbulence, flow around immersed bodies, transportation of sediment, general charcteristics of wave phenomena, gravity waves in open channels, etc. Index. Appendix of physical properties of common fluids. Frontispiece + 245 figures and photographs. xvi + 422pp. 5⅜ x 8.
S729 Paperbound **$2.25**

THE MEASUREMENT OF POWER SPECTRA FROM THE POINT OF VIEW OF COMMUNICATIONS ENGINEERING, R. B. Blackman, J. W. Tukey. This pathfinding work, reprinted from the "Bell System Technical Journal," explains various ways of getting practically useful answers in the measurement of power spectra, using results from both transmission theory and the theory of statistical estimation. Treats: Autocovariance Functions and Power Spectra; Direct Analog Computation; Distortion, Noise, Heterodyne Filtering and Pre-whitening; Aliasing; Rejection Filtering and Separation; Smoothing and Decimation Procedures; Very Low Frequencies; Transversal Filtering; much more. An appendix reviews fundamental Fourier techniques. Index of notation. Glossary of terms. 24 figures. XII tables. Bibliography. General index. 192pp. 5⅜ x 8.
S507 Paperbound **$1.85**

MICROWAVE TRANSMISSION DESIGN DATA, T. Moreno. Originally classified, now rewritten and enlarged (14 new chapters) for public release under auspices of Sperry Corp. Material of immediate value or reference use to radio engineers, systems designers, applied physicists, etc. Ordinary transmission line theory; attenuation; capacity; parameters of coaxial lines; higher modes; flexible cables; obstacles, discontinuities, and injunctions; tuneable wave guide impedance transformers; effects of temperature and humidity; much more. "Enough theoretical discussion is included to allow use of data without previous background," Electronics. 324 circuit diagrams, figures, etc. Tables of dielectrics, flexible cable, etc., data. Index. ix + 248pp. 5⅜ x 8.
S459 Paperbound **$1.50**

GASEOUS CONDUCTORS: THEORY AND ENGINEERING APPLICATIONS, J. D. Cobine. An indispensable text and reference to gaseous conduction phenomena, with the engineering viewpoint prevailing throughout. Studies the kinetic theory of gases, ionization, emission phenomena; gas breakdown, spark characteristics, glow, and discharges; engineering applications in circuit interrupters, rectifiers, light sources, etc. Separate detailed treatment of high pressure arcs (Suits); low pressure arcs (Langmuir and Tonks). Much more. "Well organized, clear, straightforward," Tonks, Review of Scientific Instruments. Index. Bibliography. 83 practice problems. 7 appendices. Over 600 figures. 58 tables. xx + 606pp. 5⅜ x 8.
S442 Paperbound **$2.85**

See also: **BRIDGES AND THEIR BUILDERS, D. Steinman, S. R. Watson; A DIDEROT PICTORIAL ENCYCLOPEDIA OF TRADES AND INDUSTRY; MATHEMATICS IN ACTION, O. G. Sutton; THE THEORY OF SOUND, Lord Rayleigh; RAYLEIGH'S PRINCIPLE AND ITS APPLICATION TO ENGINEERING, G. Temple, W. Bickley; APPLIED OPTICS AND OPTICAL DESIGN, A. E. Conrady; HYDRODYNAMICS, Dryden, Murnaghan, Bateman; LOUD SPEAKERS, N. W. McLachlan; HISTORY OF THE THEORY OF ELASTICITY AND OF THE STRENGTH OF MATERIALS, I. Todhunter,**

K. Pearson; THEORY AND OPERATION OF THE SLIDE RULE, J. P. Ellis; DIFFERENTIAL EQUATIONS FOR ENGINEERS, P. Franklin; MATHEMATICAL METHODS FOR SCIENTISTS AND ENGINEERS, L. P. Smith; APPLIED MATHEMATICS FOR RADIO AND COMMUNICATIONS ENGINEERS, C. E. Smith; MATHEMATICS OF MODERN ENGINEERING, E. G. Keller, R. E. Doherty; THEORY OF FUNCTIONS AS APPLIED TO ENGINEERING PROBLEMS, R. Rothe, F. Ollendorff, K. Pohlhausen.

CHEMISTRY AND PHYSICAL CHEMISTRY

ORGANIC CHEMISTRY, F. C. Whitmore. The entire subject of organic chemistry for the practicing chemist and the advanced student. Storehouse of facts, theories, processes found elsewhere only in specialized journals. Covers aliphatic compounds (500 pages on the properties and synthetic preparation of hydrocarbons, halides, proteins, ketones, etc.), alicyclic compounds, aromatic compounds, heterocyclic compounds, organophosphorus and organometallic compounds. Methods of synthetic preparation analyzed critically throughout. Includes much of biochemical interest. "The scope of this volume is astonishing," INDUSTRIAL AND ENGINEERING CHEMISTRY. 12,000-reference index. 2387-item bibliography. Total of x + 1005pp. 5⅜ x 8. Two volume set.
S700 Vol I Paperbound **$2.00**
S701 Vol II Paperbound **$2.00**
The set **$4.00**

THE PRINCIPLES OF ELECTROCHEMISTRY, D. A. MacInnes. Basic equations for almost every subfield of electrochemistry from first principles, referring at all times to the soundest and most recent theories and results; unusually useful as text or as reference. Covers coulometers and Faraday's Law, electrolytic conductance, the Debye-Hueckel method for the theoretical calculation of activity coefficients, concentration cells, standard electrode potentials, thermodynamic ionization constants, pH, potentiometric titrations, irreversible phenomena, Planck's equation, and much more. "Excellent treatise," AMERICAN CHEMICAL SOCIETY JOURNAL. "Highly recommended," CHEMICAL AND METALLURGICAL ENGINEERING. 2 Indices. Appendix. 585-item bibliography. 137 figures. 94 tables. ii + 478pp. 5⅝ x 8⅜.
S52 Paperbound **$2.35**

THE CHEMISTRY OF URANIUM: THE ELEMENT, ITS BINARY AND RELATED COMPOUNDS, J. J. Katz and E. Rabinowitch. Vast post-World War II collection and correlation of thousands of AEC reports and published papers in a useful and easily accessible form, still the most complete and up-to-date compilation. Treats "dry uranium chemistry," occurrences, preparation, properties, simple compounds, isotopic composition, extraction from ores, spectra, alloys, etc. Much material available only here. Index. Thousands of evaluated bibliographical references. 324 tables, charts, figures. xxi + 609pp. 5⅜ x 8. S757 Paperbound **$2.95**

KINETIC THEORY OF LIQUIDS, J. Frenkel. Regarding the kinetic theory of liquids as a generalization and extension of the theory of solid bodies, this volume covers all types of arrangements of solids, thermal displacements of atoms, interstitial atoms and ions, orientational and rotational motion of molecules, and transition between states of matter. Mathematical theory is developed close to the physical subject matter. 216 bibliographical footnotes. 55 figures. xi + 485pp. 5⅜ x 8.
S94 Clothbound **$3.95**
S95 Paperbound **$2.45**

POLAR MOLECULES, Pieter Debye. This work by Nobel laureate Debye offers a complete guide to fundamental electrostatic field relations, polarizability, molecular structure. Partial contents: electric intensity, displacement and force, polarization by orientation, molar polarization and molar refraction, halogen-hydrides, polar liquids, ionic saturation, dielectric constant, etc. Special chapter considers quantum theory. Indexed. 172pp. 5⅜ x 8.
S64 Paperbound **$1.50**

ELASTICITY, PLASTICITY AND STRUCTURE OF MATTER, R. Houwink. Standard treatise on rheological aspects of different technically important solids such as crystals, resins, textiles, rubber, clay, many others. Investigates general laws for deformations; determines divergences from these laws for certain substances. Covers general physical and mathematical aspects of plasticity, elasticity, viscosity. Detailed examination of deformations, internal structure of matter in relation to elastic and plastic behavior, formation of solid matter from a fluid, conditions for elastic and plastic behavior of matter. Treats glass, asphalt, gutta percha, balata, proteins, baker's dough, lacquers, sulphur, others. 2nd revised, enlarged edition. Extensive revised bibliography in over 500 footnotes. Index. Table of symbols. 214 figures. xviii + 368pp. 6 x 9¼. S385 Paperbound **$2.45**

THE PHASE RULE AND ITS APPLICATION, Alexander Findlay. Covering chemical phenomena of 1, 2, 3, 4, and multiple component systems, this "standard work on the subject" (NATURE, London), has been completely revised and brought up to date by A. N. Campbell and N. O. Smith. Brand new material has been added on such matters as binary, tertiary liquid equilibria, solid solutions in ternary systems, quinary systems of salts and water. Completely revised to triangular coordinates in ternary systems, clarified graphic representation, solid models, etc. 9th revised edition. Author, subject indexes. 236 figures. 505 footnotes, mostly bibliographic. xii + 494pp. 5⅜ x 8. S91 Paperbound **$2.45**